The Sedimentary Record of Meteorite Impacts

edited by

Kevin R. Evans
Department of Geography, Geology and Planning
Missouri State University
901 South National Avenue
Springfield, Missouri 65897, USA

J. Wright Horton Jr.
U.S. Geological Survey
MS 926A, 12201 Sunrise Valley Drive
Reston, Virginia 20192, USA

David T. King Jr.
Department of Geology and Geography
Auburn University
Auburn, Alabama 36849-5305, USA

Jared R. Morrow
Department of Geological Sciences
San Diego State University
5500 Campanile Drive
San Diego, California 92182-1020, USA

THE GEOLOGICAL SOCIETY OF AMERICA®

Special Paper 437

3300 Penrose Place, P.O. Box 9140 ▪ Boulder, Colorado 80301-9140 USA

2008

Published by The Geological Society of America, Inc.
3300 Penrose Place, P.O. Box 9140, Boulder, Colorado 80301-9140, USA
www.geosociety.org

Printed in U.S.A.

GSA Books Science Editors: Marion E. Bickford and Donald I. Siegel

Library of Congress Cataloging-in-Publication Data

The sedimentary record of meteorite impacts / edited by Kevin R. Evans ... [et al.].
p. cm. -- (Special paper ; 437)
Includes bibliographical references.
ISBN 978-0-8137-2437-9 (pbk.)
1. Cryptoexplosion structures. 2. Sedimentary structures. 3. Geology, Stratigraphic. 4. Meteorites. I. Evans, Kevin R.
QE613.S43 2008
551.3'97--dc22

2008000789

Front cover: Marine impact at Wetumpka, Alabama. This painting by Jerry Armstrong, noted space artist from Atlanta, Georgia, depicts the Late Cretaceous shallow marine impact event that formed the 7.6-km diameter Wetumpka impact structure in central Alabama. Witnessing the impact is the small, eastern North American tyrannosaur, *Appalachiosaurus montgomeriensis*. This view is looking south at the impact fireball rising from ground zero about 25 km offshore. Copyright © City of Wetumpka, Alabama; used with permission.

Back cover: Deep drilling in the annular trough of the Chesapeake Bay impact structure near Bayside, Mathews County, Virginia, in 2001 (see Horton et al., chapter 6). View is southward overlooking tidal wetland to Chesapeake Bay on the horizon. Photograph by E. Randolph McFarland, U.S. Geological Survey; also used as cover in McFarland, E.R., and Bruce, T.S., 2006, The Virginia Coastal Plain Hydrogeologic Framework: U.S. Geological Survey Professional Paper 1731, 118 p.; used with permission.

10 9 8 7 6 5 4 3 2 1

Contents

Preface

Nearly all known terrestrial meteorite impacts are on continents or continental shelves, and according to the *Earth Impact Database* (University of New Brunswick Planetary and Space Science Centre, 2007), approximately 70% of the 174 generally accepted terrestrial impacts involve sedimentary target rocks (see Osinski et al., this volume). Using the record of terrestrial impact as a random sampling of continental rock types strikes fairly close to the fact that 75% of the continents are covered by sedimentary cover (Encyclopædia Britannica, 2007); seemingly, this is one of the few accurately predictable aspects of terrestrial impacts.

Although the remains of ancient impacts generally are found on dry land, many, including most of the examples detailed in this volume, actually occurred in marine settings. In addition to the deformation of sedimentary target rocks, the record of meteorite impacts also includes proximal to distal ejecta and tsunami deposits, both of which are typically preserved in sedimentary successions. So, despite general agreement that impactites are metamorphic rocks because they have been exposed to shock-metamorphic pressures, many impacts and their lateral correlates are sedimentary by nature. This Special Paper explores the scope of the sedimentary record of meteorite impacts.

A topical session on impact geology at the 2004 Geological Society of America Annual Meeting in Denver, Colorado, November 7–10, and an SEPM Research Conference held in Springfield, Missouri, May 21–25, 2005, provided the impetus for joint publication of this volume. Co-editors David T. King Jr. and Jared Morrow chaired the GSA session. Co-editors Kevin Evans and J. Wright Horton Jr. were co-conveners of the SEPM Research Conference with John Warme and Mark Thompson. The organization of papers in this volume are drawn along the lines of proximal settings (chapters 1–6) and distal settings (chapters 7–12). In both sections, papers are arranged in stratigraphic order, oldest to youngest ages.

In chapter 1, Osinski, Spray, and Grieve characterize impact melt rocks derived from sedimentary target rocks, including carbonates, evaporites, and siliciclastics. Chapter 2, by Kalleson, Dypvik, and Naterstad, documents the transition from emplacement of impact-derived breccias to postimpact sediment-gravity flows within the Gardnos structure (Cambrian) of southern Norway. In chapter 3, Lindström, Ormö, and Sturkell have developed a depositional model for processes related to brecciation during the excavation and modification stages of impact cratering at the Lockne structure (Early Ordovician) in central Sweden. In chapter 4, Dulin and Elmore provide an account of their investigations of remanent paleomagnetism in the Weaubleau structure (Mississippian) of southwestern Missouri. Chapter 5, by Dypvik, Wolbach, Shuvalov, and Weaver, describes soot particles recovered from cores and outcrops in and near the Mjølnir structure (Jurassic) in the Barents Sea off the northern coast of Norway; they provide a compelling argument for the ignition and combustion of petroleum source rocks on the sea floor. Horton, Gohn, Powars, and Edwards, in chapter 6, trace the development and emplacement of breccias in the Chesapeake Bay impact structure (Eocene) in Virginia.

In chapter 7, Pinto and Warme provide an account of the proximal to distal impact realms of rocks associated with the Alamo event (Devonian) in southeastern Nevada and western Utah. In chapter 8, Banet and Fenton provide stratigraphic evidence for the age of the Avak structure of northernmost Alaska based on examination of cores from a distal setting. A breccia containing exotic mixed-age clasts, including the Lower Cretaceous Pebble shale, is found in a matrix of Upper Cretaceous (Turonian) sediments.

Chapters 9–12 concern studies of distal ejecta associated with the K-T boundary interval. In chapter 9, Keller provides evidence of possible diachroneity around the K-T boundary interval and interprets the stratigraphic record from several areas around the Gulf of Mexico and Caribbean as products of multiple impact events. King and Petruny, in chapter 10, describe the occurrence of impact spherules in a K-T boundary succession preserved in Alabama. In chapter 11, Campbell, Oboh-Ikuenobe, and Eifert document a K-T succession from southeastern Missouri that includes a megatsunami deposit. Chapter 12, by Jannett and Terry, documents the occurrence of shocked quartz and spherules associated with distal ejecta in deltaic environments of the South Dakota badlands. The complexity of these diverse structures and breccias of different ages provides valuable insight into the widely variable sedimentary record of terrestrial meteorite impacts.

Kevin R. Evans
Missouri State University
Springfield, Missouri

REFERENCES CITED

Encyclopædia Britannica, 2007, Sedimentary rock: Encyclopædia Britannica Online, http://www.britannica.com/eb/article-9109697 (July 20, 2007).

University of New Brunswick Planetary and Space Science Centre, 2007, Earth Impact Database, http://www.unb.ca/passc/ImpactDatabase/ (July 20, 2007).

The Geological Society of America
Special Paper 437
2008

Impact melting in sedimentary target rocks: An assessment

G.R. Osinski
Department of Earth Sciences/Physics and Astronomy, University of Western Ontario, London, Ontario N6A 5B7, Canada

J.G. Spray
Planetary and Space Science Centre, Department of Geology, University of New Brunswick, 2 Bailey Drive, Fredericton, New Brunswick E3B 5A3, Canada

R.A.F. Grieve
Earth Sciences Sector, Natural Resources Canada, Ottawa, Ontario K1A 0E4, Canada

ABSTRACT

Despite being present in the target sequence of ~70% of the world's known impact structures, the response of sedimentary rocks to hypervelocity impact remains poorly understood. Of particular significance is the relative importance and role of impact melting versus decomposition in carbonate and sulfate lithologies. In this work, we review experimental evidence and phase equilibria and synthesize these data with observations from studies of naturally shocked rocks from several terrestrial impact sites. Shock experiments on carbonates and sulfates currently provide contrasting and ambiguous results. Studies of naturally shocked materials indicate that impact melting is much more common in sedimentary rocks than previously thought. This is in agreement with the phase relations for calcite. A summary of the criteria for the recognition of impact melts derived from sedimentary rocks is presented, and it is hoped that this will stimulate further studies of impact structures in sedimentary target rocks. This assessment leads us to conclude that impact melting is common during hypervelocity impact into both crystalline and sedimentary rocks. However, the products are texturally and chemically distinct, which has led to much confusion in the past, particularly in terms of the recognition of impact melts derived from sedimentary rocks.

Keywords: impact cratering, impact melting, decomposition, impactites.

INTRODUCTION

An extensive literature review indicates that sedimentary rocks are present in the target sequence of ~70% of the world's known impact structures (Table 1). Despite this large percentage, the response of sedimentary rocks to hypervelocity impact remains poorly understood. Sedimentary rocks differ from most igneous and metamorphic rocks in that they are typically rich in volatiles, be it CO_2 in carbonates, SO_x in evaporites, or H_2O in hydrous minerals, such as clays. The high porosities of many sedimentary rocks and, thus, the potential to be water-saturated, and the ubiquitous presence of layering, further sets these lithologies apart from metamorphic and igneous rocks. It has been believed for some time that porosity, the presence of volatiles,

Osinski, G.R., Spray, J.G., and Grieve, R.A.F., 2008, Impact melting in sedimentary target rocks: An assessment, *in* Evans, K.R., Horton, J.W., Jr., King, D.T., Jr., and Morrow, J.R., eds., The Sedimentary Record of Meteorite Impacts: Geological Society of America Special Paper 437, p. 1–18, doi: 10.1130/2008.2437(01).

TABLE 1. LIST OF CONFIRMED TERRESTRIAL IMPACT STRUCTURES WITH THEIR IMPORTANT ATTRIBUTES (DATA FROM THE EARTH IMPACT DATABASE) AND A SUMMARY OF THE TARGET STRATIGRAPHY (THIS STUDY)

Crater name	Location	Diameter (km)	Target rock	Notes on the target stratigraphy
Acraman	Australia	90	C	Mesoproterozoic volcanics, predominantly dacite
Ames	USA	16	M	Precambrian granite, overlain by Cambrian-Ordovician dolomite
Amelia Creek	Australia	~20	M	Paleoproterozoic metasediments and metavolcanics, overlain by ~150 m of Neoproterozoic to Cenozoic sandstones
Amguid	Algeria	0.45	S	Lower Devonian sandstone
Aorounga	Chad	12.6	S	Devonian carbonate-bearing sandstone
Aouelloul	Mauritania	0.39	S	Ordovician sandstone
Araguainha	Brazil	40	M	Precambrian granite, overlain by ~1500–1800 m of Devonian sandstone, conglomerate, siltstone; Permo-Carboniferous sandstone; Permian clay/siltstone
Arkenu 1	Libya	6.8	S	Cretaceous sandstone
Arkenu 2	Libya	10	S	Cretaceous sandstone
Avak	USA	12	S	Ordovician-Cretaceous shale, sandstone, argillite
B.P Structure	Libya	2	S	Cretaceous sandstone, siltstone, conglomerate
Barringer	USA	1.186	S	Sandstone, siltstone, limestone, dolomite
Beaverhead	USA	60	M	Precambrian, crystalline; Paleozoic, sedimentary
Beyenchime-Salaatin	Russia	8	S	Cambrian limestone, minor dolomite
Bigach	Kazakhstan	8	M	Devonian and Carboniferous sandstone, andesite, basalt, dacite, tuff (all folded)
Boltysh	Ukraine	24	C	Precambrian granite, migmatite, minor gneisses
Bosumtwi	Ghana	10.5	M	2 Ga greenschist facies metasediments, overlain by minor limestone
Boxhole	Australia	0.17	C	Precambrian gneiss
Brent	Canada	3.8	C	Precambrian gneiss
Calvin	USA	8.5	S	Cambrian to Devonian limestone, dolomite, siliciclastics
Campo Del Cielo	Argentina	0.05	M	Sandy loess overlying volcanics
Carswell	Canada	39	M	Proterozoic gneiss, granite, and siliciclastics
Charlevoix	Canada	54	M	Precambrian gneiss, overlain by Paleozoic limestone (~150 m)
Chesapeake Bay	USA	90	M	Upper Proterozoic gneiss, granite, overlain by Jurassic-Triassic sandstone, shale, siltstone, and Cretaceous deltaic deposits
Chicxulub	Mexico	170	M	Paleozoic crystalline basement, overlain by ~3 km of Cretaceous limestone, dolomite, evaporites
Chiyli	Kazakhstan	5.5	S	Paleozoic-Cenozoic, sedimentary
Chukcha	Russia	6	M	Precambrian, sedimentary and crystalline
Clearwater East	Canada	26	M	Archean granodiorite, metagranite, mafic granulites, cut by dolerite dikes, overlain by Paleozoic limestone
Clearwater West	Canada	36	M	Archean granodiorite, metagranite, mafic granulites, cut by dolerite dikes, overlain by Paleozoic limestone
Cloud Creek	USA	7	S	Late Triassic sandstone
Connolly Basin	Australia	9	S	Permian to Cretaceous sandstone, siltstone, shale
Couture	Canada	8	C	Precambrian gneiss
Crawford	Australia	8.5	C-Ms	Neoproterozoic to lower Paleozoic metasediments and felsic intrusives
Crooked Creek	USA	7	S	Cambrian dolomite, limestone, minor sandstone; Ordovician dolomite, minor chert, sandstone
Dalgaranga	Australia	0.024	C	Precambrian granite
Decaturville	USA	6	M	540 m of Cambrian dolomite, limestone, sandstone Precambrian overlying granite pegmatites and schist
Deep Bay	Canada	13	C	Proterozoic pelitic gneiss, minor granite, migmatite
Dellen	Sweden	19	C	1.7–2.6 Ga gneisses
Des Plaines	USA	8	S	Lower Paleozoic dolomite, limestone, siliciclastics
Dobele	Latvia	4.5	S	750–800 m; Silurian carbonate overlain by Devonian sandstone, carbonate-evaporite, carbonate
Eagle Butte	Canada	10	S	Cretaceous–lower Tertiary shales, carbonates
Elbow	Canada	8	S	Paleozoic to Mesozoic sedimentary rocks
El'gygytgyn	Russia	18	C	Slightly folded Cretaceous volcanics (andesite, rhyolite, tuff)
Flaxman	Australia	10	C-Ms	Neoproterozoic to lower Paleozoic metasediments and felsic intrusives
Flynn Creek	USA	3.8	S	1700 m of Paleozoic limestone and dolomite
Foelsche	Australia	6	M	>1 km sandstone, dolomite, mudstone, with 85 m thick dolerite sill at ~150 m depth; other volcanics at depth
Gardnos	Norway	5	C	Precambrian granites, gneisses, minor amphibolites
Glasford	USA	4	S	Shale and dolomite, unknown thickness

(*continued*)

TABLE 1. LIST OF CONFIRMED TERRESTRIAL IMPACT STRUCTURES WITH THEIR IMPORTANT ATTRIBUTES (DATA FROM THE EARTH IMPACT DATABASE) AND A SUMMARY OF THE TARGET STRATIGRAPHY (THIS STUDY) (*CONTINUED*)

Crater name	Location	Diameter (km)	Target rock	Notes on the target stratigraphy
Glikson	Australia	~19	M	Slightly folded Neoproterozoic siliciclastics (1800 m thick) and dolerite sills/dikes
Glover Bluff	USA	8	S	Cambrian sandstone, Ordovician dolomite
Goat Paddock	Australia	5.1	S	Proterozoic sandstone, shale, limestone
Gosses Bluff	Australia	22	S	4.5 km of Ordovician sandstone, siltstone; Devonian sandstone
Gow	Canada	5	C	Precambrian, granite, quartzofeldspathic gneiss
Goyder	Australia	3	S	Mesoproterozoic sandstone, mudstone
Granby	Sweden	3	M	Precambrian, crystalline; Paleozoic, sedimentary
Gusev	Russia	3	S	Folded Carboniferous and Permian sandstone, siltstone, shales, limestone overlain by Triassic slates, siltstone
Gweni-Fada	Africa	14	S	Devonian sandstone
Haughton	Canada	23	S	1880 m of lower Paleozoic limestone, dolomite, sandstone, shale, overlying Precambrian gneiss, metagranite
Haviland	USA	0.015	S	Cenozoic, sedimentary
Henbury	Australia	0.157	S	Neoproterozoic sandstone, mudstone
Holleford	Canada	2.35	C	Precambrian, crystalline and metasediments
Ile Rouleau	Canada	4	S	Proterozoic dolomite, minor shale
Ilumetsä	Estonia	0.08	S	Paleozoic and Cenozoic, sedimentary
Ilyinets	Ukraine	8.5	M	Precambrian granite, gneiss, amphibolite, overlain by thin cover of Paleozoic siltstone, limestone
Iso-Naakkima	Finland	3	S	10 m conglomerate overlying mica gneiss
Jänisjärvi	Russia	14	C-Ms	Precambrian metasediments; amphibolite facies schists
Kaalijärv	Estonia	0.11	S	Paleozoic, sedimentary; Cenozoic, alluvium
Kalkkop	South Africa	0.64	S	Cenozoic sandstone, mudstone, shale
Kaluga	Russia	15	M	Precambrian gneiss, metagranite, schist, overlain by Precambrian siltstone (125 m) and Devonian claystone, siltstone, sandstone, sulfate-bearing carbonates
Kamensk	Russia	25	S	Folded Carboniferous and Permian sandstone, siltstone, shales, limestone overlain by Triassic slates, siltstone
Kara	Russia	65	M	5.5 km of lower Paleozoic carbonate, sandstone, intruded by rare dolerite sills
Kara-Kul	Tajikistan	52	C	Slightly metamorphosed Paleozoic sedimentary rocks, intensively folded and intruded with granites
Kärdla	Estonia	4	M	200 m of weakly consolidated sandstone, siltstone, carbonate, overlying granite, gneisses
Karikkoselkä	Finland	1.5	C	Proterozoic, granite
Karla	Russia	10	S	Carboniferous limestone, dolomite (>390 m thick); Permian limestone, dolomite (~320 m thick); Jurassic claystone, sandstone; Cretaceous claystone
Kelly West	Australia	10	C-Ms	Precambrian metasediments, mainly quartzite
Kentland	USA	13	S	Ordovician-Silurian dolomite, limestone; Devonian shale; Carboniferous shale, limestone
Keurusselkä	Finland	30	C	Paleoproterozoic granite
Kgagodi	Botswana	3.5	C	Granite, overlain and intruded by dolomite
Kursk	Russia	6	M	100 m of Devonian siltstone, limestone, sandstone overlying Precambrian gneiss, metagranite
La Moinerie	Canada	8	C	Precambrian gneiss
Lappajärvi	Finland	23	M	Precambrian, crystalline; Paleozoic, sedimentary
Lawn Hill	Australia	18	M	Folded Proterozoic sandstone, shale, siltstone, tuff, overlain by Cambrian limestone
Liverpool	Australia	1.6	S	Precambrian flat-lying sandstone
Lockne	Sweden	7.5	M	Precambrian, crystalline; Paleozoic, sedimentary
Logancha	Russia	20	M	Permian to Triassic siltstone, claystone, sandstone, basalt, tuff
Logoisk	Belarus	15	M	Precambrian gneiss overlain by Proterozoic sandstone, silt/claystone (220–340 m) and Devonian claystone, marl, dolomite (<90 m)
Lonar	India	1.83	C	Cretaceous basalt lava flows
Lumparn	Finland	9	M	1.5 Ga coarse-grained granite
Macha	Russia	0.3	S	Late Proterozoic sandstone, limestone, overlain by Quaternary sands
Manicouagan	Canada	100	M	Precambrian, crystalline; Paleozoic, sedimentary
Manson	Iowa, USA	35	M	Precambrian, crystalline and sedimentary; Paleozoic and Mesozoic, sedimentary
Maple Creek	Canada	6	S	Cretaceous sandstone, shale, siltstone
Marquez	USA	12.7	S	Unconsolidated Paleocene strata overlying Cretaceous marl, dolomite, shale
Middlesboro	USA	6	S	Paleozoic, sedimentary
Mien	Sweden	9	C	Precambrian, crystalline

(*continued*)

TABLE 1. LIST OF CONFIRMED TERRESTRIAL IMPACT STRUCTURES WITH THEIR IMPORTANT ATTRIBUTES (DATA FROM THE EARTH IMPACT DATABASE) AND A SUMMARY OF THE TARGET STRATIGRAPHY (THIS STUDY) (*CONTINUED*)

Crater name	Location	Diameter (km)	Target rock	Notes on the target stratigraphy
Mishina Gora	Russia	4	M	Proterozoic sandstone (90 m); Cambrian claystone, sandstone (100 m); Ordovician sandstone, dolomite, limestone (150 m); Devonian marl, dolomite, sandstone (200 m), overlying Precambrian gneiss
Mistastin	Canada	28	C	Precambrian, crystalline
Mizarai	Lithuania	5	C	Precambrian amphibolite, gneiss, granite
Mjølnir	Norway	40	S	Mesozoic, sedimentary
Montagnais	Canada	45	S	Phanerozoic, sedimentary
Monturaqui	Chile	0.46	C	Mesozoic and Cenozoic, crystalline
Morasko	Poland	0.1	S	Cenozoic, soil
Morokweng	South Africa	70	C	Precambrian, crystalline
Mount Toondina	South Australia	4	S	>1 km Mesozoic and Paleozoic sandstone, shale
Neugrund	Estonia	8	S	~100 m siliciclastic sediments overlying Precambrian basement
New Quebec	Canada	3.44	C	Precambrian, crystalline
Newporte	USA	3.2	M	~150 m of Cambrian-Ordovician sandstone, shale, overlying Precambrian metasediments, schists, granite
Nicholson	Canada	12.5	M	Precambrian, crystalline; Paleozoic, sedimentary
Oasis	Libya	18	S	Cretaceous sandstone, siltstone, conglomerate
Obolon	Ukraine	20	M	250–350 m of Carboniferous claystone, sandstone, limestone, Triassic sandstone, overlying Precambrian gneiss, granite
Odessa	USA	0.168	S	Cenozoic and Mesozoic, sedimentary
Ouarkziz	Algeria	3.5	S	Carboniferous limestone, shale
Paasselkä	Finland	10		
Piccaninny	Australia	7	S	Devonian sandstone and conglomerate
Pilot	Canada	6	C	Precambrian, crystalline
Popigai	Russia	100	M	Precambrian gneiss, schist, granite, overlain by 1–1.5 km of Proterozoic and Cambrian quartzite, dolomite, limestone, Permian sandstone, siltstone, Triassic tuff, dolerite sills
Presqu'ile	Canada	24	C	Precambrian, crystalline
Puchezh-Katunki	Russia	80	M	~2 km of Devonian to Triassic limestone, dolomite, sandstone, shale, overlying crystalline basement
Ragozinka	Russia	9	M	Folded Devonian-Carboniferous carbonate, volcanics, sandstone, shale, overlain by 100–200 m of Jurassic-Cretaceous sandstone, coal, siltstone
Red Wing	USA	9	S	Silurian-Triassic limestone, dolomite, with minor sandstone, evaporite
Riachao Ring	Brazil	4.5	S	Paleozoic, sedimentary
Ries	Germany	24	M	Paleozoic and older crystalline; Mesozoic and Cenozoic, sedimentary
Rio Cuarto	Argentina	1 by 4.5	M	Cenozoic, sedimentary and crystalline
Rochechouart	France	23	C	Precambrian, crystalline
Rock Elm	USA	6	S	Paleozoic sedimentary
Roter Kamm	Namibia	2.5	C	Metasediments (schist, marble, quartzite) overlying crystalline basement
Rotmistrovka	Ukraine	2.7	C	Precambrian granite
Sääksjärvi	Finland	6	M	Precambrian, crystalline and sedimentary
Saarijärvi	Finland	1.5	C	Metagranite cut by dolerite dikes
Saint Martin	Canada	40	M	Precambrian, crystalline; Paleozoic, sedimentary
Serpent Mound	Ohio, USA	8	S	Cambrian-Ordovician to Carboniferous dolomite, limestone, shale, sandstone
Serra da Cangalha	Brazil	12	S	Devonian to Permian sandstone, shale
Shoemaker (formerly Teague Ring)	Australia	30	M	Archean granite, syenite, overlain by Proterozoic siltstone, shale, limestone
Shunak	Kazakhstan	2.8	C	Folded Devonian volcanics, mainly rhyolite
Sierra Madera	USA	13	S	Permian limestone, dolomite, overlain by Cretaceous sandstone, carbonates
Sikhote Alin	Russia	0.027	C	Mesozoic volcanics
Siljan	Sweden	52	M	Precambrian, crystalline; Paleozoic, sedimentary
Slate Islands	Canada	30	C	Archean (2.7 Ga) pyroclastics, flows; Proterozoic metavolcanics, metasediments
Sobolev	Russia	0.053	M	Mesozoic and Cenozoic, volcanic and sedimentary
Söderfjärden	Finland	5.5	C	Granite
Spider	Australia	13	S	Precambrian sandstone, siltstone
Steen River	Canada	25	M	1.4 km of Devonian-Carboniferous carbonate, evaporite, shale, overlying Precambrian granitic gneiss
Steinheim	Germany	3.8	S	1180 m of limestone, marls, sandstone

(*continued*)

TABLE 1. LIST OF CONFIRMED TERRESTRIAL IMPACT STRUCTURES WITH THEIR IMPORTANT ATTRIBUTES (DATA FROM THE EARTH IMPACT DATABASE) AND A SUMMARY OF THE TARGET STRATIGRAPHY (THIS STUDY) (*CONTINUED*)

Crater name	Location	Diameter (km)	Target rock	Notes on the target stratigraphy
Strangways	Australia	25	M	Precambrian metagranite overlain by Mesoproterozoic shale, siltstone, sandstone, Cambrian sedimentary rocks and volcanics
Suavjärvi	Russia	16	C-Ms	Precambrian metasediments (pelite, quartzite)
Sudbury	Canada	250	C	Precambrian gneiss, granite, metasediments
Suvasvesi N	Finland	4	C	Precambrian granite, migmatite, schist
Tabun-Khara-Obo	Mongolia	1.3	C	Paleozoic mica schists
Talemzane	Algeria	1.75	S	Mesozoic and Cenozoic limestone, shale
Tenoumer	Mauritania	1.9	M	Precambrian gneisses, granites; Cenozoic carbonates
Ternovka	Ukraine	11	C	Precambrian schist, quartzite
Tin Bider	Algeria	6	S	>1 km Cretaceous claystone, sandstone, limestone
Tookoonooka	Australia	55	M	Ordovician metasediments, overlain by Jurassic-Cretaceous sandstone, siltstone
Tswaing (formerly Pretoria Saltpan)	South Africa	1.13	C	2.05 Ga metagranite
Tvären	Sweden	2	M	Precambrian, crystalline; Paleozoic, sedimentary
Upheaval Dome	USA	10	S	Paleozoic and Mesozoic, sedimentary
Vargeao Dome	Brazil	12	M	Jurassic-Triassic sandstone, Cretaceous basalt
Veevers	Australia	0.08	S	Laterite
Vepriai	Lithuania	8	S	Vendian sandstone, Cambrian sandstone, siltstone, claystone, Ordovician carbonate, claystone, Silurian marl, dolomite, Devonian sandstone, siltstone, carbonates
Viewfield	Canada	2.5	S	Lower Paleozoic limestone, siltstone, sandstone, evaporite
Vista Alegre	Brazil	9.5		Jurassic-Triassic sandstone, Cretaceous basalt
Vredefort	South Africa	300	M	Archean gneiss, Archean-Paleoproterozoic supracrustal rocks
Wabar	Saudi Arabia	0.116	S	Cenozoic, loosely consolidated sand
Wanapitei	Canada	7.5	C	Precambrian, crystalline and metasediments
Wells Creek	USA	12	S	Ordovician and Silurian limestone, dolomite; Devonian shale, Carboniferous limestone
West Hawk	Canada	2.44	C	Archean basaltic-andesite metavolcanics
Wetumpka	USA	6.5	M	Precambrian gneiss, overlain by unconsolidated Cretaceous sands, silts, marls
Wolfe Creek	Australia	0.875	S	Devonian sandstone, laterite
Woodleigh	Australia	40	M	Precambrian granite, gneiss, overlain by Ordovician to Devonian sandstone
Yarrabubba	Australia	30	C	Precambrian granite, greenstones
Zapadnaya	Ukraine	3.2	C	Precambrian granite, migmatite, gneiss
Zelenv Gai	Ukraine	2.5	C	Precambrian granite, migmatite, gneiss
Zhamanshin	Kazakhstan	14	M	Folded Silurian metasediments (quartzite, schist), Carboniferous andesite, tuff, overlain by ~300 m of Cretaceous claystone, sandstone, siltstone

Note: See Earth Impact Database (2007) for geographic coordinates and age of the listed impact structures. C—crystalline target; C-Ms—metasedimentary target; M—mixed target (i.e., sedimentary strata overlying crystalline basement); S—sedimentary target (i.e., no crystalline rocks affected by the impact event).

and the heterogeneity of layered targets exert a considerable influence on the details of processes and, ultimately, the products of hypervelocity impact (e.g., Kieffer and Simonds, 1980). An understanding of the response of sedimentary rocks to impact is, therefore, needed in order to assess the effect of the cratering process on these lithologies and the potential environmental influences such impacts may have had on Earth.

For example, it is widely held that carbonates and evaporites decompose during impact and release massive amounts of CO_2 (e.g., O'Keefe and Ahrens, 1989) and sulfur-bearing gases (e.g., Pope et al., 1994), respectively. Impacts into sedimentary targets may, therefore, be more environmentally damaging than impacts of the same size into igneous or metamorphic terrains. This is widely assumed to be the case with the ~180 km diameter Chicxulub impact structure, Mexico, which impacted into a thick sequence of carbonates and evaporites (e.g., Penfield and Camargo, 1981). Chicxulub coincides with the Cretaceous-Tertiary (K-T) boundary, and quantification of the amount of CO_2 and sulfur-bearing gases released during the impact process—from the decomposition of carbonates and evaporites, respectively—is required to assess their role in the resultant K-T mass extinction event (Alvarez et al., 1980). It is notable that the relative climatic importance of CO_2 and sulfur-bearing gases (e.g., SO_2 and SO_3) in the wake of the Chicxulub impact event is still actively debated (Pierazzo et al., 1998).

One of the outstanding questions in impact cratering studies is the relative importance and role of impact melting versus decomposition for impacts into sedimentary rocks. This question cannot be completely addressed through experimentation in the laboratory, which is limited to impact velocities generally below that required for extensive melting (Grieve and Cintala, 1992). The duration of the shock state is also much shorter in experiments than in nature. Numerical and computer-based modeling may offer some important information; however, as Pierazzo et al. (1998) note, "there is no good model for melt production from impact craters in sedimentary targets." Very few detailed, systematic studies of naturally shocked sedimentary rocks have been carried out to date, yet such studies offer the only true ground-truth data on the response of sedimentary rocks to impact. The aim of this paper is to provide an up-to-date assessment of the importance of impact melting versus decomposition during impacts into sedimentary target rocks, based on studies carried out by the authors and a review of the existing literature.

PHYSICS OF IMPACT MELT GENERATION

Theoretical considerations of the impact process have revealed important results regarding the generation of impact melt:

1. It is widely understood that impact melting occurs upon decompression. Shock compression deposits a large amount of energy in the target, which remains as heat following decompression. If the shock is strong enough and sufficient heat remains, the released material may be left in the form of a liquid (i.e., melt) or vapor (Melosh, 1989, p. 42–44).
2. A large amount of compression and shock heating occurs in porous target rocks (Kieffer, 1971; Ahrens and Cole, 1974). High porosity significantly increases the amount of pressure-volume work in the target rocks resulting from the shock wave, which results in greater amounts of postshock waste heat, raising temperatures. However, the crushing of pore space reduces the overall shock pressures in the target.
3. The volumes of target material shocked to pressures sufficient for melting are not significantly different in sedimentary or crystalline rocks (Kieffer and Simonds, 1980).
4. Model calculations indicate that both wet and dry sedimentary rocks yield greater volumes of melt on impact than crystalline targets (Kieffer and Simonds, 1980).

Thus, impacts into sedimentary targets should produce as much melt as do impacts into crystalline targets. However, it has been a generally accepted observation that impact melt rocks are not generated in impact structures formed in sedimentary targets (e.g., Dressler and Reimold, 2001). This anomaly has been attributed to the formation and expansion of enormous quantities of sediment-derived vapor (e.g., H_2O, CO_2, SO_2), resulting in an unusually wide dispersion of shock-melted sedimentary rocks (Kieffer and Simonds, 1980).

METEORITE IMPACT INTO CARBONATES

Carbonates are present in the target rocks of approximately one third of the world's known impact structures (Table 1). Despite the many uncertainties regarding the response of carbonates to impact, it is commonly accepted that these lithologies decompose after pressure release due to high residual temperatures and that subsequent fast back-reactions trap a significant part of the gaseous species (Kieffer and Simonds, 1980; Martinez et al., 1994a; Agrinier et al., 2001). It should be noted that decomposition is a chemical reaction in which a compound or molecule breaks down into smaller, simpler compounds, molecules, or elements. Vaporization, on the other hand, is the process of converting a substance from a liquid or solid state to the gaseous (i.e., vapor) state.

Shock Experiments and Thermodynamic Calculations

A compilation by Agrinier et al. (2001), and updated here, reveals that shock experiments provide contrasting and ambiguous results regarding the onset of decomposition of carbonates (Table 2). This may be due, in part, to differences between experimental techniques (single shock versus reverberation), and/or properties of the sample material (e.g., porosity) (Martinez et al., 1994a), and/or the duration of the shock state. To complicate matters, substantial differences exist between experimental observations and thermodynamic calculations (e.g., Martinez et al., 1994a; Agrinier et al., 2001; Skála et al., 2002, and references therein). Early experimental studies suggested that calcite undergoes significant decomposition (>10%–50%) at pressures as low as 10–20 GPa (e.g., Lange and Ahrens, 1986). However, recent shock experiments, coupled with optical and X-ray diffraction studies, suggest that decomposition of calcite and dolomite only occurs at pressures >65 GPa and >70 GPa, respectively (Gupta et al., 2002; Langenhorst et al., 2000; Skála et al., 2002). Furthermore, recent dynamic loading and fast unloading experiments have produced complete shock melting of $CaCO_3$ at pressures of ~25 GPa and temperatures of ~2700 K (Langenhorst et al., 2000).

Phase Relations of $CaCO_3$

The phase diagram for $CaCO_3$ has recently been reevaluated in relation to shock compression and decompression by Ivanov and Deutsch (2002) (Fig. 1). The main outcome of their work has been the extension of the liquid field of $CaCO_3$. For example, isentropic release paths for calcite shocked to a pressure of >10 GPa first enter the liquid field, with decomposition only possible after pressure has dropped to <0.003 GPa (30 bar) at temperatures of ~1500 K (Fig. 1). Decomposition is terminated at temperatures of <1200 K at atmospheric pressure (Ivanov and Deutsch, 2002). Thus, the phase relations of $CaCO_3$ suggest that the expected result of hypervelocity impact into calcite is melting, with decomposition only occurring during postshock cooling. Indeed, the phase relations would suggest that $CaCO_3$, shocked to pressures >1 GPa and temperatures >1500–2000 K, should undergo melting (Fig. 1). Ivanov and Deutsch (2002) also note that complex reaction kinetics, in particular the rate of diffusion, may limit the amount of CO_2 released during impact events.

TABLE 2. COMPILATION OF EXPERIMENTAL AND MODELING DATA ON THE SHOCK BEHAVIOR OF CARBONATES

Pressure (GPa)	Material	Experimental method(s)	Observed effect(s)	Reference
10–42	Cc., single crystal	Shock reverberation; sample recovery	10%–67% devolatilization, nonhomogeneous	Lange and Ahrens, 1986
10	Chalk, 49% porosity	Impedance matching; particle velocity measurements	90% devolatilization	Tyburczy and Ahrens, 1986
17	Arag., single crystal, 1.7% porosity	Impedance matching; particle velocity measurements	Onset of devolatilization	Vizgirda and Ahrens, 1982
18	Cc., single crystal	Impedance matching; trapping of gas	Onset (~0.3 mol%)	Boslough et al., 1982
20	Cc., single crystal	Shock reverberation; sample recovery	~50% devolatilization	Lange and Ahrens, 1986
25	Lst., polycryst.	Impedance matching; particle velocity measurements	20%–30% devolatilization	Tyburczy and Ahrens, 1986
25	Cc., single crystal	Multi-anvil	Melting	Langenhorst et al., 2000
30	Lst., polycryst.	Impedance matching; single shocks; sample recovery	13% devolatilization	Kotra et al., 1983
>30	Dol. rock, polycryst.	Impedance matching; sample recovery	Possible formation of periclase	Skála et al., 1999
33	Cc., nonporous	Calculation of postshock entropy; isentropic release	Onset of devolatilization	Vizgirda and Ahrens, 1982
35–45	Cc.	Shock reverberation; sample recovery	Onset to complete devolatilization	Martinez et al., 1995
40	Cc., single crystal	Impedance matching; color *T* measurements	High-*T* shear band regions with *T* ~ 2200 K	Kondo and Ahrens, 1983
40–50	Cc., Arag., nonporous	Calculations	Onset of devolatilization	Pasternak et al., 1976*
45	Cc.	Calculation of postshock energy; Hugoniot	Onset of devolatilization	Kieffer and Simonds, 1980
55	Arag., nonporous	Calculation of postshock entropy; isentropic release	Onset of devolatilization	Vizgirda and Ahrens, 1982
55–65	Dol.	Shock reverberation; sample recovery	Onset to complete devolatilization	Martinez et al., 1995
<60	Lst., Dol., solid disks	Shock reverberation; sample recovery	Stable	M.S. Bell et al., 1998
<60	Cc., single crystal	Impedance matching; sample recovery	Minor, if any, devolatilization	Kotra et al., 1983
<60	Dol. rock, polycryst. 4%–6% porosity	Shock reverberation; sample recovery	Decrease in grain size	Martinez et al., 1995
60	Cc., single crystal	Shock reverberation; thermal gravimetric analysis	7% CO_2 loss	Bell, 2001
60	Dol. rock, polycryst. 4%–6% porosity	Shock reverberation; preheating; sample recovery	nm-sized periclase crystals	Martinez et al., 1994b
60–68	Dol. rock, polycryst.	Impedance matching; sample recovery	Relatively stable; <1 µm-sized periclase crystals	Skála et al., 2002
61	Cc.	Shock reverberation; sample recovery	Recrystallized Cc.; probably from a melt	Badjukov et al., 1995
>70	Dol. rock, polycryst.	Impedance matching; sample recovery	Disintegration of sample containers apparently due to extensive vapor production	Skála et al., 2002

Note: Modified and updated from Agrinier et al. (2001). Cc.—calcite; Lst.—limestone; Dol.—dolomite; Arag.—aragonite; polycryst.—polycrystalline.

*Unpublished work cited in Boslough et al. (1982).

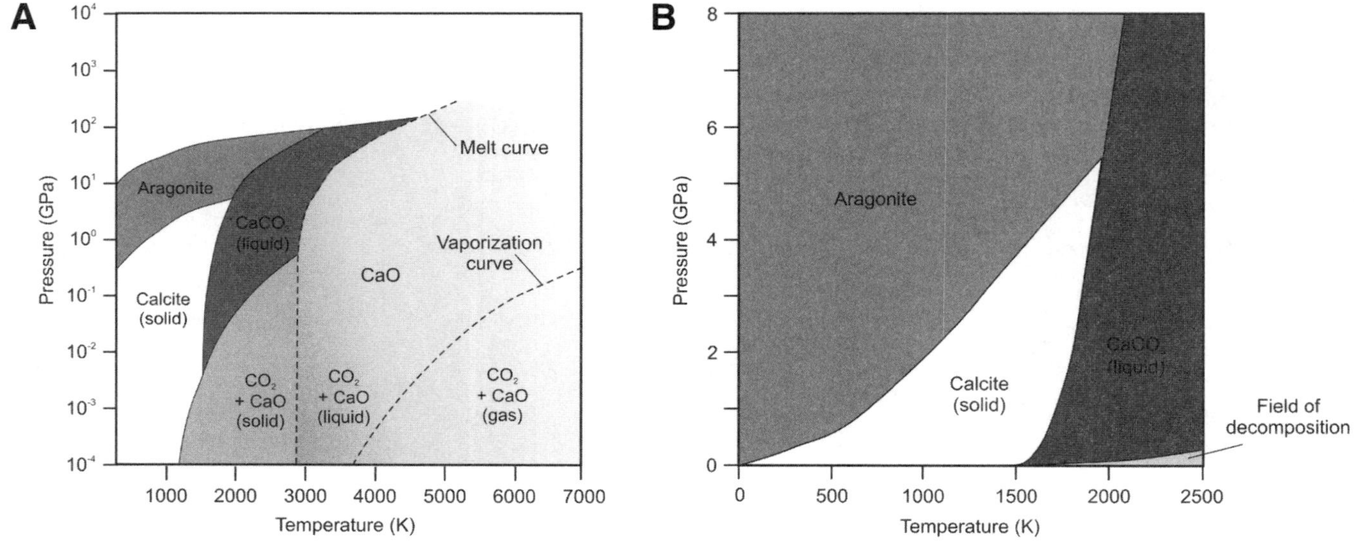

Figure 1. A: Phase diagram for $CaCO_3$ in the temperature-log (pressure) plane. B: Expanded left part of the $CaCO_3$ phase diagram shown in A. Note that this diagram is plotted in linear coordinates to reveal the real geometry of the phase boundaries. Ivanov and Deutsch (2002).

Decomposition of Carbonates During Impact: The Record in the Rocks

The decomposition of carbonates releases CO_2 and produces residual solid oxides: CaO and MgO from the decomposition of calcite and dolomite, respectively. Thus, if carbonates decompose to any great degree during impact, we would expect to detect CaO and MgO in impactites within and around terrestrial impact structures. However, CaO has not been documented from any terrestrial impact structure to date (Martinez et al., 1994a; Osinski and Spray, 2001). For impacts into limestone (i.e., predominantly $CaCO_3$), this anomaly has been attributed to subsequent rapid back-reactions of CO_2 with the initially produced CaO (e.g., Martinez et al., 1994a; Agrinier et al., 2001). While such reactions are possible, it is important to note that 100% reconversion to calcite was not achieved in any of the experiments of Agrinier et al. (2001). Reconversion is also dependent on grain size, with a reduction in efficiency with increasing grain size (Agrinier et al., 2001). Small amounts of CaO may, therefore, be expected to survive. A single case of evidence for back-reactions at terrestrial impact structures has been presented. Martinez et al. (1994a) described unshocked calcite present within vesicles and holes in silicate impact glasses in naturally shocked rocks from the Haughton impact structure as evidence for back-reactions. However, the evidence presented by these authors has been reinterpreted as liquid immiscible textures between coexisting carbonate and silicate melts (Graup, 1999; Osinski and Spray, 2001). Thus, the character of some impactites from Haughton should no longer be quoted as displaying unequivocal evidence for back-reactions.

Evidence for very minor decomposition of carbonates has currently only been noted at the Ries impact structure, Germany. Baranyi (1980) noted that a <1.5 mm thick orange-brown crust or rim, comprising clays and X-ray amorphous material, surrounds 3%–10% of the limestone clasts within the Ries suevites. Baranyi (1980) inferred that the rims were formed by the thermal decomposition of carbonate clasts during postimpact contact metamorphism and that the CaO was removed by H_2O circulating in the suevite: CaO is unstable in the presence of H_2O, rapidly reacting to form calcium hydroxide ($Ca(OH)_2$), which itself can dissolve in H_2O to release Ca^{2+} and OH^- ions (Chang et al., 1998). Recent studies have confirmed the findings of Baranyi (1980). In particular, it is clear that the decarbonated rims neither are cut by nor diffuse into the suevite groundmass, indicating that decomposition occurred after deposition of the suevite at temperatures of >900 °C (Osinski et al., 2004). Preliminary data, only available in abstract form, suggest the presence of similar clasts at the Chicxulub impact structure, Mexico (Deutsch et al., 2003).

Dolomite ($CaMg(CO_3)_2$) is an important component of the target stratigraphy of many terrestrial impact sites, such as Haughton. Unlike CaO, MgO is a stable mineral (i.e., periclase), which should be preserved in impactites. Indeed, Agrinier et al. (2001) note that "similar experiments with dolomite and magnesite show that residual Mg oxides do not react significantly at the 10^3 s timescale and may, therefore, survive as a witness of degassing in impact breccias." To the knowledge of the authors, periclase has been documented at two terrestrial impact sites. However, it is clear, in these cases, that the this mineral was formed during contact metamorphism of carbonates, present either as clasts within a silicate impact melt layer (Clearwater Lake; Rosa, 2004) or in target rocks directly beneath a melt sheet (Manicouagan; Spray, 2006). In other words, the decomposition of carbonates at these two sites was a postshock phenomenon due to the high temperatures of juxtaposed silicate impact melt (cf. limestone at the Ries structure). For example, at Manicouagan, the carbonates beneath the melt sheet attained temperatures of >900 °C, resulting in a series of contact-metamorphic minerals: diopside + labradorite + periclase + brownmillerite + spurrite + perovskite + mayenite (Spray, 2006).

Melting of Carbonates During Impact: The Record in the Rocks

Evidence for the impact melting of carbonates has now been recognized at five terrestrial impact structures (Table 3), as well as in distal impact ejecta from undetermined source craters in West Greenland (Jones et al., 2005), and the Western Desert of Egypt (Osinski et al., 2007). Based on our observations of impactites from several impact sites, and a review of the existing literature, three main occurrences of carbonate impact melts are distinguished: (1) groundmass-forming phases within impact melt-bearing crater-fill and proximal ejecta deposits, (2) globules and irregularly shaped masses within impact glass clasts from proximal ejecta deposits, and (3) individual particles and spherules within the proximal and distal ejecta deposits. Textural and chemical evidence for the impact melting of carbonates during hypervelocity impact is provided by:

1. *Liquid immiscible textures.* Liquid immiscibility is an important mechanism of magmatic differentiation in some igneous systems (Roedder, 1978). It is important to note that liquid immiscibility *sensu stricto* describes the process whereby an initially homogeneous melt reaches a temperature where it can no longer exist stably and so unmixes into two liquids of markedly different composition and density (Roedder, 1978). Textural evidence for the presence of immiscible impact-generated melts has been documented at several terrestrial impact sites (Table 3). This evidence includes ocellar or emulsion textures of globules of carbonate in silicate glass, sharp menisci and budding between silicate and carbonate glasses, and deformable and coalescing carbonate spheres within silicate glass (Figs. 2A–2D). These provide unequivocal evidence for carbonates and silicate glasses being in the liquid state at the same time. However, in contrast to magmatic systems, it is possible, and is to be expected, that impact-generated melts from target rocks at different stratigraphic levels will not completely mix or homogenize. This is particularly true for melt derived from a crystal-

TABLE 3. IMPACT SITES WHERE EVIDENCE FOR THE MELTING OF CARBONATES HAS BEEN REPORTED

Impact site	Impactite	Setting of carbonate	Textures and geochemistry						References
			Liq Imm	Quen	Sph	Euh Carb	Ca–Mg–rich silicates	Ca–Mg–CO_2–rich glass	Carb Chem	
Chicxulub	Proximal ejecta	Individual particles	N	Y	N	N	N	N	n.a.	Jones et al., 2000
	Distal ejecta	Individual particles	N	N	Y	N	N	N	n.a.	Jones et al., 2000
	Proximal ejecta	Embedded in glass clasts	Y	N	N	N	N	N	n.a.	Dressler et al., 2004; Kring et al., 2004; Schmitt et al., 2004; Stöffler et al., 2004; Tuchscherer et al., 2004
	Proximal ejecta	Within groundmass	Y	N	N	?	N	N	n.a.	Tuchscherer et al., 2004
	Proximal ejecta	Within groundmass	N	N	N	?	N	Y	n.a.	Nelson and Newsom, 2006
Haughton	Crater-fill	Groundmass	Y	N	Y	Y	N	Y	Y	Osinski and Spray, 2001; Osinski et al., 2005
	Crater-fill	Embedded in glass clasts	Y	N	N	Y	N	Y	Y	Osinski and Spray, 2001; Osinski et al., 2005
	Proximal ejecta	Groundmass	Y	N	N	N	N	N	N	Osinski and Spray, 2001; Osinski et al., 2005
Meteor Crater	Proximal ejecta	Embedded in glass clasts	N	N	Y	Y	Y	Y	Y	Osinski et al., 2003
Ries	Proximal ejecta	Embedded in glass clasts	Y	N	Y	N	N	N	n.a.	Graup, 1999; Osinski, 2003
	Proximal ejecta	Individual particles	N	Y	Y	N	N	N	n.a.	Graup, 1999
	Proximal ejecta	Groundmass	Y	N	N	N	N	Y	Y	Osinski et al., 2004
	Crater-fill	Glass clasts	Y	N	Y	N	N	N	Y	Graup, 1999
Tenoumer	Proximal ejecta	Glass clasts	Y	N	N	N	N	N	n.a.	Pratesi et al., 2005

Note: Liq Imm—liquid immiscible textures; Quen—quench-textured feathery carbonate; Sph—spherule; Euh Carb—euhedral carbonate crystals; Carb Chem—unusual carbonate chemistry; Y—yes, N—no; n.a.—no analyses presented.

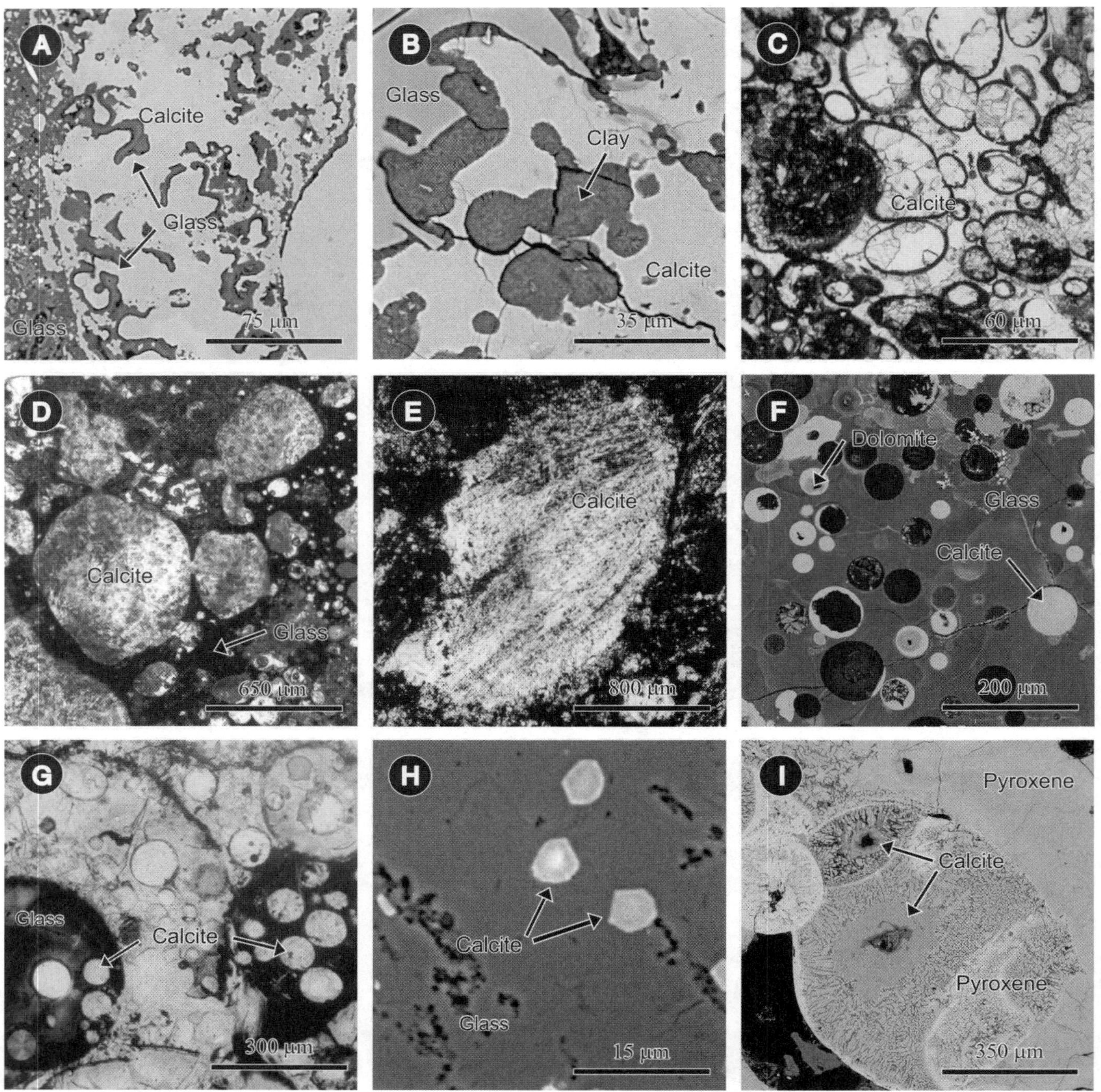

Figure 2. Backscattered electron (BSE) images (A, B, F, H, I) and plane-polarized light photomicrographs (C–E, G) showing the various textural associations of carbonate melt phases within impactites from terrestrial impact structures. A: Immiscible intergrowth of groundmass-forming calcite and MgO-rich silicate glass from crater-fill deposits at the Haughton structure (Osinski and Spray, 2001). B: Globules and irregular patches of clay minerals (originally hydrous silicate glass) within calcite and silicate glass in surficial suevites from the Ries structure (Osinski et al., 2004). C: Globules of calcite within a silicate glass–calcite groundmass in surficial suevites from the Ries structure (Osinski et al., 2004). D: Immiscible globules of calcite within silicate glass from the Tenoumer impact crater (Pratesi et al., 2005). E: Feathery carbonate from the Chicxulub structure (Jones et al., 2000). F: Calcite spherules embedded in CO_2-rich impact glass from the ballistic ejecta at Meteor Crater. Note that some calcite spherules are cored by dolomite (Osinski et al., 2003). G: Silicate glass spherules from Disko Bay, Greenland, which themselves contain calcite spherules (Jones et al., 2005). H: Impact glass clast from crater-fill deposits at the Haughton structure showing the well-developed euhedral form of the zoned calcite crystals. EDS (energy-dispersive spectrometry) analyses reveal that these calcites can contain up to ~7 wt% Al_2O_3 and ~2 wt% SiO_2. Given that the enclosing glass is pure SiO_2 (+ H_2O), the Al_2O_3 has clearly been incorporated into the calcite (Osinski et al., 2005). I: Calcite associated with Ca-rich clinopyroxene and Mg-rich olivine in ballistic ejecta from Meteor Crater (Osinski et al., 2003).

line basement with overlying sedimentary rocks. Thus, it is suggested that the term "carbonate-silicate liquid immiscibility" be avoided unless there is unequivocal evidence for the unmixing of an originally homogeneous impact melt.

2. *Quench textures.* Fragments of calcite displaying a distinctive feathery texture are common in proximal ejecta at Chicxulub (Fig. 2E) (Jones et al., 2000). These are well-understood quench crystal morphologies, which indicate rapid crystallization from a melt (Jones et al., 2000).
3. *Carbonate spherules.* Individual calcite spherules are present within a variety of impactites at several impact structures (Table 3; Figs. 2F and 2G). As noted by French (1998), the production of spherules appears to be a typical process in impact events. Carbonate spherules may be misinterpreted as secondary vesicle fillings; however, in several cases there is clear evidence indicating that these features are quenched droplets of carbonate melt (Figs. 2F and 2G). This evidence includes the presence of carbonate spherules embedded in unaltered impact glass and in samples where no other secondary carbonate minerals are present (Jones et al., 2005; Osinski and Spray, 2001).
4. *Euhedral calcite crystals within impact glass clasts.* Micrometer-sized euhedral crystals of calcite have been described from impact glass clasts within crater-fill impactites from the Haughton structure (Osinski et al., 2005). Their euhedral nature indicates that these crystals grew while the surrounding silicate glass was still in a fluid state.
5. *Carbonates intergrown with CaO-MgO-rich silicates.* Figure 2I shows euhedral crystals and pockets of calcite-dolomite within a groundmass of pyroxene in glassy clasts from the proximal ejecta at Meteor Crater. These textures are difficult to reconcile with an origin through alteration but are consistent with an impact melt origin for the carbonates (Osinski et al., 2003). The melting of carbonates at Meteor Crater is also supported by the unusual composition of associated pyroxenes and olivines: coexisting Ca-rich pyroxene (diopside/wollastonite) and Mg-rich olivine (forsterite) are common in carbonatitic igneous rocks (Barker, 2001; Gittins, 1989). The assemblage of Ca-Mg-rich clinopyroxene (diopside/augite) + Mg-rich olivine (forsterite) + calcite ± dolomite has also been produced in experiments using the simplified system of CaO-MgO-SiO_2-CO_2 (Lee et al., 2000; Lee and Wyllie, 2000).
6. *CaO-MgO-CO_2-rich glasses.* Unusual silicate glasses have been documented at several terrestrial impact sites (Table 3). At the Haughton structure, groundmass-forming glasses possess extremely high MgO contents of up to ~35 wt% (Osinski et al., 2005). The only MgO-rich target lithology at Haughton is dolomite (~21 wt% MgO and ~31 wt% CaO), which indicates that this lithology must have been a major component in the melt zone. It is also notable that many glasses at Haughton yield consistently low EDS (energy-dispersive spectrometry) analytical totals, typically ranging from ~50 wt% to ~65 wt%. Qualitative analyses suggest that the predominant volatile species in these glasses is CO_2. Similar glasses are present at Meteor Crater (Figs. 3A and 3B).
7. *Carbonate chemistry.* It has become clear from work at the Ries and Haughton structures that calcite displaying textures indicative of an impact melt origin is also often distinctly different in composition than carbonates from the target rocks and postimpact hydrothermal and/or diagenetic settings, at the same structure. For example, groundmass-forming calcite in crater-fill impact melt breccias at Haughton displays higher MgO, FeO, SO_3, Al_2O_3, and SiO_2 contents than the calcite developed in sedimentary target rocks and postimpact hydrothermal products (Osinski et al., 2005). The relatively high Al_2O_3 and SiO_2 component is particularly notable as carbonatitic (i.e., igneous) calcites are the only other known carbonates to contain elevated levels of SiO_2 and Al_2O_3. The possibility that these anomalous compositions are due to problems during analysis has also been ruled out (Osinski, 2005a; Osinski et al., 2005) (e.g., the euhedral calcite crystals in Figure 2H contain up to 8 wt% Al_2O_3 and 2 wt% SiO_2; no Al_2O_3 is present in the host glass). Experiments have shown that rapid crystallization (quenching) of a high-temperature SiO_2-rich carbonate melt can produce SiO_2-rich carbonates (~3–10 wt% SiO_2; Brooker, 1998). This is not to say that secondary carbonates do not occur at these sites; however, due consideration of textures and chemistry allows the discrimination of igneous hydrothermal calcite, even within individual clasts (e.g., Figs. 3C–3E).

METEORITE IMPACT INTO EVAPORITES

Despite the suggested environmental consequences of the release of sulfur species during impact events (Pierazzo et al., 1998), the response of sulfates to hypervelocity impact has been little studied. Previous work focused primarily on experimental studies and computer-based simulations, with a view to estimating the threshold for the vaporization and decomposition of sulfate minerals (Chen et al., 1994; Gupta et al., 2001; Ivanov et al., 1996; Pierazzo et al., 1998; Yang and Ahrens, 1998). It is apparent that shock experiments currently provide equivocal results, with estimates for incipient vaporization ranging from ~32.5 ± 2.5 GPa (Gupta et al., 2001) to >85 GPa (Langenhorst et al., 2003). Preliminary attempts at producing a phase diagram and equation of state for anhydrite in relation to impact events were presented by Ivanov et al. (2004). These authors noted that non-porous anhydrite has two solid-solid polymorphic phase transitions at high pressure and should theoretically undergo melting at >80–90 GPa, with incipient decomposition at pressures of ~60–70 GPa, for venting of gas products, and 100–110 GPa for gas

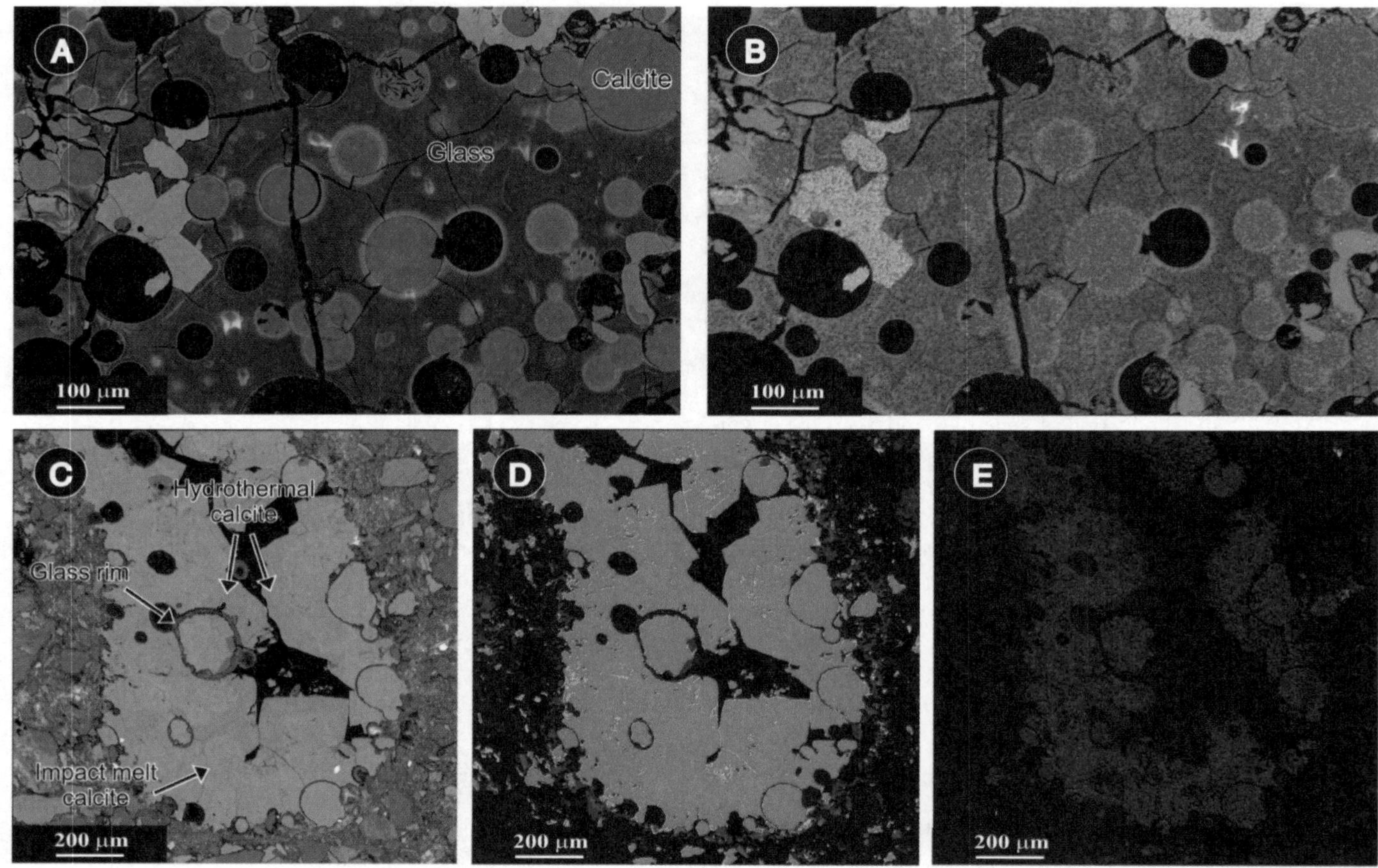

Figure 3. A and B: Backscattered electron (BSE) image and element map, respectively, of CO_2-rich glasses from Meteor Crater (Osinski et al., 2006). Red—C; blue—Si; green—Mg. C: Backscattered electron (BSE) image showing a vesiculated calcite melt clast with embedded glass-coated calcite spherules, with element maps of Ca (D) and Mn (E) (Osinski, 2005a). Note the vuggy calcite infilling the void at the center of the glass clast. The late-stage vuggy calcite is poorer in Fe and Mg and richer in Ca than the impact melt calcite.

products of the reaction in equilibrium. Little is known about the phase relations for porous anhydrite and gypsum, which would be more applicable for impact into natural sedimentary targets.

Evaporites, comprising predominantly gypsum ($CaSO_4 \cdot 2H_2O$) and anhydrite ($CaSO_4$), are present at only a handful of terrestrial impact craters (Table 1), the majority of which are buried or eroded. The only published study available of naturally shocked sulfate rocks and minerals from a terrestrial impact structure is the recent work on the Haughton structure (Osinski and Spray, 2003). Only the main observations and conclusions will be discussed here.

Detailed SEM (scanning electron microscope) studies reveal that in addition to shock-melted calcite and silicate glass (Osinski and Spray, 2001), anhydrite also constitutes an important component of the groundmass of the crater-fill melt breccias at Haughton. Textural and chemical evidence presented in Osinski and Spray (2003) indicates that the anhydrite represents a primary impact melt phase. Evidence for this includes (1) liquid immiscible textures developed between celestite and/or anhydrite and silicate glass and/or calcite (Figs. 4B and 4C), (2) possible quench-textured groundmass-forming anhydrite, (3) the presence of carbonate melt globules within anhydrite (Fig. 4A), (4) flow textures developed between anhydrite and silicate glasses, (5) elevated (trapped) concentrations of Si, Al, Mg, and H_2O in groundmass-forming anhydrite, and (6) clasts of anhydrite-quartz lithologies exhibiting evidence for (partial) shock melting (Figs. 4D–4F). Evidence from clasts of shocked anhydrite-quartz lithologies in the crater-fill deposits at Haughton suggests that anhydrite undergoes shock melting at pressures of >50–60 GPa (Figs. 4D–4F). Preliminary evidence from Chicxulub suggests that clasts of anhydrite within impact melt rocks and impact melt breccias underwent postshock annealing, melting, and possible decomposition (Deutsch et al., 2003; Langenhorst et al., 2003).

METEORITE IMPACT INTO TERRIGENOUS CLASTIC ROCKS

Terrigenous clastic sedimentary rocks, including sandstones, conglomerates, and mudrocks (i.e., siltstone, claystone, shale), are present at many impact sites (Table 1). Unlike carbonates and

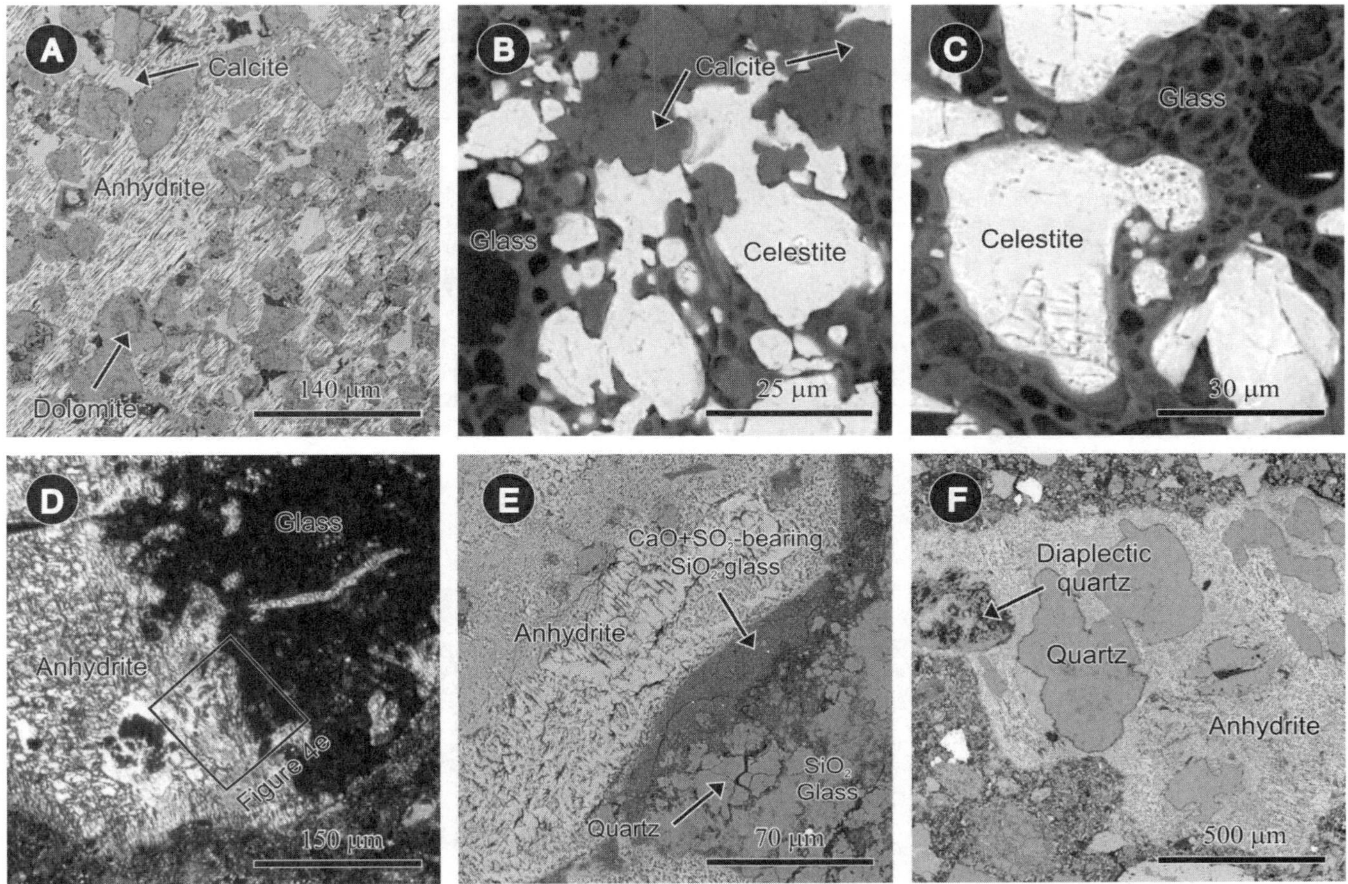

Figure 4. Backscattered electron (BSE) images (A–C, E, F) and cross-polarized light photomicrograph (D) showing the various textural associations of sulfate impact melt phases within impactites from the Haughton impact structure (Osinski and Spray, 2003). A: Anhydrite-dominant groundmass with calcite melt blebs and strongly shocked dolomite clasts. B and C: Immiscible globules of celestite and calcite within SiO_2 glass. Note the sharp contact between calcite and celestite. D: A highly shocked anhydrite-quartz clast in which quartz has been transformed into glass and is now isotropic. E: Inset from Figure 4D. BSE imagery reveals that this clast is composed of relic anhydrite and quartz enclosed by fine-grained anhydrite and SiO_2 glass, respectively. Glass of a different composition (dark) occurs at the contact between these latter two phases, suggesting mixing between two shock-melted phases. F: Anhydrite-quartz clast in which anhydrite preserves its original pinacoidal habit. Three irregularly shaped quartz grains have been transformed into diaplectic glass. Note that planar deformation features are not developed in the nondiaplectic quartz grains.

evaporites, there has been a limited acceptance that these lithologies can melt during the impact process, although it is generally assumed that the majority of the resultant melts are dispersed due to the expansion of large amounts of H_2O originally present in these lithologies (Kieffer and Simonds, 1980).

Fragments of shock-melted sandstones in ejecta were first studied in detail at Meteor Crater, Arizona (Kieffer, 1971; Kieffer et al., 1976) (Table 4). These studies revealed the dramatic effects of porosity, grain characteristics, and volatiles on the response of quartz to impact in sedimentary targets. For example, in crystalline rocks, quartz will typically be transformed to diaplectic glass at 32–50 GPa, with melting at >50–60 GPa (Grieve et al., 1996). At Meteor Crater, however, diaplectic glass is present in rocks shocked to pressures as low as ~5.5 GPa, with whole-rock melting occurring at >30–35 GPa (Kieffer, 1971; Kieffer et al., 1976). Shock-melted sandstones have now been recognized at a number of terrestrial impact structures in a variety of different stratigraphic settings (Fig. 5; Table 4). Shock-melted shales have also been recognized from the Haughton structure (Redeker and Stöffler, 1988; Osinski et al., 2005). The high SiO_2 contents of the Libyan Desert glass (>95 wt% SiO_2; Weeks et al., 1984) and urengoites from Siberia (89–96 wt% SiO_2; Deutsch et al., 1997) also indicate an origin through the impact melting of sandstones, although no source craters for these two glass types have been found to date.

Impact breccias at the Gosses Bluff (Milton et al., 1996) and Goat Paddock (Milton and Macdonald, 2005) impact structures, comprise a groundmass of predominantly silica glass, indicating that considerable amounts of sandstone-derived impact melts can also be preserved within structures developed in sandstone-rich

TABLE 4. TERRESTRIAL IMPACT STRUCTURES THAT PRESERVE UNEQUIVOCAL EVIDENCE FOR SHOCK-MELTED SANDSTONES

Impact site	Impactite setting			Reference(s)
	Crater-fill	Proximal	Distal	
Aouelloul	–	CL	–	Koeberl et al., 1998
Goat Paddock	CL, GR	–	–	Milton and Macdonald, 2005
Gosses Bluff	CL, GR	–	–	Milton et al., 1996
Haughton	CL, GR	CL, GR	–	Martinez et al., 1994a; Metzler et al., 1988; Osinski et al., 2005; Redeker and Stöffler, 1988
Meteor Crater	–	CL	–	Kieffer, 1971; Kieffer et al., 1976
Ries	–	CL, GR	CL	Engelhardt et al., 1987; Osinski, 2003; Osinski et al., 2004

Note: CL—occurrence as clasts; GR—occurrence in groundmass (i.e., emplaced in a molten state).

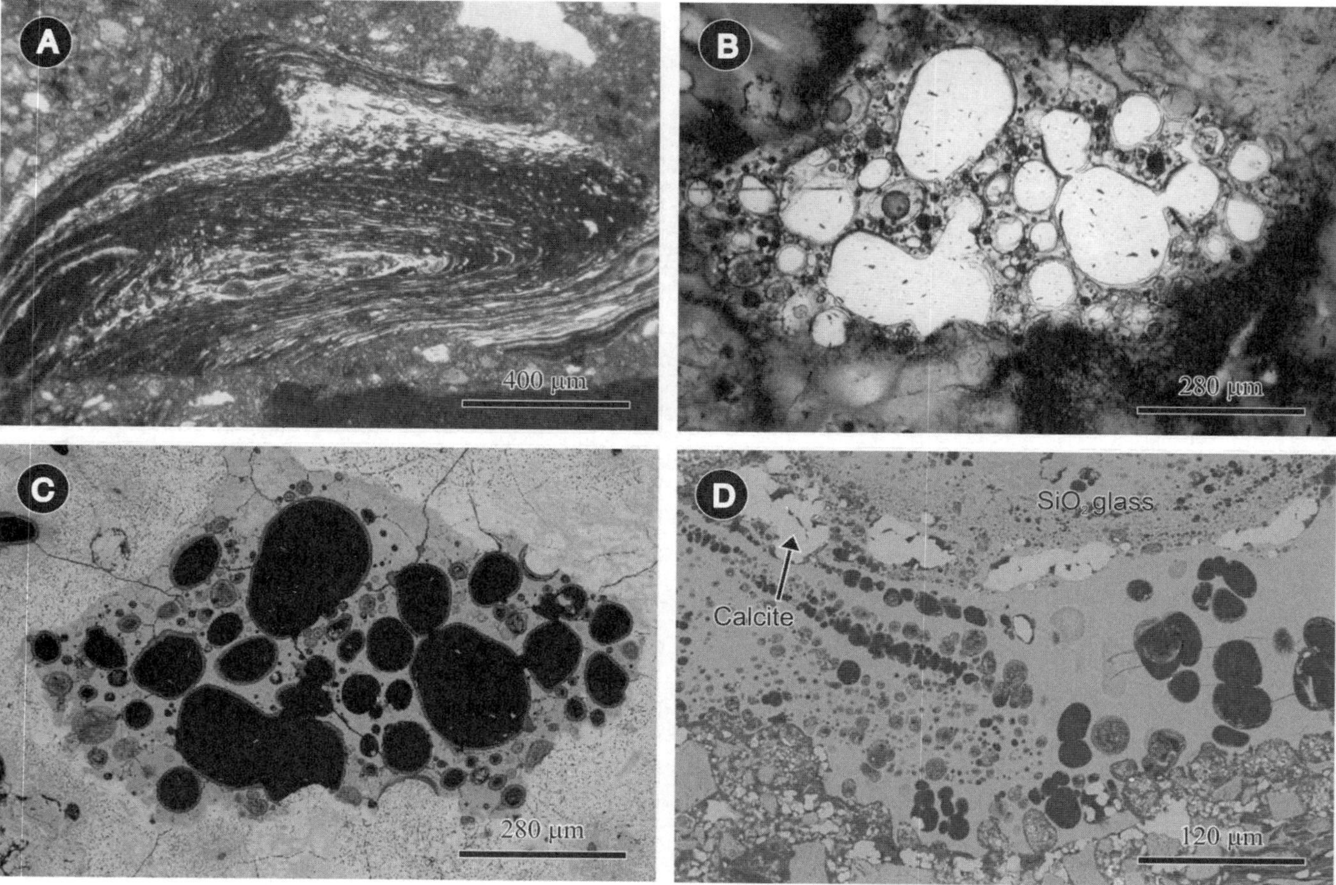

Figure 5. Plane-polarized light photomicrographs (A and B) and backscattered electron (BSE) images (C and D) showing SiO_2-rich glasses derived from shock-melted sandstones. A: Glass clast within impact melt breccias from the Haughton structure (Osinski et al., 2005). Note the well-developed flow banding, which does not reflect any internal difference in composition. B and C: Clast of SiO_2-rich glass within a glass clast from surficial suevites from the Ries structure (Osinski et al., 2004). D: Vesiculated glass clast (vesicles appear black) containing globules of calcite (upper three-quarters of the image) (Osinski et al., 2005).

targets. It has also been shown that sandstone-derived impact melts form a volumetrically minor, yet ubiquitous component of the groundmass of Ries surficial suevites (Osinski et al., 2004).

DISCUSSION AND CONCLUSIONS

It is apparent that impact melting in sedimentary targets is much more common than previously realized. In terms of carbonates, this is consistent with the recently reappraised phase relations of $CaCO_3$ (Fig. 1), which indicate that melting is the dominant response of carbonates to hypervelocity impact. Very high postshock temperatures close to the point of impact will result in vaporization of carbonates, as is the case for all rock types; however, this is not the same as chemical decomposition. The phase relations of $CaCO_3$ suggest that limited decomposition from $CaCO_3$ *melt* may be possible following decompression, although evidence for this has not yet been observed in naturally shocked rocks. For impact into limestones, this absence of evidence may be due, in part, to the recombination of CO_2 and CaO during fast back-reactions (e.g., Agrinier et al., 2001). However, Ivanov and Deutsch (2002) note that complex reaction kinetics, in particular the rate of diffusion of CO_2, may limit the amount of this gas released from carbonates during impact events.

It is also important to note that decomposition of $CaCO_3$ can only occur after pressure release due to high residual temperatures during postshock cooling (>1200 K; Ivanov and Deutsch, 2002). It is suggested that this thermal constraint is important in determining if, and when, decomposition can occur (cf. Deutsch et al., 2003). For example, it is notable that all silicate melt phases in the groundmass of the crater-fill deposits at Haughton were quenched to glasses (Osinski et al., 2005), indicating rapid cooling that would have inhibited the decomposition of $CaCO_3$ melt and clasts. At the Ries impact structure, however, postimpact temperatures in the suevite deposits were sufficiently high, in some regions, to result in minor decomposition at the edge of limestone clasts (Baranyi, 1980). This is consistent with preliminary results from Chicxulub, which suggest that minor decomposition of limestone clasts occurred only in impact melt rocks and impact melt breccias, and not in the lower-temperature lithic breccias or melt-bearing breccias (Deutsch et al., 2003). Thus, this decomposition is a postimpact contact-metamorphic process, which also occurs in igneous rocks. In other words, the decomposition of limestone during the impact process is not, therefore, due to shock *sensu stricto*, but is governed by the postimpact temperature of the melt-clast mixture: rapid quenching and/or low postshock temperatures will inhibit carbonate decomposition.

Despite the growing body of evidence for the melting of carbonates during hypervelocity impact, the pressure-temperature conditions at which this occurs are poorly understood. Appraisal of Figure 1 predicts that calcite, shocked to >0.1 GPa and postshock temperatures >1500 K, should undergo impact melting during decompression (Ivanov and Deutsch, 2002) (i.e., the melting of calcite requires low pressures but high temperatures). In this respect, it is interesting to note that calcite underwent melting at >10 to <20 GPa in porous sandstones at the Haughton structure (Osinski, 2005b), corresponding to postshock temperatures of >1250 K (Kieffer, 1971). Thus, temperature is obviously the limiting factor and explains why melting of calcite did not occur in the Haughton sandstones at pressures <10 GPa (i.e., the corresponding shock temperatures were not high enough). Further studies are required to assess the pressure-temperature conditions at which calcite undergoes melting and the influence of porosity.

With respect to dolomite, the lack of periclase (MgO)—which unlike CaO is stable—in impactites suggests that the decomposition of dolomite is not an important process during impact events. Complications also arise because accurate data for the melting of dolomite are not available (i.e., there are no known binary equilibria in which dolomite does not form solid solutions [Treiman, 1989]). It is also interesting to note that primary dolomite (i.e., impact melt phases) is extremely rare in terrestrial impact structures. This is compatible with observations of carbonatites, the vast majority of which are calcitic (K. Bell et al., 1998). Indeed, the phase relations of the systems CaO-MgO-CO_2-H_2O (Lee et al., 2000) and CaO-MgO-SiO_2-CO_2-H_2O (Otto and Wyllie, 1993) indicate that calcite is the liquidus phase for a wide range of compositions and pressure-temperature conditions. Thus, for impacts into dolomite-rich target rocks, a CaO-MgO-rich melt will be generated and calcite will typically be the first phase to crystallize out of the melt, with dolomite only forming at lower temperatures upon slow cooling. This is consistent with observations from the Haughton structure in which the groundmass of crater-fill impact melt breccias comprises calcite and MgO-rich glasses (Osinski et al., 2005). This also appears to be the case for impact melt breccias from Chicxulub, although the primary MgO-rich glass has been altered to MgO-rich clay (Nelson and Newsom, 2006). These observations also suggest that for impacts into carbonate-dominated targets, such as Chicxulub and Haughton, calcite and MgO-rich glasses are the typical product. For target sequences richer in siliciclastic sedimentary rocks, such as at Meteor Crater, it appears that SiO_2-CaO-MgO-rich melts are produced, from which a variety of silicate phases can crystallize (e.g., Ca-rich clinopyroxene and Mg-rich olivine) (Osinski et al., 2003) in addition to, or in place of, MgO-rich glasses.

To complicate matters further, it has been known for decades that natural systems are several orders of magnitude more complex than the simple one- or two-component systems used in experiments and theoretical calculations. This is the norm in natural carbonates that are typically "impure" (e.g., containing quartz, alkali feldspar, clays, etc.). Thus, impact melting of large volumes of carbonate strata will yield melts that are not pure $CaCO_3$ or $CaMg(CO_3)_2$. To the knowledge of the authors, there are no studies that have addressed this problem with respect to decomposition. The very existence of carbonatite lava flows indicates that "impurities" play a major role in carbonate melts (Gittins and Jago, 1991). According to the phase relations of $CaCO_3$, the existence of carbonatite lava flows is not possible. However, it has been shown that the presence of "impurities" in the melt, such as H_2O (Wyllie and Tuttle, 1960) and F (Gittins and Jago,

1991), dramatically lowers the minimum melting temperature of calcite, which allows carbonate melts to be stable at atmospheric pressure at temperatures <900 °C (Gittins and Jago, 1991).

In comparison to carbonates, fewer studies have been conducted into the response of sulfates to hypervelocity impact. Studies of naturally shocked rocks and minerals from the Haughton structure suggest that the anhydrite undergoes shock melting at pressures of >50–60 GPa. This is lower than preliminary estimates of >80–90 GPa derived from the phase relations of Ivanov et al. (2004); however, this estimate is for nonporous anhydrite, and it is widely known that melting of minerals in porous targets occurs at much lower pressures (e.g., Kieffer, 1971). These results are consistent with recent shock experiments, which suggest that anhydrite does not decompose below pressures of ~85 GPa (Langenhorst et al., 2003).

Impact-generated melt occurs in two main forms within terrestrial impact structures (Grieve and Cintala, 1992): (1) as crystallized and/or glassy coherent impact melt sheets or layers, within allochthonous crater-fill deposits, and (2) as discrete glassy clasts within so-called "suevites" in allochthonous crater-fill deposits and ejecta deposits. It has long been accepted that the volume of impact melt recognized in sedimentary and in mixed sedimentary-crystalline targets is about two orders of magnitude less than for crystalline targets in comparably sized impact structures (Kieffer and Simonds, 1980). However, recent work has shown that the crater-fill deposits at Haughton are impact melt breccias or clast-rich impact melt rocks (Osinski et al., 2005) and that they are stratigraphically equivalent to coherent impact melt layers developed at craters in crystalline targets (Grieve, 1988; Osinski et al., 2005). Importantly, the present and probable original volumes (~7 km^3 and ~22.5 km^3, respectively) of the crater-fill impact melt breccias at Haughton are analogous to characteristics of coherent impact melt sheets developed in comparably sized structures formed in crystalline targets (Osinski et al., 2005). For example, the volume of crater-fill impact melt rocks and melt-bearing breccias at the 24 km diameter Boltysh and 28 km diameter Mistastin impact structures are ~11 and 20 km^3, respectively (Grieve, 1975; Grieve et al., 1987).

It should be noted that the clast content of crater-fill impact melt breccias at Haughton (up to ~40–50 vol%; Osinski et al., 2005) is higher than in comparably sized structures developed in crystalline targets (e.g., ~20–30 vol% at Mistastin; Grieve, 1975). Based on the work of Kieffer and Simonds (1980), Osinski et al. (2005) suggested that this observation might be explained by the effect of mixing "wet" sediments or carbonates into a melt as opposed to dry crystalline rocks. The enthalpies of H_2O-bearing and carbonate systems are such that a much smaller proportion of admixed sedimentary rocks than of anhydrous crystalline rock is required to quench the melt to subsolidus temperatures (Kieffer and Simonds, 1980). Thus, a lower percentage of sedimentary rocks will be assimilated than crystalline rocks, before a melt is quenched, resulting in higher final clast contents for melts derived from impacts into sedimentary as opposed to crystalline targets (Osinski et al., 2005).

In summary, synthesizing observations from terrestrial impact structures with recent experimental results, computer simulations, and phase relations, it is clear that previous assumptions about the response of sedimentary rocks during hypervelocity impact events are inaccurate. Impact melting appears to be the dominant response of hypervelocity impact into carbonates, evaporites, and terrigenous sedimentary rocks. Limited decomposition from the melt phase may be possible following decompression if the melt remains at high temperatures long enough for this to occur. However, in the limited cases where decomposition of carbonates has been recognized, this was clearly a postimpact contact-metamorphic process, which is governed by the postimpact temperature of the melt-clast mixture. The apparent "anomaly" between the volumes of impact melt generated in sedimentary versus crystalline targets in comparably sized impact structures may, therefore, be due to difficulties in recognizing impact melts derived from sedimentary rocks.

ACKNOWLEDGMENTS

The authors thank the many individuals and institutions that have facilitated and funded fieldwork at various terrestrial impact structures, which forms the basis for this study. This study is based, in part, on the Discussion chapter of the Ph.D. thesis of G.R.O. and was funded by the Natural Sciences and Engineering Research Council of Canada (NSERC) through research grants to J.G.S. G.R.O. was supported by Canadian Space Agency (CSA) Space Science Research Project 05P-07. Giovanni Pratesi and Adrian Jones are thanked for supplying several of the images in Figure 2. Jayanta Kumar Pati and Kai Wünnemann are thanked for their helpful and constructive reviews. Planetary and Space Science Centre contribution 54. Contribution from the Earth Sciences Sector 2006519.

REFERENCES CITED

Agrinier, P., Deutsch, A., Schärer, U Anhydrite EOS and phase diagram in relation to shock decomposition., and Martinez, I., 2001, Fast back-reactions of shock-released CO_2 from carbonates: An experimental approach: Geochimica et Cosmochimica Acta, v. 65, p. 2615–2632, doi: 10.1016/S0016-7037(01)00617-2.

Ahrens, T.J., and Cole, D.M., 1974, Shock compression and adiabatic release of lunar fines from Apollo 17: Geochimica et Cosmochimica Acta, v. 3, p. 2333–2345.

Alvarez, L.W., Alvarez, W., Asaro, F., and Michel, H.V., 1980, Extraterrestrial cause for the Cretaceous/Tertiary extinction: Science, v. 208, p. 1095–1108, doi: 10.1126/science.208.4448.1095.

Badjukov, D.D., Dikov, Y.P., Petrova, T.L., and Pershin, S.V., 1995, Shock behavior of calcite, anhydrite, and gypsum: Lunar and Planetary Science Conference, 26th, p. 63–64.

Baranyi, I., 1980, Untersuchungen über die veränderungen von sedimentgesteinseinschlüssen im suevit des Nördlinger Rieses: Beiträge zur naturkundlichen Forschung Südwest-Deutschlands, v. 39, p. 37–56.

Barker, D.S., 2001, Calculated silica activities in carbonatite liquids: Contributions to Mineralogy and Petrology, v. 141, p. 704–709.

Bell, K., Kjarsgaard, B.A., and Simonetti, A., 1998, Carbonatites—Into the twenty-first century: Journal of Petrology, v. 39, no. 11/12, p. 1839–1845.

Bell, M.S., 2001, Thermal gravimetric analysis of experimentally shocked calcite [abs.]: Meteoritics and Planetary Science, v. 36, p. A17.

Bell, M.S., Hörz, F., and Reid, A., 1998, Characterization of experimental shock effects in calcite and dolomite by X-ray diffraction: Lunar and Planetary Science Conference, 29th, abstract 1422.

Boslough, M.B., Ahrens, T.J., Vizgirda, J., Becker, R.H., and Epstein, S., 1982, Shock-induced devolatilization of calcite: Earth and Planetary Science Letters, v. 61, p. 166–170, doi: 10.1016/0012-821X(82)90049-8.

Brooker, R.A., 1998, The effect of CO_2 Saturation on immiscibility between silicate and carbonate liquids: An experimental study: Journal of Petrology, v. 39, p. 1905–1915, doi: 10.1093/petrology/39.11.1905.

Chang, L.L.Y., Howie, R.A., and Zussman, J., 1998, Rock-forming minerals, volume 5B (2nd edition): Non-silicates: Sulphates, carbonates, phosphates, halides: London, Geological Society, 383 p.

Chen, G., Tyburczy, J.A., and Ahrens, T.J., 1994, Shock-induced devolatilization of calcium sulfate and implications for K-T extinctions: Earth and Planetary Science Letters, v. 128, p. 615–628, doi: 10.1016/0012-821X(94)90174-0.

Deutsch, A., Ostermann, M., and Masaitis, V.L., 1997, Geochemistry and neodymium-strontium isotope signature of tektite-like objects from Siberia (urengoites, South Ural glass): Meteoritics and Planetary Science, v. 32, p. 679–686.

Deutsch, A., Langenhorst, F., Hornemann, U., and Ivanov, B.A., 2003, On the shock behavior of anhydrite and carbonates—Is post-shock melting the most important effect? Examples from Chicxulub: Third International Conference on Large Meteorite Impacts, abstract 4080.

Dressler, B.O., and Reimold, W.U., 2001, Terrestrial impact melt rocks and glasses: Earth-Science Reviews, v. 56, p. 205–284, doi: 10.1016/S0012-8252(01)00064-2.

Dressler, B.O., Sharpton, V.L., Schwandt, C.S., and Ames, D.E., 2004, Impactites of the Yaxcopoil-1 drilling site, Chicxulub impact structure: Petrography, geochemistry, and depositional environment: Meteoritics and Planetary Science, v. 39, p. 857–878.

Earth Impact Database, 2007, http://www.unb.ca/passc/ImpactDatabase/ (September 2007).

Engelhardt, W. von, Luft, E., Arndt, J., Schock, H., and Weiskirchner, W., 1987, Origin of moldavites: Geochimica et Cosmochimica Acta, v. 51, p. 1425–1443, doi: 10.1016/0016-7037(87)90326-7.

French, B.M., 1998, Traces of catastrophe: Handbook of shock-metamorphic effects in terrestrial meteorite impact structures: Houston, Lunar and Planetary Institute, 120 p.

Gittins, J., 1989, The origin and evolution of carbonatite magmas, *in* Bell, K., ed., Carbonatites: Genesis and evolution: London, Unwin Hyman, p. 580–600.

Gittins, J., and Jago, B.C., 1991, Extrusive carbonatites: Their origins reappraised in the light of new experimental data: Geological Magazine, v. 128, p. 301–305.

Graup, G., 1999, Carbonate-silicate liquid immiscibility upon impact melting: Ries Crater, Germany: Meteoritics and Planetary Science, v. 34, p. 425–438.

Grieve, R.A.F., 1975, Petrology and chemistry of impact melt at Mistastin Lake crater, Labrador: Geological Society of America Bulletin, v. 86, p. 1617–1629, doi: 10.1130/0016-7606(1975)86<1617:PACOTI>2.0.CO;2.

Grieve, R.A.F., 1988, The Haughton impact structure: Summary and synthesis of the results of the HISS project: Meteoritics, v. 23, p. 249–254.

Grieve, R.A.F., and Cintala, M.J., 1992, An analysis of differential impact melt-crater scaling and implications for the terrestrial impact record: Meteoritics, v. 27, p. 526–538.

Grieve, R.A.F., Reny, G., Gurov, E.P., and Ryabenko, V.A., 1987, The melt rocks of the Boltysh impact crater, Ukraine, USSR: Contributions to Mineralogy and Petrology, v. 96, p. 56–62, doi: 10.1007/BF00375525.

Grieve, R.A.F., Langenhorst, F., and Stöffler, D., 1996, Shock metamorphism of quartz in nature and experiment, II: Significance in geoscience: Meteoritics and Planetary Science, v. 31, p. 6–35.

Gupta, S.C., Ahrens, T.J., and Yang, W., 2001, Shock-induced vaporization of anhydrite and global cooling from the K/T impact: Earth and Planetary Science Letters, v. 188, p. 399–412, doi: 10.1016/S0012-821X(01)00327-2.

Gupta, S.C., Love, S.G., and Ahrens, T.J., 2002, Shock temperature in calcite ($CaCO_3$) at 95–160 GPa: Earth and Planetary Science Letters, v. 201, p. 1–12, doi: 10.1016/S0012-821X(02)00685-4.

Ivanov, B.A., and Deutsch, A., 2002, The phase diagram of $CaCO_3$ in relation to shock compression and decompression: Physics of the Earth and Planetary Interiors, v. 129, p. 131–143, doi: 10.1016/S0031-9201(01)00268-0.

Ivanov, B.A., Badukov, D.D., Yakovlev, O.L., Gerasimov, M.V., Dikov, Y.P., Pope, K.O., and Ocampo, A.C., 1996, Degassing of sedimentary rocks due to Chicxulub impact: Hydrocode and physical simulations, *in* Ryder, G., et al., eds., The Cretaceous-Tertiary event and other catastrophes in Earth history: Geological Society of America Special Paper 307, p. 125–139.

Ivanov, B.A., Langenhorst, F., Deutsch, A., and Hornemann, U., 2004, Anhydrite EOS and phase diagram in relation to shock decomposition: Lunar and Planetary Science Conference, 35th, abstract 1489.

Jones, A.P., Claeys, P., and Heuschkel, S., 2000, Impact melting of carbonates from the Chicxulub Crater, *in* Gilmour, I., and Koeberl, C., eds., Impacts and the early Earth: Lecture Notes in Earth Sciences, 91: Berlin, Springer-Verlag, p. 343–361.

Jones, A.P., Kearsley, A.T., Friend, C.R.L., Robin, E., Beard, A., Tamura, A., Trickett, S., and Claeys, P., 2005, Are there signs of a large Paleocene impact, preserved around Disko Bay, West Greenland? Nuussuaq spherule beds origin by impact instead of volcanic eruption?, *in* Kenkmann, T., et al., eds., Large meteorite impacts, III: Geological Society of America Special Paper 384, p. 281–298.

Kieffer, S.W., 1971, Shock metamorphism of the Coconino Sandstone at Meteor Crater, Arizona: Journal of Geophysical Research, v. 76, p. 5449–5473.

Kieffer, S.W., and Simonds, C.H., 1980, The role of volatiles and lithology in the impact cratering process: Reviews of Geophysics and Space Physics, v. 18, p. 143–181.

Kieffer, S.W., Phakey, P.P., and Christie, J.M., 1976, Shock processes in porous quartzite: Transmission electron microscope observations and theory: Contributions to Mineralogy and Petrology, v. 59, p. 41–93, doi: 10.1007/BF00375110.

Koeberl, C., Reimold, W.U., and Shirey, S.B., 1998, The Aouelloul crater, Mauritania: On the problem of confirming the impact origin of a small crater: Meteoritics and Planetary Science, v. 33, p. 513–517.

Kondo, K.I., and Ahrens, T.J., 1983, Heterogeneous shock-induced thermal radiation in minerals: Physics and Chemistry of Minerals, v. 9, p. 173–181, doi: 10.1007/BF00308375.

Kotra, R.K., See, J.H., Gibson, E.K.J., Hörz, F., Cintala, M.J., and Schmidt, R.M., 1983, Carbon dioxide loss in experimentally shocked calcite and limestone: Lunar and Planetary Science Conference, 14th, p. 401–402.

Kring, D.A., Hörz, F., Zurcher, L., and Urrutia Fucugauchi, J., 2004, Impact lithologies and their emplacement in the Chicxulub impact crater: Initial results from the Chicxulub Scientific Drilling Project, Yaxcopoil, Mexico: Meteoritics and Planetary Science, v. 39, p. 879–897.

Lange, M.A., and Ahrens, T.J., 1986, Shock-induced CO_2 loss from $CaCO_3$: Implications for early planetary atmospheres: Earth and Planetary Science Letters, v. 77, p. 409–418, doi: 10.1016/0012-821X(86)90150-0.

Langenhorst, F., Deutsch, A., Ivanov, B.A., and Hornemann, U., 2000, On the shock behavior of $CaCO_3$: Dynamic loading and fast unloading experiments – modeling – mineralogical observations: Lunar and Planetary Science Conference, 31st, abstract 1851.

Langenhorst, F., Deutsch, A., Hornemann, U., Ivanov, B.A., and Lounejava, E., 2003, On the shock behaviour of anhydrite: Experimental results and natural observations: Lunar and Planetary Science Conference, 34th, abstract 1638.

Lee, W.J., and Wyllie, P.J., 2000, The system CaO-MgO-SiO_2-CO_2 at 1 GPa, metasomatic wehrlites, and primary carbonatite magmas: Contributions to Mineralogy and Petrology, v. 138, p. 214–228, doi: 10.1007/s004100050558.

Lee, W.J., Huang, W.L., and Wyllie, P.J., 2000, Melts in the mantle modeled in the system CaO-MgO-SiO_2-CO_2 at 2.7 GPa: Contributions to Mineralogy and Petrology, v. 138, p. 199–213, doi: 10.1007/s004100050557.

Martinez, I., Agrinier, P., Schärer, U., and Javoy, M., 1994a, A SEM-ATEM and stable isotope study of carbonates from the Haughton impact crater, Canada: Earth and Planetary Science Letters, v. 121, p. 559–574, doi: 10.1016/0012-821X(94)90091-4.

Martinez, I., Schörer, U., Guyot, F., Deutsch, A., and Hornemann, U., 1994b, Experimental and theoretical investigation of shock-induced outgassing of dolomite: Lunar and Planetary Science Conference, 25th, p. 839.

Martinez, I., Deutsch, A., Guyot, F., Ildefonse, P., Guyot, F., and Agrinier, P., 1995, Shock recovery experiments on dolomite and thermodynamical calculations of impact-induced decarbonation: Journal of Geophysical Research, v. 100, p. 15,465–15,476, doi: 10.1029/95JB01151.

Melosh, H.J., 1989, Impact Cratering: A Geologic Process: New York: Oxford University Press, 245 p.

Metzler, A., Ostertag, R., Redeker, H.J., and Stöffler, D., 1988, Composition of the crystalline basement and shock metamorphism of crystalline and sedimentary target rocks at the Haughton impact crater, Devon Island, Canada: Meteoritics, v. 23, p. 197–207.

Milton, D.J., and Macdonald, F.A., 2005, Goat Paddock, Western Australia: An impact crater near the simple-complex transition: Australian Journal of Earth Sciences, v. 52, p. 689–697, doi: 10.1080/08120090500170435.

Milton, D.J., Glikson, A.Y., and Brett, R., 1996, Gosses Bluff—A latest Jurassic impact structure, central Australia, part 1: Geological structure, stratigraphy, and origin: Journal of Australian Geology and Geophysics, v. 16, p. 453–486.

Nelson, M.J., and Newsom, H.E., 2006, Yaxcopoil-1 impact melt breccias: Silicate melt clasts among dolomite melt and implications for deposition: Lunar and Planetary Science Conference, 37th, abstract 2081.

O'Keefe, J.D., and Ahrens, T.J., 1989, Impact production of CO_2 by the Cretaceous/Tertiary extinction bolide and the resultant heating of the Earth: Nature, v. 338, p. 247–249, doi: 10.1038/338247a0.

Osinski, G.R., 2003, Impact glasses in fallout suevites from the Ries impact structure, Germany: An analytical SEM study: Meteoritics and Planetary Science, v. 38, p. 1641–1668.

Osinski, G.R., 2005a, Hydrothermal activity associated with the Ries impact event, Germany: Geofluids, v. 5, p. 202–220, doi: 10.1111/j.1468-8123.2005.00119.x.

Osinski, G.R., 2005b, Shock-metamorphosed and shock-melted $CaCO_3$-bearing sandstones from the Haughton impact structure, Canada: Melting of calcite at ~10–20 GPa: Lunar and Planetary Science Conference, 36th, abstract 2038.

Osinski, G.R., and Spray, J.G., 2001, Impact-generated carbonate melts: Evidence from the Haughton structure, Canada: Earth and Planetary Science Letters, v. 194, p. 17–29, doi: 10.1016/S0012-821X(01)00558-1.

Osinski, G.R., and Spray, J.G., 2003, Evidence for the shock melting of sulfates from the Haughton impact structure, Arctic Canada: Earth and Planetary Science Letters, v. 215, p. 357–370, doi: 10.1016/S0012-821X(03)00420-5.

Osinski, G.R., Bunch, T.E., and Wittke, J., 2003, Evidence for shock melting of carbonates from Meteor Crater, Arizona: Münster, Germany: Meteoritical Society Meeting, 66th, abstract 5070.

Osinski, G.R., Grieve, R.A.F., and Spray, J.G., 2004, The nature of the groundmass of surficial suevites from the Ries impact structure, Germany, and constraints on its origin: Meteoritics and Planetary Science, v. 39, p. 1655–1684.

Osinski, G.R., Spray, J.G., and Lee, P., 2005, Impactites of the Haughton impact structure, Devon Island, Canadian High Arctic: Meteoritics and Planetary Science, v. 40, p. 1789–1812.

Osinski, G.R., Schwarcz, H.P., Smith, J., Kleindienst, M.R., Haldemann, A.F.C., and Churcher, C.S., 2007, Evidence for a 100–200 ka meteorite impact in western Egypt: Earth and Planetary Science Letters, v. 253, p. 378–388, doi: 10.1016/j.epsl.2006.10.039.

Otto, J.W., and Wyllie, P.J., 1993, Relationships between silicate melts and carbonate-precipitating melts in $CaO-MgO-SiO_2-CO_2-H_2O$ at 2 kbar: Mineralogy and Petrology, v. 48, p. 343–365, doi: 10.1007/BF01163107.

Penfield, G.T., and Camargo, Z.A., 1981, Definition of a major igneous zone in the central Yucatán platform with aeromagnetics and gravity: Society of Exploration Geophysicists 51st Annual Meeting, Los Angeles, Society of Exploration of Geophysicists, p. 37.

Pierazzo, E., Kring, D.A., and Melosh, H.J., 1998, Hydrocode simulation of the Chicxulub impact event and the production of climatically active gases: Journal of Geophysical Research, v. 103, p. 28,607–28,625, doi: 10.1029/98JE02496.

Pope, K.O., Baines, K.H., Ocampo, A.C., and Ivanov, B.A., 1994, Impact winter and the Cretaceous/Tertiary extinctions: Results of a Chicxulub asteroid impact model: Earth and Planetary Science Letters, v. 128, p. 719–725, doi: 10.1016/0012-821X(94)90186-4.

Pratesi, G., Morelli, M., Rossi, A.P., and Ori, G.G., 2005, Chemical compositions of impact melt breccias and target rocks from the Tenoumer impact crater, Mauritania: Meteoritics and Planetary Science, v. 40, p. 1653–1672.

Redeker, H.J., and Stöffler, D., 1988, The allochthonous polymict breccia layer of the Haughton impact crater, Devon Island, Canada: Meteoritics, v. 23, p. 185–196.

Roedder, E., 1978, Silicate liquid immiscibility in magmas and in the system $K_2O-FeO-Al_2O_3-SiO_2$: An example of serendipity: Geochimica et Cosmochimica Acta, v. 42, p. 1597–1617, doi: 10.1016/0016-7037(78)90250-8.

Rosa, D.F., 2004, Marble enclaves in the melt sheet at the West Clearwater Lake impact crater, northern Quebec [M.S. thesis]: Montreal, McGill University, 146 p.

Schmitt, R.T., Wittmann, A., and Stöffler, D., 2004, Geochemistry of drill core samples from Yaxcopoil-1, Chicxulub impact crater, Mexico: Meteoritics and Planetary Science, v. 39, p. 979–1001.

Skála, R., Hörz, F., and Jakes, P., 1999, X-ray powder diffraction study of experimentally shocked dolomite: Lunar and Planetary Science Conference, 30th, abstract 1327.

Skála, R., Ederová, J., Matìjka, P., and Hörz, F., 2002, Mineralogical investigation of experimentally shocked dolomite: Implications for the outgassing of carbonates, *in* Koeberl, C., and MacLeod, K.G., eds., Catastrophic events and mass extinctions: Impacts and beyond: Geological Society of America Special Paper 356, p. 571–586.

Spray, J.G., 2006, Ultrametamorphism of impure carbonates beneath the Manicouagan impact melt sheet: Evidence for superheating: Lunar and Planetary Science Conference, 37th, abstract 2385.

Stöffler, D., Artemieva, N.A., Ivanov, B.A., Hecht, L., Kenkmann, T., Schmitt, R.T., Tagle, R., and Wittmann, A., 2004, Origin and emplacement of the impact formations at Chicxulub, Mexico, as revealed by the ICDP deep drilling at Yaxcopoil-1 and by numerical modeling: Meteoritics and Planetary Science, v. 39, p. 1035–1067.

Treiman, A.H., 1989, Carbonatite magma: Properties and process, *in* Bell, K., ed., Carbonatites: Genesis and evolution: London, Unwin Hyman, p. 89–104.

Tuchscherer, M.G., Reimold, W.U., Koeberl, C., Gibson, R.L., and de Bruin, D., 2004, First petrographic results on impactites from the Yaxcopoil-1 borehole, Chicxulub structure, Mexico: Meteoritics and Planetary Science, v. 39, p. 899–931.

Tyburczy, J.A., and Ahrens, T.J., 1986, Dynamic compression and volatile release of carbonates: Journal of Geophysical Research, v. 91, p. 4730–4744.

Vizgirda, J., and Ahrens, T.J., 1982, Shock compression of aragonite and implications for the equation of state for carbonates: Journal of Geophysical Research, v. 87, p. 4747–4758.

Weeks, R.A., Underwood, J.R., and Giegengack, R., 1984, Libyan Desert glass: A review: Journal of Non-Crystalline Solids, v. 67, p. 593–619, doi: 10.1016/0022-3093(84)90177-7.

Wyllie, P.J., and Tuttle, O.F., 1960, The system $CaO-CO_2-H_2O$ and the origin of carbonatites: Journal of Petrology, v. 1, p. 1–46.

Yang, W., and Ahrens, T.J., 1998, Shock vaporization of anhydrite and global effects of the K/T bolide: Earth and Planetary Science Letters, v. 156, p. 125–140, doi: 10.1016/S0012-821X(98)00006-5.

Manuscript Accepted by the Society 10 July 2007

The Geological Society of America
Special Paper 437
2008

Postimpact sediments in the Gardnos impact structure, Norway

E. Kalleson
Natural History Museum, University of Oslo, PO 1172, Blindern, NO-0318 Oslo, Norway

H. Dypvik
Department of Geosciences, University of Oslo, PO 1047, Blindern, NO-0316 Oslo, Norway

J. Naterstad
Skoleveien 7, NO-1389 Heggedal, Norway

ABSTRACT

The Gardnos structure in Hallingdal, Norway, is a 5 km in diameter, eroded impact crater of probably late Precambrian age. Within the structure, both impactites and postimpact crater sediments are preserved. The sediments comprise a wide range of siliciclastics: (1) postimpact breccias, (2) coarse conglomerates, (3) conglomeratic sandstones, (4) sandstones, and (5) interbedded fine sandstones, siltstones, and shales.

The sedimentary succession reveals a shifting depositional environment. The postimpact breccias covering the crater floor were deposited by rock avalanches. Directly after impact, loose rock debris slid down the crater wall and the central uplift, settling on top of the newly formed impactites (suevite and lithic breccia known as Gardnos Breccia). The overlying conglomeratic and sandy sequence shows significant local thickness variations, consistent with coalescing fan-shaped deposits along the lower crater wall and on the crater floor. Probably the resurging water breached the crater rim at its weakest parts, initiating series of screes and debris flows, which built out into an eventually water-filled crater. Sand-enriched density flows then dominated in the water-filled crater basin. Above fine-grained sandstones, siltstones and shales were deposited, representing the reestablishment of quiet conditions, possibly similar to the preimpact depositional conditions. Carbon-enrichments in the impactites, and partly deformed clasts of sedimentary origin in the strata just above, suggest that the crystalline basement of the target area was covered by a thin layer of organic-rich sediments. This supports a scenario with a target area in shallow, stagnant water.

Keywords: Gardnos, infill history, breccia, conglomerates, sandstones.

Kalleson, E., Dypvik, H., and Naterstad, J., 2008, Postimpact sediments in the Gardnos impact structure, Norway, *in* Evans, K.R., Horton, J.W., Jr., King, D.T., Jr., and Morrow, J.R., eds., The Sedimentary Record of Meteorite Impacts: Geological Society of America Special Paper 437, p. 19–41, doi: 10.1130/2008.2437(02).

INTRODUCTION

The Gardnos impact structure is located in Hallingdal, 150 km northwest of Oslo, approximately at 500,700 m E, 6,722,800 m N UTM (Fig. 1). Being deeply eroded, the Gardnos structure now is represented by outcrops of impact-produced breccia and postimpact sediments within a roughly circular area of 5 km in diameter (Fig. 1). Broch (1945) explained the brecciated rocks as the product of explosive volcanism, though he found the absence of other volcanic material peculiar. In the early 1990s, petrographic studies revealed the existence of planar deformation features in quartz, and an impact origin of the structure was proposed (Dons and Naterstad, 1992; Naterstad and Dons, 1994). Further studies detected an extraterrestrial signature from the projectile in the breccias (osmium-isotope analyses), and mixing models demonstrated that the breccias could have been produced by mixing the exposed target rocks (French et al., 1997).

Generally the impact breccias contain significant amounts of carbon (typically 0.2–1.0 wt%), some five to ten times the amount present in the target rocks (Gilmour et al., 2003). The Sudbury crater in Canada is the only other known impact structure with presence of comparable amounts of carbon (French, 1968; Bunch et al., 1999; Heymann et al., 1999). The presence of carbon has been explained by a target area where the crystalline basement was covered by a thin, organic-rich sedimentary layer (Gilmour et al., 2003). However, the paleoenvironmental setting at the time of impact is still poorly constrained. This work therefore concentrates on the sedimentary rocks, their composition, structure, and distribution, related to the infill history. The rocks have been studied in the field, in core, and by laboratory analysis.

GEOLOGICAL SETTING

The Precambrian basement consists mostly of granitic gneisses and quartzites, along with some banded gneisses, mafic rocks, and local pegmatites (Nordgulen et al., 1997). French et al. (1997) describes five lithologies of impact rocks: black quartzite,

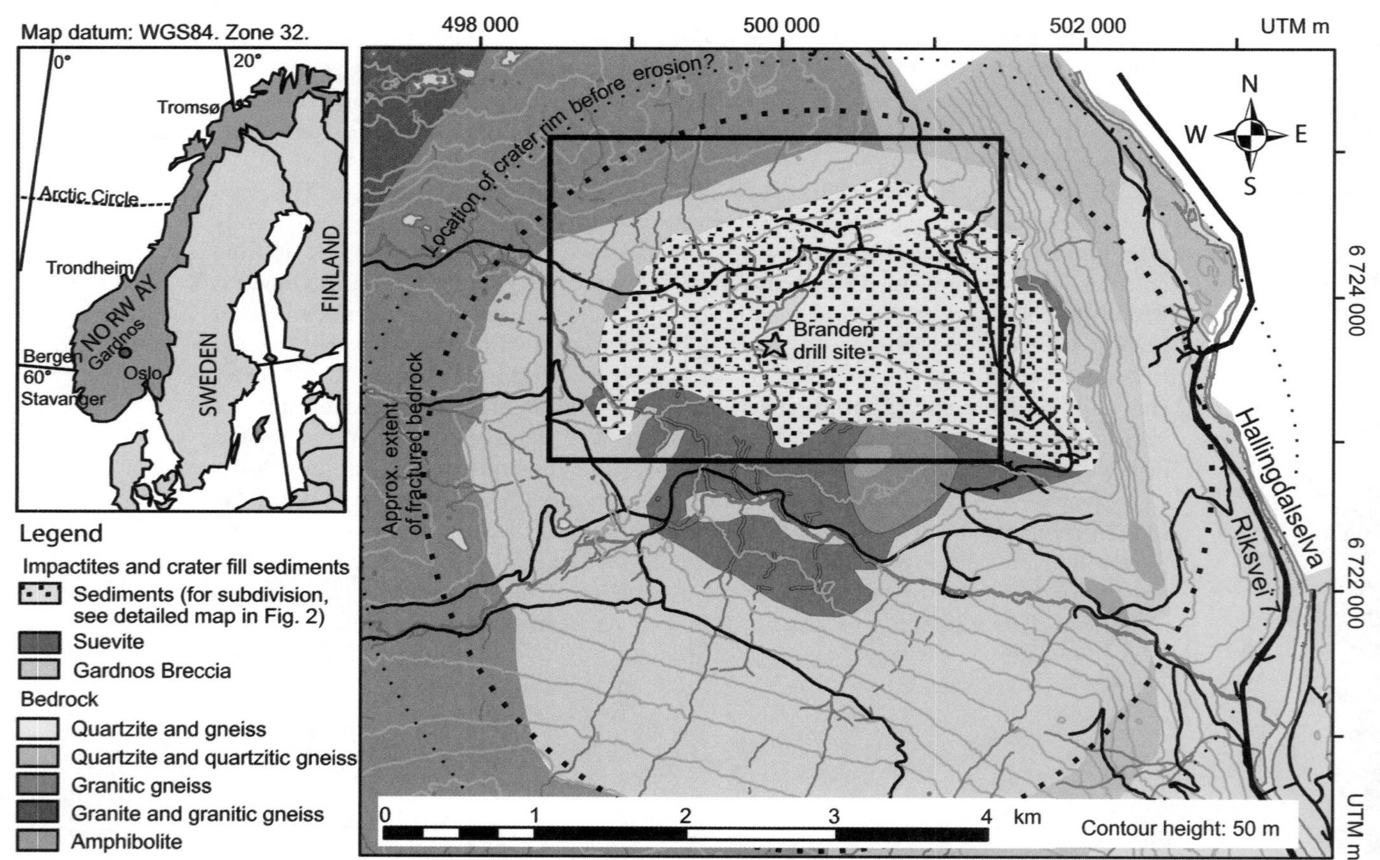

Figure 1. Geographic location and geological map of the Gardnos structure. Dotted circular line on geological map marks approximate outer boundary of bedrock fractured by the impact. Original location of the crater rim is indicated. Rectangle within the map marks detailed map presented in Figure 2, and approximately the maps in Figures 12, 14, and 18.

Gardnos Breccia, black-matrix breccia, suevite, and melt-matrix breccia. The shocked, black quartzite is only present at one small outcrop, and is here excluded from the map. The lithic breccias are divided into "Gardnos Breccia" and "black-matrix breccia" (French et al., 1997). However, in the field these breccias are often observed closely associated, without any clear boundaries. Likely, the two lithologies are only subdivisions of autochthonous breccias produced during impact by fracturing of target rock during the shock wave propagation. The amounts of black matrix probably reflect the degree of internal abrasion (grinding) and marginal thermal softening. In the typical Gardnos Breccia, angular white granitic or quartzitic clasts are embedded in a dark matrix. Exposures, however, show the whole range from breccias where black matrix appears only as millimeter-thick zones dividing the bedrock into separate fragments, still in place, to matrix-dominated breccias where the clasts are few and small, floating in the matrix and even somewhat rounded along the edges.

French et al. (1997) also divide the melt-bearing breccias into suevite and melt-matrix breccia. The suevite is an allochthonous breccia, where angular rock fragments of different sizes and lithologies are floating in a fine-grained clastic matrix. The suevite is distinguished by centimeter-sized dark fragments of melted rocks, which in the field is best seen on weathered surfaces. In the melt-matrix breccia, rock and mineral clasts float in a matrix of crystallized melt. Melt-matrix breccia is exposed only at one outcrop. As the impactites are not within the scope of this study, the maps and models presented here only show the two main lithologies of the impactites: the Gardnos Breccia and the suevite.

The Gardnos Breccia makes up large parts of the crater floor. In central parts of the crater, a sheet of suevite is found above the Gardnos Breccia. In places the bedrock consists of amphibolitic intrusions. These have responded to the shock waves in a different way from that of the granitic and quartzitic rocks. Typically, the amphibolitic rocks do not appear as breccias, and they lack the characteristic black matrix. They seem to be more resistant to erosion, and often appear as small topographic highs. Most of the present exposures of the central peak in the Gardnos structure are made up of amphibolitic rocks.

The age of the Gardnos structure is not certain. So far, attempts at paleontological and radiometric dating of the crater have failed, but locally occurring pegmatite intrusions in the target rock have been dated at 900 Ma, giving a maximum age for crater formation (Naterstad and Dons, 1994). $^{40}Ar/^{39}Ar$ dating of feldspar and devitrified glass from suevite gave ages ca. 385 Ma, possibly reflecting Caledonian overprint (Grier et al., 1999). Biogenic microstructures preserved in clasts from suevite indicate a late Precambrian age of the impact. A possible core imprint of a brachiopod (http://www.nhm.uio.no/palmus/palvenn/index.htm) has been found at an outcrop near Branden.

The crater structure was exposed through Tertiary uplift and subsequent weathering and erosional processes. During Quaternary time the area was repeatedly glaciated and parts of the structure were covered by moraine. There are, however, good exposures of impactites at steep hillsides and along riverbeds. In addition, important new information was obtained in 1993, when the moat of the Gardnos structure was cored, resulting in a 400 m long drill core of both impactites (lower 244 m) and postimpact sediments (upper 156 m) (Naterstad and Dons, 1994). In the central part of the structure, a central peak is represented by exposures of deformed bedrock. In the northern half of the structure, crater-fill sediments are preserved, whereas in the southern half only impactites are exposed (Figs. 1 and 2).

Most volume calculations of crater dimensions rely on the "rim diameter" (Melosh, 1989), a feature not preserved in the Gardnos crater. A tentative reconstruction of the pristine crater is here presented (Fig. 3). Crater parameters are defined as recommended by Turtle et al. (2005). No impactite exposures have been observed above 830 masl. The present peneplain of the crystalline target rocks is at ~1000 masl, representing the absolute minimum for the upper level of target rock at the time of impact. Extrapolation of the crater structure up to at least the level of today's peneplain suggests a minimum apparent crater diameter (D_A) of 5.5 km. Further extrapolation to the top of an imaginary rim gives an estimated rim diameter (D) for the original Gardnos crater of ~6 km.

Possible rim height can be calculated by $h_R = 0.36 \times D^{1.014}$, where all units are in meters (Pike, 1977). With $D = 6000$ m, the rim height is estimated at 240 m above original ground surface (in the case of very shallow water setting, the seafloor may be regarded as the base level).

The stratigraphical uplift in the center can be estimated by $h_{SU} = 0.06 \times D^{1.1}$, where all units are in kilometers (Grieve et al., 1981). With a rim to rim diameter D of 6 km, the stratigraphic uplift is estimated at 430 m. This is very difficult to confirm by field observations in Gardnos, as there is no horizontal layering of the massive bedrocks in the area. The precollapse height of the central peak is uncertain.

The central peak diameter is calculated by $D_{CP} = 0.22 \times D$ (Pike, 1985). This gives a basal central peak diameter of 1302 m, compared to an estimate of about 1000 m based on field observations.

DATA COLLECTION

The crater-fill breccias and conglomerates were studied in detail by clast counts. Twenty-eight squares of 0.5 m × 0.5 m were selected on exposures (sub)parallel to bedding planes. All clasts equal to or larger than 1 cm were counted and registered by size, type, and roundness. In the core, such statistics were done for intervals of 0.5 m throughout the conglomeratic sequence. The relative proportions of matrix and clasts were estimated by field and core observations, and the matrix compositions were further studied by thin-section microscopy. Thin sections (116 total) from core and field samples were studied with optical and scanning electron microscopy, and modal compositions were determined in 29 thin sections (Table 1). Fifteen bulk samples were analyzed by XRD (X-ray diffraction) (Philips X'Pert MPD

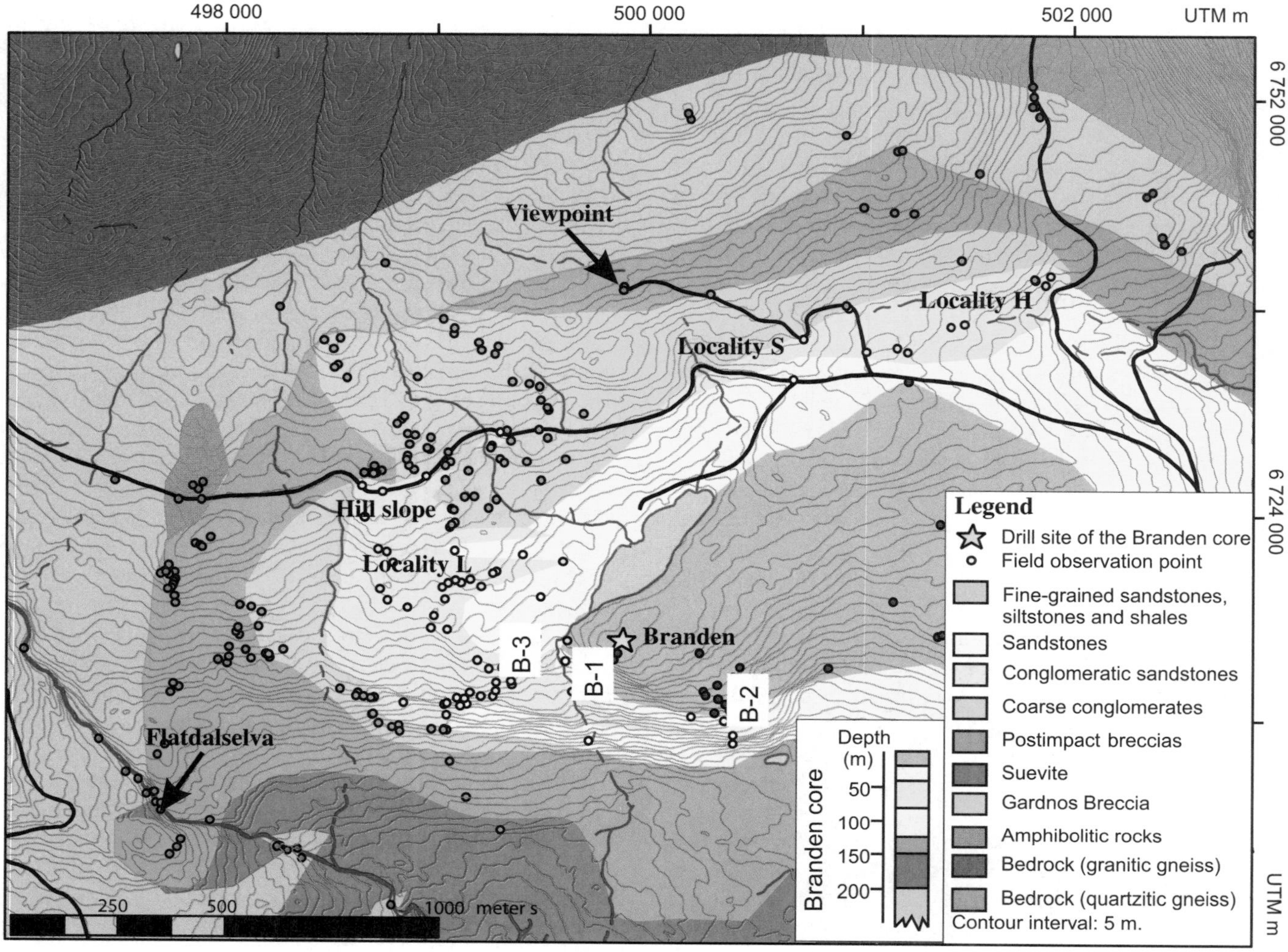

Figure 2. Detailed map (rectangle in Fig. 1) showing distribution of the sedimentary facies in the northwestern part of the Gardnos structure. B-1, B-2, and B-3 are location labels referring to the main field sections.

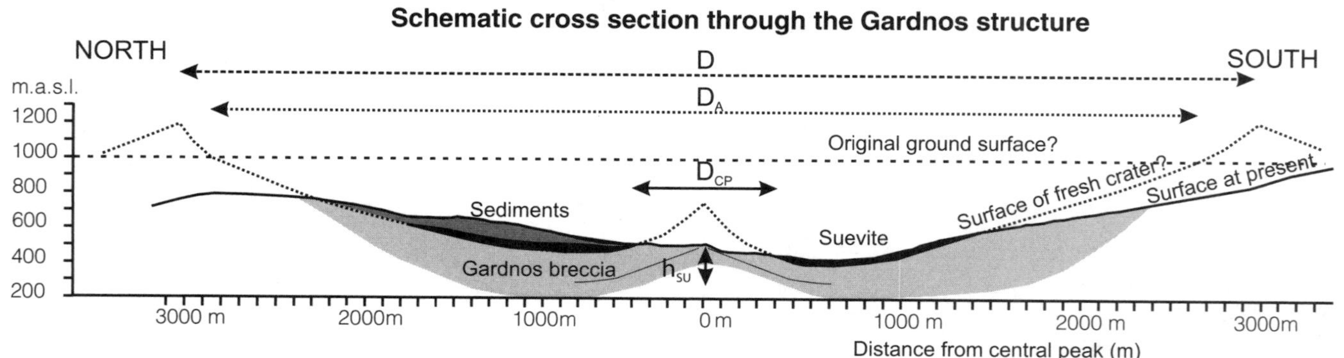

Figure 3. A cross section through a geometrical reconstruction of the Gardnos structure based on field observations. It forms the basis for the calculations of the crater rim and central peak (compare text).

TABLE 1. THIN-SECTION COUNTS

Mineral	Branden core depth (m)										
	127.56 (%)	119.5 (%)	113.15 (%)	102 (%)	101.85 (%)	98.25 (%)	85.9 (%)	83.95 (%)	79.77 (%)	60.3 (%)	23.38 (%)
Monocrystalline quartz	13.5	28.9	19.0	26.3	18.4	21.4	33.2	23.3	29.6	37.5	40.6
Polycrystalline quartz	3.8	1.0	1.7	1.3	0.5	0.0	1.0	0.5	0.0	0.0	1.0
Feldspar	30.5	35.7	53.1	45.2	47.7	48.3	43.7	40.8	46.4	45.6	46.8
Rock fragments	17.6	11.0	8.9	5.3	7.1	15.1	7.5	7.5	5.8	5.1	0.8
Heavy minerals	0.8	0.0	0.0	0.0	0.0	0.0	0.0	0.0	0.0	0.0	0.0
Mica	4.3	0.2	2.2	0.5	1.6	1.0	0.3	1.7	0.0	0.0	0.0
Chlorite	7.4	12.2	8.4	12.6	10.1	4.7	7.7	6.3	2.0	1.5	0.0
Carbonate	5.1	6.0	1.0	0.8	2.2	0.8	0.3	0.0	0.3	1.0	0.5
Glass / melted fragments	0.8	0.2	0.0	2.5	0.0	0.5	0.5	0.2	1.5	0.3	0.0
Matrix	11.5	1.7	3.7	3.5	12.3	3.1	4.1	16.3	7.5	2.8	9.0
Organic matter	4.8	3.0	2.0	2.0	0.0	5.0	1.8	3.4	7.0	6.3	1.3
Pyrite	0.0	0.0	0.0	0.0	0.0	0.0	0.0	0.0	0.0	0.0	0.0
Amphibolitic minerals	0.0	0.0	0.0	0.0	0.0	0.0	0.0	0.0	0.0	0.0	0.0
Total grains counted	393	401	405	396	365	383	389	412	399	395	387

Mineral	Location H		Location S		Location B-3		Location B-2		Location B-1		
	H-1A-1 (%)	H-1B-1 (%)	S-1-1 (%)	S-1-2 (%)	B-1A-1 (%)	B-1A-2 (%)	B-1B-1 (%)	B-2-6 (%)	B-1A-1 (%)	B-1A-2 (%)	B-1B-1 (%)
Monocrystalline quartz	32.3	28.0	53.0	47.1	73.8	68.1	68.3	45.8	73.8	68.1	68.3
Polycrystalline quartz	0.0	0.0	0.0	0.0	0.7	0.7	0.0	0.5	0.7	0.7	0.0
Feldspar	34.5	27.8	26.0	32.3	20.4	27.2	25.7	43.8	20.4	27.2	25.7
Rock fragments	4.1	4.3	4.0	8.7	0.5	0.0	0.0	2.5	0.5	0.0	0.0
Heavy minerals	0.0	0.0	0.0	0.0	0.7	0.0	0.0	0.2	0.7	0.0	0.0
Mica	1.0	5.1	0.5	0.5	0.0	1.0	0.7	0.2	0.0	1.0	0.7
Chlorite	9.0	3.1	1.8	0.0	0.0	0.0	0.0	0.0	0.0	0.0	0.0
Carbonate	0.0	0.0	4.0	0.0	0.0	0.0	0.0	0.2	0.0	0.0	0.0
Glass / melted fragments	0.0	0.0	0.0	0.0	0.0	0.0	0.0	0.0	0.0	0.0	0.0
Matrix	19.2	26.6	10.6	11.5	3.2	2.2	4.1	5.4	3.2	2.2	4.1
Organic matter	0.0	2.9	0.0	0.0	0.5	0.7	1.2	1.2	0.5	0.7	1.2
Pyrite	0.0	2.2	0.0	0.0	0.0	0.0	0.0	0.0	0.0	0.0	0.0
Amphibolitic minerals	0.0	0.0	0.0	0.0	0.0	0.0	0.0	0.0	0.0	0.0	0.0
Total grains counted	412	414	396	393	401	405	413	404	401	405	413

Mineral	Various outcrops in the L area					
	sst-6 (%)	sst-4 (%)	sst-5 (%)	sst-2 (%)	sst-1 (%)	sst-7 (%)
Monocrystalline quartz	31.0	33.7	34.6	23.0	26.9	33.9
Polycrystalline quartz	0.2	0.0	0.0	0.2	0.0	0.0
Feldspar	47.5	56.6	62.5	62.0	55.7	55.0
Rock fragments	3.9	1.9	1.7	2.5	5.0	6.6
Heavy minerals	0.0	0.0	0.0	0.0	0.0	0.0
Mica	4.4	1.5	0.2	1.0	2.1	0.0
Chlorite	0.0	0.0	0.0	0.0	0.7	0.0
Carbonate	0.0	0.0	0.0	0.0	0.0	0.0
Glass / melted fragments	0.0	0.0	0.0	0.0	0.0	0.0
Matrix	9.1	2.4	0.2	3.7	5.5	3.7
Organic matter	3.7	3.9	0.7	7.4	2.6	0.0
Pyrite	0.0	0.0	0.0	0.0	0.0	0.0
Amphibolitic minerals	0.0	0.0	0.0	0.2	1.4	0.7
Total grains counted	412	414	396	393	411	395

TABLE 2. XRD-RESULTS

Sample numbers		Peak height (counts)			Ratio quartz/feldspar
		Quartz*	K-feldspar†	Plagioclase§	
Branden core depth (m)	0.6	2670	313	849	2.30
	22.97	3587	532	471	3.58
	25.09	383	398	616	0.38
	30.29	938	329	709	0.90
	37.3	1098	383	606	1.11
	79.7	2918	4126	2013	0.48
	83.95	778	3027	2060	0.15
	85.9	2281	3018	2339	0.43
	98.25	1694	2858	2745	0.30
	101.85	1378	2248	3089	0.26
	102	1172	2084	2605	0.25
	113.15	1768	2397	3372	0.31
	119.5	1406	1725	2377	0.34
	124.5	1192	1401	1909	0.36
	127.56	1868	2088	2497	0.41
Field sample	SK-KGL	2035	793	1555	0.87

*Quartz is identified by Miller indices (101) and quantified by the peak at d = 4.26 Å (interplanar spacing).

†K-feldspar is identified by Miller indices (-200) (002) (040) and quantified by the peak at d = 3.24 (3.25) Å (interplanar spacing).

§Plagioclase is identified by Miller indices (002) (040) (220) and quantified by the peak at d = 3.18 (3.19) Å (interplanar spacing).

at the University of Oslo), and the main mineral content was semiquantitatively determined (Table 2).

In the clast analysis in the field, the smallest and darkest grains (amphibolites and some clasts of metasediments) were, in spite of intensive cleaning operations to remove organic matter, very difficult to observe. Efforts, however, were taken to find good exposures and minimize this and related observational difficulties. The numbers of counted clasts in the core and in the field are not directly comparable, as field sections were cut almost parallel to bedding, and the core represents a vertical section (e.g., almost normal to bedding) through the sediments. Within these coarsest sediments, neither clear trends of upward fining or upward coarsening nor preferred fabric orientations were seen. It should therefore be possible to compare the relative distributions of clast size, type, and roundness.

The cylindrical shape and limited surface area present in the core may have influenced the textural interpretation of larger clasts that appeared too well rounded. In some core positions the supporting mechanism of the conglomerates is poorly documented, due to lack of contact observations between larger clasts. Postdepositional diagenesis has altered both minerals and impact melt. Dissolution and precipitation of the main components such as quartz and feldspar are common and partly mask original textures. Grain size classes applied are according to Wentworth (1922).

OBSERVATIONS

Transition from Impactites to Sediments

The boundaries between suevite and postimpact sediments have only been observed in one well-exposed location at Flatdalselva and in the Branden core. In both places, thin (decimeter-scale) zones of fine-grained sediments occur at the boundary, and detailed examination of the boundary reveals significant differences at these occurrences (Fig. 4).

In the Flatdalselva riverbed, postimpact breccias overlie the suevite. The suevite becomes depleted in large rock fragments toward its upper boundary, where it consists almost entirely of dark clastic matrix with millimeter-sized black shards of altered melt. Directly above the suevite follow two beds of conglomerates/breccias, each of ~40 cm thickness. The lower parts of the beds are partly clast-supported, consisting of angular clasts (typically 2–7 cm in size) in a dark matrix. Granitic gneiss and amphibolite clasts dominate, reflecting the local bedrock in the western part of the Gardnos structure. Toward the top of the beds, dominant clast size decreases to 1–2 cm, grading into sandstone with only scattered granule-sized granitic grains in the uppermost 10 cm. The light gray granitic clasts are easily distinguished, in contrast to the components of the gray to dark gray matrix. Amphibolitic clasts are generally dark, often blending in with the matrix when observed from a distance. Breccia clasts can be seen sinking into

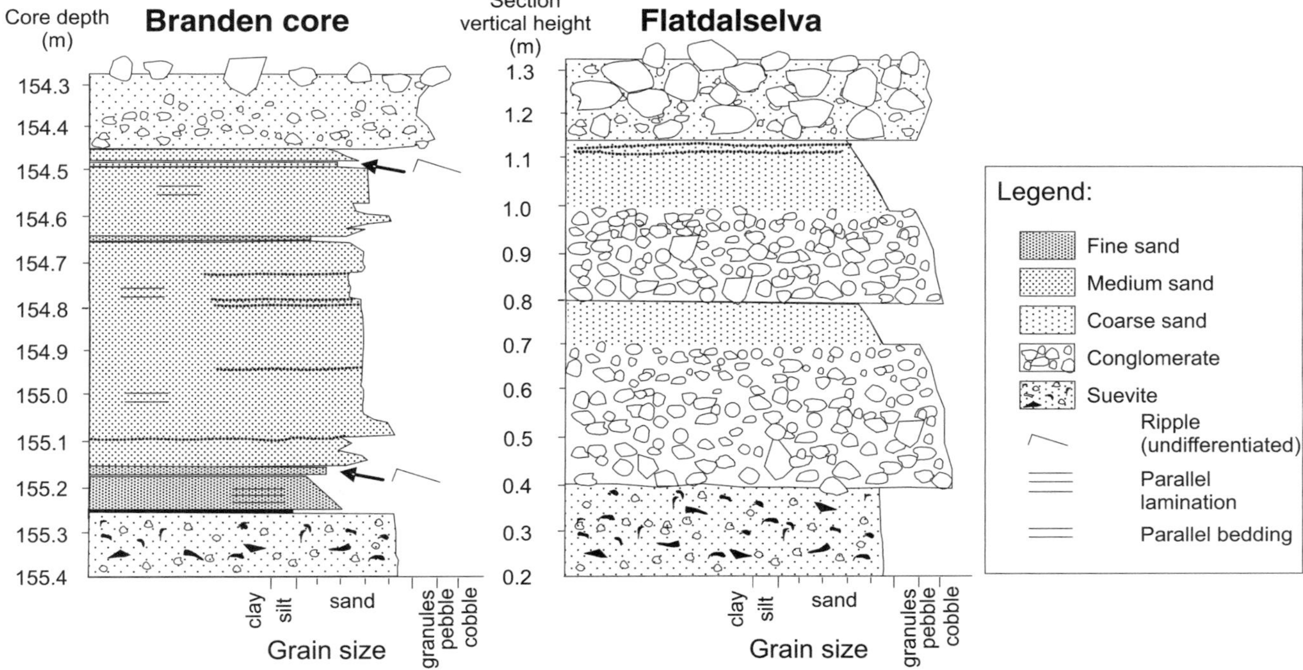

Figure 4. Detailed logs across the boundary between suevite and postimpact sediments in the Branden core (left) and at Flatdalselva (right) (location in Fig. 2).

the suevite at the boundary, which indicate that the suevite was still unconsolidated when breccia deposition started (Fig. 5).

In the Branden core, similar to the contact at the Flatdalselva locality, the uppermost meter of suevite is poor in large (>5 cm) rock fragments. The overlying postimpact breccia is mainly clast-supported and dominated by angular to subangular clasts of different lithologies (see the next section). In the Branden core the suevite–postimpact breccia boundary includes a transitional one-meter-thick interval (155.3–154.3 m core depth) of well-sorted, dominantly fine- to medium-grained sandstone (Fig. 6). Parallel bedding is common throughout the transitional interval, with dispersed centimeter-thick zones of coarse-grained sand. However, the lowermost 10 cm and the topmost couple of centimeters contains more fine-grained sediments such as partly laminated, fine-grained sandstones that grade upward into dark gray siltstones and claystones. At the top of each of these two very fine-grained intervals occurs a 1–2 cm bed of fine-grained sand with a small ripple-like structure (Fig. 6).

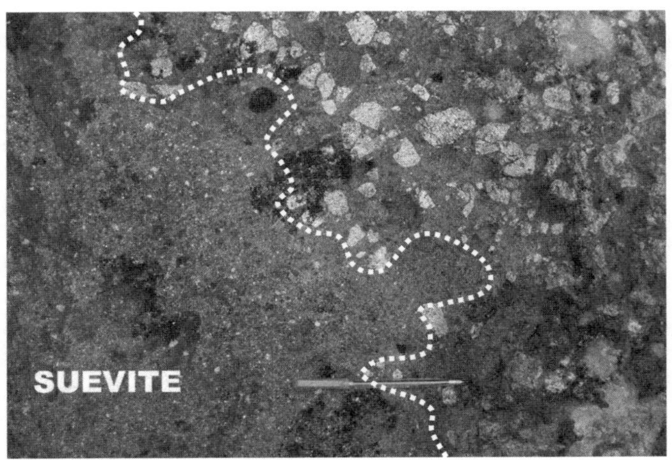

Figure 5. The boundary between suevite and postimpact sediments as exposed at Flatdalselva (Fig. 2). The exposed surface has a gentle slope that cuts through the stratigraphy at an angle of ~10°. Ballpoint pen in picture is 15 cm in length.

In both postimpact breccia sections described above, individual grains generally have an angular appearance and consist of quartz, plagioclase, K-feldspar, amphiboles, mica, and chlorite. Traces of possible impact-altered material, such as microcrystalline quartz and remnants of molten fragments, are also present. The main constituents are similar in the two boundary sections, whereas the core section is generally finer-grained and better sorted than the Flatdalselva section.

Within the fine-grained interval separating suevite from postimpact breccias, several centimeter-thick beds with lighter, millimeter-sized "spots" (Fig. 7) are common. The "spots" consist of slightly coarser sand grains. The light color is due to less fines than in the surrounding sand, in addition to some carbonate present. The "spots" are irregular rather than concentric, with a

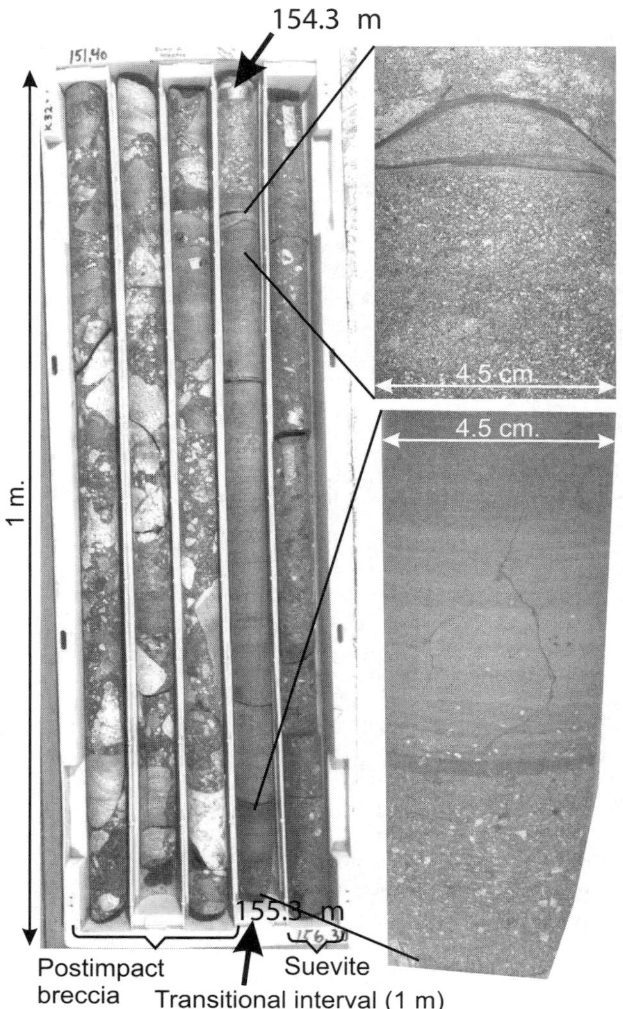

Figure 6. In the Branden core a transitional interval of fine-grained sediments occurs between the suevite and the postimpact breccia. Bedding, lamination, and a ripple structure are present in these fine-grained transitional sediments. This relation has not been observed in the few outcrops across this boundary (Figs. 4 and 5).

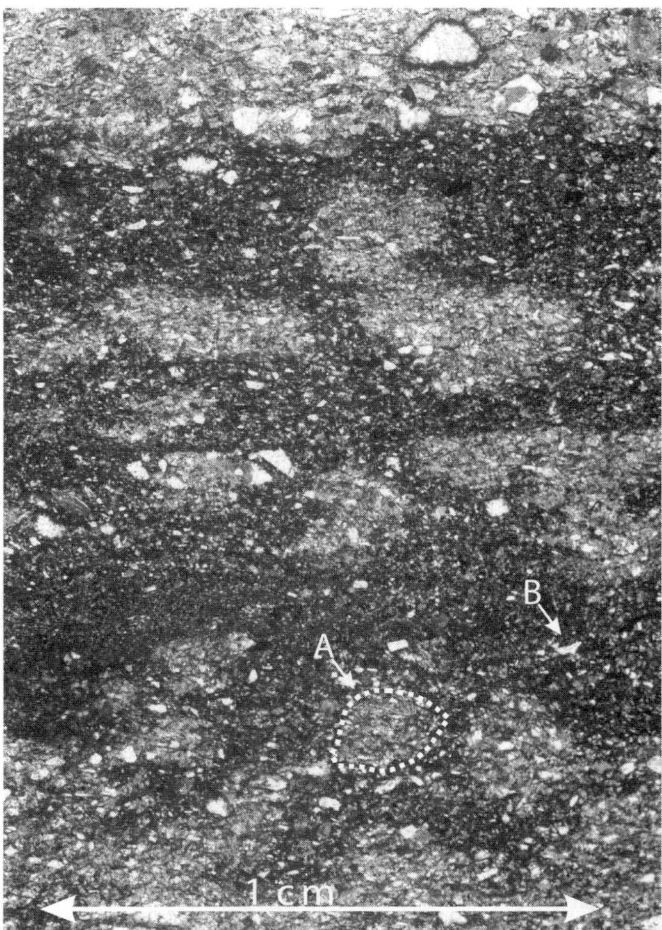

Figure 7. A thin-section scan of "spots" in the fine-grained transition interval between the suevite and overlying coarse clastics at the Flatdalselva location. Example of a "spot" is marked with A, whereas the white dot marked with B is just a light grain.

homogeneous interior and a diffuse outer zone. Thus, they do not resemble accretionary lapilli as described from other impact structures such as Ries (Graup, 1981) and Chicxulub (Pope et al., 1999). The "spots" in the Gardnos deposits are being analyzed, but their origin is still not known.

The postimpact sediments are described in terms of different facies. The term "facies" (lithofacies) is applied to bodies of sedimentary rock based on descriptive features such as composition, grain size, bedding characteristics, and sedimentary structures. The term "facies" can, however, also be used also in an interpretive sense, emphasizing specific depositional processes, such as a debris flow facies.

The type section of the postimpact sedimentary succession in the Gardnos crater is presented in the Branden core (Fig. 8). The sediments in the Branden area follow on top of the suevite and represent the more basin-central part of the postimpact deposits. The sediments in the area called the "Hillslope" (Fig. 2) are located closer to the original crater wall than the sediments at Branden, and may rest directly on Gardnos Breccia. The mapped postimpact sediments can be subdivided into five main facies: (1) postimpact breccias, (2) coarse conglomerates, (3) conglomeratic sandstones, (4) sandstones, and (5) interbedded fine sandstones, siltstones, and shales (facies distribution shown on the map in Fig. 2; characteristics summarized in Fig. 9).

Postimpact Breccias—Facies 1

The breccias of this facies are extremely clast-rich. Clasts, mainly resting on each other, make up a framework, with sandy matrix filling interstices (clast-supported texture). The clasts are generally angular to subangular. Pebble-sized clasts

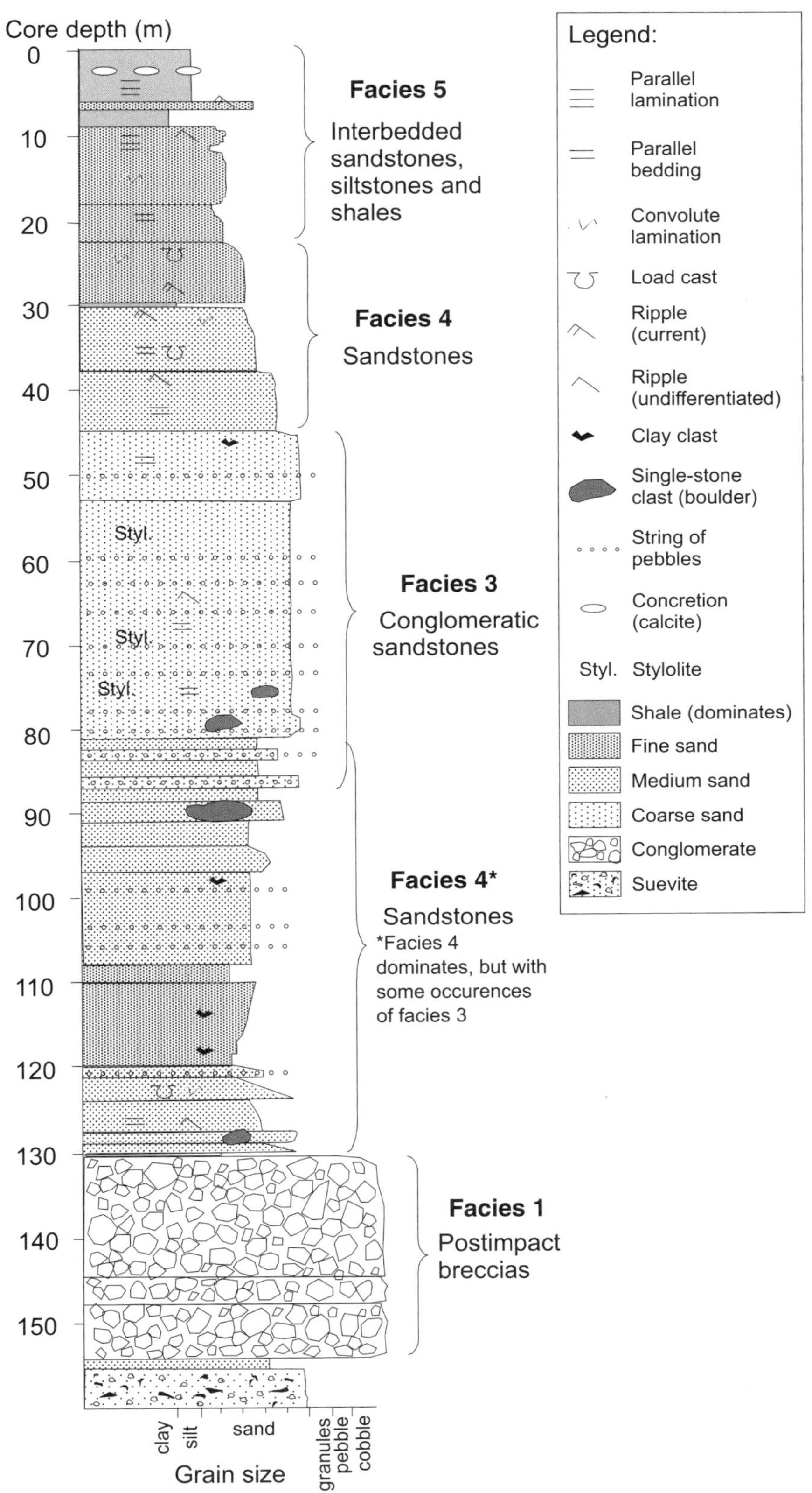

Figure 8. Schematic presentation of the sedimentary upper part of the Branden core.

Facies 5:
Interbedded fine sandstones, siltstones and shales

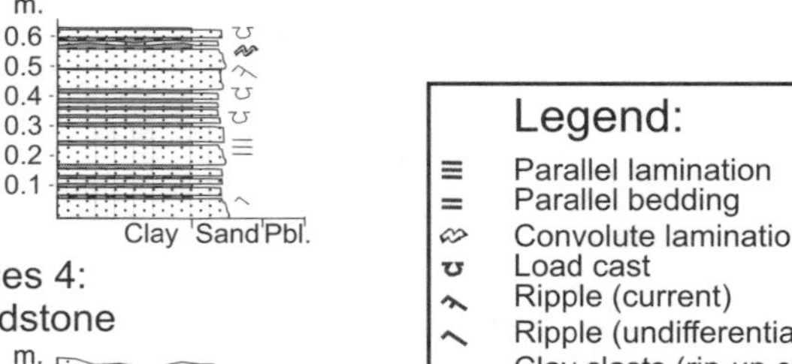

Facies 4:
Sandstone

m.
0.6
0.5
0.4
0.3
0.2
0.1
Clay Sand Pbl.

Facies 3:
Conglomeratic sandstone

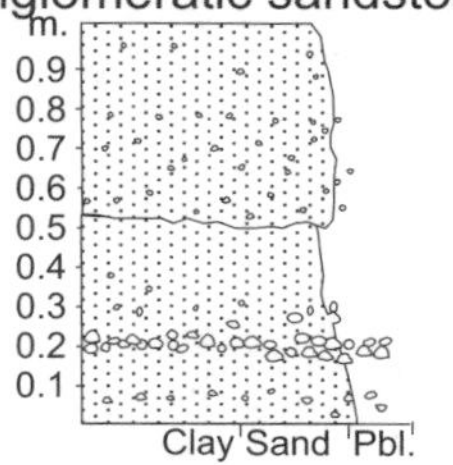

Facies 2:
Coarse conglomerate

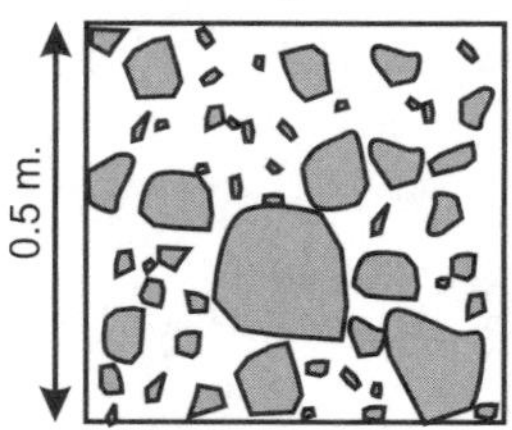

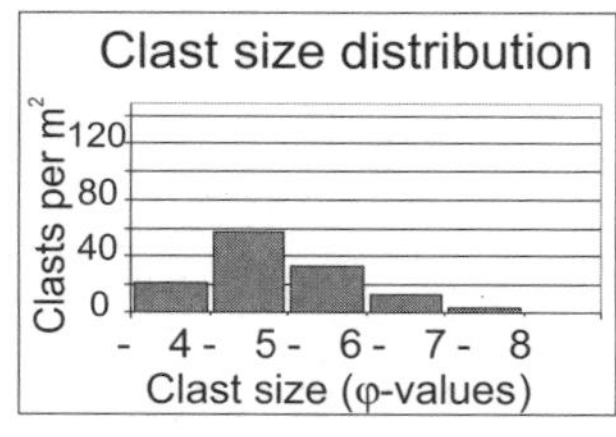

Facies 1:
Postimpact breccia

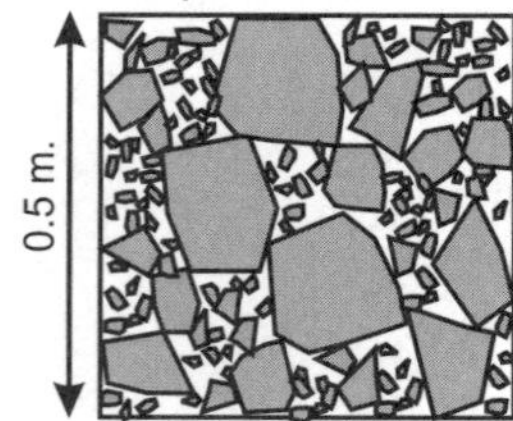

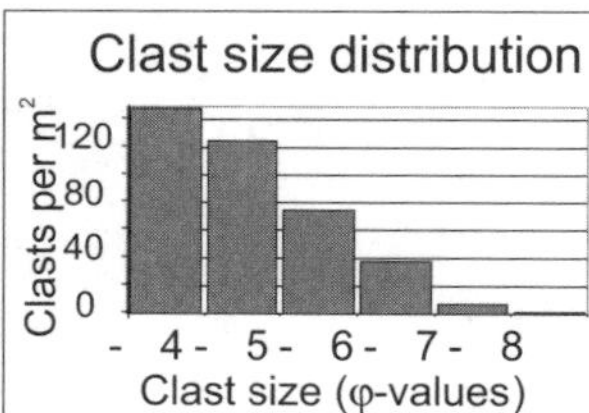

Figure 9. Summary of typical appearance and textures for the main sedimentary facies. In facies 5 individual sandstone beds (fine- to very fine-grained) are thin (<5 cm) and alternate with shales. Sandstone beds of medium-grained sand in facies 4 are generally 20–100 cm thick, commonly separated by clay seams, and occasionally contain clay clast conglomerates at the base. The conglomeratic sandstones of facies 3 consist of coarse-grained sand to granules, sometimes with pebble-sized clasts concentrated in stringers. The conglomerates of facies 2 are matrix-supported and poorly sorted, with dominating cobble pebble-sized clasts of subangular to subrounded shape. Most clasts are large pebbles. In clast-supported facies 1, the clasts are generally densely packed and angular to subangular, and small pebbles are much more common than in facies 2.

dominate, with a few cobbles, and some occasional boulders. Facies 1 is observed at several field locations and in the Branden core, and cover most of the crater walls and floor, resting directly above the impactites.

In general the matrix of the breccias appears gray to dark gray in color, whereas the clasts display different colors according to rock type. Fragments of granite/granitic gneiss, quartzite, pegmatite, amphibolites, and some metasediments are commonly recognized. Along the western side of the crater, granitic clasts dominate, whereas quartzitic clasts are often seen in the east (Fig. 2). At one location west of the Hillslope, there is a bedrock outcrop of amphibolite, and at the adjacent exposures of facies 1, amphibolitic clasts are common (20%–30% of clast volume). This indicates that local bedrocks are important sources for the clasts of facies 1. A few clasts of original soft sediments (presently not correlated to any known local bedrock) were observed in several field exposures. The clasts were partly folded around other fragments, but in thin section these sediment clasts display preserved internal structures are still intact (only slightly folded). Both studied samples of sediment fragments are finely laminated sediments, one rich in clay minerals and with a carbon content of 0.4%, the other dominantly comprising very fine-grained quartz sand, dark gray to black in color due to a carbon content of 0.8 wt%.

In a few cases the rock clasts are clearly associated with each other, as if clasts were broken and fell to pieces at the moment of deposition, more or less in place (Fig. 10). At the Viewpoint locality (Fig. 2), the clasts are packed extremely densely, and the matrix is barely visible. The matrix is dominated by lithic fragments and single grains of quartz and feldspar. Thin-section studies show that the mafic minerals of the amphibolites are commonly altered to chlorite. Other phyllosilicates only occur in minor amounts. The matrix of rock fragments and single mineral grains resemble the larger clasts compositionally, and are most likely derived from those (or the same precursor lithologies). The breccias have a homogeneous appearance, generally lacking internal erosion surfaces, shear zones, and depositional gaps.

In the Branden core a 25 m thick package of facies 1 directly overlies the fine-grained sandstones above the suevite. The facies 1 breccia contains clasts of granites, quartzites, amphibolites, and some metasediments, varying in grain size. The distribution patterns (sizes and composition) are remarkably uniform throughout the conglomeratic section in the core (Fig. 11). It is worth noting that the lowermost 15 cm of this section contains no clasts larger than 1 cm. From just one core observation, it is not possible to know whether this is a general development, or whether the core just happened to penetrate a pocket of finer-grained material between/below large boulders. At two levels (a 10 cm interval at 148 m depth and a 5 cm interval at 145 m depth), parallel bedding is present. These intervals lack large clasts, but it is not clear whether they mark the top or the bottom of a depositional unit, or could be the result of internal shear.

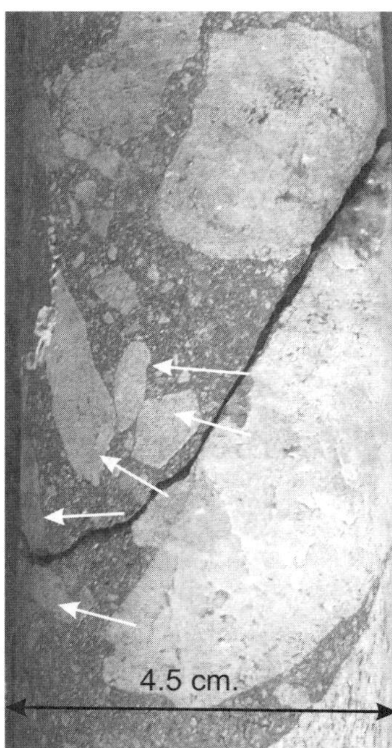

Figure 10. Detail of postimpact breccia (facies 1) in the Branden core at 152 m depth (compare Fig. 8). White arrows point at small clasts of the same type as the larger clast to the right in the core. The smaller clasts are probably fragments derived from larger ones.

Coarse Conglomerates—Facies 2

Conglomerate facies 2 has considerably fewer clasts than facies 1, and clasts generally float in matrix (matrix-supported texture). Facies 2 clasts are mainly small cobbles, subangular or angular in shape, though subrounded clasts also frequently occur. The unsorted matrix is sand-dominated, with only a minor clay fraction. Mineral grains are generally angular, and their compositions reflect the composition of the clasts. Where present, facies 2 stratigraphically succeeds the postimpact breccias of facies 1. Outcrops of facies 2 conglomerates occur in a belt along most of the north and northwestern crater wall (Fig. 2), but the facies is absent in the core. The total thickness of the unit varies from zero to a maximum estimate of 50–100 m.

The clast composition of both the postimpact breccias (facies 1) and the coarse conglomerates (facies 2) varies geographically and stratigraphically (Fig. 12). Along a section from the crater wall to the deeper part of the basin, matrix-supported conglomerates show stratigraphic changes in composition (Fig. 12). In these outcrops, no bedding planes were seen separating the compositionally distinct deposits.

In the stratigraphically uppermost part of facies 2, bedding planes were occasionally observed. These beds vary in thickness from 10 cm to several meters, and occasionally display trends of upward-fining or -coarsening development, but most commonly with a random fabric. Erosional channels occur in the upper part, where the facies becomes sandier and grades into the conglomeratic sandstones of facies 3.

Conglomeratic Sandstones—Facies 3

The transition from facies 2 to facies 3 is gradual. The conglomeratic sandstones of facies 3 typically appear in 0.5–2.5 m thick beds. They are dominated by very coarse sandstones, with pebble-sized clasts floating in a sandy matrix, often forming bedding-parallel stringers. The clasts have subrounded to rounded shapes and are dominated by quartzite and granite. A couple of shallow erosional channels and one groove cast (at the side of one channel) are sedimentological features that indicate sediment transport directions (Fig. 13).

Along the crater wall (Fig. 2) the conglomeratic sandstones of facies 3 succeed the coarse conglomerates (facies 2), though at one place conglomeratic sandstone overlies the postimpact breccia of facies 1. In the drill core, a different development has been observed: about 40 m of conglomeratic sandstones (facies 3) occur above a 45 m thick interval dominated by well-sorted sandstones here classified as facies 4. In other areas of the crater structure unit thickness could not be estimated properly, due to deep glacial erosion or limited exposure. The mineral composition of these gray to dark gray sandstones is dominated by quartz and feldspar. The sandstones, in contrast to larger, light gray quartz and granite clasts, weather to a reddish color. Well-developed stylolites are abundant in several beds, commonly with color shifts from lighter to darker gray across the stylolite surface, which may be coated by black carbonaceous material. Maximum stylolite amplitudes measured are on the order of 3 cm.

Sandstones—Facies 4

The relatively well-sorted sandstones of facies 4 are made up of individual beds, typically from 20 cm to 1 m in thickness. These units are normally upward-fining or uniform in grain size composition. The grain size ranges from very fine to very coarse sand, with a clear dominance of medium-grained sand. Locally, clay laminae/beds are found on top of the sand layers, and fresh rip-up clasts of clay occur in the sandstone beds. A couple of large, cobble- to boulder-sized clasts of granitic gneiss occur in the sand.

The best field exposures of sandstones are found along the steep slopes in the Branden area (Fig. 2). In the drill core, the sandstones occur in two intervals: (1) a 45 m thick interval between the clast-supported conglomerate and the conglomeratic sandstone, and (2) as a part of a fining-upward sequence above the conglomeratic sandstone interval (Fig. 8), in a gradual transition to overlying fine sandstones and shales of facies 5.

The sandstones display parallel bedding, tool marks, groove casts, one flute cast, and rare loading structures. Ripples are present in the drill core but were not observed in the field. The directional structures indicate a wide range of possible sediment transport directions (Fig. 13).

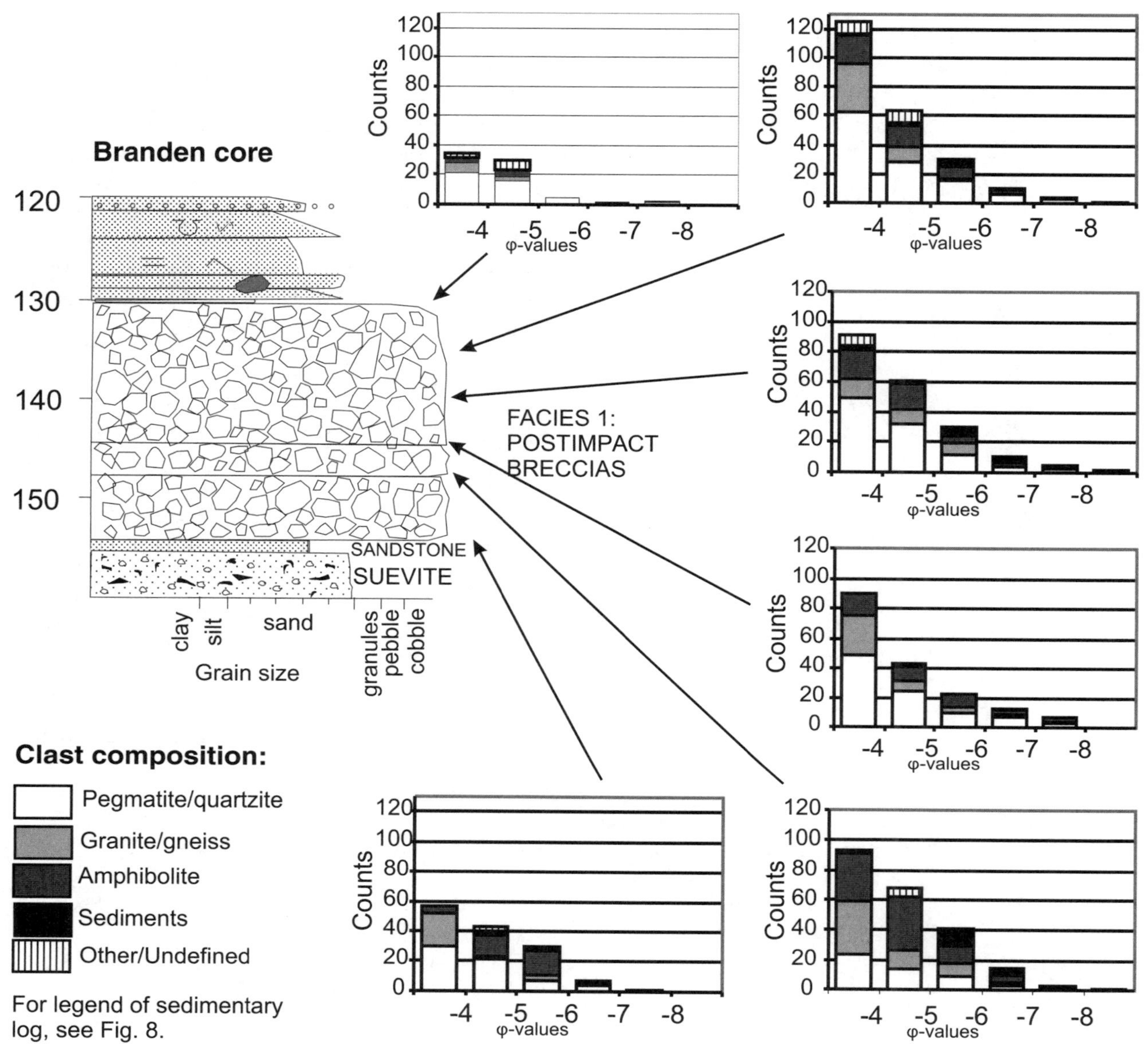

Figure 11. In the postimpact breccia of facies 1 (Branden core), clasts were counted and identified by size and rock type. The results illustrate the modest variations in clast size and composition throughout the section.

The mineralogical composition is relatively constant throughout facies 3 and 4 in the Branden area, as indicated by the quartz/feldspar ratios determined on thin sections and by XRD (Fig. 13). The marked increase in quartz content relative to feldspar in the upper parts of the Branden profiles coincides with the transition to facies 5. Within facies 3 and 4, the quartz/feldspar ratio shows local geographical variations (Fig. 14). The ratio is rather constant in the Hillslope and Branden sections, whereas the quartz content is significantly higher at two localities northeast of Branden (marked with "S" and "H" on the map of Fig. 14).

Interbedded Sandstones, Siltstones, and Shale—Facies 5

The uppermost part of the sedimentary sections of Gardnos is well exposed in the field and in the Branden drill core (Fig. 2). Sandstone beds are typically 1–5 cm thick, consisting of fine sandstone and separated by thin shale/siltstone layers. The relatively quartz-rich sandstone layers are generally light gray in color. The individual, well-sorted, sand beds may be mixed with clay beds/lamina due to loading effects and soft-sediment deformation (Fig. 15).

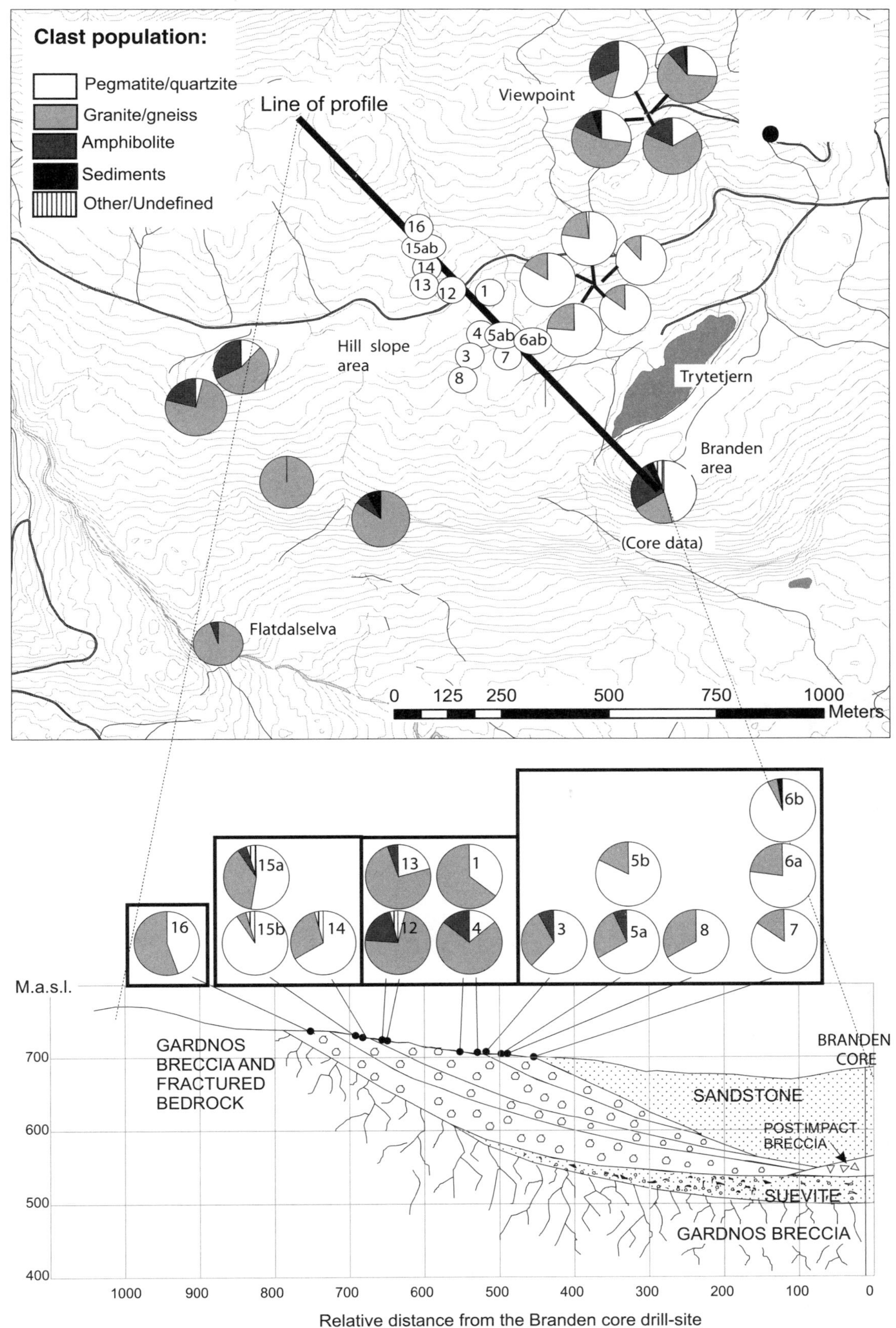

Figure 12. Variation in clast composition in the breccias and conglomerates of facies 1 and 2 (upper figure) and for only facies 2 conglomerates along a profile at Hillslope (lower figure).

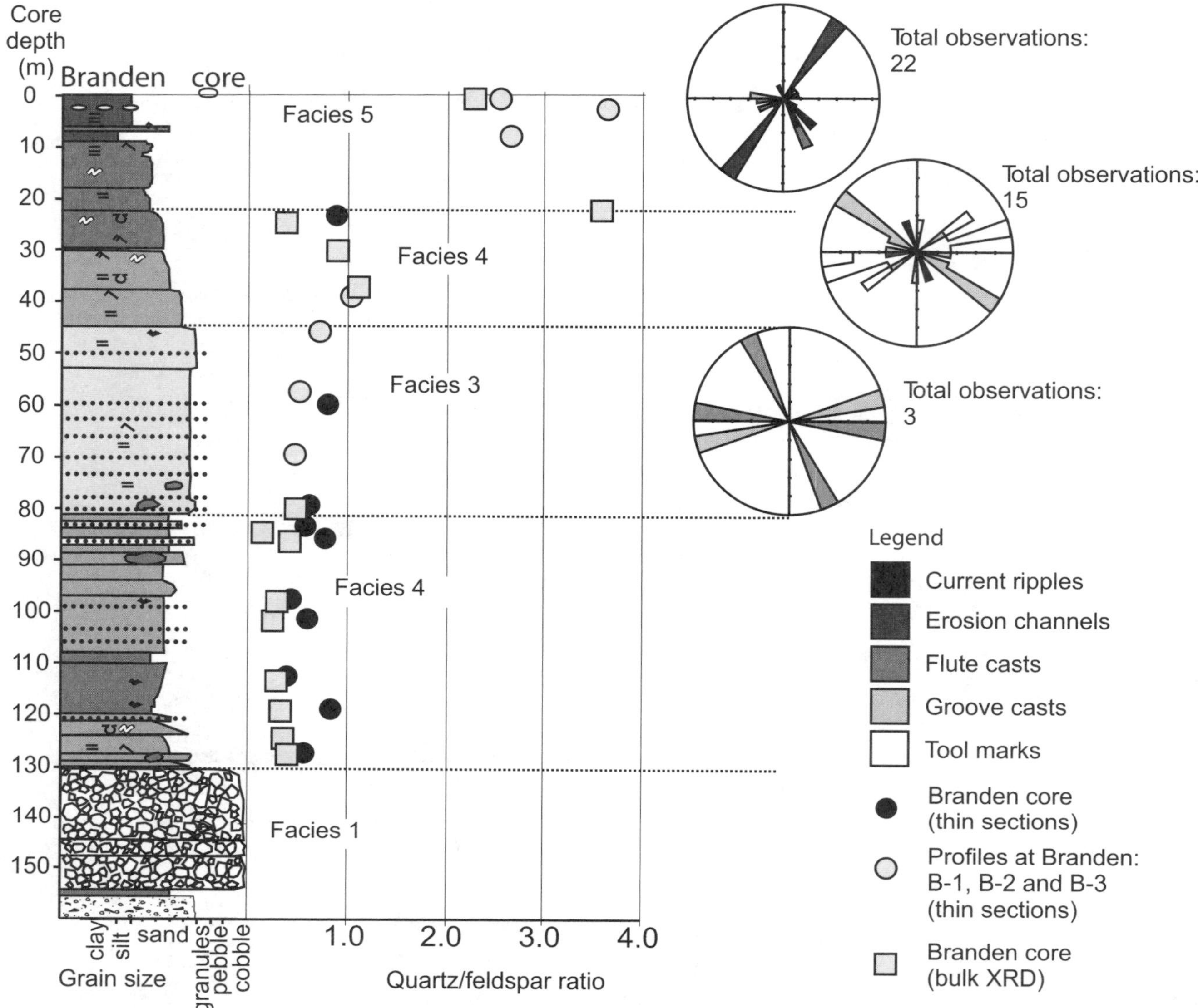

Figure 13. Stratigraphic variation in the quartz/feldspar ratio through the Branden profile (drill core and field samples) based on point counting in thin sections and bulk XRD (X-ray diffraction) analysis. Rose diagrams for directional measurements of sedimentary structures within facies 3, 4, and 5. All structures were measured in the field at the Branden area. Note the very low number of directional observations in facies 3.

Individual sand and clay layers appear sheet-like and are disrupted by shallow (<50 cm deep) erosional scour channels. The thickness of the clay beds increases upward, and the uppermost 5 m are dominated by shale with occasional carbonate-cemented sand lenses.

Ripples were seen both in the field and in the drill core, and in the field, flute casts and one groove cast were also observed. Directional measurements of these features are included in Figure 13. The current ripple and flute cast measurements show common sediment transportation toward the southeast, whereas west-southwest transport directions are less common. The majority of erosional channels point to the southwest, whereas the ripple and flute cast directions are more scattered. Compared to the sandstones of facies 4, this might indicate a shift in transport direction toward the top of facies 5.

At Branden, sand-filled fractures in shale are found. The cracks may be V-shaped and penetrate at least 2 cm into the underlying shale, but they may also appear smaller and more irregular (Fig. 16). The cracks may be single-standing or intersecting, but are always concentrated in specific beds.

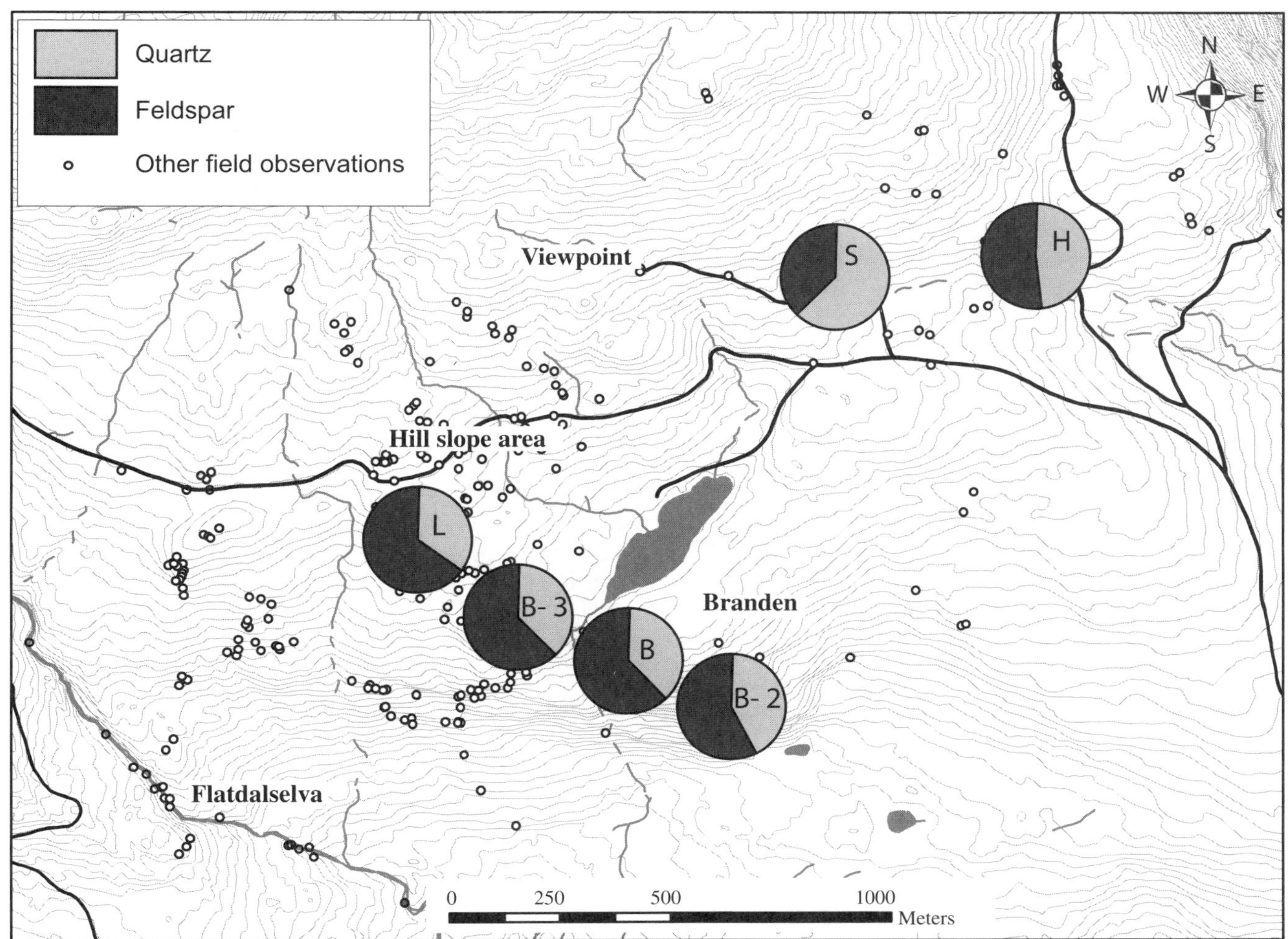

Figure 14. The quartz/feldspar ratio of the conglomeratic and well-sorted sandstones of facies 3 and 4. Quartz and feldspar contents are based on grain counts on thin sections (Table 1) and XRD-data (Table 2). Labels S, H, L, and B (Branden) refer to localities in Figure 2. The quartz/feldspar ratios show the same pattern for samples from the Hillslope and Branden areas, with the highest values at the two more easterly locations.

INTERPRETATION

The various postimpact sedimentary rocks at Gardnos reflect different mechanisms of sediment transport and local depositional systems. Steep crater walls and large amounts of newly excavated impact material favor gravity-controlled sedimentation in the first phase of deposition. In addition, suspension deposition of fine-grained material took place.

There has been considerable discussion of density flow nomenclature (e.g., Shanmugam, 1996; Mulder and Alexander, 2001; Gani, 2004). Gani (2004) classified gravity flows according to four main parameters (sediment concentration, sediment-support mechanism, flow state, and rheology), and he emphasized how the different parameters affect the classification. In this study, gravity flows are classified according to Mulder and Alexander (2001), by combined consideration of physical flow properties and sediment-support mechanism.

Transition from Impactites to Sediments

The 1 m thick, fine-grained bed between the suevite and postimpact breccias, as seen in the drill core, may represent the last fallout stage from the ejecta plume. However, a fallout deposit would most likely have resulted in a fining-upward sequence, whereas the observed succession generally lacks gradation. Together with the parallel bedding/lamination, presence of ripple structures, and repeated grain size variations, this points toward shifting flow regimes. The resurge of seawater would be expected to return as a mixture of water and debris of most grain sizes as water rushed down the steep crater walls covered with impact debris. These well-sorted, fine-grained sediments, therefore, are not likely to be resurge deposits. The distribution of the bed is uncertain due to lack of exposures. The deposition was most likely very fast, in order to precede the quick deposition of overlying conglomerates. Ongoing detailed studies will hopefully give more clues to its origin.

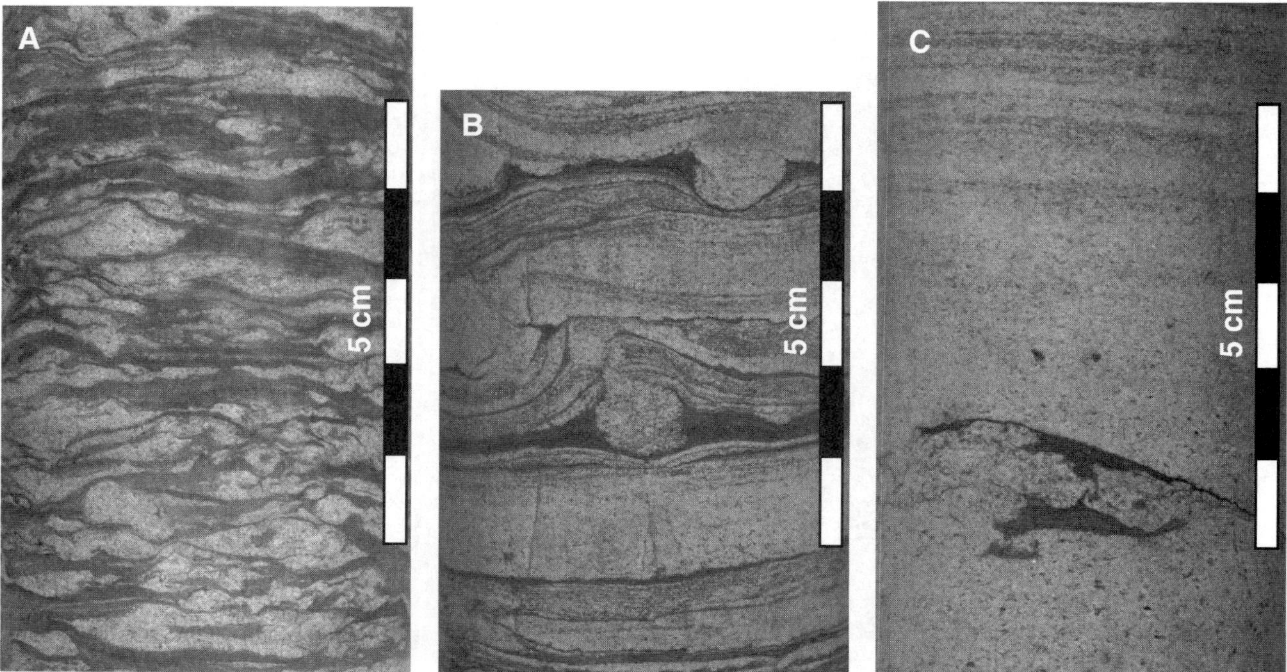

Figure 15. Core photos showing examples of sedimentary structures in the interbedded sandstones, siltstones, and shales of facies 5. A: Thin sand and clay bed disrupted and partly mixed after deposition. B: Sand-dominated interval, with examples of loading and soft-sediment deformation structures where sand penetrates into the thin clay beds. C: Rip-up clasts in a clean sandstone unit. The sandstone bed is slightly upward-fining, with some parallel lamination in the upper part.

Figure 16. Left: An example of sand-filled cracks in shale at Branden. The drill core photo to the right shows the cross section of a smaller, irregular sand-filled crack.

Postimpact Breccias—Facies 1

The dominance of coarse, angular clasts of local origin indicates very short transport distance for these particles. Few soft-deformed sedimentary clasts were seen, and the general absence of clay minerals in the poorly sorted deposits reflects a sediment source deficient of clay. This indicates very freshly crushed, unweathered, source rocks, as does the high abundance of feldspar relative to quartz.

Kessler and Bédard (2000) described avalanches as large rock masses that move rapidly down steep slopes, generating poorly sorted deposits as a mixture of megablocks, boulders and cobble-sized fragments, gravel, sand, and mud. Due to collisional breakage during transport, individual clasts are generally very angular and often form a jigsaw puzzle breccia, well matching the features observed in facies 1 breccias of Gardnos.

The coarsest parts of the excavated material have been found along the crater rim and most likely represent some of the first beds deposited after impact. The steep crater walls and the central peak area were unstable and susceptible to rockfalls. Blikra and Nemec (1998) studied recent postglacial colluvium in Norway and concluded that fans dominated by rockfall debris are generally much steeper and shorter than those formed by debris flows. Facies 1 has a texture and structure corresponding to rockfall avalanches, but the question is how these deposits could be sourced in the more distal basinal location of the Branden core. The uniform distribution of clast types throughout the conglomeratic formation in the Branden core indicates a single source area for the unit in that location. The Branden location represents the deepest part (moat) of the basin, which had the potential of receiving material from both the crater walls and the central uplift (see the discussion section). The wide distribution of rockfall and avalanche deposits may have been caused by the combination of several local avalanches, developing a continuous apron. The local origin of the avalanches is also supported by the geographic variations in clast content, reflecting local bedrock composition.

Deposition of facies 1 may have taken place in subaerial or subaquatic conditions. No direct indications of subaerial exposure have been found (such as iron oxide coatings), but with the expected high sedimentation rates such weathering features would not have had time to develop. The preservation of fine-grained sediments beneath the succeeding conglomerates (facies 2) in the deeper part of the basin is in accordance with a configuration where facies 1 was deposited as avalanches before water reentered the crater.

Coarse Conglomerates—Facies 2

The matrix-supported and poorly sorted conglomerates of facies 2 are consistent with en masse deposition characteristics for cohesive debris flows (Mulder and Alexander, 2001). Sporadic clusters of rock fragments (pegmatite and quartzite) may represent original larger rock fragments that were fractured during the initiation of the flow, and kept close together during later transport in accordance with the cohesive behavior of the flow.

The matrix of facies 2 conglomerates is dominated by angular sand grains of feldspars, quartz, some amphibolitic minerals, and only minor amount of clay minerals. The mineralogical compositions of the matrix reflect the local bedrock, indicating a fresh source of debris generated by the impact. The volume fraction of cohesive solids (silt and clay compounds) required to produce cohesive behavior is different in subaerial and subaqueous conditions. Water aids subaerial flow by adding additional mass, and the pore pressure helps to hold the flow together by surface tension effects, reducing the need for clay mineral content (Mulder and Alexander, 2001). This may support a subaerial explanation of the very first sediments in the crater basin.

The debris flow deposits of facies 2 occur in most places along the crater wall, whereas occurrences on the central crater floor are few. The absence of this facies in the Branden core indicates a discontinuous distribution pattern, possibly controlled by variations in local relief. The debris flows of facies 2 may have been triggered when seawater resurged into the crater. Breaching of the rim most likely took place at weak and lower locations. The resurging water eroded gullies in the crater wall and created initial centers of deposition along the inner periphery of the rim wall. These sediments were farther spread out to the more central parts of the crater during subsequent stages of sedimentation.

The debris flow facies sedimentologically represents several events, as seen in their shifting composition (Fig. 12). This episodic behavior is to be expected in a setting of a highly breached rim. The incoming, powerful, and partly sediment-laden water could continuously erode and widen (and deepen) the inlets, leading to repeated rim collapses and renewed activity in the resurge gullies.

The transition between facies 1 and 2 matches a downslope development from avalanches to debris flows. Kessler and Bédard (2000) proposed such development as a consequence of clast crushing and grinding during transport, introducing the term "proto–debris flow" for the intermediate stage between avalanche and debris flow. The division between facies 1 and facies 2 conglomerates reflects such a transition.

Conglomeratic Sandstones—Facies 3

The shift from matrix-supported conglomerates (facies 2) to the conglomeratic sandstones (facies 3) may correspond to the change from cohesive to noncohesive flow. According to the Mulder and Alexander (2001) classification, hyperconcentrated density flows (noncohesive debris flows, grain flows) are characterized by lack of grain size grading, bed forms, and sedimentary structures, and occurrence of only a few erosional features such as flutes and scours. The conglomeratic sandstone (facies 3) in Gardnos matches these criteria well.

By entraining water, a debris flow could develop into a cohesionless flow. Both Mulder and Alexander (2001) (theoretically) and Felix and Peakall (2006) (experimentally) described such transformation of density flows toward turbidity currents. If the Gardnos crater was filled with water during deposition of facies 2, it is possible that parts of facies 3 (and 4) could represent more distal formations of diluted debris flows, as they entered the rising water in the crater basin. This corresponds well with the field observations at Gardnos, where thick conglomerates of facies 2 are found at the Hillslope area close to the crater wall, but are replaced by the sandier facies 3 at the distal Branden location (Fig. 2).

Sandstones—Facies 4

Sandstones of facies 4 have either uniform or fining-upward developments. The sedimentary structures (flute and groove casts) within the facies are typical of turbulent flows, and a concentrated density flow explanation could be applied (Mulder and Alexander, 2001). These flows of facies 4 may have been part of a downslope evolution of hyperconcentrated density flows, with facies 4 deposited as a continuation of facies 3.

Crater Lake (Oregon, USA) is a possible recent depositional analogue of a small, steep, deep basin. Nelson et al. (1986) described how debris flows evolved to sheet flow turbidity currents in Crater Lake, depositing finer-grained materials on the basin floor. Part of the facies 4 sediments probably was deposited separate from the other facies, as a result of reworking by local currents. The large variations in measured transport directions for facies 4 (Fig. 13) most likely reflect this reworking.

Thin clay lamina, commonly separating the sand units, may reflect short and quiet periods, with ample time for clay sedimentation out of suspension. The intermixed zones of clay clast conglomerates show renewed erosion of these clay beds and clay clast deposition just after clay consolidation.

The depositional mechanisms of the occasional boulders appearing in the facies 4 sandstones are not clear. The smaller ones may have been part of thin debris flows, but at ~90 m depth in the Branden core a more than 40 cm diameter granitic gneiss clast occurs within an interval of repeated 20–40 cm thick fining-upward sand units. These units may have been deposited by turbidity currents, but the boulder is too large for such transport. Granitic gneisses are common bedrock in the western region of Gardnos, and the boulders could be derived from rockfall. In steep terrain, boulders may settle far beyond the base of the main rockfall talus. The concept of rockfall shadow (Evans and Hungr, 1993) is based on a shadow angle from a horizontal line drawn from the apex of the talus with typical shadow angles between 20° and 25°. Due to severe erosion of the Gardnos structure, the apex of any possible talus is hard to pinpoint. Based on the reconstructed crater morphology, the Branden area was too far away (more than 2 km from the crater rim) to receive rockfalls from the crater rim walls. The central peak or local topographic highs may have been possible sources for the observed rockfall material. A possible glacial dropstone explanation cannot be ruled out, as the general geological setting/timing of the impact is poorly constrained. Tillites of the Varanger ice age are known from the Mjøsa region, 100 km east of Gardnos (Spjeldnæs, 1959; Nystuen, 1976; Bjørlykke and Nystuen, 1981). The tillites are of Vendian age (Bjørlykke and Nystuen, 1981) and thus within the possible time span for the Gardnos impact event. However, no other possible glacial indications have yet been found.

Lateral variations in the quartz/feldspar ratios within facies 3 and 4 may indicate subareas representing separate depositional systems and different source areas. The Hillslope and Branden areas (Fig. 14) seem to be dominated by one such system, with continuous sedimentation developed from coarse-grained conglomerates to sandstones. The deposits display a fan-shaped geometry (Figs. 17 and 18), which suggests an initial progradational phase, succeeded by retreat.

Interbedded Fine Sandstones, Siltstones, and Shale—Facies 5

Repeated cycles of sandstones and shales in facies 5 reflect periodicity in the depositional processes. The sandstones were probably deposited from concentrated density flows or turbidity currents, though complete Bouma sequences have not been observed. The shales were most likely formed from clay-sized material that settled from suspension in quiet periods between the sand depositional events. The abundant shallow channels have a southeasterly direction and promoted transport of sediments toward the central part of the basin. Directional measurements of sedimentary structures, such as flute casts and current ripples, commonly deviate from directions outlined by the channels (Fig. 13), which may be due to sheet-like spillover deposition from the many channels. The abundance of loading and water escape structures indicate rapid deposition of water-enriched sediment in the lower sections of facies 5.

Overall upward-fining trends in the sedimentary units of the upper part of facies 5 reflect decrease in the clastic input to the basin. The marked mineralogical change toward quartz enrichments, in combination with evidence for shift in sediment transport directions (Fig. 13), support changes in depositional conditions. The single preimpact sediment sample analyzed has relatively high quartz content, and the increasing quartz/feldspar ratio upward in facies 5 may reflect a return to preimpact sedimentary conditions.

The brachiopod imprint found at Branden (http://www.nhm.uio.no/palmus/palvenn/index.htm) indicates a marine depositional environment, but an uncertain age for this fossil does not help in constraining the timing of impact.

The sand-filled cracks (Fig. 16) may have several explanations: subaerial desiccation, synaeresis near the water-sediment interface, or interstratal cracking at shallow burial depth. Plummer and Gostin (1981) reviewed previous literature on desiccation and synaeresis cracks, and they proposed a classification scheme based on field appearance. They do, however, emphasize the ambiguity of the interpretations, and warn against the use of such cracks in environmental analysis without looking at other sedimentological structures. A later review by Astin and Rogers (1991) stated that there is no evidence for subaqueously formed cracks being preserved in rocks. Likewise, both Tanner (1998) and Pratt (1998) found the existence of synaeresis cracks in natural systems unlikely. Tanner (1998) proposed interstratal

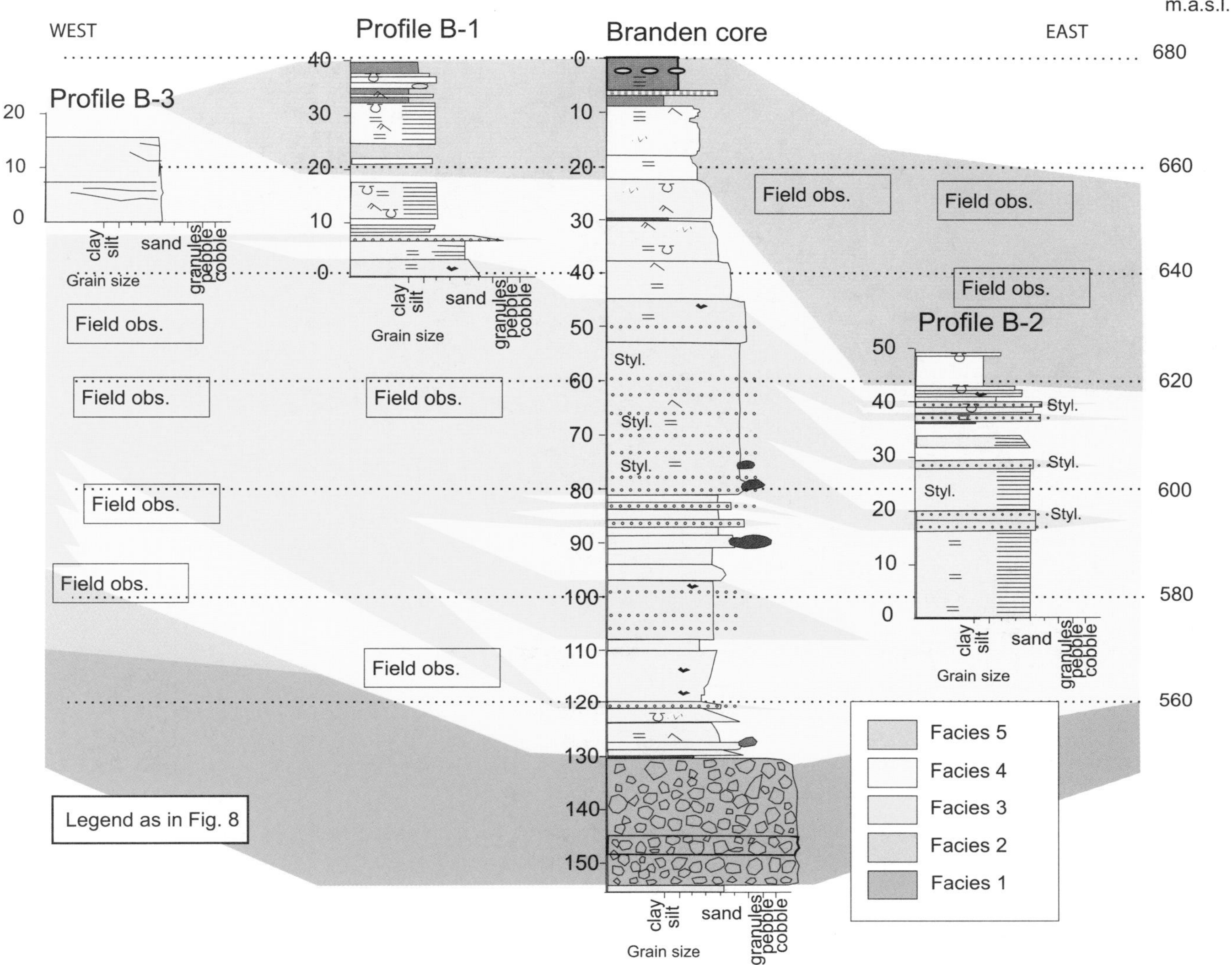

Figure 17. Correlation of drill core and field sections from the Branden area, supplemented by other field observations (marked "field obs."). The constructed cross section is approximately NW-SE, intersecting the main direction of sediment transport toward the SE.

cracking caused by layer-parallel contraction accompanied by water escape as important mechanisms for development of sediment-filled cracks in sedimentary rocks. According to Pratt (1998), cracking at very shallow burial depth could be initiated by earthquakes. In the sedimentary section at Branden, no evidence of subaerial exposure/deposition, such as oxidized red beds or evaporitic minerals, have been observed. Interstratal cracking is therefore our favored interpretation, and the sand-filled cracks found in the Branden section should not be used as evidence of depositional environment. Water escape, loading effects, and soft-sediment deformation structures are consistent with subaqueous deposition. Initiation of the cracking might have been a seismic event, or maybe the compaction of the rapidly deposited sediments in the crater basin was sufficient to disturb the sedimentary beds.

DISCUSSION

The sediment composition of the Gardnos structure is dominated by original fresh, unweathered material. To demonstrate the importance of impact-generated debris as sources of the postimpact sediments, some tentative volume estimations are presented. All calculations are in km^3, neglecting differences in porosity and density between crystalline target rock and postimpact sediments. Severe compaction and diagenesis in the sediments due to deep burial during and after the Caledonian orogeny minimized porosity and density contrasts between sedimentary and crystalline rocks.

The total ejected volume includes material excavated only down to about one-third of the crater's transient depth, or one-tenth of its transient diameter (Melosh, 1989), e.g., ~400 m for the Gardnos crater. The diameter of the excavated cavity corresponds

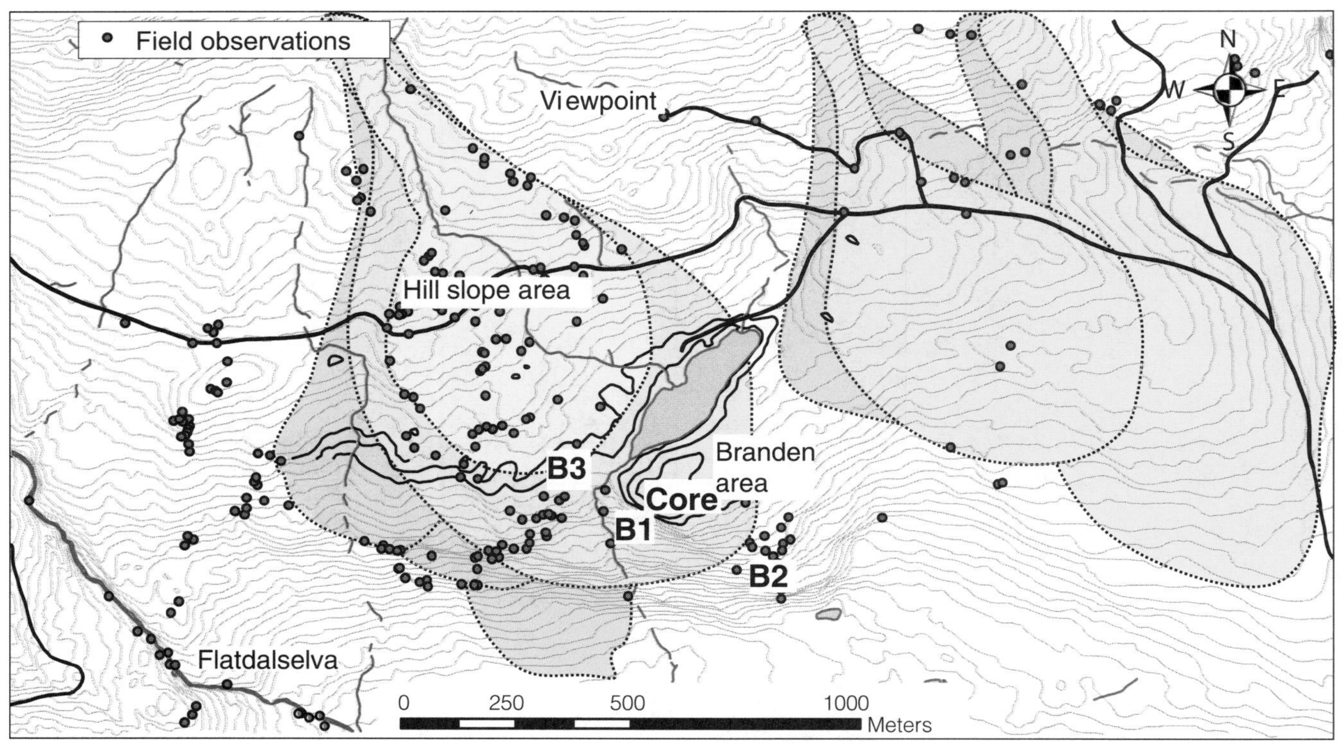

Figure 18. Interpretation of depositional patterns for facies 3 and 4.

to the transient crater diameter, measured at the preimpact surface (Turtle et al., 2005). The Gardnos structure has a central peak, but no evidence of large-scale faulting or slumping has been observed along the crater walls, and the final diameter is probably only slightly modified from the value for the transient diameter. The suggested apparent diameter (D_A = 5.5 km) is therefore used as the diameter of the excavated cavity, corresponding to a volume of ~4.8 km^3. Some of this excavated material made up the crater rim, some fell back into the crater as suevitic deposits, and some were deposited as an ejecta blanket around the crater. Theoretically 50% of the ejecta were deposited between one and two radii from the crater center, 17% between two and three radii, 8% between three and four radii (Melosh, 1989), and ~5% are assumed deposited even farther away. The estimated volumes of melted and vaporized target rocks are only 0.06 km^3 and 0.10 km^3, respectively (based on Melosh, 1989).

About half the rim height may be crushed, uplifted, and possibly inverted bedrock rather than loose debris (Melosh, 1989). Still, the rim, as outlined in Figure 3, contains a total debris volume of 1.5 km^3. The volume of the central peak is here approximated by 1000 m for the base diameter of the central peak and a central peak height of 300 m. This gives an original central peak volume close to 0.08 km^3, which probably collapsed immediately or soon after impact, being only a minor, local sediment source. This demonstrates the crater rim as the major source. In addition, fractured bedrock of the crater walls may have been an easily erodible target.

The sedimentary sections at Gardnos are proposed to consist of impact-excavated material reentering the crater basin, except for the material of the final few meters at the very top. This indicates a major depositional change at that time. The total thickness of the present postimpact sedimentary column of Gardnos does not exceed 150 m, and all sediments are found within 2 km from the crater center. To fill the crater with sediments up to 150 m above the crater floor, ~1 km^3 of sediment is needed. This is an absolute minimum estimation of sediment volume. Probably even more of the, at present, deeply eroded crater walls were covered with sediments. Assuming an additional 50 m thickness of sediments accumulating in the area from 2 to 2.5 km distance from the crater center, the total volume of suggested impact-related debris in the crater basin adds up to 1.4 km^3.

Though very rough, the above calculations demonstrate how impact-excavated debris originally deposited as part of the crater rim and nearby ejecta are sufficient sources for the postimpact sedimentary infill in the crater.

The apparently local origin and short transport of facies 1 clasts along the crater wall indicate fractured bedrock from the upper part of the crater wall as an important sediment source. The contribution from the central peak is expected only to be of local importance, and most important immediate after impact, as large parts of the peak collapsed. The Branden core represents the moat of the basin, close to the central peak, consequently with the potential of having received erosional products from the central high. The coarse conglomerates of facies 1 in the Branden core contain a variability of clasts, including a considerable amount of amphibolitic rocks, in accordance with the central peak acting as a sediment source.

Even before water reentered the crater basin, the first clastic deposits were sourced from the collapsing central peak, most likely as rockfalls. As water filled and eroded the high-relief areas, debris flows were generated. Height to length ratios of rockfalls (referred to as "apparent coefficient of friction") depend on rockfall volume and are smaller for large falls than for small ones (Hsü, 2004). The size of the supposed avalanches only derived from the central peak in the Gardnos crater is not known, but they obviously were less than the total volume of the peak. In this calculation we suppose that rock debris from the central peak moved in ten (arbitrary number) major avalanches, and typical height to length ratios would be expected within the range 0.3–0.6. If sedimentation started where the main erosion terminated (i.e., at the point where the central peak diameter was 1000 m), and most material was deposited within the range of 500 m of further transport, sediment distribution would likely have been within a 3 km^3 ring-formed area. This corresponds to run-out distances with height to length ratios in the range 0.3–0.6, as expected for this type of deposits. An even distribution of available material from the collapsing central peak gives a thickness of 25 m. The Branden core, which is located within this area, has an observed thickness of facies 1 conglomerate of 25 m. A considerable uncertainty is the timing of the collapse of the central peak. Did it collapse instantly so that large parts were incorporated into the accumulating suevite? Did it break down soon after impact with the potential to source the facies 1 deposits seen in the Branden core? Or was it even more resistant to collapse, standing up for a longer time? Presently we do not know.

Facies 2 conglomerates display large variations in clast composition, and the large fraction of large subrounded clasts indicates some distance of transportation from source to depositional site. The smaller clasts and matrix sand grains are angular, but naturally much less vulnerable to rounding than large grains, when exposed to otherwise equal conditions. The large clasts of facies 2, due to both texture and stratigraphic position, represent deposition immediately after impact. The crater rim and associated ejecta deposits were possible sediment sources.

The sandstones at Branden (facies 3 and 4) have relative uniform mineralogical composition throughout the profile, comparable to the sandstones at the Hillslope closer to the crater wall. They were probably part of the same depositional system, fed from a single source area (Figs. 14, 17, and 18). Other possible sediment depositional systems (Figs. 17 and 18) have also been identified, each probably fed from different source areas. Differences in mineral composition of the sandstones may mirror varied original preimpact bedrock composition of the crater, from granites to more quartz-dominated rocks, going from west to east. Reworking of material excavated during impact is the main sediment source, implying that the excavated material was not thoroughly mixed and became ejected out to different parts of the crater. The rocks from the central parts of the crater are expected to have experienced the greatest forces and ejected the farthest, while rocks from deeper horizons and closer to the crater rim traveled shorter distances. The crater rim itself may partly consist of a flap of overturned local rocks (Melosh, 1989). Compositional resemblance between the ejecta close to the crater and the local bedrock is consequently to be expected.

Within facies 5 there is a change from feldspar-rich (arkoses) to quartzitic sandstones in the uppermost 25 m, indicating a shift in sediment sourcing. This may mark the return to preimpact sediment sources. Quartz is more resistant to weathering than most other minerals, and a relative increase in the quartz content with time could be expected, reflecting increased sediment maturation in the crater-fill sediments. The abrupt and pronounced (about four times increase in quartz relative to feldspar content through just a few meters of the sedimentary column; Fig. 13) change in mineralogical composition makes this not very likely. The change in quartz content approximately coincides with an indicated shift in sediment transport direction (Fig. 13), supporting a shift to a different sediment source area. The comparable mineralogical composition of the topmost sandstones of facies 5 and the preimpact sediment analyzed support a return to preimpact conditions.

The paleoenvironment in the target area at the time of impact is still poorly constrained, as is the age of the Gardnos crater structure (French et al., 1997). The discovery of soft-sediment deformed and folded clasts in the oldest Gardnos conglomerates indicates the presence of a young, only partly consolidated sedimentary cover on top of the crystalline bedrock at the time of impact. The relatively high carbon content of the impactites and occurrence of impact-generated diamonds are further evidence of carbon-rich sedimentary target deposits (Gilmour et al., 2003). Accumulation and preservation of these sediments are very unlikely to have occurred subaerially, and this supports a preimpact scenario with water covering the area.

The diameter of the Brent crater in Canada is close to 4 km and thus comparable to the Gardnos crater. According to Lozej and Beales (1975), the Brent crater was filled with impact breccia, leaving only a shallow depression with an almost flat crater floor, close to mean sea level. Lozej and Beales (1975) describe two distinct phases of succeeding sediment infilling. The lowermost part is a 116 m thick sequence of dolostones, arkosic siltstones, and evaporites formed in a perimarine tidal-flat setting, in an arid to semiarid climate, subjected to periodic flooding. The overlying 30 m of silty arkoses marks a transgression to a subtidal, quiet-water environment, where the crater rim no longer was an obstacle to direct communication with the sea. The sediment infill of the Brent crater differs from the succession at Gardnos in several respects. In the Gardnos crater there are no evaporites or red beds to indicate subaerial deposition, and the first infill deposits are dominated by sedimentary breccias and coarse conglomerates, probably mainly derived from the collapsing crater rim. The dissimilarities in sedimentary infill suggest differences between the target areas at the time of impact. The freshly formed Brent crater was probably situated at the margin of an epicontinental sea, close to sea level, and the rim protected the crater depression from much external input (Lozej and Beales, 1975). In the Gardnos case, seawater seems to have played an important role in eroding the crater rim and filling the crater basin soon after impact.

Ormö and Lindstöm (2000) summarized data from five other craters in Baltoscandia (Lockne, Kärdla, Granby, Hummeln, and Tvären) supposed to have been formed in comparable settings, i.e., relatively thin sedimentary cover above crystalline basement, in an epicontinental sea. In their cases, marine settings were established by the presence of similar marine facies before and immediately after impact. In the Swedish Tvären structure, preimpact sedimentary rocks are only preserved as fragments within the resurge deposits, supposed to have originated from the shattered sedimentary succession outside the crater (Ormö, 1994). Tvären has a suggested crater diameter of 2 km, an upward-fining "graded resurge unit" of 58 m, and an estimated water depth at time of impact of 100–150 m (Lindström et al., 1994). In Gardnos only a few preimpact rock fragments have been found. In the Tvären case impact-related resurge was almost instantaneous and not slowed down by a crater rim, while the filling of Gardnos was more complex and most likely involved crater rim erosion.

The assumed presence of a crater rim in Gardnos may promote the Estonian Kärdla crater as a possible analogue. The Kärdla crater, as described by Puura and Suuroja (1992), is 4 km in diameter and positioned in crystalline basement. The target water depth was probably a few tens of meters. Outside the crater, close to the rim, the sedimentary preimpact cover is almost missing but becomes gradually more complete farther away from the crater. Rather than interpreting this impact-related erosion as an outer crater, similar to the brim modeled for the Lockne crater (Lindström et al., 2005), Puura and Suuroja (1992) suggested that this zone was possibly formed by a ground surge associated with the ejecta curtain during the cratering process. Proposing a shallow target water depth and a ground surge process related to the Gardnos impact, a thin sedimentary cover could have been almost entirely removed in the vicinity of the crater. This could explain the sparse occurrence of preimpact sedimentary rock fragments in the postimpact crater-fill deposits.

SUMMARY

The relatively large amount of carbon in the impactites and appearance of preimpact sedimentary clasts in the postimpact infill deposits suggest a Gardnos target area covered by water. In contrast to the infill succession in the Brent crater, the postimpact sedimentary deposits in Gardnos demonstrate that the crater was filled with water shortly after impact. The sedimentary succession does not display the same distribution as inferred resurge deposits in craters such as Lockne and Tvären. The Gardnos succession appears more complex and deposited through several events over some time. The preferred interpretation is therefore a setting were the crater rim dammed a surrounding shallow sea, the rim being able to keep the water out for a very short time, before multiple breaching, channeling, and subsequent sedimentation. This interpretation gives an upper constraint on paleo–water depth to considerably less than rim height, i.e., less than 240 m. With Kärdla as a possible analogue, target water depth was probably not more than several tens of meters. Eroded sediments from both rim and ejecta blanket were the most likely main source for the postimpact sediments, with only a minor component from the central peak.

The laminated, fine-grained deposits directly overlying the suevite on the crater floor most likely were deposited immediately after impact, just before deposition of the rockfalls and avalanches of facies 1, which were rushing down the crater walls and the collapsing central peak. The succeeding debris flow (facies 2) and hyperconcentrated flow deposits (facies 3) have more confined distribution, radially building out in fan-shaped geometries (Fig. 18), indicating deposition through a limited number of established inlets. These inlets may be classified as resurge gullies, developed as seawater breached the rim and flowed back into the crater. If the sea really breached the rim at several locations, the crater would be filled with water in a matter of hours. Most of the facies 2 conglomerates were probably deposited during this stage. Even after water filling, new debris flows were generated in the crater by further erosion of the remains of the crater rim. The main deposition of facies 3 most likely took place after the basin was water-filled, as did the deposition of sandstones of facies 4 in the deepest part of the basin. The facies 4 sandstones apparently represent distal continuations of facies 3 or reworked material brought farther into the basin by local currents. Deposition of facies 4 obviously took place over a considerably longer time (years?), which is evident by the presence of repeated clay seams and clay clast conglomerates. Deposition of facies 3 and 4 in the Hillslope and Branden localities seem to represent progradational phases. In an early phase the sedimentary system built out with high sedimentation rates, but thereafter retreated as sedimentation rates decreased. This sedimentary event ended in an upward-fining development in the interbedded fine sandstones and shales of facies 5. A shift in sediment transport directions supports a major change in sedimentation environment, probably reflecting the reestablishment of preimpact conditions.

ACKNOWLEDGMENTS

This paper forms a part of a Ph.D. thesis carried out by the first author. Elen Roaldset (Natural History Museum, University of Oslo) has organized the investigations and the financial support. During extensive fieldwork in the Gardnos area, Tom Jahren (Nes community) has been of great support. Salahalldin Akhavan (Natural History Museum, University of Oslo) provided more than 100 thin sections and was of general help with sample preparation, and Berit Løken Berg (Department of Geoscience, University of Oslo) assisted with the XRD investigations. The manuscript greatly benefited from reviews by Uwe Reimold and Erik Sturkell.

REFERENCES CITED

Astin, T.R., and Rogers, D.A., 1991, Subaqueous shrinkage cracks in the Devonian of Scotland reinterpreted: Journal of Sedimentary Petrology, v. 61, p. 851–859.

Bjørlykke, K., and Nystuen, J.P., 1981, Late Precambrian tillites of South Norway, *in* Hambrey, M.J., and Harland, W.B., eds., Earth's pre-Pleistocene glacial record: Cambridge, UK, Cambridge University Press, p. 624–628.

Blikra, L.H., and Nemec, W., 1998, Postglacial colluvium in western Norway: Depositional processes, facies and paleoclimatic record: Sedimentology, v. 45, p. 909–959, doi: 10.1046/j.1365-3091.1998.00200.x.

Broch, O.A., 1945, Gardnosbreksjen i Hallingdal: Norsk Geologisk Tidsskrift, v. 25, p. 16–25.

Bunch, T.E., Becker, L., Des Marais, D., Tharpe, A., Schultz, P.H., Wolbach, W., Glavin, D.P., Brinton, K.L., and Bada, J.L., 1999, Carbonaceous matter in the rocks of the Sudbury Basin, Ontario, Canada, *in* Dressler, B.O., and Sharpton, V.L., eds., Large meteorite impacts and planetary evolution: Geological Society of America Special Paper 339, p. 331–343.

Dons, J.A., and Naterstad, J., 1992, The Gardnos impact structure, Norway: Meteoritics, v. 27, p. 215.

Evans, S.G., and Hungr, O., 1993, The assessment of rockfall hazard at the base of talus slopes: Canadian Geotechnical Journal, v. 30, p. 620–636.

Felix, M., and Peakall, J., 2006, Transformation of debris flows into turbidity currents: Mechanisms inferred from laboratory experiments: Sedimentology, v. 53, p. 107–123, doi: 10.1111/j.1365-3091.2005.00757.x.

French, B.M., 1968, Sudbury structure, Ontario: Some petrographic evidence for an origin by meteorite impact, *in* French, B.M., and Short, N.M., eds., Shock metamorphism of natural minerals: Baltimore, Maryland, Mono Book, p. 383–412.

French, B.M., Koeberl, C., Gilmour, I., Shirley, S.B., Dons, J.A., and Naterstad, J., 1997, The Gardnos impact structure, Norway: Petrology and geochemistry of target rocks and impactites: Geochimica et Cosmochimica Acta, v. 61, p. 873–904, doi: 10.1016/S0016-7037(96)00382-1.

Gani, M.R., 2004, From turbid to lucid: A straightforward approach to sediment gravity flows and their deposits: The Sedimentary Record, v. 2, no. 4, p. 4–8.

Gilmour, I., French, B.M., Franchi, I.A., Abbott, J.I., Hough, R.M., Newton, J., and Koeberl, C., 2003, Geochemistry of carbonaceous impactites from the Gardnos impact structure, Norway: Geochimica et Cosmochimica Acta, v. 67, p. 3889–3903, doi: 10.1016/S0016-7037(03)00213-8.

Graup, G., 1981, Terrestrial chondrules, glass spherules and accretionary lapilli from the suevite, Ries Crater, Germany: Earth and Planetary Science Letters, v. 55, p. 407–418, doi: 10.1016/0012-821X(81)90168-0.

Grier, J.A., Swindle, T.S., Kring, D.A., and Melosh, H.J., 1999, $^{40}Ar/^{39}Ar$ dating of samples from the Gardnos impact structure: Norway: Meteoritics and Planetary Science, v. 34, p. 803–808.

Grieve, R.A.F., Robertson, P.B., and Dence, M.R., 1981, Constraints on the formation of ring impact structures, based on terrestrial data, *in* Schultz, P.H., and Merrill, R.B., eds., Multiring basins: Proceedings Lunar and Planetary Science Conference, 12A, p. 37–57.

Heymann, D., Dressler, B.O., Knell, J., Thiemens, M.H., Buseck, P.R., Dunbar, R.B., and Mucciarone, D., 1999, Origin of carbonaceous matter, fullerenes, and elemental sulfur in rocks of the Whitewater Group, Sudbury impact structure, Ontario, Canada, *in* Dressler, B.O., and Sharpton, V.L., eds., Large meteorite impacts and planetary evolution: Geological Society of America Special Paper 339, p. 345–360.

Hsü, K.J., 2004, Physics of sedimentology: Textbook and reference (2nd edition): Berlin, Springer, 240 p.

Kessler, L.G., and Bédard, J.H., 2000, Epiclastic volcanic debrites—Evidence of flow transformations between avalanche and debris flow processes, Middle Ordovician, Baie Verte Penisula, New Foundland, Canada: Precambrian Research, v. 101, p. 135–161, doi: 10.1016/S0301-9268(99)00086-8.

Lindström, M., Floden, T., Grahn, T., and Kathol, B., 1994, Post-impact deposits in Tvären, a marine Middle Ordovician crater south of Stockholm, Sweden: Geological Magazine, v. 131, no. 1, p. 91–103.

Lindström, M., Ormö, J., Sturkell, E., and von Dalwigk, I., 2005, The Lockne Crater: Revision and reassessment of structure and impact stratigraphy, *in* Koeberl, C., and Henkel, H., eds., Impact tectonics: Berlin, Heidelberg, Springer-Verlag, p. 357–388.

Lozej, G.P., and Beales, F.W., 1975, The unmetamorphosed sedimentary fill of the Brent meteorite crater, southeastern Ontario: Canadian Journal of Earth Sciences, v. 12, no. 1, p. 606–628.

Melosh, H.J., 1989, Impact cratering: A geological process: New York, Oxford University Press, 245 p.

Mulder, T., and Alexander, J., 2001, The physical character of subaqueous sedimentary density flows and their deposits: Sedimentology, v. 48, p. 269–299, doi: 10.1046/j.1365-3091.2001.00360.x.

Naterstad, J., and Dons, J.A., 1994, Unpublished notes and geological map for excursion to the Gardnos impact structure during the European Science Foundation Workshop "The Identification and Characterization of Impacts," Östersund, 31 May–5 June, 1994.

Nelson, C.A., Meyer, A.W., Devin, T., and Larsen, M., 1986, Crater Lake, Oregon: A restricted basin with base-of-slope aprons of nonchannelized turbidites: Geology, v. 14, p. 238–241, doi: 10.1130/0091-7613(1986)14<238:CLOARB>2.0.CO;2.

Nordgulen, Ø., Riiber, K., and Bargel, T.H., 1997, Nes county, Norway: Geological map M: Adresseavisen, Trondheim, NGU (Geological Survey of Norway), scale 1:100,000, 1 sheet.

Nystuen, J.P., 1976, Facies and sedimentation of the late Precambrian Moelv Tillite in the eastern part of the sparagmite region, southern Norway: Norges Geologiske Undersøkelse, v. 329, Bulletin 40.

Ormö, J., 1994, The pre-impact Ordovician stratigraphy of the Tvären Bay impact structure, SE Sweden: GFF, v. 116, no. 3, p. 139–144.

Ormö, J., and Lindström, M., 2000, When a cosmic impact strikes the sea bed: Geological Magazine, v. 137, no. 1, p. 67–80, doi: 10.1017/S0016756800003538.

Pike, R.J., 1977, Size dependence in the shape of fresh impact craters on the moon, *in* Roddy, D.J., et al., eds., Impact and explosion cratering: New York, Pergamon Press, p. 489–509.

Pike, R.J., 1985, Some morphologic systematics of complex impact structures: Meteoritics, v. 20, p. 49–68.

Plummer, P.S., and Gostin, V.A., 1981, Shrinkage cracks: Desiccation or synaeresis?: Journal of Sedimentary Petrology, v. 51, p. 1147–1156.

Pope, K.O., Ocampo, A.C., Fischer, A.G., Alvarez, W., Fouke, B.W., Webster, C.L., Vega, F.J., Smit, J., Fritsche, A.E., and Claeys, P., 1999, Chicxulub impact ejecta from Albion Island, Belize: Earth and Planetary Science Letters, v. 170, p. 351–364, doi: 10.1016/S0012-821X(99)00123-5.

Pratt, B.R., 1998, Syneresis cracks: Subaqueous shrinkage in argillaceous sediments caused by earthquake-induced dewatering: Sedimentary Geology, v. 117, p. 1–10, doi: 10.1016/S0037-0738(98)00023-2.

Puura, V., and Suuroja, K., 1992, Ordovician impact crater at Kärdla, Hiiumaa Island, Estonia: Tectonophysics, v. 216, p. 143–156, doi: 10.1016/0040-1951(92)90161-X.

Shanmugam, G., 2000, 50 years of the turbidite paradigm (1950s–1990s): Deep-water processes and facies models—A critical perspective: Marine and Petroleum Geology, v. 17, p. 285–342, doi: 10.1016/S0264-8172(99)00011-2.

Spjeldnæs, N., 1959, Traces of an Eocambrian orogeny in southern Norway: Norsk Geologisk Tidsskrift, v. 39, p. 83–86.

Tanner, P.W.G., 1998, Interstratal dewatering origin for polygonal patterns of sand-filled cracks: A case study from Late Proterozoic metasediments of Islay, Scotland: Sedimentology, v. 45, p. 71–89, doi: 10.1046/j.1365-3091.1998.00135.x.

Turtle, E.P., Pierazzo, E., Collins, G.S., Osinski, G.R., Melosh, H.J., Morgan, J.V., and Reimold, W.U., 2005, Impact structures: What does crater diameter mean?, *in* Kenkmann, T., et al., eds., Large meteorite impacts, III: Geological Society of America Special Paper 384, p. 1–24.

Wentworth, C.H., 1922, A scale of grade and class terms for clastic sediments: Journal of Geology, v. 30, no. 5, p. 377–392.

Manuscript Accepted by the Society 10 July 2007

The Geological Society of America
Special Paper 437
2008

Water-blow and resurge breccias at the Lockne marine-target impact structure

Maurits Lindström*
Department of Geology and Geochemistry, Stockholm University, 10691 Stockholm, Sweden

Jens Ormö*
Centro de Astrobiología (CSIC/INTA), Instituto Nacional de Técnica Aeroespacial, Ctra de Torrejón a Ajalvir, km 4, E-28850 Torrejón de Ardoz, Madrid, Spain

Erik Sturkell*
Nordic Volcanological Center, Institute of Earth Sciences, University of Iceland, Sturlugata 7, 101 Reykjavik, Iceland

ABSTRACT

The Ordovician (early Sandbian) Lockne impact crater in central Sweden formed in a sea at least 500 m deep. The structure and impact stratigraphy are sufficiently well preserved to permit detailed analysis of the cratering process. The target seabed consisted of a partly lithified, 75–80-m-thick sediment cover resting on continental crystalline basement. An over 7-km-wide inner crater formed in the basement. The surrounding sediment cover was almost completely removed within about 2 km from the rim of the inner crater. At 2.5–8.5 km from the rim, the thickness of preserved preimpact sediment increases with the distance to the inner crater. The top of the preserved sediment is mostly limestone breccia. Brecciation was probably driven by extremely forceful flow of water that was charged with sediment to the limit of its carrying capacity, and therefore stirred deeper than it would erode. The resulting lithology, for which we suggest the term "water-blow breccia," is monomictic, with all clasts deriving from the underlying parent rock. Great volumes of crystalline ejecta were emplaced during and immediately after water-blow brecciation. The resurge, connected with the subsequent collapse of the water crater, deposited a mixed breccia of transported clasts that includes eroded limestone of various local provenances, as well as crystalline ejecta. The resurge breccia occurs extensively in the region, even outside the area characterized by occurrences of water-blow breccia. Its thickness and clast sizes decrease away from the crater.

Keywords: impacts, water crater, water-blow breccia, resurge breccia, Ordovician.

---

*Lindström: maurits.lindstrom@geo.su.se; Ormö: ormo@inta.es; Sturkell: sturkell@hi.is

Lindström, M., Ormö, J., and Sturkell, E., 2008, Water-blow and resurge breccias at the Lockne marine-target impact structure, *in* Evans, K.R., Horton, J.W., Jr., King, D.T., Jr., and Morrow, J.R., eds., The Sedimentary Record of Meteorite Impacts: Geological Society of America Special Paper 437, p. 43–54, doi: 10.1130/2008.2437(03). For permission to copy, contact editing@geosociety.org.

INTRODUCTION

The Lockne impact structure in central Sweden (Fig. 1; centered at 63°00′08″N, 14°49′03″E) formed at sea early in the time of deposition of the Upper Ordovician (lower Sandbian) Dalby Limestone (Lindström et al., 2005a). The Dalby Limestone is represented by the youngest deposits formed at the site before the impact as well as by the oldest deposit formed afterwards. The essential parts of the structure are exposed in good and relatively complete preservation.

Lockne has a clearly defined crater about 7.5 km wide that was excavated into crystalline basement rock of Proterozoic age; it will be referred to as inner crater in the following text. There is also an over 15 km wide outer crater that is less precisely delimited. It formed through the excavation of 75–80 m thick sedimentary cover rock of Cambrian to earliest Late Ordovician age. The lower part of the cover rock consisted of poorly lithified bituminous mud (that is preserved as black shale in areas outside the central crater and the immediately surrounding area) of Middle to Late Cambrian age, whereas the upper 50 m was Ordovician, mainly well-lithified limestone. This cover was completely removed in most places within 1–2 km from the inner crater, but the thickness of it that was spared increases with the distance to the inner crater.

Under the assumption of an oblique impact and a water depth of about 500 m, Shuvalov et al. (2005) found that the inner crater could have been surrounded by a 10 km wide water crater toward the end of the crater excavation stage. Surrounding the inner crater, the bottom of the water crater consisted of crystalline basement that was essentially stripped of sedimentary cover but instead became covered by a blanket of crystalline ejecta (described as "brim" by Lindström et al., 2005a). The name coined for this lithologic unit is Tandsbyn Breccia after the principal village where it occurs (Lindström et al., 1996).

The uppermost parts of the sedimentary cover were removed even beyond the reach of the water crater as it is given by the computer simulation. This denuded area extends to about 5 km from the margin of the inner crater. When the water crater collapsed, the sea that surged back toward the central crater was charged with particles of mud to block size that were dumped on the denuded surface as resurge deposits. The coarsest-grained resurge deposit was the first to form. It has been named Lockne Breccia (Lindström et al., 1983) after the administrative parish in which it occurs. It consists of a mixture of clasts of gravel to block size in which all target lithologies are represented, although Ordovician limestone tends to predominate.

The arenite overlying the Lockne Breccia is called Loftarstone, which is a very old local name for a graywacke-like and carbonate-rich sandstone to siltstone. The grains represent the same lithologies that form the Lockne Breccia.

In addition to the Lockne Breccia there is another sedimentary breccia, the existence of which was suggested by von Dalwigk and Ormö (2001). This breccia was formally distinguished by Lindström et al. (2005a), who named it Ynntjärnen Breccia after a small lake in the forest where it is typically exposed. In the early papers on the Lockne crater, its outcrops were taken to represent the Lockne Breccia, from which it differs by having an essentially monomictic clast population of local limestone parent rock. The clasts frequently fit together even if they are more or less rotated.

It is the aim of this contribution to elaborate on these sedimentary breccias and their probable significance.

THE YNNTJÄRNEN BRECCIA

In the initial geological studies of the Lockne crater (Lindström and Sturkell, 1992; Lindström et al., 1996), the breccia outcrops referred to under this heading were mapped as Lockne Breccia, the typical outcrops of which were recognized to represent a resurge deposit. They shared with the typical Lockne Breccia a seemingly chaotic arrangement of abundant clasts of Ordovician limestone.

However, the Ynntjärnen Breccia contrasts with the Lockne Breccia by being essentially monomictic, very poorly sorted, and everywhere matrix-supported. At most outcrops of the Ynntjärnen Breccia, the clasts represent only the autochthonous limestone beneath the breccia. Some outcrops show slabs of the immediately underlying bed caught in the process of becoming part of the breccia. In some places—e.g., Kajan (coordinates 6985530/1442180) (Figs. 2 and 3; see also Simon, 1987, p. 52) and Sved (coordinates 6986775/1441800) (Fig. 4)—a few isolated clasts of granitic ejecta in the size range 3–10 cm occur in monomictic, very poorly sorted, and matrix-supported calcareous breccia, the clasts and matrix of which are derived from the uppermost part of the orthoceratite limestone (conodont Zones of *Scalpellodus gracilis* to *Baltoniodus gerdae*). The autochthonous beds that immediately underlie the breccia in these outcrops belong to the Zone of *Scalpellodus gracilis* (Simon, 1987). The thickness of the involved conodont Zones in the autochthonous orthoceratite limestone of the adjacent Brunflo area is about 20 m, whereas the estimated thickness of the monomictic breccia at Kajan and Sved is 10–20 m. We regard these occurrences as Ynntjärnen Breccia and are dealing further with them in the Discussion.

Clast sizes in the breccia vary greatly within an outcrop. As extreme examples, the size ranges in the outcrops at the parish community center of Brunflo and at Sved (Fig. 4) are from a few millimeters to over 100 m^3. Some clasts are strongly rotated. The affected stratigraphic levels are rich in clay that either is interlayered between thicker limestone beds or contains abundant limestone nodules. Transitions between continuous beds of solid limestone and nodular layers are frequent in these parent lithologies. The gradual disintegration and passage into nodular breccia is particularly well displayed in the Brunflo outcrop. Brecciation ultimately resulted in a mortar-like rock in which angular fragments of bedded limestone and abundant nodules are mixed in chaotic fashion in a marly matrix.

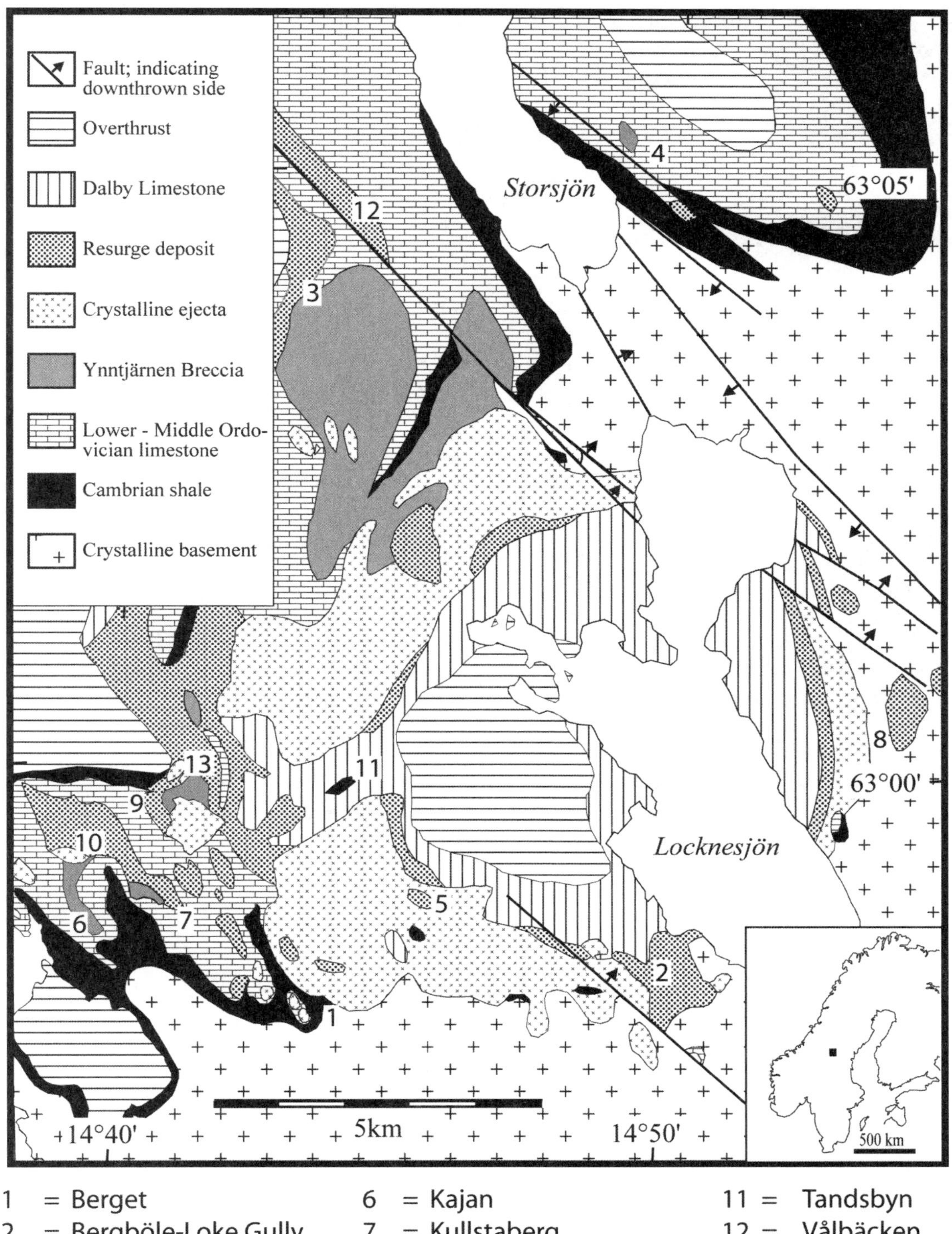

1 = Berget	6 = Kajan	11 = Tandsbyn
2 = Bergböle-Loke Gully	7 = Kullstaberg	12 = Vålbäcken
3 = Brännsmyren	8 = road to Skettmyrtjärnen	13 = Ynntjärnen
4 = Brunflo	9 = Skutetjärnen	
5 = road to Handsjö lakes	10 = Sved	

Figure 1. Geological overview of the Lockne crater.

Figure 2. Water-blow breccia. Ynntjärnen Breccia. Monomictic, with reoriented clasts of Middle Ordovician limestone. Kajan outcrop (coordinates 6985530/1442180). Photo by J. Ormö.

Nodular breccia of the described kind occurs as vertical as well as horizontal fillings of fractures in autochthonous limestone outside the crater (Simon, 1987). Sturkell and Ormö (1997) demonstrated that the injected fracture fillings at Kullstaberg (coordinates 6986090/1442560) were derived from nodular beds immediately overlying fractured limestone. Subsequent investigation of the outcrop by the authors has shown that the affected limestone beds are in their turn overlain by slabs of crystalline ejecta. As demonstrated through mapping by Lindström et al. (2005a), the Ynntjärnen Breccia can also be overlain by major masses of crystalline ejecta.

The Ynntjärnen Breccia occurs between 6 km and 12 km from the center of the inner crater (Fig. 1). The known occurrences nearest to the center are at the outer margin of the ejecta brim (Lindström et al., 2005a) that surrounds the crater. They formed in the lowermost part of the Ordovician orthoceratite limestone, about 40 m above the sub-Cambrian peneplain (Sturkell and Lindström, 2004) that defines the base of the lower Paleozoic marine sediments in the Lockne area. This peneplain was cut into the crystalline basement of the Baltica craton before the craton was inundated by the Early Cambrian sea (Lindström et al., 2005a). The remotest occurrences of Ynntjärnen Breccia rest on sediments only slightly older than the impact, about 75 m above the peneplain. At intermediate distances, the thickness of sediments between the breccia and the peneplain is between 40 and 75 m.

THE LOCKNE BRECCIA

The Lockne Breccia (Lindström et al., 1983) is typically polymictic, clast-supported, and more or less well-sorted. As a rule, Ordovician limestone from different stratigraphic horizons is by far the most frequent lithology. There are crystalline rocks among the clasts, although they are not always common. The upper part of the breccia tends to be dominated by gravel and may pass into polymictic and carbonate-rich Loftarstone arenite

Figure 3. Water-blow breccia. Ynntjärnen Breccia, on autochthonous Middle Ordovician limestone. Road section near Kajan, Tandsbyn (coordinates 6985530/1442180). Photo by J. Ormö.

(Lindström et al., 1996). The Loftarstone arenite, which occurs either as a finer-grained upward continuation of the Lockne Breccia (Fig. 5) or with a sharp erosional boundary against the latter, contains numerous grains of shocked quartz (Therriault and Lindström, 1995).

Cambrian bituminous limestone (for which the Scandinavian name is "orsten") is sporadically occurring as cobbles in a number of breccia localities. Of these, Berget (coordinates 6984390/1445280) and Handsjövägen (coordinates 6986020/1447030) are close to outcrops of autochthonous Upper Cambrian, whereas the other outcrops are instead rather distant from possible sources of cobbles of Cambrian limestone. Such localities are Vålbäcken (coordinates 6996730/1445375) (Thorslund, 1940), Brännsmyren (coordinates 6994930/1444970), and Skutetjärnen (coordinates 6987600/1441800) (Simon, 1987), all at distances of several kilometers from where the Cambrian limestone might have come, and Hallen (coordinates 6953450/1422700) (Thorslund, 1940; Simon, 1987), 40 km southwest of Lockne.

Just outside the southern margin of the inner crater there is an area of about 1 km^2 that is referred to as the Bergböle-Loke Gully (Lindström et al., 1996). Extensive outcrops in this area have been mapped as Lockne Breccia. Much of this outcrop is now overgrown by shrubby, planted forest, but much was recorded while there was still opportunity, and some still remains to be seen in the field. There are two kinds of breccia, one of which consists of fairly well-sorted pebbles to gravel. It is overlain with either gradual transition or erosional discontinuity by Loftarstone arenite (Lindström et al., 2005a). The other kind of breccia consists of large boulders and blocks of diverse red

Figure 4. Water-blow breccia. Ynntjärnen Breccia. Reoriented major blocks of local Middle Ordovician limestone. Sved, Tandsbyn (coordinates 6986725/1441800) (Lindström et al., 1983).

and gray lithologies of Lower to Middle Ordovician limestone. Some of these large clasts are turned upside down (Lindström et al., 1983, their Pl. 2 and Fig. 1). They are surrounded by matrix-supported, marly-nodular breccia of the same lithologies. This breccia rests on either crystalline ejecta or autochthonous, crystalline basement. It is clear that only the first-mentioned lithology agrees with the updated definition of Lockne Breccia. As for the second kind of breccia, it cannot be referred to the Ynntjärnen Breccia, because it is neither monomictic nor related to the immediately underlying rock unit.

The remotest known occurrence of Lockne Breccia in the vicinity of the Lockne crater is at Torvalla, south of Östersund (coordinates 7004250/1444825) and 16.5 km from the crater center. This outcrop contains limestone cobbles as well as numerous cobbles of upper Proterozoic Åsby Dolerite, the nearest in situ outcrops of which are near the rim of the inner crater (coordinates 6995300/1451355) and 12 km from Torvalla. Another occurrence, described by Thorslund (1940) and Sturkell et al. (2000), is at Hallen, 40 km southwest of the crater center. At this outcrop the clasts of limestone and granitoids do not form a continuous bed but rest either in clusters or isolated on the basal beds of the Dalby Limestone. They are overlain by a thin layer of silty Loftarstone above which there are further beds of Dalby Limestone. The Lockne Breccia rests on different lithologies. Where it is underlain by autochthonous Ordovician limestone, the distance to the basement varies like that of the Ynntjärnen Breccia. East of the crater it rests directly on the sub-Cambrian peneplain.

Figure 6 shows an outcrop of Lockne Breccia that was deposited on the crystalline ejecta of the western "brim" of the crater. This is a rare case of a major fraction of crystalline rocks among the clasts. Sedimentary clasts are mainly weathered dark brown in the pictured part of the outcrop. At some localities, such as Sved and Ynntjärnen (Fig. 7), Lockne Breccia rests on Ynntjärnen Breccia.

DISCUSSION

The Ynntjärnen Breccia and the Lockne Breccia belong to the class of sedimentary-clast breccias on which Witzke and Anderson (1996) presented a thorough and lucid discussion. Although their discussion principally concerned the Manson structure, it is highly relevant for Lockne as well. Of the three local breccia formations at Manson, it is the so-called "Phanerozoic-clast breccia" that is of particular interest in connection with Lockne. It is a polymictic breccia with mainly sedimentary clasts and a gray, marly matrix with sand-sized grains, most of which are quartz. The size range of clasts is millimeters to 100 m. The clasts are neither sorted nor consistently oriented. Fragments of crystalline basement and impact melt do occur, although they are not very common.

The Phanerozoic-clast breccia of the Manson structure very much resembles the Bunte Breccia of the Ries crater (Hörz et al., 1983; Hüttner and Schmidt-Kaler, 1999). However, the distribution of the Bunte Breccia is limited to a peripheral zone of the structural crater and large areas outside it, whereas the

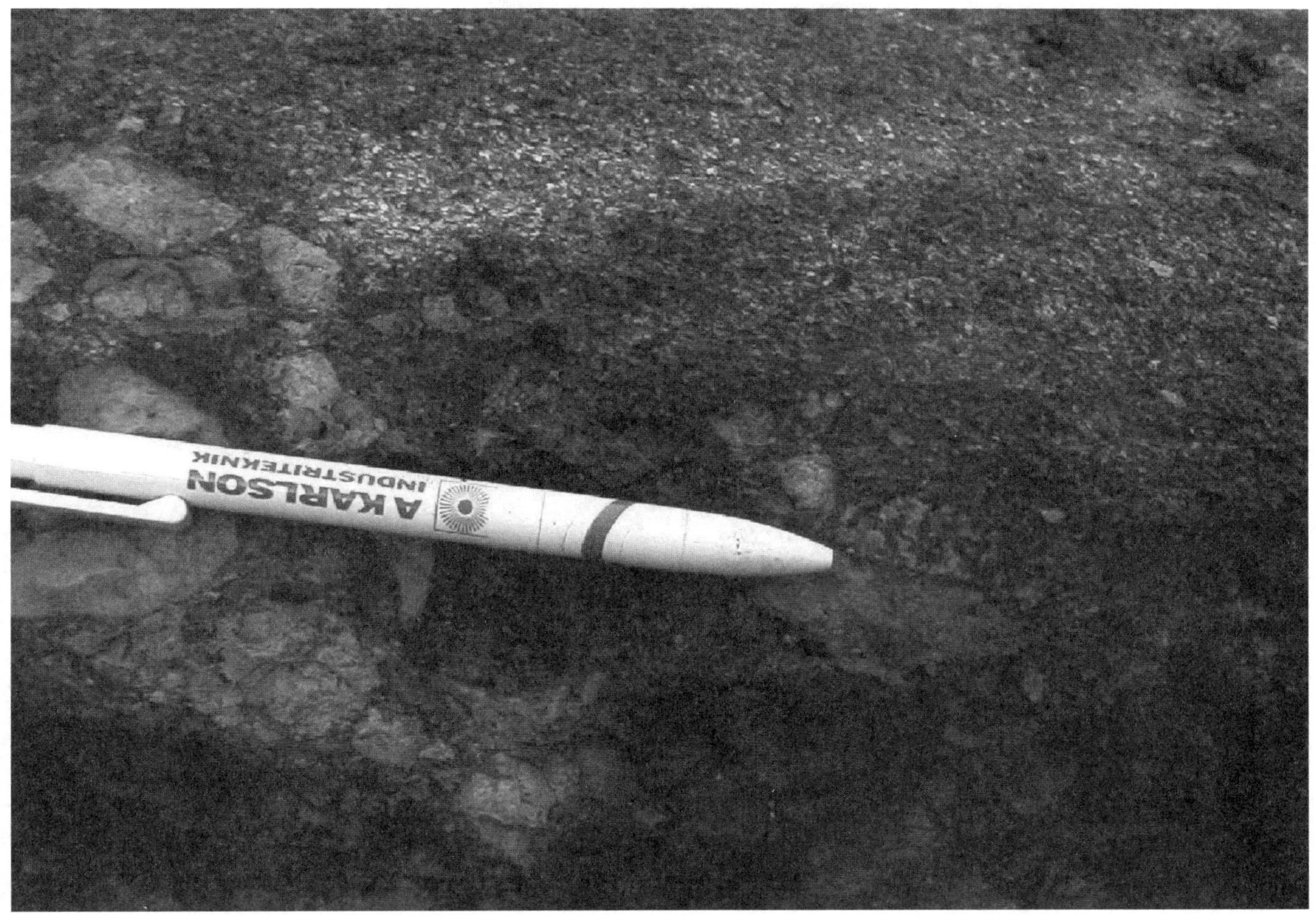

Figure 5. Gravelly Lockne Breccia transitional upward into coarsely sandy Loftarstone. Forestry road to Skettmyrtjärnen (coordinates 6988500/1454950). Photo by J. Ormö.

Phanerozoic-clast breccia occurs throughout the Manson crater as well as in areas outside it.

The Bunte Breccia formed through horizontal outward movement by what Witzke and Anderson (1996) refer to as "ground-hugging debris surge," and they assume that an identical process initiated the formation of the Phanerozoic-clast breccia. As pointed out by Witzke and Anderson, the occurrence of thick deposits of the Phanerozoic-clast breccia in many places throughout the interior of the Manson crater would seem to contradict this idea, but it is hypothetically explained through transport by resurge in shallow seawater. However, this very tentatively assumed transport by water produced neither sorting, nor grading, nor sedimentary layering. Furthermore, the authors were careful in pointing out that the impact occurred in a regressive episode during which the area might have been above sea level.

The Phanerozoic-clast breccia differs from the Lockne Breccia through the circumstances that its clasts are everywhere matrix-supported and apparently nowhere sorted. These circumstances could lead to the conclusion that transport in the case of the Phanerozoic-clast breccia took place as a high-velocity debris flow and not in watery suspension, which apparently was the transporting medium in the case of the Lockne Breccia. An important difference between the Phanerozoic-clast breccia and the Ynntjärnen Breccia is that its clasts are relatively far-transported and of mixed origin. For the following reasons neither the Ynntjärnen Breccia nor the Lockne Breccia is likely to have formed through impact surge (Knauth et al., 2005).

Impact surges have essentially the same kinematic background as pyroclastic surges, and the deposits are expected to resemble one another structurally. Fisher (1979) distinguishes two kinds of pyroclastic surge: ground surge and base surge. Ground surge could produce deposits with the poor sorting and omnipresent matrix support that characterize the Ynntjärnen Breccia, but unlike the case with the Ynntjärnen Breccia the clasts would be relatively far-traveled and of greatly mixed origin. On the other hand, base surge is of phreatomagmatic origin, and thus requires

Figure 6. Resurge breccia. Lockne Breccia, dominated by crystalline ejecta clasts. Private road to Handsjön lakes, Tandsbyn (coordinates 6985959/1447030). Photo by J. Ormö.

Figure 7. Resurge-generated Lockne Breccia (the fairly well-sorted clasts of which appear in three dimensions owing to easily weathered matrix) overlying water-blow breccia (Ynntjärnen Breccia) (monomictic limestone breccia with clasts visible only in two dimensions). Railroad section at Ynntjärnen, Tandsbyn (coordinates 6988050/1443800). Photo by J. Ormö.

moisture. It creates deposits with structures that may look identical to those laid down by water (Fisher, 1977; Knauth et al., 2005). This would fit the description of the Lockne Breccia and the overlying Loftarstone but not the position of these members in the impact stratigraphy of Lockne, because they overlie the ejecta sheet and are transitional, inside the crater as well as outside of it, to the postimpact secular, marine sediments.

In places, the Ynntjärnen Breccia is covered by crystalline ejecta (Figs. 8A and 8B) (Lindström et al., 2005a). This circumstance relates it to the excavation stage, whereas the Lockne Breccia belongs to the modification stage (Melosh, 1989) (Figs. 8B and 8C).

The mix of clasts in the Lockne Breccia is evidence of some distance of transport. Although a part of this transport may have been by debris flow, particularly near the margin of the inner crater and in the outer reaches of the water crater, the predominant transport was by traction and suspension in water. The evidence consists of sorting (generally moderate to good; Figs. 5–7), clast-supported fabric (Figs. 5–7), and more or less gradual transitions to water-laid arenite (Lindström et al., 2005a). Lack of lamination and the occurrence of cobbles in somewhat finer-grained lithologies are probable consequences of combined bed flow and flow in critically charged suspension.

By contrast, the clasts of the Ynntjärnen Breccia moved within their own parent lithology. This movement could be a matter of several meters, as it demonstrably was in the case of dike injections, or when a major slab was rotated (Fig. 4), but it was frequently less than the diameter of a clast. Because the attitude of the bedding in the whole region that immediately surrounds the crater does not deviate significantly from the horizontal, and no Ordovician tectonism is known to have occurred in the area before the impact, gravity would be improbable as a driving force of brecciation.

Outside the apparent crater, the strongest force exerted on the seabed came with the flow of water connected with the growth of the water crater. The velocity of this flow was far greater than required for the mere erosion of the uppermost level of sediment.

Shuvalov et al. (2005) performed both two-dimensional simulation (vertical impact) and three-dimensional simulation (45° oblique impact) of the Lockne impact event. The two-dimensional simulation focused on ejecta distribution and water

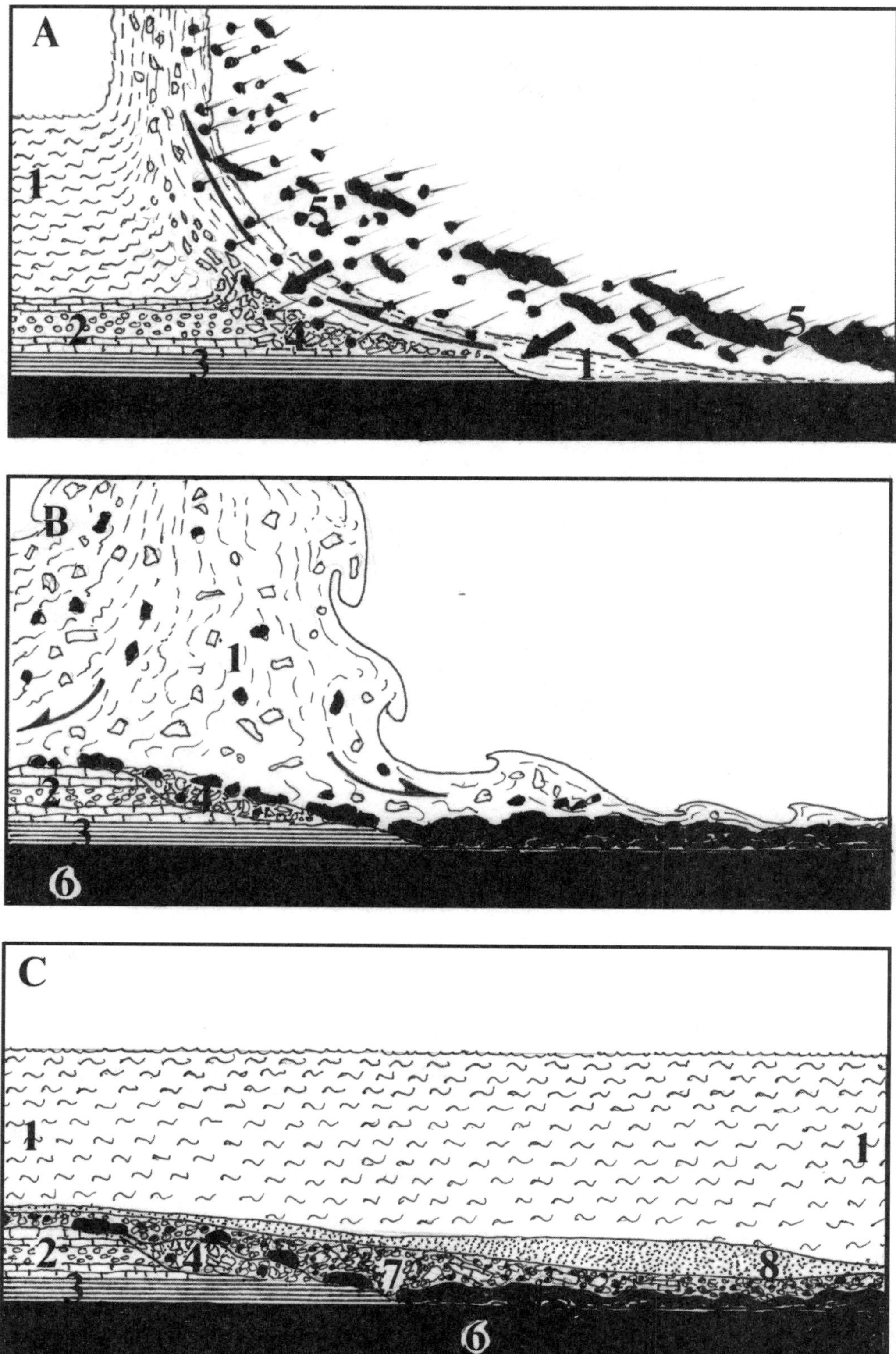

Figure 8. Sketched origin of the two discussed breccias. A: Formation of the Ynntjärnen Breccia by water blow in connection with the excavation of the water crater. B: Collapse of the water crater. The Ynntjärnen Breccia is partly covered by crystalline ejecta. Ejecta and other clasts are carried by the cascading water masses. C: Lockne Breccia and Loftarstone, deposited by resurge on top of ejecta and Ynntjärnen Breccia, form the postcratering seabed. 1—seawater; 2—Lower to Middle Ordovician limestone, bedded as well as nodular; 3—Cambrian bituminous shale; 4—Ynntjärnen Breccia; 5—ejecta; 6—crystalline basement; 7—Lockne Breccia; 8—Loftarstone; half arrows—direction of shear; full arrows—direction of greatest principal stress.

velocities in connection with the cratering. The impact velocity was set to 20 km/s for a stony asteroid with 300 m radius into an 800 m deep sea. The erosive power of outflowing water was found to be very significant even as far as 15 km from the crater center. The velocity of the outgoing flow 5 km from the center culminated at about 125 m/s about 10 s after impact. This marks the approximate end of growth of the water crater. The three-dimensional simulation showed, however, that the best fit between the geological observations and the numerical model should be sought for slightly shallower water.

Because a marked east-west asymmetry of the Lockne crater was revealed by a number of new outcrops (Lindström et al., 2005a), Lindström et al. (2005b) did further refinements to the oblique impact alternative. The impact velocity was reduced to the more conservative 15 km/s. Because the assumption of obliquity brought with it a longer passage through water, it became clear that the water depth had to be less than 800 m. The diameter and other properties of the impactor were assumed to be the same as in the previous model.

Two alternative depths of sea were investigated, 500 m and 700 m. At 500 m there was a significant production of ejecta from the inner (basement) crater, in agreement with conditions found at Lockne, and the water crater was wider than the crater in bedrock, though not by more than about 1 km. At 700 m depth there was a wider water crater, but the basement crater, which in this case was due mainly to downward push, did not yield any appreciable ejecta. At 500 m depth the downrange (westward) spread of an ejecta sheet >1 m thick reached about 10 km from the crater center. The water crater had already collapsed when distal ejecta were emplaced beyond its range.

All known occurrences of the Ynntjärnen Breccia are outside the reach of the modeled water crater. However, as far from the crater center as 7 km the velocity of outgoing flow of water calculated by Shuvalov et al. (2005) reached 40 m/s about 25 s after impact. It was 15 m/s about a minute later at 15 km from the center. One would expect a sedimentary seabed to have been stirred if assaulted by water flowing at these velocities.

The ingoing flow 1.2 minutes later was almost as strong. However, the timing of ejecta emplacement, after the Ynntjärnen Breccia had formed, indicates that the brecciation was caused by the outgoing flow. Ejecta were still in the air 40 s after the impact (Lindström et al., 2005b), but excavation of the crater had ceased by then, and all ejecta within 15 km from the crater center had been emplaced by the time the ingoing flow was culminating.

We are proposing the new term "water-blow breccia" for the Ynntjärnen Breccia. The noun "blow" refers to a sudden, frequently destructive, transmission of kinetic energy (whereas the verb "blow" most frequently refers to perceptible movement in the gaseous state). We conclude that the breccia formed under momentary and exceptionally strong rise of liquid flow velocity (Fig. 8A). At some localities the kinetic impulse transmitted from the water may have been greatly enhanced by the emplacement of the crystalline ejecta masses. The evidence from the Kajan locality indicates that some of these ejecta penetrated into the Ynntjärnen Breccia and became part of it. Water-blow breccias may have formed mainly in parent rock that was water-saturated and partly plastic or semilithified, such as limestone-mud alternations on the seabed.

The Lockne Breccia occurs much more extensively because it could form wherever ejecta above sand size met with resurge-related bottom currents strong enough to transport coarse sediment, and it was not dependent on the presence of any particular preimpact lithology. Thus, its most voluminous occurrence is within the inner crater. Bottom current velocities were over 15 m/s for minutes after the impact as far as 10 km from the crater center (Shuvalov et al., 2005).

SUMMARY AND CONCLUSIONS

The Lockne impact took place in a sea at least 500 m deep. The seabed consisted of 75–80 m thick sedimentary rocks, the upper 50 m of which was lithified limestone with intervals of more plastic, marly-nodular limestone. The lower 25–30 m was poorly lithified Cambrian bituminous mud. The rigid basement beneath these sedimentary cover rocks consisted of magmatic and metamorphic rocks of Late Proterozoic age.

The early Late Ordovician impact excavated a 7–7.5 km wide inner crater in the basement and a somewhat wider water crater. Outside the inner crater, most of the sedimentary cover was swept away along the floor of the expanding water crater. The outgoing water flow was strong enough to remove Ordovician limestone even outside the water crater. This erosion became less effective with decreasing peak flow velocity away from the crater.

The top of the remaining limestone was transformed into monomictic breccia. The brecciation took place in situ. We propose the term "water-blow breccia" for rocks that form through this kind of process. The local water-blow breccia at Lockne carries the name Ynntjärnen Breccia.

Major masses of crystalline ejecta from the inner crater were emplaced either after the Ynntjärnen Breccia had formed or, possibly in some cases, while this breccia was still in the process of forming. The local geological name for these ejecta masses is Tandsbyn Breccia.

In addition to the breccias formed during and immediately after the excavation stage, there is a widely distributed rock that is known under the local formation name Lockne Breccia. It formed by resurge connected with the collapse of the water crater; thus it is a product of the modification stage. It consists of clasts of limestone eroded from different levels and localities and ejecta of different crystalline lithologies. Sorting can be moderate to good, and the fabric is frequently clast-supported. The Lockne Breccia can provide measures of the local erosive and carrying capacity of the resurge flow. Among the clasts one can occasionally find Ynntjärnen Breccia and Tandsbyn Breccia.

ACKNOWLEDGMENTS

Christian Koeberl and Alexander Deutsch provided essential and highly constructive criticism. Eve Arnold gave helpful comments about terminology. The work of M.L. was carried out at the Department of Geology and Geochemistry of Stockholm University. It was a great help to have the support of this institution. The work by J.O. was supported by the Spanish Ministry for Science and Technology (references AYA2003-01203 and CGL2004-03215/BTE).

REFERENCES CITED

Fisher, R.V., 1977, Erosion by volcanic base-surge density currents: U-shaped channels: Geological Society of America Bulletin, v. 88, p. 1287–1297.

Fisher, R.V., 1979, Models for pyroclastic surges and pyroclastic flows: Journal of Volcanology and Geothermal Research, v. 6, p. 305–318.

Hörz, F., Ostertag, R., and Rainey, D.A., 1983, Bunte Breccia of the Ries: Continuous deposits of large impact craters: Reviews of Geophysics and Space Physics, v. 21, p. 1667–1725.

Hüttner, R., and Schmidt-Kaler, H., 1999, Die Geologische Karte des Rieses 1:50,000 (2, überarbeitete Auflage): Geologica Bavarica, v. 104, p. 7–76.

Knauth, L.P., Burt, D.M., and Wohletz, K.H., 2005, Impact origin of sediments at the Opportunity landing site on Mars: Nature, v. 438, p. 1123–1128.

Lindström, M., and Sturkell, E.F.F., 1992, Geology of the early Paleozoic Lockne impact structure, central Sweden: Tectonophysics, v. 216, p. 169–185.

Lindström, M., Simon, S., Paul, B., and Kessler, K., 1983, The Ordovician and its mass movements in the Lockne area near the Caledonian margin, central Sweden: Geologica et Palaeontologica, v. 17, p. 17–27.

Lindström, M., Sturkell, E.F.F., Törnberg, R., and Ormö, J., 1996, The marine impact crater at Lockne, central Sweden: GFF, v. 118, p. 193–206.

Lindström, M., Ormö, J., Sturkell, E., and von Dalwigk, I., 2005a, The Lockne crater: Revision and reassessment of structure and impact stratigraphy, *in* Koeberl, C., and Henkel, H., eds., Impact tectonics: Berlin, Heidelberg, Springer-Verlag, p. 357–388.

Lindström, M., Shuvalov, V., and Ivanov, B., 2005b, Lockne crater as a result of marine-target oblique impact: Planetary and Space Science, v. 53, p. 803–815.

Melosh, H.J., 1989, Impact cratering: A geological process: New York, Oxford University Press, 245 p.

Shuvalov, V., Ormö, J., and Lindström, M., 2005, Hydrocode simulation of the Lockne marine target impact event, *in* Koeberl, C., and Henkel, H., eds., Impact tectonics: Berlin, Heidelberg, Springer-Verlag, p. 403–422.

Simon, S., 1987, Stratigraphie, Petrographie und Entstehungsbedingungen von Grobklastika in der autochthonen, ordovizischen Schichtenfolge Jämtlands (Schweden): Sveriges Geologiska Undersökning C 815, 156 p.

Sturkell, E., and Lindström, M., 2004, The target peneplain of the Lockne impact: Meteoritics and Planetary Science, v. 39, no. 10, p. 1721–1731.

Sturkell, E.F.F., and Ormö, J., 1997, Impact-related clastic injections in the marine Ordovician Lockne impact structure, central Sweden: Sedimentology, v. 44, p. 793–804.

Sturkell, E., Ormö, J., Nölvak, J., and Wallin, Å., 2000, Distant ejecta from the Lockne marine-target impact crater, Sweden: Meteoritics and Planetary Science, v. 35, p. 929–936.

Therriault, A.M., and Lindström, M., 1995, Planar deformation features in quartz grains from the resurge deposit of the Lockne structure, Sweden: Meteoritics and Planetary Science, v. 30, no. 6, p. 700–703.

Thorslund, P., 1940, On the Chasmops Series of Jemtland and Södermanland (Tvären): Sveriges Geologiska Undersökning C 436, 191 p.

von Dalwigk, I., and Ormö, J., 2001, Formation of resurge gullies at impacts at sea: The Lockne Crater, Sweden: Meteoritics and Planetary Science, v. 36, p. 359–369.

Witzke, B.J., and Anderson, R.R., 1996, Sedimentary-clast breccias of the Manson impact structure, *in* Koeberl, C., and Anderson, R.R., The Manson impact structure, Iowa: Anatomy of an impact crater: Geological Society of America Special Paper 302, p. 115–144.

Manuscript Accepted by the Society 10 July 2007

Printed in the USA

The Geological Society of America
Special Paper 437
2008

Paleomagnetism of the Weaubleau structure, southwestern Missouri

Shannon Dulin
R.D. Elmore
School of Geology and Geophysics, University of Oklahoma, 100 E. Boyd Street, Norman, Oklahoma 73019, USA

ABSTRACT

The Weaubleau structure consists of a 19-km circular feature that contains deformed Mississippian limestones. The age of the structure is stratigraphically constrained between deposition of the deformed Osagean limestones and the overlying undeformed Pennsylvanian (Desmoinesian) units. Paleomagnetic samples were collected from tilted Burlington-Keokuk Limestone (undivided), a polymict breccia inside the structure, and undeformed Burlington-Keokuk Limestone outside of the structure. Stepwise thermal and alternating-field demagnetization of tilted limestone samples reveals a characteristic remanent magnetization (ChRM) with southeasterly declinations and shallow positive inclinations with maximum unblocking temperatures of 475 °C. The ChRM is post-tilting and resides in magnetite. The pole is 30.2°N, 135.4°E (d_p = 4.1°, d_m = 7.9°), which lies on the Late Mississippian portion of the apparent polar wander path. The breccia samples only contain a present-day field component. Many of the samples from outside the structure contain a present-day field component residing in magnetite, although some contain a poorly defined component with southeasterly declinations and moderate positive inclinations. The ChRM is apparently localized within the deformation feature. Since the ChRM is post-tilting, the age of the deformation has been constrained better than the stratigraphic age. The post-deformational ChRM is not a shock magnetization and is interpreted as a chemical remanent magnetization (CRM). One hypothesis for the origin of the CRM is hydrothermal fluids that were activated as a result of the impact. This hypothesis is consistent with $^{87}Sr/^{86}Sr$ values in the deformed limestones, which suggest alteration by radiogenic fluids.

Keywords: remagnetization, Weaubleau, supposed impact, paleomagnetism.

INTRODUCTION

The Weaubleau structure is an 8 km circular feature bounded by a 19 km rim denoted by drainage patterns. The structure is located in southwestern Missouri (Fig. 1) and contains deformed Mississippian (Osagean) limestones within the Burlington-Keokuk Limestone (undivided). The age of the deformation is stratigraphically constrained between deposition of the deformed Osagean limestones and the overlying undeformed Pennsylvanian (Desmoinesian) Cherokee Group sandstones. The origin of the structure and associated deformation is unresolved, with some authors suggesting deformation by meteorite impact (Evans et al., 2003; Rampino and Volk, 1996), whereas others suggest deformation by cryptovolcanic processes (Snyder and Gerdemann,

Dulin, S., and Elmore, R.D., 2008, Paleomagnetism of the Weaubleau structure, southwestern Missouri, *in* Evans, K.R., Horton, J.W., Jr., King, D.T., Jr., and Morrow, J.R., eds., The Sedimentary Record of Meteorite Impacts: Geological Society of America Special Paper 437, p. 55–64, doi: 10.1130/2008.2437(04). For permission to copy, contact editing@geosociety.org.

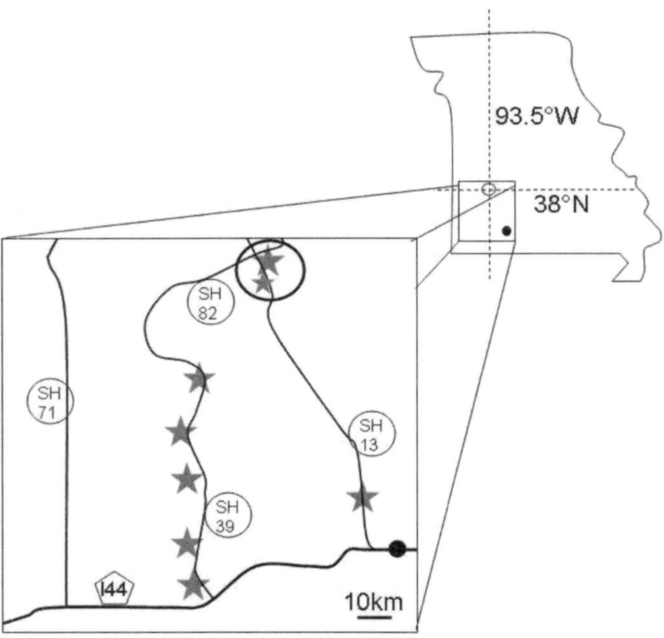

Figure 1. Map of the study area with an inset map of Missouri. The open circle shows the extent of the Weaubleau structure. The closed circle shows the town of Springfield. The stars represent sampling locations. The star located in the north-central part of the structure shows the approximate location of the Ash Grove Aggregates Quarry, whereas the smaller star is the location of the sampling site for the "Weaubleau Breccia." SH—State Highway; I-44—Interstate Highway 44.

1965; Luczaj, 1998). At present, the absence of shatter cones and indexed shocked quartz at the Weaubleau structure indicates that it can only be classified as a proposed impact crater.

The purpose of this study is to test if a paleomagnetic signature is localized within the deformed limestones of the Weaubleau structure, and to test if this signature can be used to constrain the origin and timing of the deformation. The paleomagnetic signatures of impacts in carbonate rocks have been related to processes such as shock effects, heating, and chemical alteration (e.g., Cisowski and Fuller, 1978; Jackson and Van der Voo, 1986; Pesonen et al., 1992; Pilkington and Grieve, 1992). These previous studies provide a background and guided our study of the Weaubleau structure. Samples were collected from both inside and outside the structure to test if the magnetization is localized within the structure. The deformed limestones were sampled for a tilt test to constrain the timing of the magnetization relative to the deformation. Petrographic and geochemical results were used to investigate the possibility of fluid alteration that could be related to a remagnetization mechanism.

GEOLOGIC BACKGROUND

In the late 1940s, Thomas Beveridge from the University of Iowa became the first geologist to describe the complex folding, faulting, and brecciation seen in the Burlington-Keokuk Limestone and surrounding strata in southwestern Missouri (Evans et al., 2003). The structure lies ~100 km north of Springfield, Missouri, near the intersection of State Highways 82 and 13 (Fig. 1). The feature consists of an 8 km circular structure surrounded by an outer ring defined by annular drainage patterns that measures 19 km in diameter (Evans et al., 2005). As previously stated, the age of the structure is stratigraphically constrained between deposition of the deformed Osagean limestones and the overlying undeformed Pennsylvanian (Desmoinesian) Cherokee Group sandstones.

The Burlington-Keokuk Limestone is Late Mississippian (Osagean) in age and comprises most of the disturbed beds associated with the Weaubleau deformation feature (Fig. 2). It is a coarse-grained crinoid grainstone with sparse chert nodules and paleokarst features (Evans et al., 2003). Within the Ash Grove Aggregates Quarry (38.007°N, 93.634°W), the Burlington-Keokuk Limestone is highly faulted and folded.

At the top of the Burlington-Keokuk Limestone (Fig. 3) within the deformation zone is an ~8.8 m thick polymict, unbedded, graded breccia called the "Weaubleau Breccia" (Miller et al., 2005). The breccia contains a fine- to medium-grained carbonate matrix with clasts of underlying Middle Mississippian units, such as the Northview Formation, as well as Lower Ordovician and exotic granitic clasts (Evans et al., 2003). Evans et al. (2003) describe several types of breccia (e.g., fallback and basement breccias) within the "Weaubleau Breccia" and interpret the upper breccia unit as a fallback breccia, which was modified and exhumed by karst processes. Fossil debris is present throughout the breccia. The fossils are of varying ages, though none are younger than Osagean (Miller et al., 2005). The conodont alteration index of conodonts in the breccia is ~1 (Miller et al., 2005).

The Burlington-Keokuk Limestone is overlain unconformably by the Pennsylvanian (Cherokee Group) channel sandstones and shales. None of the Pennsylvanian strata were incorporated into the breccia, nor are the strata deformed.

Evans et al. (2003) suggest an impact origin for the Weaubleau structure based on the presence of the unusual breccia types and intense deformation bounded by a circular drainage pattern. Approximately 600 feet of core was collected within the structure, revealing a granitic breccia and apparent "melt rocks" (Evans et al., 2003). Planar deformation features have also been reported from quartz grains within the "Weaubleau Breccia" but have not been indexed (Evans et al., 2003). The lack of indexing and the absence of shatter cones leave the origin of the structure in question.

The Weaubleau structure, along with seven other structures, lies along the 38th parallel. These structures have been described as either a serial impact (Rampino and Volk, 1996) or all cryptovolcanic in nature (Snyder and Gerdemann, 1965; Luczaj, 1998), or a mixture of the two (Melosh, 1998). The case for serial impact has been refuted based on the prediction of the occurrence of serial impact chains as extraordinarily low (Bottke et al., 1997) and also on the fact that impact theory cannot explain the

Figure 2. Wall of deformed Burlington-Keokuk Limestone within the quarry.

geometry and physical processes that would produce such a crater chain (Melosh, 1998). Ages for most of the structures are not well defined (Luczaj, 1998; Rampino et al., 1999). The Decaturville and Crooked Creek structures are well-documented impacts (Offield and Pohn, 1979; Hendriks, 1954) due to the presence of shock features such as shatter cones and shocked quartz.

Previous Paleomagnetic Studies of Impacts in Carbonate Rocks

Numerous studies have been undertaken to isolate remanent magnetizations associated with impacts in carbonate target rocks. Cisowski and Fuller (1978) report that a small percentage of samples in the Kaibab Limestone at Meteor Crater, Arizona, hold both primary and secondary remanences that were acquired pre-deformationally. The secondary remanence is interpreted as a shock remanent magnetization formed by resetting of the low-coercivity fraction of remanence during the passing of a low-pressure shock wave. The Coconino Sandstone at Meteor Crater is interpreted to carry a thermal remanent magnetization set by high temperature associated with shock (Cisowski and Fuller, 1978).

A post-deformational remanent magnetization was found within Middle Ordovician carbonate rocks at Kentland, Indiana (Jackson and Van der Voo, 1986), a 4 km radius impact structure that Dietz (1947) identified as an impact. This normal-polarity, post-deformational magnetization has relatively low unblocking temperatures (~300 °C) and was interpreted to be Late Cretaceous in age. It was interpreted as a thermoviscous remanent magnetization acquired during prolonged cooling of the impact.

Recent studies from a core at Chicxulub suggest that melt-rich suevites contain a magnetization with multivectorial characteristics (Soler-Arechalde et al., 2003; Urrutia-Fucugauchi et al., 2004). A reversed component is interpreted as the primary remanence, which is consistent with the acquisition at the Cretaceous-Tertiary boundary. The authors interpreted the magnetic characteristics as resulting from the heterogeneous nature of the breccias as well as hydrothermal activity.

The Crow Creek Member (fine-grained marl) of the Pierre Shale in Iowa is proposed by Steiner and Shoemaker (1996) as a tsunami deposit associated with the Manson impact. A Cretaceous normal direction is present in the shale and is interpreted as a depositional remanent magnetization acquired just after the impact (Steiner and Shoemaker, 1996).

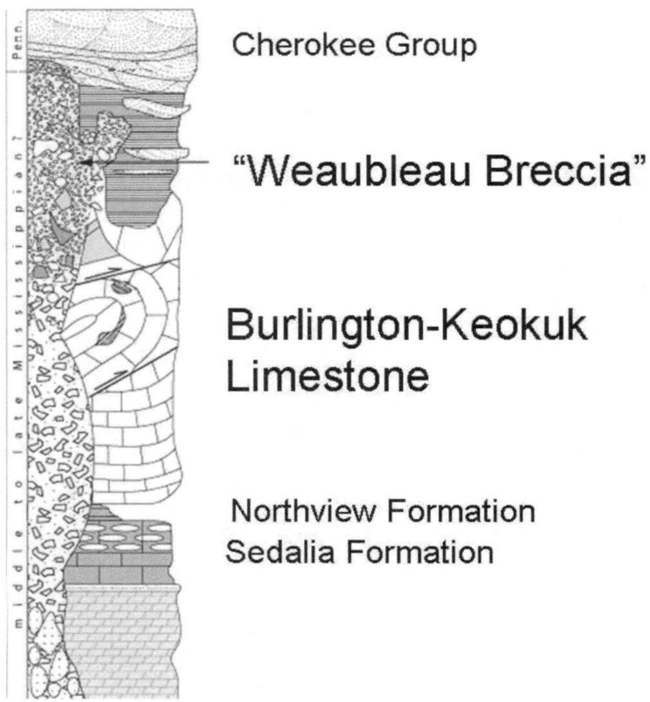

Figure 3. Stratigraphic column showing extent of deformation within the Burlington-Keokuk Limestone and surrounding units (from Evans et al., 2003).

At the Decaturville, Missouri, impact, Offield and Pohn (1979) describe polymict breccias containing numerous carbonate lithologies that are Cambrian-Ordovician in age. Paleomagnetic results suggest that breccia clasts in one polymict breccia contain a Middle Permian post-depositional magnetization in magnetite (Elmore and Dulin, 2007).

PALEOMAGNETIC AND OTHER METHODS

Eighteen sites (six to eight specimens per site) were sampled from tilted beds within the deformed Burlington-Keokuk Limestone in the Ash Grove Aggregates Quarry, which lies in the north-central section of the structure (Fig. 1). The "Weaubleau Breccia" was also sampled along SH 13, south of the quarry inside the structure (Fig. 1). Six sites were collected in the undeformed Burlington-Keokuk Limestone along SH 13, north of Springfield, west on SH 82, and south of the structure along SH 39 to I-44 (Fig. 1). The samples were collected using a gasoline-powered portable drill and were oriented using a clinometer and Brunton compass.

The cores were cut into standard 2.2 cm length specimens. The natural remanent magnetizations (NRMs) were measured with a three-axis 2G Enterprises cryogenic magnetometer with DC squids in a magnetically shielded room. Representative specimens of the Burlington-Keokuk Limestone (deformed and undeformed) and the "Weaubleau Breccia" were subjected to alternating-field (AF) demagnetization (10 mT steps from NRM to 120 mT) in a 2G Enterprises AF demagnetizer. Most specimens were thermally demagnetized in a stepwise fashion (100 °C, 200 °C, 250 °C, 300 °C, and 25 °C steps from 300 to 680 °C) using an ASC thermal demagnetizer. The demagnetization data were analyzed for magnetic directions using the Super IAPD program (http://www.geodynamics.no/software.htm). The data were plotted on Zijderveld diagrams, which are orthogonal projections representing the horizontal and vertical components of the magnetization in the sample (Zijderveld, 1967). Magnetic directions were determined using principal component analysis (Kirschvink, 1980) with mean angular deviations of less than 17°, though most mean angular deviations were 11° or less. Fisher (1953) statistics were used to compute the mean directions. The directional correction (Enkin, 2003) and the Watson and Enkin (1993) tilt tests were performed.

The AF demagnetization results and isothermal remanent magnetization (IRM) acquisition and decay measurements were used to identify the magnetic mineralogy. An IRM was imparted by an impulse magnetizer to representative samples in a stepwise process from 0 to 2500 mT. The samples were then subjected to a second AF demagnetization regime. Subsequently, three perpendicular IRMs (120 mT, 500 mT, and 2500 mT) were applied and the specimens were thermally demagnetized to give the triaxial IRM decay curves (Lowrie, 1990).

Petrographic analysis was performed on polished thin sections using reflected and transmitted light to determine the magnetic mineralogy. A scanning electron microscope and backscattered electron imaging were employed to aid in identification of the magnetic mineralogy. Dunham's (1962) classification scheme was used to classify the rocks.

Strontium isotopic analysis was performed at the University of Texas, Austin, following the methods described by Gao et al. (1992). Three samples of deformed Burlington-Keokuk Limestone within the quarry were analyzed to determine if the rocks were altered. The $^{87}Sr/^{86}Sr$ values were normalized relative to NBS 987 = 0.71014 and then compared to coeval seawater values for Mississippian limestones (Denison et al., 1994).

RESULTS AND INTERPRETATIONS

Paleomagnetism

Thermal demagnetization removed a present-day viscous remanent magnetization (VRM) at low temperatures (<300 °C) and a ChRM with southeasterly declinations and shallow inclinations at temperatures from 300 to 475 °C (Figs. 4A–4C). Maximum unblocking temperatures were 475 °C. At higher temperatures the magnetic intensities increase, probably because magnetite is being created from pyrite. The AF demagnetization of ten Burlington-Keokuk Limestone specimens from the quarry removed a component with northerly declinations and steep down inclinations at low field strengths, interpreted as the present-day VRM, and the ChRM with southeasterly declinations and

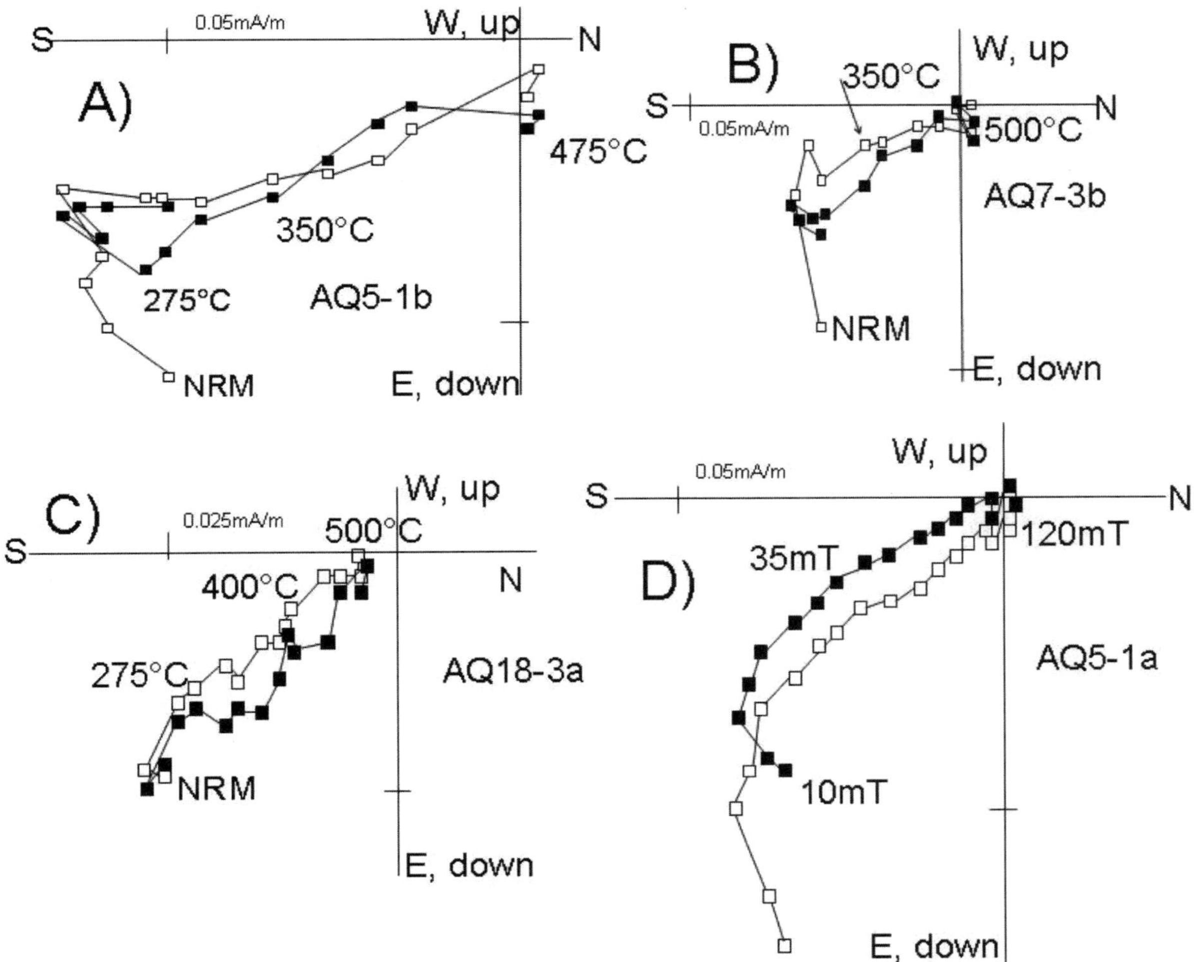

Figure 4. Zijderveld diagrams showing decay of the magnetization within the limestones. Closed symbols represent the horizontal component, open symbols the vertical component. A–C: Thermal demagnetization of three representative deformed Burlington-Keokuk Limestone samples. D: Alternating-field (AF) demagnetization for a deformed Burlington-Keokuk Limestone within the quarry. The trajectory is slightly curved, suggesting vector addition of magnetic components. The sample has fully decayed by 120 mT. For A–C, thermal demagnetization steps at higher temperatures were removed because they show an increase in intensity, which is interpreted to have been caused by creation of new minerals (e.g., magnetite) at high temperatures. The natural remanent magnetization (NRM) intensities are as follows: 0.040 mA/m for A, 0.028 mA/m for B, 0.021 mA/m for C, and 0.046 mA/m for D. The noise level for the cryogenic magnetometer is ~0.002 mA/m.

moderate down inclinations at higher field strengths (Fig. 4D; Table 1). The median destructive field for the ChRM is 20–30 mT. During AF treatment the demagnetization trajectories are curved and the directions have steeper positive inclinations compared to the ChRM removed by thermal demagnetization (Table 1). The component removed by AF treatment is interpreted to be contaminated by overlap with a present-day VRM, and thus the AF results were not used to determine directional information (Table 1).

Fifty-eight thermally demagnetized specimens from nine sites within the deformed Burlington-Keokuk Limestone at the quarry contain the ChRM (Table 1). The specimen directions in the sites have good groupings, with k values ranging from 26.1 to 61.9 and α_{95} values below 15.3°. Nine sites within the quarry were not included in the statistics. Specimens from three of these sites carry the southeasterly direction but have mean angular deviations >20°. Four sites have weak unstable magnetizations, and two sites are dominated by a present-day VRM.

The site means group best in geographic rather than stratigraphic coordinates (Fig. 5; Table 1), suggesting a post-deformational ChRM. The geographic mean of the ChRM has declination of 138.6° and inclination of 15.6° (k = 45.3, N = 9 sites,

TABLE 1. PALEOMAGNETIC SITE DATA FROM WEAUBLEAU

Site	Strike (°)	Dip (°)	Geographic				Stratigraphic		
			N/N_o	Dec_g (°)	Inc_g (°)	k	α_{95} (°)	Dec_s (°)	Inc_s (°)
AQ1	0	0	5/6	137.9	4.8	26.1	15.3	137.9	4.8
AQ4	22.5	15	5/10	154	4.8	31.5	13.8	154.1	–6.5
AQ5	202.5	78	8/8	134.4	16.5	61.9	7.1	217.1	68.3
AQ6	202.5	78	7/8	132.1	24.9	54.6	8.2	240.1	67.4
AQ7	22.5	15	7/7	140.7	19.4	38.6	9.8	139.1	6.1
AQ9	327.5	32	5/9	141.5	10.7	42.7	11.8	136.8	5.9
AQ11	259.5	14	9/9	148.4	8.1	40.8	8.2	147	21.1
AQ14	205.5	15	5/10	130.5	20.1	29.8	14.2	132.7	34.5
AQ18	265.5	60	7/8	128.5	29.7	41.8	9.4	73.7	49.5
AF mean			10/10	144.1	26	21.8	10.6	149.1	35.4
Tilt Test									
DC test –14.5% ± 16.9% untilting									
Geographic			9/9	138.9	15.6	45.3	7.7		
Stratigraphic			9/9	142.8	31.9	4.1	29.2		

Note: N/N_o—number of specimens with direction versus number of specimens demagnetized; Dec—declination; Inc—inclination; k—precision parameter; Stratigraphic—tilt-corrected direction; α_{95}—cone of 95% confidence. Tilt test does not include alternating-field (AF) data because orthogonal projections displayed curved trajectories, which indicates overlapping components. DC—directional correction test of Enkin (2003).

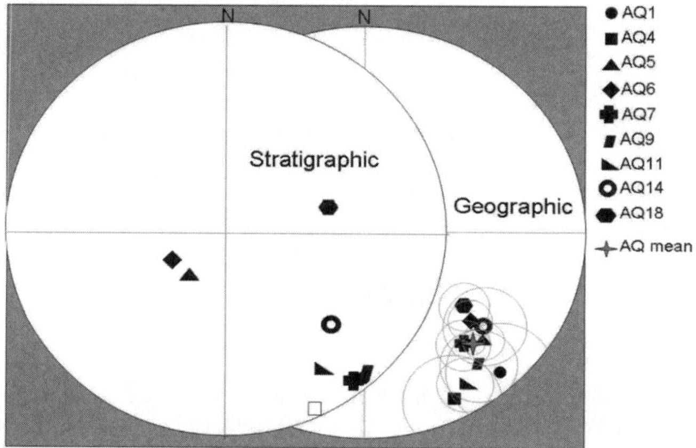

Figure 5. Equal-area plots showing site means in geographic and stratigraphic coordinates. The α_{95} for each site is also shown in geographic coordinates. Open symbols represent negative inclinations; closed symbols are positive inclinations.

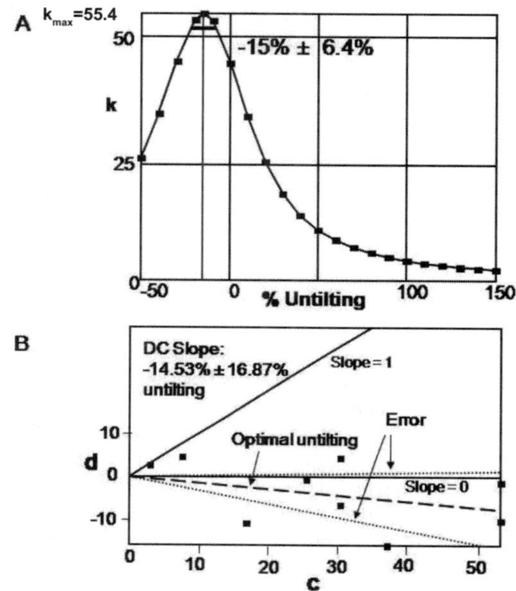

Figure 6. A: The Watson and Enkin (1993) tilt test plots k, which is a measure of grouping, versus untilting from –50% to 150%. The best grouping (largest k value) is at –15% untilting, with an error of 6.4%, shown by the bar. B: The directional correction (DC) test (Enkin, 2003) plots c versus d, where c is the angle between the geographic mean direction and the tilt-corrected site mean direction that is back-rotated by the angular relationship between the two directions, and d is a projection of the arc between geographic site mean directions and the geographic mean of means direction onto the arc used to calculate the c value (Enkin, 2003). The c and d values are closely correlated if the best grouping is at 100% untilting (DC slope = 1.0). The value of the slope of the data (optimal untilting, dashed line) represents the optimal degree of untilting, with error (dotted lines). Since this line lies close to the slope = 0 line, this means the magnetization was acquired post-tilting (or 0% untilting).

and α_{95} = 7.7°). Both the Watson and Enkin (1993) tilt test and a directional correction test (Enkin, 2003) were performed on the beds of different attitudes within the quarry. The Watson and Enkin (1993) test shows that the best grouping for the tilted beds is at –15% ± 6.4% untilting (Fig. 6A). The error calculations for this test yield an unrealistically small error (Enkin, 2003), so the directional correction test was also performed. The directional correction test is negative (Fig. 6B) with a best estimate of –14.5% ± 16.9% untilting. This overlaps the 0% untilting interval, consistent with a post-deformational magnetization.

The age of the ChRM in the quarry was determined by plotting the pole position of the post-tilting magnetization on the apparent polar wander path for North America (Fig. 7). The pole (30.2°N, 135.4°E; $d_p = 4.1°$, $d_m = 7.9°$ [semiaxes of the 95% ellipse of confidence around the pole]) lies on the Mississippian portion of the apparent polar wander path. This age of deformation is supported by a fossil analysis of the breccia, which reported no deformed/broken fossils younger than Osagean (Miller et al., 2005).

Most undeformed Burlington-Keokuk Limestone outside the impact structure either contains a present-day VRM (6.8° declination, 73° inclination) at low temperatures or did not hold a stable ancient magnetization. At higher temperatures the magnetization is very weak and does not display a stable direction. A few specimens contain a poorly defined component with southeasterly declinations and moderate positive inclinations. The "Weaubleau Breccia" contains a component with northerly declinations and shallow inclinations that is interpreted to be a present-day VRM.

Rock Magnetism

The AF decay of the NRM for a representative sample of the deformed Burlington-Keokuk Limestone showed that the magnetic intensity was reduced to 10%–15% of the NRM by 60 mT (Fig. 8A). A similar decay pattern was seen within the undeformed Burlington-Keokuk Limestone. There was little to no AF decay in the "Weaubleau Breccia," suggesting the presence of a higher-coercivity mineral.

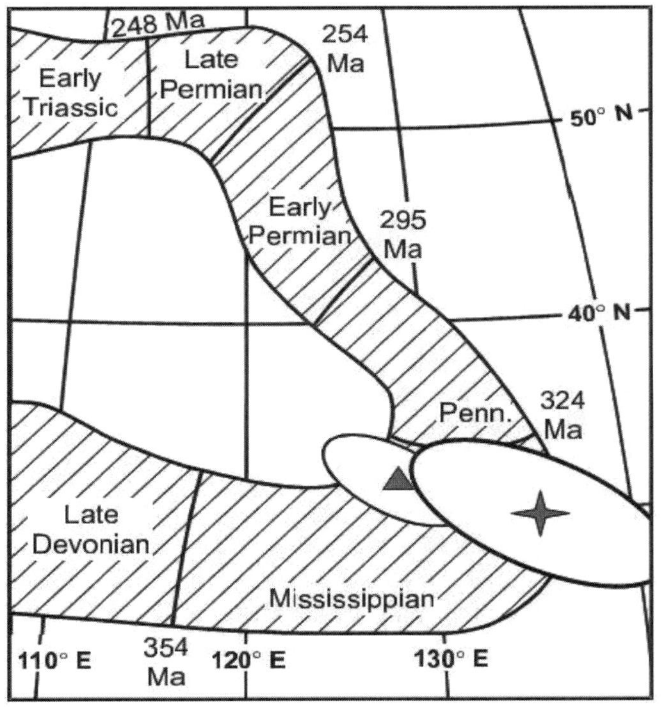

Figure 7. The apparent polar wander path for North America for the middle Paleozoic to the Triassic (modified from Symons et al., 2005). The star represents the pole position of the magnetization found at the Weaubleau deformation structure. The triangle represents the D-component pole from Symons et al. (2005). The ellipses represent the error. Mean poles for the apparent polar wander path are from Van der Voo (1993).

An IRM was imparted on the sample after decay of the NRM (Fig. 8B). The sample is nearly saturated by 300 mT, which is suggestive of a low-coercivity phase such as magnetite. There is a slight rise above 300 mT, indicating that a higher-coercivity mineral, such as hematite, may be present in the rock. The AF decay of the IRM showed that the magnetic intensity was reduced to ~20% of the IRM by 120 mT (Fig. 8C). Triaxial decay of the IRM shows that a lower-coercivity mineral dominates and decays by 580 °C, suggesting that the magnetization resides in magnetite (Fig. 8D).

Petrographic Results

The Burlington-Keokuk Limestone samples collected are classified as fossiliferous grainstones and packstones. Crinoids are the most abundant allochem, followed by brachiopods and bryozoans. Many crinoids within the deformed Burlington-Keokuk Limestone appear sheared or altered (Fig. 9A). Small amounts of dolomite are present in some samples. Pyrite is abundant in all samples (Fig. 9B). In many samples, the pyrite is partly replaced by an iron oxide (Fig. 9C). Hematite is present in some of the replacement rims, and magnetite may also be present (Fig. 9C).

Geochemistry

The $^{87}Sr/^{86}Sr$ analysis of samples from three of the deformed limestones (AQ4: 0.7079; AQ6: 0.7082; AQ7: 0.7084) within the quarry indicate coeval (0.7075–0.7080; Denison et al., 1994) to slightly radiogenic values. The slight enrichment of strontium within these samples indicates that radiogenic fluids may have altered the limestones within the quarry.

DISCUSSION

The deformed carbonates in the Weaubleau structure contain a post-tilting Mississippian remagnetization interpreted to reside in magnetite. This ChRM is also apparently localized within the deformed zone. There are several possibilities for the origin of this ChRM.

A thermoviscous remagnetization can be ruled out because maximum unblocking temperatures (~475 °C) are too high considering that the rocks were only heated to low temperatures (<100 °C for a CAI of 1) and based on experimental evidence in Kent (1985). A shock remanent magnetization is pre-deformational (Cisowski and Fuller, 1978; Halls, 1979) and is therefore ruled out as the origin of the ChRM.

It should be noted that a low conodont alteration index for rocks associated with impacts is common (e.g., Jackson and Van der Voo, 1986) and may not be an accurate record of temperature associated with impact. The instantaneous nature of heating associated with impact may not be sustained for periods of time long

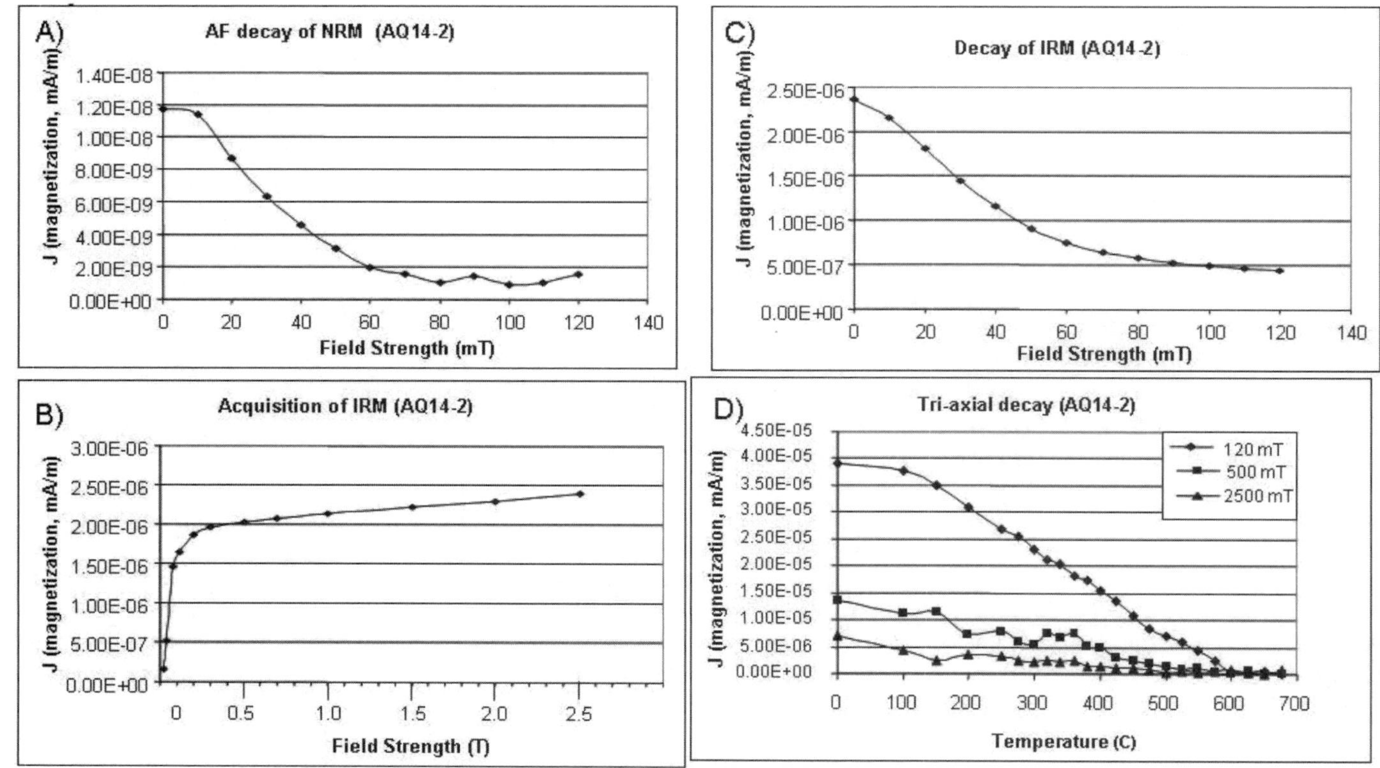

Figure 8. A: Alternating-field (AF) decay curve of natural remanent magnetization (NRM) of representative sample of Burlington-Keokuk Limestone within the quarry. B: Acquisition of isothermal remanent magnetization (IRM) curve. C: AF decay of the IRM. D: Triaxial thermal decay of an imparted IRM. Diamonds—120 mT; squares—500 mT;—triangles 2500 mT.

enough to alter conodonts (J. Morrow, 2004, personal commun.), and therefore a low conodont alteration index should not immediately rule out the possibility of impact.

Another possible remagnetization mechanism is a post-depositional remanent magnetization, where the magnetic grains became realigned in fluid-filled pores after deformation. This mechanism would be viable if the deformation occurred soon after deposition prior to lithification, when the pores were still filled with marine water.

A chemical remanent magnetization (CRM) is also a possibility for the origin of the ChRM, and several chemical mechanisms may be responsible. For example, Symons et al. (2005) report a component interpreted as a CRM acquired during either deposition or compaction in Middle to Late Mississippian (Meramecian-Chesterian) fine-grained limestones in the Tri-State Mining District, Missouri. The limestones are slightly younger than the Burlington-Keokuk Limestone. This CRM has a paleopole position of 32.2°N, 128.4°E ($d_p = 4.5°$, $d_m = 8.6°$), which is interpreted to be Mississippian (Fig. 7). The component is interpreted as being related to "recrystallization of carbonate from cryptocrystalline to very fine grained" (Symons et al., 2005; Pan et al., 1990). It is not clear that this mechanism could apply to the deformed Burlington-Keokuk Limestone because it is a coarse-grained unit. The component is not present in the undeformed limestones around the Weaubleau structure.

Another possible CRM mechanism could be externally derived fluids. The strontium isotopic analysis suggests that the Burlington-Keokuk Limestone was probably altered by radiogenic fluids, and such fluids could be an agent of remagnetization. Other studies report that hydrothermal alteration can lead to the acquisition of a CRM within impact structures (e.g., Pilkington and Grieve, 1992). The ChRM is apparently localized in the Weaubleau structure. Fractures and faults produced by the impact could provide conduits for fluids that could produce localized alteration. However, deformation by other processes could also produce fractures that may localize fluids and cause alteration. The fluids that may be responsible for the CRM could be related to impact, being driven through the rocks shortly after the instantaneous deformation that occurred, or the CRM may be related to a later fluid flow event. If a later event is responsible for the CRM, then it is probably not related to Mississippi Valley-type (MVT) mineralization in southwestern Missouri, which is younger (Symons et al., 2005).

The case for serial impact to explain the eight 38th parallel structures can only be dispelled when more accurate dates for

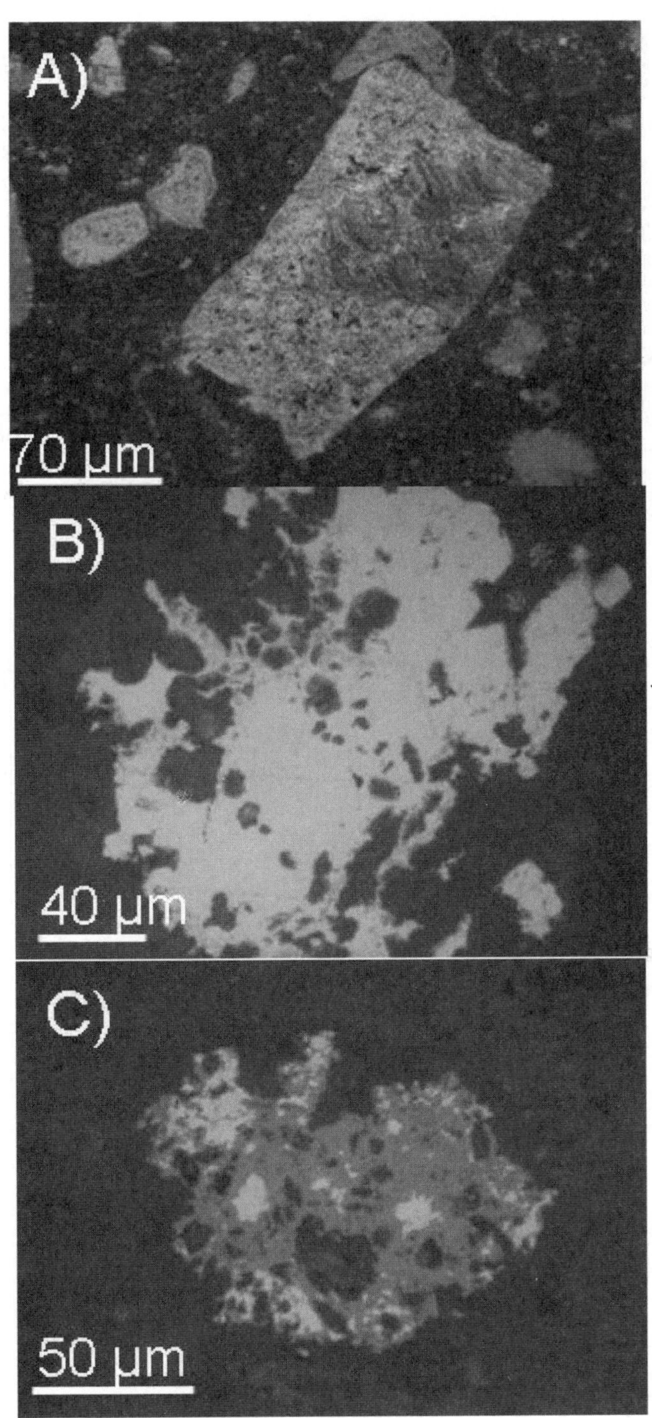

Figure 9. A: Transmitted light photomicrograph of sheared or altered crinoid grain in the deformed limestone. B: Reflected light photomicrograph of pyrite from the deformed limestone. C: Reflected light photomicrograph of iron oxide (magnetite?) partially replacing pyrite from the deformed limestone.

the structures are known. Regardless of the origins of the ChRM at Weaubleau, the deformation can be no younger than the Late Mississippian, since the magnetization is post-deformational. This is a better-defined timing of deformation compared to the stratigraphic constraints. Further studies dating the impacts and other structures are necessary to test the hypothesis that a serial impact was responsible for the structures along the 38th parallel.

CONCLUSIONS

The Weaubleau deformation structure contains a ChRM, interpreted to reside in magnetite, that is localized in the deformed limestones within the structure. The age of the magnetization is Late Mississippian, which constrains the timing of the deformation better than the stratigraphic constraints and is consistent with the paleontologic interpretation for the age of the structure. The origin of the magnetization is not conclusively known, but it is likely a CRM related to fluids. The results of this study show that paleomagnetic methods can be used to help constrain the timing of deformation, either in addition to or in the absence of other dating methods.

ACKNOWLEDGMENTS

We thank the NASA Oklahoma Space Grant Consortium and the Society for Sedimentary Geology (SEPM) for grant support. I thank Kathryn Gardner and Ivy Graham for help in the field. We also thank Kevin Evans and Jim Miller for field orientation and for many discussions.

REFERENCES CITED

Bottke, W.F., Richardson, D.C., and Love, S.G., 1997, Can tidal disruption of asteroids make crater chains on the Earth and Moon?: Icarus, v. 126, p. 470–474, doi: 10.1006/icar.1997.5685.

Cisowski, S.M., and Fuller, M., 1978, The effect of shock on the magnetism of terrestrial rocks: Journal of Geophysical Research, v. 83, p. 3441–3458.

Denison, R.E., Koepnick, R.B., Burke, W.H., Hetherington, E.A., and Fletcher, A., 1994, Construction of a Mississippian, Pennsylvanian and Permian seawater $^{87}Sr/^{86}Sr$ curve: Chemical Geology, v. 112, p. 145–167, doi: 10.1016/0009-2541(94)90111-2.

Dietz, R.S., 1947, Meteorite impact suggested by the orientation of shatter-cones at the Kentland, Indiana disturbance: Science, v. 105, p. 42–43, doi: 10.1126/science.105.2715.42.

Dunham, R.J., 1962, Classification of carbonate rocks according to depositional texture: American Association of Petroleum Geologists Memoir 1, p. 108–121.

Elmore, R.D., and Dulin, S., 2007, New paleomagnetic age constraints on the Decaturville impact structure and Weaubleau structure along the 38th parallel in Missouri (North America), Geophysical Research Letters, v. 34, L13308, doi:10.1029/2007GL030113.

Enkin, R.J., 2003, The direction-correction tilt test: An all-purpose tilt/fold test for paleomagnetic studies: Earth and Planetary Science Letters, v. 212, p. 151–166, doi: 10.1016/S0012-821X(03)00238-3.

Evans, K.R., Mickas, K., Rovey, C., II, and Davis, G., 2003, The Weaubleau-Osceola structure: Evidence of a Mississippian meteorite impact in southwestern Missouri, *in* Plymate, T., ed., Association of Missouri Geologists, 50th Annual Meeting, Guidebook, p. 1–26.

Evans, K.R., Mulvany, P.S., Miller, J.F., Mickus, K.L., and Davis, G.H., 2005, Prologue: Possible, probable, and accepted impact structures of southern Missouri, *in* Field Trip Guidebook, SEPM Research Conference, The

Sedimentary Record of Meteorite Impacts, Springfield, Missouri, May 21–23, 2005, p. 5–11.

Fisher, R.A., 1953, Dispersion on a sphere: Proceedings of the Royal Society of London, ser. A, Mathematical and Physical Sciences, v. 217, p. 295–305.

Gao, G., Elmore, R.D., and Land, L.S., 1992, Geochemical constraints on the origin of calcite veins and associated limestone alteration, Ordovician Viola Group, Arbuckle Mountains, Oklahoma, USA: Chemical Geology, v. 98, p. 257–269, doi: 10.1016/0009-2541(92)90188-B.

Halls, H.C., 1979, The Slate Island meteorite impact site: A study of shock remanent magnetization: Geophysical Journal of the Royal Astronomical Society, v. 59, p. 553–591.

Hendriks, H.E., 1954, The geology of the Steelville quadrangle, Missouri: Missouri Division, Geologic Survey and Water Resources, v. 36, 88 p.

Jackson, M., and Van der Voo, R., 1986, A paleomagnetic estimate of the age and thermal history of the Kentland, Indiana, cryptoexplosion structure: Journal of Geology, v. 94, p. 713–723.

Kent, D.V., 1985, Thermoviscous remagnetization in some Appalachian limestones, Geophysical Research Letters, v. 12, 805–808.

Kirschvink, J.L., 1980, The least-squares line and plane and the analysis of paleomagnetic data: Geophysical Journal of the Royal Astronomical Society, v. 62, p. 699–718.

Lowrie, W., 1990, Identification of ferromagnetic minerals in a rock by coercivity and unblocking temperature properties: Geophysical Research Letters, v. 17, p. 159–162.

Luczaj, J., 1998, Argument supporting explosive igneous activity for the origin of "cryptoexplosion" structures in the mid-continent, United States: Geology, v. 26, p. 295–298, doi: 10.1130/0091-7613(1998)026<0295:ASEIAF>2.3.CO;2.

Melosh, H.J., 1998, Craters unchained: Nature, v. 394, p. 222–223.

Miller, J.F., Bolyard, S., Evans, K.R., Ausich, W.I., Ethington, R.L., Thompson, T.L., and Waters, J.A., 2005, Implications of fossils in the ejecta breccia associated with the Weaubleau-Osceola structure, St. Clair County, Missouri, *in* Proceedings, SEPM Research Conference, The Sedimentary Record of Meteorite Impacts, Springfield, Missouri, May 21–23, 2005, p. 26.

Offield, T.W., and Pohn, H.A., 1979, Geology of the Decaturville impact structure, Missouri: U.S. Geological Survey Professional Paper 1042, 48 p.

Pan, H., Symons, D.T.A., and Sangster, D.F., 1990, Paleomagnetism of the Mississippian Valley-type ores and host rocks in the northern Arkansas and Tri-State districts: Canadian Journal of Earth Sciences, v. 27, p. 923–931.

Pesonen, L.J., Marcos, N., and Pipping, F., 1992, Palaeomagnetism of the Lappajarvi impact structure, western Finland: Tectonophysics, v. 216, p. 123–142, doi: 10.1016/0040-1951(92)90160-8.

Pilkington, M., and Grieve, R.A.F., 1992, The geophysical signature of terrestrial impact craters: Review of Geophysics, v. 30, p. 161–181.

Rampino, M.R., and Volk, T., 1996, Multiple impact events in the Paleozoic: Collision of a string of comets or asteroids: Geophysical Research Letters, v. 23, p. 49–53, doi: 10.1029/95GL03605.

Rampino, M.R., Glikson, A., Koeberl, C., Reimond, W.U., and Luczaj, J., 1999, Argument supporting explosive igneous activity for the origin of "cryptoexplosion" structures in the mid-continent, United States: Comment: Geology, v. 27, p. 279–285, doi: 10.1130/0091-7613(1999)027<0279:ASEIAF>2.3.CO;2.

Snyder, F.G., and Gerdemann, P.E., 1965, Explosive igneous activity along an Illinois-Missouri-Kansas axis: American Journal of Science, v. 263, p. 465–493.

Soler-Arechalde, A.M., Urrutia-Fucugauchi, J., Rebolledo-Vieyra, M., and Vera-Sanchez, P., 2003, Paleomagnetic and rock magnetic study of the Yaxcopoil-1 impact breccia sequence, Chicxulub impact crater: Eos (Transactions, American Geophysical Union), v. 84, p. F538.

Steiner, M.B., and Shoemaker, E.M., 1996, A hypothesized Manson impact tsunami: Paleomagnetic and stratigraphic evidence in the Crow Creek Member, Pierre Shale, *in* Koeberl, C., and Anderson, R.R., eds., The Manson Impact Structure, Iowa: Anatomy of an Impact Crater: Geological Society of America Special Paper 302, p. 419–432.

Symons, D.T.A., Panalal, S.J., Coveney, R.M., Jr., and Sangster, D.F., 2005, Paleomagnetism of late Paleozoic strata and mineralization in the Tri-State lead-zinc ore district: Society of Economic Geologists, v. 100, p. 295–309.

Urrutia-Fucugauchi, J., Soler-Arechalde, A.M., Rebolledo-Vieyra, M., and Vera-Sanchez, P., 2004, Paleomagnetic and rock magnetic study of the Yaxcopoil-1 impact breccia sequence, Chicxulub impact crater, Mexico: Meteoritics and Planetary Science, v. 39, p. 843–856.

Van der Voo, 1993, Paleomagnetism of the Atlantic, Tethys, and Iapetus Oceans: New York, Cambridge University Press, 411 p.

Watson, G.S., and Enkin, R.J., 1993, The fold test in paleomagnetism as a parameter estimation problem: Geophysical Research Letters, v. 20, p. 2135–2137.

Zijderveld, J.D.A., 1967, A.C. demagnetization of rocks: Analysis of results, *in* Collison, D.E., et al., eds., Methods in paleomagnetism: New York, Elsevier Science, p. 254–286.

Manuscript Accepted by the Society 10 July 2007

Printed in the USA

The Geological Society of America
Special Paper 437
2008

Did the Mjølnir asteroid impact ignite Barents Sea hydrocarbon source rocks?

Henning Dypvik
Department of Geosciences, University of Oslo, P.O. Box 1047, Blindern, No 0316 Oslo, Norway

Wendy S. Wolbach
Department of Chemistry, DePaul University, 1036 W. Belden Avenue, Chicago, Illinois 60614, USA

Valery Shuvalov
Institute of Geosphere Dynamics, Russian Academy of Sciences, 38 Leninsky Prospect, Building 1, 119334 Moscow, Russia

Susanna L. Widicus Weaver
Departments of Chemistry and Astronomy, University of Illinois, Roger Adams Lab 164, Box 23-5, Urbana, Illinois 61801, USA

ABSTRACT

Organic-rich shales of Late Jurassic age make up the main source rock for oil and gas in large parts of the Arctic. These sediments, which locally may contain more than 15% total organic carbon (TOC), covered the target area of the Mjølnir impact. We suggest that the extreme richness of organic matter and highly volatile components in the target rock resulted in colossal and intense fires in the impact area, both in the air and on the seafloor. This hypothesis is supported by numerical simulations and explains the large quantities of soot that have been found in samples associated with the Mjølnir impact.

Keywords: Mjølnir, Barents Sea, Jurassic, Hekkingen, impact, soot, numerical simulation, petroleum, source rocks, kerogen, pyrolysis, fire.

INTRODUCTION

The Mjølnir impact crater is located in the Barents Sea (Fig. 1) beneath 50–150 m of sediments and 350 m of water. It was formed in the Late Jurassic (142 ± 2.6 Ma) by the impact of a 1.5–2 km diameter asteroid (Gudlaugsson, 1993; Dypvik et al., 1996; Tsikalas et al., 1998). Geochemical analysis of Dypvik and Attrep (1999) and Robin et al. (2001) indicate a possible iron-nickel-dominated asteroid composition. During this encounter, the crater was formed, ejecta were widely spread, and a tsunami was generated (Shuvalov et al., 2002; Shuvalov and Dypvik, 2004). Simulations indicate that normal marine sea level was reestablished in and around the crater after a subaerial exposure of ~20 min (Shuvalov et al., 2002).

In the Late Jurassic, the paleo–Barents Sea formed the south-eastern part of a wide, shallow epicontinental sea, with possible openings toward the south (the proto–North Atlantic) and the north (present Arctic Basin) (Fig. 1). The Mjølnir crater formed ~500 km from the paleo-Scandinavian landmass in the south and 300 km from the western coastline (the present North Greenland)

Dypvik, H., Wolbach, W.S., Shuvalov, V., and Weaver, S.L.W., 2008, Did the Mjølnir asteroid impact ignite Barents Sea hydrocarbon source rocks?, *in* Evans, K.R., Horton, J.W., Jr., King, D.T., Jr., and Morrow, J.R., eds., The Sedimentary Record of Meteorite Impacts: Geological Society of America Special Paper 437, p. 65–72, doi: 10.1130/2008.2437(05). For permission to copy, contact editing@geosociety.org.

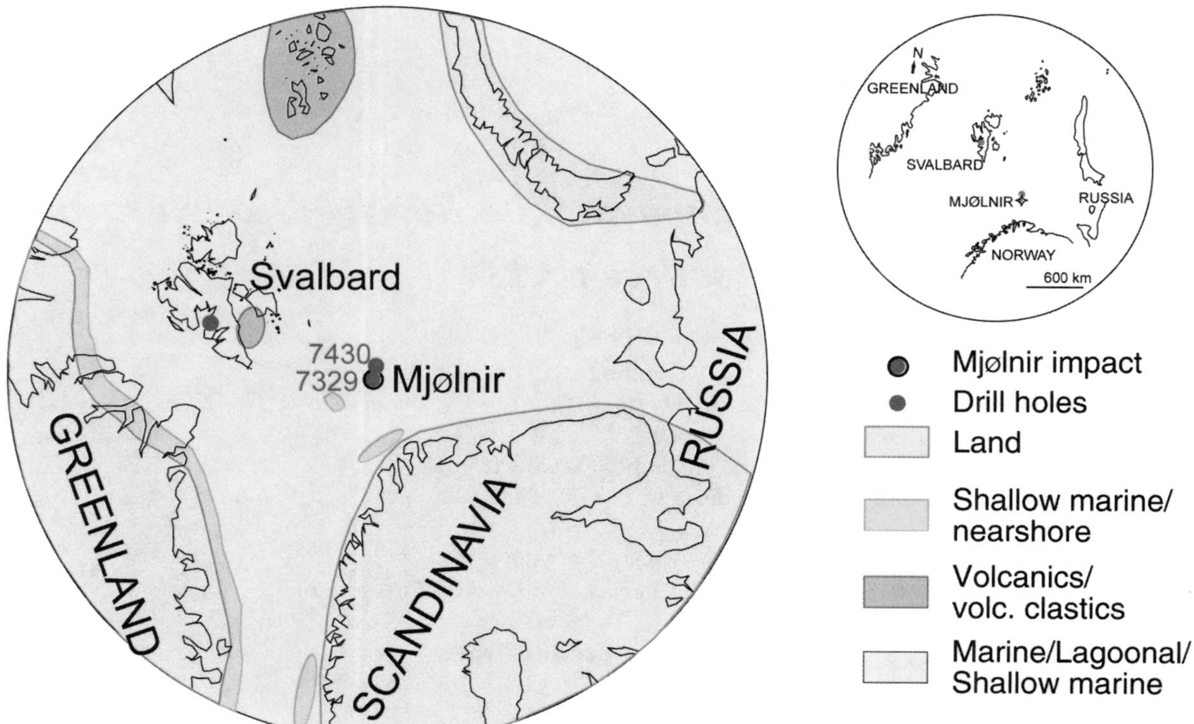

Figure 1. Recent map (right) and a paleogeographic setting (left, 150 Ma, Kimmeridgian, absolute frame) of the studied region. Two shallow cores from the Barents Sea (7430/10-U-01 and 7329/03-U-01) and a section on Svalbard (Janusfjellet) are marked by red dots.

of the paleo–Barents Sea (Fig. 1). Well-developed flora and fauna were probably established around the paleo–Barents Sea with its low topographic relief and paleoposition ~50–55 °N.

The Late Jurassic sediments of the Barents Sea were deposited during hypoxic conditions, dominated by sedimentation of organic-rich clays in wide areas (Leith et al., 1993). These dark gray to black shales (Hekkingen Formation, Oxfordian-Berriasian) currently have an average thickness of ~100 m (Figs. 1 and 2) (max. 150 m; Tsikalas and Faleide, 2004) (Worsley et al., 1988; Smelror et al., 2001; Bugge et al., 2002). The black shales of the Hekkingen Formation can be correlated to the Agardhfjellet Formation on Svalbard (Fig. 2). On average, these shales contain ~8% TOC (total organic carbon), locally up to 15% TOC, have an average kerogen composition of type II/III, and currently form a major source rock for oil and gas in the Barents Sea. In the study area, the Jurassic shales presently are moderately mature (about R_0 0.5%, up to 1%) and carry high concentrations of extractable organic matter (Larsen et al., 1993; Johansen et al., 1993; Leith et al., 1993).

The Cretaceous-Tertiary (K-T) impact 65 m.y. ago at Chicxulub (Mexico) has been associated with widespread wildfires, based on the discovery of soot in the K-T boundary at over a dozen locations around the world (Wolbach et al., 1985, 1988, 1990; Wolbach and Anders, 1989). The fires were likely ignited by atmospheric heating caused by the reaccretion of high-velocity ejecta (Melosh et al., 1990), vapor-rich plume material (Kring and Durda, 2002; Durda and Kring, 2004), or some other cause unrelated to this study. The fuel source (biomass) for these fires is considered to originate from outside the crater (Venkatesan and Dahl, 1989). When the Mjølnir impact was discovered, we decided to investigate the site for evidence of fires (Wolbach et al., 2001). Because the Mjølnir crater diameter is only roughly 1/5 that of Chicxulub, and since Mjølnir has a different target rock composition, soot was not necessarily expected in any great quantity. Unlike Chicxulub, there were no forests around the Mjølnir impact site. However, the uppermost layer of the Mjølnir target rock consisted of thick, black clays very rich in algal organic matter. Additional organic-rich Middle Triassic formations are located a few hundred meters stratigraphically below these Late Jurassic horizons.

Shock heating on impact could transform these sediments by pyrolysis and gasification into a fuel source for combustion and subsequent formation of soot. In this paper we test this idea both by electron-microscopy studies of field and core-sample residues and by numerical simulations.

METHODS

Samples were taken from cores (7430/10-U-01 and 7329/03-U-01) and surface sections on Svalbard. Drill hole 7329/03-U-01

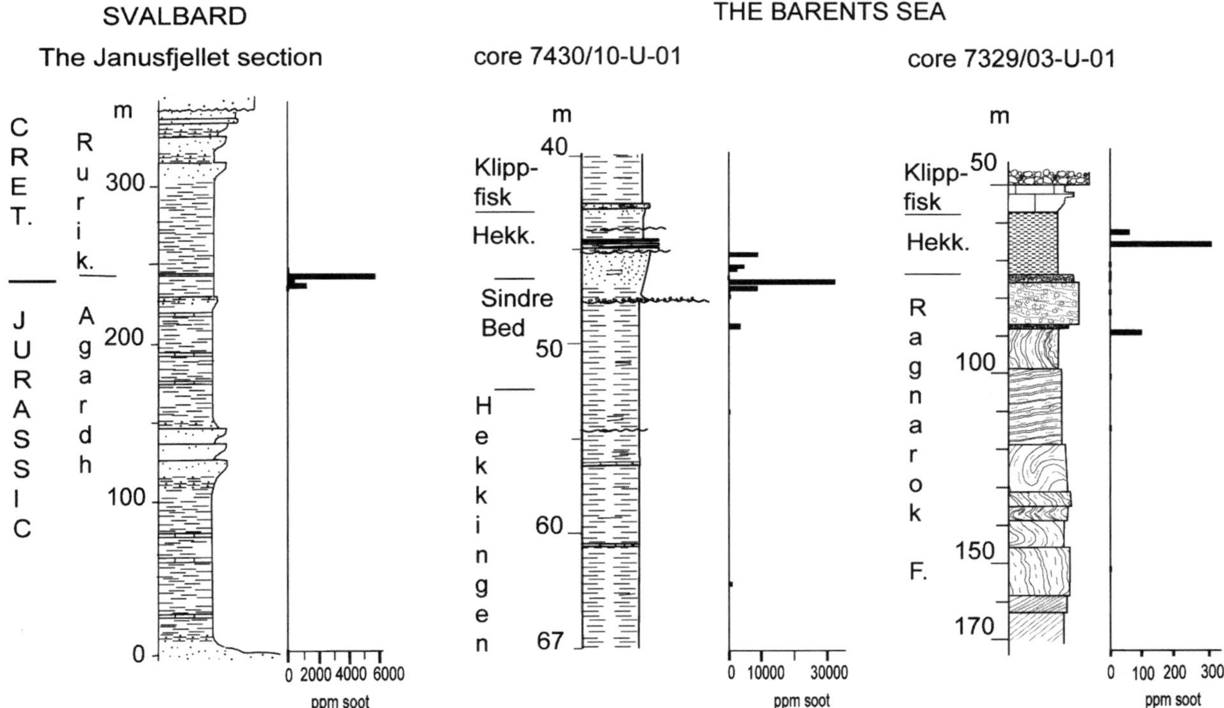

Figure 2. The soot distribution in three stratigraphic sections (Table 1) along with the general stratigraphic subdivision of the area. Note that the horizontal and vertical scales are different in the three sections. Hekk.—Hekkingen Formation, Rurik.—Rurikfjellet Formation, Agardh—Agardhfjellet Formation. The Agardhfjellet and Rurikfjellet Formations of Svalbard can be correlated to the Hekkingen Formation of the Barents Sea. The Ragnarok Formation is a wedge-shaped unit within the Hekkingen Formation. In the Mjølnir crater, the impactites of the Ragnarok Formation consist of different conglomerates/breccias, while a mappable ejecta bed, the Sindre Bed, is found outside the crater (Dypvik et al., 2004). The Sindre Bed in core 7430/10-U-01 is found as a bed within the Hekkingen Formation. Detailed geological studies from the region make regional stratigraphic correlations and paleogeographic reconstructions possible from this area (Brekke et al., 2001; Dallmann et al., 1999; Dypvik et al., 2006). Standard lithological symbols are used in the columns, with sand- and clay-sized particles dominating in the two left successions. In the very right column, conglomeratic beds occur between 75 m and 87 m, while highly folded and deformed sedimentary formations appear below 87 m (Dypvik et al., 2006).

(171 m deep) was drilled in the central peak of the crater, while 7430/10-U-01 (67.6 m deep) was drilled 50 km northeast of the center of the crater. Svalbard is located ~500 km west of the center of the crater (Figs. 1 and 2).

Lightly mortared samples were dissolved in HCl and HF. After dissolution, elemental carbon was separated from organic carbon by acidic dichromate oxidation. The elemental carbon of interest (aciniform soot) was identified and characterized using SEM (scanning electron microscope) imaging and quantified by weighing and particle size analysis (Wolbach and Anders, 1989; Wolbach et al., 2001) (Table 1; Fig. 3). No elemental or isotopic analyses have been executed on the soot particles at this stage since biomass will have the same isotopic signature as those carbon-rich fuel sources that derive from biomass (Deines, 1980) and the probable iron-nickel bolide (Dypvik and Attrep, 1999; Robin et al., 2001) would not provide significant enough carbon to account for the observed amount of soot.

Numerical simulations were performed according to the SOVA multimaterial hydrocode (Shuvalov, 1999). The two-dimensional and three-dimensional SOVA simulations are used to obtain initial ejecta mass and velocity distributions (they terminate at the end of the excavation stage, when ejection velocity falls below 100 m/s, ~5 s after impact). The flight and ultimate deposition of the ejecta have been calculated using ballistic approximation. Further discussion of the method can be found in Shuvalov and Dypvik (2004). The input parameters were a granite bolide 800 m in radius, a 45° oblique impact from southwest at 20 km/s, calcite target area, and a 400 m deep paleo–Barents Sea. A discussion of the selected parameters is provided later (see "Results," "Numerical Simulation").

Ignition temperatures used in the calculations are 673 K and 443 K. Shafizadeh (1984) claims 443 K to be sufficient for combustion/pyrolysis of cellulosic components. Gilmour and Pillinger (1985) did isotopic analyses by stepped combustion on organic matter from the Green River Shale, which contains algae-rich kerogen, partly comparable to the Hekkingen kerogen of the Barents Sea. Gilmour and Pillinger (1985) showed that all carbon had been liberated by 673 K, which is a common upper temperature limit used for pyrolysis and combustion experiments (e.g., Shafizadeh, 1984). Gilmour and Pillinger (1985), however, claim

TABLE 1. SOOT DISTRIBUTION IN THE THREE SECTIONS, WITH A STANDARD DEVIATION OF 10% MAXIMUM

Svalbard		Core 7430/10-U-01		Core 7329/03-U-01	
Sample number	soot (ppm)	Sample number	soot (ppm)	Sample number	soot (ppm)
37.54	0	45.09	8811	62.27	66
36.7	5775	45.73	4784	64.93	314
35.3	0	46.15	3414	71.1	0
34.32	0	46.57	0	73.72	0
33.70	0	46.79	32,189	75.43	0
33.5	3	47.4	0	75.68	0
32.6	39	47.52	463	79.27	0
30.9	989	47.8	15,838	85.48	0
33.9	0	48.85	3733	87.93	100
33.3	9	53.5	0	101.05	0
33.55	5	62.8	536	115.09	0
33.75	519			152	0
34	0				

Note: The Svalbard sample numbers give meters above a stratigraphical marker (Dorsoplanites Bed), while the sample numbers from the two cores give depth (m) below seafloor.

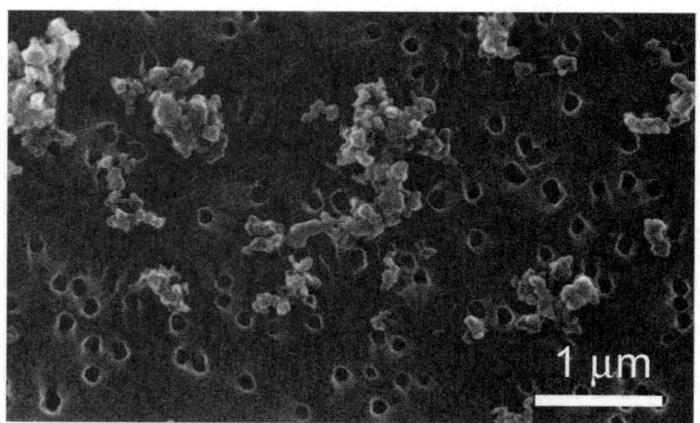

Figure 3. SEM micrograph of aciniform soot in sample 36.7 from Svalbard. The soot particles are mounted on a Nuclepore polycarbonate filter.

that kerogen with high maturity and low H/C ratios will combust at temperatures as high as 870 K.

In the Mjølnir case, the temperature span (443 K to 673 K) used in this modeling covers the pyrolysis of type II/III kerogen. During pyrolysis, lighter, more easily combustible components are formed. The kerogen of the Hekkingen Formation in addition contains varying amounts of types II and I kerogen, which makes it even more combustible (Tissot and Welte, 1984). The pyrolysis could, however, be somewhat retarded due to the presence of inorganic material (Shafizadeh, 1984). In beds with higher organic maturity and more marine and algae-dominated kerogen composition (type I/II), the water and CO_2 contents are reduced and the relative concentrations of more reactive and volatile components are higher. Consequently the suggested temperature span used in this modeling, along with low organic maturity and large original water content in the target sediments, should give a conservative value for soot production.

RESULTS

Soot

Soot particles have been identified in several samples from various localities and stratigraphic positions (Table 1). The soot morphology is similar in all the locations, with smooth particle surfaces and homogeneous particle size distribution. The average single soot particle is ~0.05 µm in diameter (Fig. 3). This morphology is more consistent with soot formed by combustion of hydrocarbon material versus soot formed by the combustion of biomass (Fernandes et al., 2003). The Mjølnir soot looks very much like Fernandes et al.'s (2003) diesel soot. It should be mentioned that soot-like particle can also form by pyrolysis.

The 7329/03-U-01 core consists of sediments that were deposited along the central peak as a slump complex consisting of various density flows (the Ragnarok Formation). These postimpact sediments are succeeded by the youngest units of the dark gray to black shales of the Hekkingen Formation, marls of the Klippfisk Formation, and ~50 m of Quaternary glacial sediments (Fig. 2) (Dypvik et al., 2004). Twelve samples from the 7329/03-U-01 core were analyzed for soot, but only three samples contained measurable amounts (Fig. 3; Table 1). In two samples the concentrations are fairly low (≤100 ppm), while the sample from the Hekkingen Formation, 10 m above the Ragnarok Formation, contains 310 ppm. The soot content in these crater-core samples is much less than found in the 7430/10-U-01 core and on Svalbard (Figs. 2 and 3; Table 1).

The ejecta bed (Sindre Bed) within the Hekkingen Formation of core 7430/10-U-01, consists of dark gray shales with some coarser-grained sandstone beds in its uppermost part. The dark gray shales of the ejecta bed (Sindre Bed) contain smectite (alteration from glass), grains of shocked quartz, and iridium anomalies (Dypvik et al., 1996; Dypvik and Ferrell, 1998). Eleven samples, several carrying high soot concentration, have been analyzed from this core (Figs. 2 and 3; Table 1). The soot-enriched samples are located in the uppermost part of the Hekkingen Formation, partly in the Sindre Bed. The highest enrichment is found in sample 46.79 (32,000 ppm), with elevated values in samples above and below. The extreme soot concentrations are accompanied by shocked quartz and enrichments of Ir, V, Cr, and Ni (Dypvik and Attrep, 1999).

Thirteen samples have been analyzed from the Svalbard section. Three of those display fairly high concentrations, and one sample in particular (sample 36.7) contains 5800 ppm (Figs. 2 and 3; Table 1). This sample is located ~3 m above an iridium-enriched sample (33.70) from the section (Dypvik et al., 2006) and 4.1 m above occurrences of possible Ni-rich iron oxides (Dypvik et al., 2006; Robin et al., 2001). Lesser soot enrichments are also found in sample 33.75 (520 ppm) and in sample 30.9 (990 ppm). In his study Salvigsen (2004) found the 36.7 m level to be characterized by smectitic clay minerals and weak enrichments in Cr, V, and Ni.

Detailed correlation of stratigraphical soot distribution and impact signals (e.g., shocked quartz, Ir anomalies) between the different sections studied is complex and difficult. Consequently due to the original partly chaotic sedimentation, uncertain degrees of bioturbation, and few samples available, we end up with a rather crude correlation.

Numerical Simulations

Within the first 2 s after impact, the target material was heated due to shock propagation; later the temperature of ejected particles likely increased due to interaction with hot gas (i.e., plume). Both processes were included in our numerical simulations. In particular, Figure 4 shows the results for shock (2 s after the impact) and total heating (including heating in the plume between 2 and 10 s).

According to two-dimensional simulations, the material heated to more than 673 K reached heights above 22 km after 60 s, while the material heated close to 443 K reached ~7.5 km heights within the same time. Target particles close to the impact site experienced stronger heating and were ejected with higher velocity. The more realistic three-dimensional experiments indicate that ejecta heated to 673 K reached a height of 28 km in ~5 s (Fig. 5). The oblique Mjølnir impact explains the asymmetrical ejecta distribution (Shuvalov and Dypvik, 2004).

In our simulations (two-dimensional and three-dimensional), temperature calculations have been essential. They have shown that the upper 100 m layer within a circular area of at least 6–8 km in diameter was heated by impact shock (within 1–2 s) above pyrolysis temperature (~500 K). This means that 5×10^{11} to 10^{12} kg of shale was "shock-ignited", while according to the three-dimensional simulations, within 10 s, 2×10^{13} kg (443 K) or 1.3×10^{13} kg (673 K) was heated above the indicated temperatures (Table 2) (Fig. 4). In two-dimensional numerical experiments, 3.8×10^{13} kg (443 K) or 1.1×10^{13} kg (673 K) was calculated after 60 s.

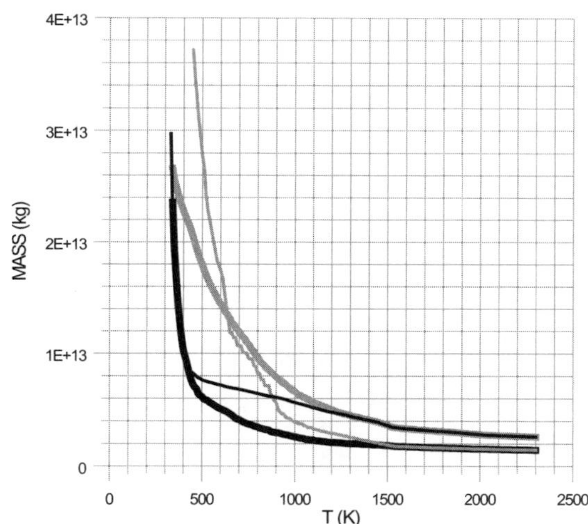

Figure 4. The mass of ejecta from the top 100 m target layer heated above combustion temperature. Thick black line shows shock heating within first 2 s after a vertical impact. Thin gray line estimates total heating within first 60 s after a vertical impact. Thin black line estimates the shock heating for a 45° oblique impact within first 2 s. Thick gray line estimates total heating within 10 s after the oblique impact.

In the simulations, calcite was used to represent target rock and granite bolide composition because these were the only model data available for use in this simulation. However, these substitutions both display similar heat capacities to shales and sandy shales (claystones 0.92–0.94 kJ/kgK and sand 0.80 kJ/kgK). We also tried granite as target material and obtained similar results (to within 10%–15%). Therefore, substituting a more realistic equation of state for actual target rock composition (shale) would not appreciably change our modeling results.

The ejecta distribution from the top 100 m thick target layer is presented in Figure 5. The structure of the downrange soot distribution depends on both water depth and impact angle. The soot distribution would approximately mirror the distribution of ejected shale of Figure 5, reduced in amount to 8% TOC and further, since only a fraction of the carbon is converted to soot. The asymmetrical distribution indicates most of the soot distribution in the direction of impact, toward the north.

DISCUSSION AND CONCLUSIONS

We consider that the soot formation occurred just after impact (due to shock heating of the target material and probably

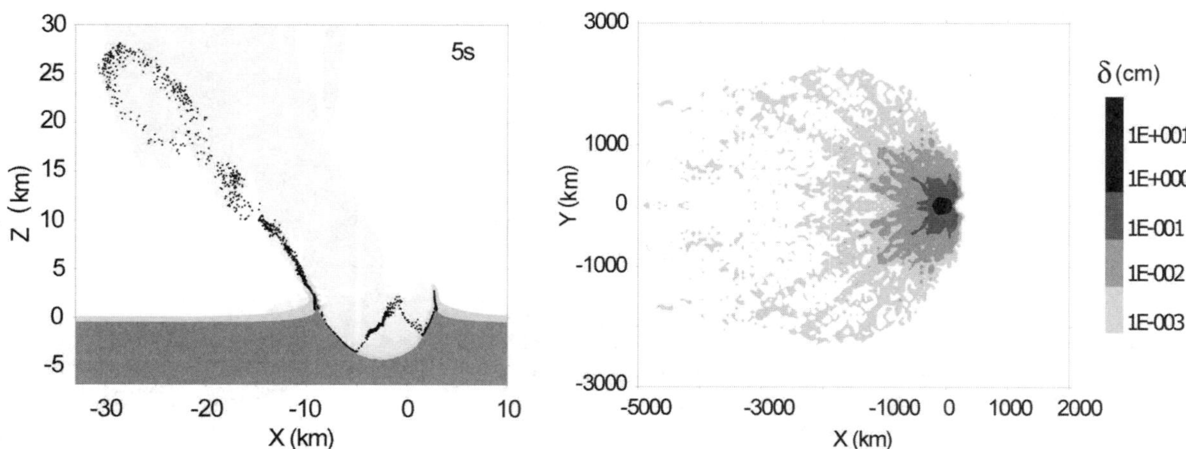

Figure 5. The numerical simulation of the ejecta distribution derived from the 100 m upper layer, which reach a temperature above 443 K. During this oblique impact, an asymmetric distribution is clearly developed toward the north (Shuvalov and Dypvik, 2004). Distance scale in km and thickness of ejecta deposited (δ) in cm.

TABLE 2. ESTIMATE OF THE SOOT PRODUCTION FROM TWO EXTREME RESULTS CALCULATED

	Radius of fire area (km)	Amount of shale pyrolyzed (kg)	Amount of organic material burned (kg)	Mass of soot formed (kg)	Volume of soot (cm^3)	Area of 1-mm-thick soot (km^2)
Model 1	3	5×10^{11}	4×10^{10}	1.6×10^{9}	2×10^{12}	2×10^{3}
Model 2	4	1.3×10^{13}	1×10^{12}	4.2×10^{10}	5.3×10^{13}	5.3×10^{4}

Note: Model 1 is based on calculations from a vertical impact, 3 km in radius of target region heated to 673 K, after 2 s. Model 2 is based on vertical impact, 4 km in radius of target region heated to 673 K, after 60 s. Area in km^2 covered by 1-mm-thick soot lamina after impact is given in the very right column.

later due to ejecta heating in the impact plume) and most likely had an initial distribution similar to that of other ejecta. The soot generation may be tied to both an early phase of pyrolysis and succeeding combustion phases, but both the size and distribution of the soot are consistent with a combustion origin (I. Gilmour, 2005, personal commun.). Winds and turbulence may to some extent make the distribution of this fluffy material less clear. This is supported by the soot occurrences in the three sites analyzed: Svalbard 500 km west of the crater, core 7430/10-U-01 30 km north of the crater, and core 7329/03-U-01 from the central peak.

At the time of impact, the paleogeographic position of the impact site was hundreds of kilometers from the closest forest, making wildfires on surrounding land not very probable. Spontaneous ignition of vegetation would also need higher temperatures. An extraterrestrial carbon source seems less likely, since the geochemical analysis indicates an iron-nickel-rich impactor (Dypvik and Attrep, 1999; Robin et al., 2001). We consequently find it most likely that the soot particles came from pyrolysis and combustion heating of the organic-rich, partly volatile, dark clays of the seabed. This heating occurred during shock wave propagation through the target sediments (1–2 s), due to ejecta interaction with the impact plume (fireball) and due to interaction between the plume and the exposed seabed. The heat- and pressure-controlled alteration and soot formation may have been complex, with a possible early pyrolysis phase when most of the atmosphere was blown away. A rapid, incomplete combustion stage succeeded after the return of the atmosphere. We presently have no information to distinguish between soot particles of different origins, but the soot grains inspected so far have an appearance comparable to those of combustion origin from oils or hydrocarbons. The last combustion phase happened during the 20 min dry seabed period, before preimpact sea level returned (Shuvalov et al., 2002). The fires in the air and on the seafloor must have started immediately at impact and may have been on and off in periods, partly controlled by the wash and backwash into the crater. Consequently the soot formation may have lasted for a long time (~20 min), and the beds contain soot of various origins and mechanisms of formation. At the present stage we are not able to separate out the different kinds of soot, but we hope future morphological and isotope chemical analyses may help in this respect.

The soot particles were widely distributed in the area, presently occurring in sediments that are associated with Ir enrichments, smectitic clays, Ni-rich iron oxides, shock-metamorphic grains of quartz, and other ejecta material.

A rough estimation has been done of the amounts of oil that could have been generated from immature kerogen pyrolyzed and combusted during impact under normal Norwegian-shelf petroleum geological conditions (e.g., an about average nC10 oil composition, 30 API gravity, 1%–5% kerogen converted and migrated to oil in place). The estimate shows that the 20 min impact/postimpact Mjølnir period pyrolyzed and combusted source rocks capable of forming ~30 million standard m^3 oil in place. This represents about a peak-year production from some of the larger Norwegian oil fields (e.g., Statfjord). Today the tiny grains of soot are the only traces found of this conflagration.

ACKNOWLEDGMENTS

The shallow cores from the Barents Sea were drilled in 1988 (7430/10-U-01) and in 1998 (7329/03-U-01) by Institutt for kontinentalsokkelundersøkelser (IKU), now SINTEF Petroleum Research. The Norwegian Research Council supported this research program. The European Science Foundation-IMPACT programme is thanked for arranging meetings and establishing environments for fruitful scientific discussions. Claire M. Belcher, Iain Gilmour, Dag Karlsen, Filippos Tsikalas, and the Mjølnir research group are thanked for valuable discussions. The comments of Wright Horton Jr., David Rajmon, and an anonymous reviewer on an earlier draft of the manuscript are highly appreciated.

REFERENCES CITED

Brekke, H., Sjulstad, H.I., Magnus, C., and Williams, R.W., 2001, Sedimentary environments offshore Norway—An overview, *in* Martinsen, O.J., and Dreyer, T., eds., Sedimentary environments offshore Norway—Palaeozoic to Recent: Amsterdam, Elsevier, Norwegian Petroleum Society Special Publication 10, p. 7–37.

Bugge, T., Elvebakk, G., Fanavoll, S., Mangerud, G., Smelror, M., Weiss, H.M., Gjelberg, J., Kristensen, S.E., and Nilsen, K., 2002, Shallow stratigraphic drilling applied in hydrocarbon exploration of the Nordkapp Basin, Barents Sea: Marine and Petroleum Geology, v. 19, p. 13–37, doi: 10.1016/S0264-8172(01)00051-4.

Dallmann, W.K., Gjelberg, J.G., Harland, W.B., Johannessen, E.P., Keilen, H.B., Lønøy, A., Nilsson, I., and Worsley, D., 1999, Upper Palaeozoic lithostratigraphy, *in* Dallmann, W.K., ed., Lithostratigraphic lexicon of Svalbard: Review and recommendations for nomenclature use: Upper Palaeozoic to Quaternary bedrock: Tromsø, Norsk Polarinstitutt, p. 127–214.

Deines, P., 1980, The isotopic composition of reduced organic carbon, *in* Fritz, P., and Fortes, J.Ch., eds., Handbook of Environmental Isotope Geochemistry, Volume 1 of The Terrestrial Environment, A: New York, Elsevier Scientific Publishing, p. 329–406.

Durda, D.D., and Kring, D.A., 2004, Ignition threshold for impact-generated fires: Journal of Geophysical Research, v. 109, E08004, p. 1–14, doi: 10.1029/2004JE002279.

Dypvik, H., and Attrep, M., Jr., 1999, Geochemical signals of the Late Jurassic, marine Mjølnir impact: Meteoritics and Planetary Science, v. 34, p. 393–406.

Dypvik, H., and Ferrell, R.E., Jr., 1998, Clay mineral alteration associated with a meteorite impact in the marine environment (Barents Sea): Clay Minerals, v. 33, p. 51–64, doi: 10.1180/000985598545426.

Dypvik, H., Gudlaugsson, S.T., Tsikalas, F., Attrep, M., Jr., Ferrell, R.E., Jr., Krinsley, D.H., Mørk, A., Faleide, J.-I., and Nagy, J., 1996, Mjølnir structure: An impact crater in the Barents Sea: Geology, v. 24, p. 779–782, doi: 10.1130/0091-7613(1996)024<0779:MLSAIC>2.3.CO;2.

Dypvik, H., Sandbakken, P.T., Postma, G., and Mørk, A., 2004, Post-impact sedimentation in the Mjølnir crater: Sedimentary Geology, v. 168, p. 227–247, doi: 10.1016/j.sedgeo.2004.03.009.

Dypvik, H., Smelror, M., Sandbakken, P.T., Salvigsen, O., and Kalleson, E., 2006, Traces of the marine Mjølnir impact event: Palaeogeography, Palaeoclimatology, Palaeoecology, v. 241, p. 621–636, doi: 10.1016/j.palaeo.2006.04.013.

Fernandes, M.B., Skjemstad, J.O., Johnson, B.B., Wells, J.D., and Brooks, P., 2003, Characterization of carbonaceous combustion residues, I: Morphological, elemental and spectroscopic features: Chemosphere, v. 51, p. 785–795, doi: 10.1016/S0045-6535(03)00098-5.

Gilmour, I., and Pillinger, C.T., 1985, Stable carbon isotopic analysis of sedimentary organic matter by stepped combustion: Organic Geochemistry, v. 8, p. 421–426, doi: 10.1016/0146-6380(85)90020-8.

Gudlaugsson, S.T., 1993, Large impact crater in the Barents Sea: Geology, v. 21, p. 291–294, doi: 10.1130/0091-7613(1993)021<0291:LICITB>2.3.CO;2.

Johansen, S.E., Ostisty, B.K., Birkeland, Ø., Federovsky, Y.F., Martirosjan, V.N., Bruun Christensen, O., Cheredeev, S.I., Ignatenko, E.A., and Margulis, L.S., 1993, Hydrocarbon potential in the Barents Sea region: Play distribution and potential, *in* Vorren, T.O., et al., eds., Arctic geology and petroleum potential: Amsterdam, Elsevier, Norwegian Petroleum Society Special Publication 2, p. 273–320.

Kring, D.A., and Durda, D.D., 2002, Trajectories and distribution of material ejected from the Chicxulub impact crater: Implications for post impact wildfires: Journal of Geophysical Research, v. 107, E8, p. 6-1 to 6-21, doi:10.1029/2001JE001532

Larsen, R.M., Fjæran, T., and Skarpnes, O., 1993, Hydrocarbon potential of the Norwegian Barents Sea based on recent well results, *in* Vorren, T.O., et al., eds., Arctic geology and petroleum potential: Amsterdam, Elsevier, Norwegian Petroleum Society Special Publication 2, p. 321–331.

Leith, T.L., Weiss, H.M., Mørk, A., Århus, N., Elvebakk, G., Embry, A.F., Brooks, P.W., Stewart, K.R., Pchelina, T.M., Bro, E.G., Verba, M.L., Danyushevskaya, A., and Borisov, A.V., 1993, Mesozoic hydrocarbon source-rocks of the Arctic region, *in* Vorren, T.O., et al., eds., Arctic geology and petroleum potential: Amsterdam, Elsevier, Norwegian Petroleum Society Special Publication 2, p. 1–25.

Melosh, H.J., Schneider, N.M., Zahnle, K.J., and Lathan, D., 1990, Ignition of global wildfires at the Cretaceous/Tertiary boundary: Nature, v. 343, p. 251–254, doi: 10.1038/343251a0.

Robin, E., Rocchia, R., Siret, D., and Dypvik, H., 2001, Discovery of nickel-iron particles in the ejecta-bearing strata of the latest Jurassic Mjølnir meteorite impact (Barents Sea): Norwegian Geological Society, Abstract series, v. 1, p. 67–68.

Salvigsen, O., 2004, Spor etter Mjølnirnedslaget i Janusfjellprofilet på Svalbard [M.S. thesis]: Oslo, Norway, University of Oslo, 141 p. (in Norwegian).

Shafizadeh, F., 1984, The chemistry of pyrolysis and combustion, *in* Rowell, R., ed., The chemistry of solid wood: Washington, D.C., American Chemical Society, Advances in Chemistry series, v. 207, p. 489–529.

Shuvalov, V.V., 1999, Multi-dimensional hydrodynamic code SOVA for interfacial flows: Application to thermal layer effects: Shock Waves, v. 9, p. 381–390, doi: 10.1007/s001930050168.

Shuvalov, V.V., and Dypvik, H., 2004, Ejecta formation and crater development of the Mjølnir impact: Meteoritics and Planetary Science, v. 39, p. 467–479.

Shuvalov, V.V., Dypvik, H., and Tsikalas, F., 2002, Numerical simulations of the Mjølnir marine impact crater: Journal of Geophysical Research, v. 107, E7, p. 1-1 to 1-13, doi: 10.1029/2001JE001698.

Smelror, M., Kelly, R.A., Dypvik, H., Mørk, A., Nagy, J., and Tsikalas, F., 2001, Mjølnir (Barents Sea) meteorite impact ejecta offers a Volgian-Ryazanian boundary marker: Newsletter in Stratigraphy, v. 38, p. 129–140.

Tissot, B., and Welte, D., 1984, Petroleum formation and occurrence: Heidelberg, Springer-Verlag, 699 p.

Tsikalas, F., and Faleide, J.I., 2004, Near-field erosional features at the Mjølnir impact crater: The role of marine sedimentary target, *in* Dypvik, H., et al., eds., Impacts in marine environments and on ice: Heidelberg, Springer-Verlag, p. 39–55.

Tsikalas, F., Gudlaugsson, S.T., and Faleide, J.I., 1998, The anatomy of a buried complex impact structure: The Mjølnir structure, Barents Sea: Journal of Geophysical Research, v. 103, p. 30,469–30,484, doi: 10.1029/97JB03389.

Venkatesan, M.I., and Dahl, J., 1989, Organic geochemical evidence for global fires at the Cretaceous/Tertiary boundary: Nature, v. 338, p. 57–60, doi: 10.1038/338057a0.

Wolbach, W.S., and Anders, E., 1989, Elemental carbon in sediments: Determination and isotopic analysis in the presence of kerogen: Geochimica et Cosmochimica Acta, v. 53, p. 1637–1647, doi: 10.1016/0016-7037(89)90245-7.

Wolbach, W.S., Lewis, R.S., and Anders, E., 1985, Cretaceous extinctions: Evidence for wildfires and search for meteoritic material: Science, v. 230, p. 167–170, doi: 10.1126/science.230.4722.167.

Wolbach, W.S., Gilmour, I., Anders, E., Orth, C.J., and Brooks, R.R., 1988, Global fire at the Cretaceous-Tertiary boundary: Nature, v. 334, p. 665–669, doi: 10.1038/334665a0.

Wolbach, W.S., Gilmour, I., and Anders, E., 1990, Major wildfires at the Cretaceous-Tertiary boundary, *in* Sharpton, V.L., and Ward, P., eds., Global catastrophes in Earth history: Geological Society of America Special Paper 247, p. 391–400.

Wolbach, W.S., Widicus, S., and Dypvik, H., 2001, A preliminary search for evidence of impact-related burning near the Mjølnir impact structure, Barents Sea: Lunar and Planetary Science Conference, XXXII, Abstract 1332, Lunar and Planetary Institute, Houston, CD-ROM.

Worsley, D., Johansen, R., and Kristensen, S.E., 1988, The Mesozoic and Cenozoic succession of Tromsøflaket, *in* Dalland, A., et al., eds., A lithostratigraphic scheme for the Mesozoic and Cenozoic succession offshore mid and northern Norway: Norwegian Petroleum Directorate Bulletin, v. 4, p. 42–65.

Manuscript Accepted by the Society 10 July 2007

Printed in the USA

The Geological Society of America
Special Paper 437
2008

Origin and emplacement of impactites in the Chesapeake Bay impact structure, Virginia, USA

J. Wright Horton Jr.
Gregory S. Gohn
David S. Powars
Lucy E. Edwards
U.S. Geological Survey, MS 926A, 12201 Sunrise Valley Drive, Reston, Virginia 20192, USA

ABSTRACT

The late Eocene Chesapeake Bay impact structure, located on the Atlantic margin of Virginia, may be Earth's best-preserved large impact structure formed in a shallow marine, siliciclastic, continental-shelf environment. It has the form of an inverted sombrero in which a central crater ~40 km in diameter is surrounded by a shallower brim, the annular trough, that extends the diameter to ~85 km. The annular trough is interpreted to have formed largely by the collapse and mobilization of weak sediments. Crystalline-clast suevite, found only in the central crater, contains clasts and blocks of shocked gneiss that likely were derived from the fragmentation of the central-uplift basement. The suevite and entrained megablocks are interpreted to have formed from impact-melt particles and crystalline-rock debris that never left the central crater, rather than as a fallback deposit. Impact-modified sediments in the annular trough include megablocks of Cretaceous nonmarine sediment disrupted by faults, fluidized sands, fractured clays, and mixed-sediment intercalations. These impact-modified sediments could have formed by a combination of processes, including ejection into and mixing of sediments in the water column, rarefaction-induced fragmentation and clastic injection, liquefaction and fluidization of sand in response to acoustic-wave vibrations, gravitational collapse, and inward lateral spreading. The Exmore beds, which blanket the entire crater and nearby areas, consist of a lower diamicton member overlain by an upper stratified member. They are interpreted as unstratified ocean-resurge deposits, having depositional cycles that may represent stages of inward resurge or outward anti-resurge flow, overlain by stratified fallout of suspended sediment from the water column.

Keywords: Chesapeake, impact, crater, impactite, breccia, suevite, resurge.

Horton, J.W., Jr., Gohn, G.S., Powars, D.S., and Edwards, L.E., 2008, Origin and emplacement of impactites in the Chesapeake Bay impact structure, Virginia, USA, *in* Evans, K.R., Horton, J.W., Jr., King, D.T., Jr., and Morrow, J.R., eds., The Sedimentary Record of Meteorite Impacts: Geological Society of America Special Paper 437, p. 73–97, doi: 10.1130/2008.2437(06). For permission to copy, contact editing@geosociety.org.

INTRODUCTION

Impact structures formed in "wet" targets of all kinds, including water-rich sediments and rocks as well as oceanic targets, are poorly understood relative to those formed in continental "dry" targets (Ormö and Lindström, 2000; Dypvik and Jansa, 2003; French, 2004). Major differences between the two types of impact structures are attributed to target properties (including water depth and target strength), resurge of seawater back into oceanic craters, and postimpact depositional environments (French, 2004; Evans et al., 2005).

The late Eocene Chesapeake Bay impact structure (CBIS) was formed in a siliciclastic, continental-shelf environment. The purpose of this synthesis is to provide general insights into the origin and emplacement of impactites produced by this large oceanic impact event. The structure has no surface outcrops and can be sampled only by drilling. Consequently, this paper is based mainly on the results of five scientific test holes drilled during 2000–2004. The partially cored Cape Charles test hole, and the continuously cored Bayside, Langley, North, and Watkins School core holes (Appendix), are respectively located ~1, 25, 36, 39, and 43 km from the center of impact, as shown on the map in Figure 1. The term "impactite" is used to designate all rocks produced during an impact event, including shock-metamorphosed target rocks, breccias, and impact-melt rocks (French, 1998).

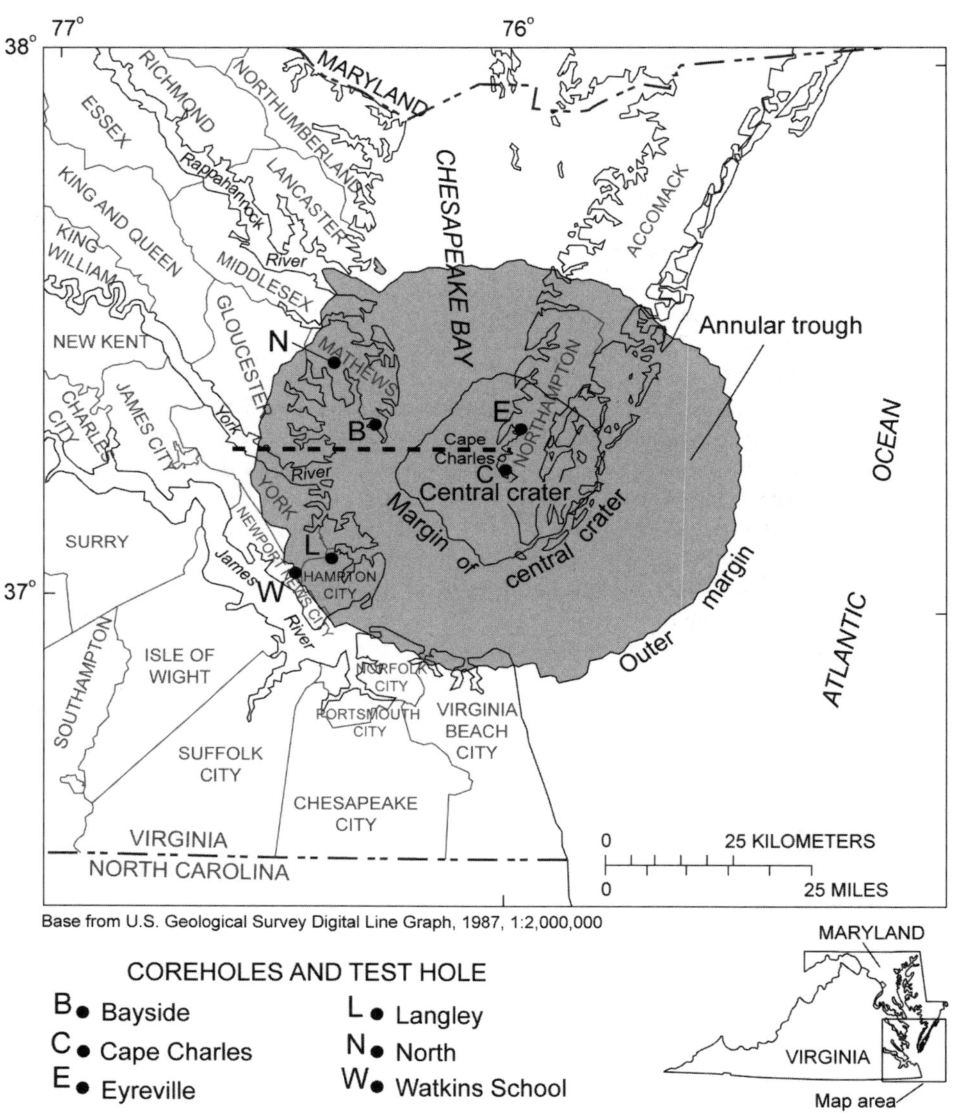

Figure 1. Map showing the location of the Chesapeake Bay impact structure and core holes in southeastern Virginia. Dashed line indicates approximate location of interpretive cross section in Figure 2. Locations of the central crater and outer margin are from Powars and Bruce (1999). Modified from Figure A1 of Horton et al. (2005d).

The buried CBIS (Fig. 1) formed ca. 35.5 Ma when an asteroid or a comet nucleus struck near what is now the mouth of Chesapeake Bay along the Atlantic margin of Virginia (Poag et al., 1994, 1999, 2004; Koeberl et al., 1996; Powars and Bruce, 1999; Powars, 2000; Horton and Izett, 2005; Horton et al., 2005a, 2005d; Poag and Norris, 2005). The impactor struck a layered target of seawater <340 m deep, weak clastic sediments >400 m in thickness, and crystalline basement rocks (Horton et al., 2005d). The clastic sediments were mostly poorly consolidated and unlithified at the time of impact; very few beds in this section could be described as sedimentary rocks.

The resulting central crater is ~40 km in diameter and is surrounded by a relatively shallow structural domain known as the "annular trough," which has an outer margin that is ~85 km in diameter (Figs. 1 and 2). The shape of the structure commonly is compared to an inverted sombrero (e.g., Powars and Bruce, 1999). The central crater was excavated through the unconsolidated sedimentary layer into the crystalline basement, but deformation in the annular trough was confined largely to the sediment layer, where concentric collapse structures (Powars et al., 2003, 2004) and a peripheral ring graben (Poag et al., 2004) are major extensional structures. A donut-shaped gravity low delineates the moat (deepest part) of the central crater and surrounds a relative gravity high that corresponds to the central uplift (Shah et al., 2005).

The general formative stages of the CBIS, based on conceptual and numerical models (e.g., Collins and Wünnemann, 2005; Horton et al., 2005d), are as follows: (1) The projectile was largely vaporized on contact and produced a shock wave in the target. (2) Expansion of the shock wave excavated a complex transient crater in the target and produced shock metamorphism, melt, ejecta, and an ejecta curtain; the upper part of the target sedimentary section was removed from the inner part of the annular trough adjacent to the transient crater by spallation at the sediment-water interface. (3) Crystalline rocks directly below the impact were pushed down, adding to the depth of the transient crater, and subsequently rebounded to form the central uplift. (4) Gravity-driven collapse of the transient crater was accompanied by inward collapse of poorly consolidated sediments that surrounded it, forming the annular trough and expanding the diameter of the structure to ~85 km. Debris-laden seawater surged back into the collapsing transient crater. (5) After this resurge abated, normal marine sedimentation resumed. The impact structure subsequently was blanketed by and preserved beneath postimpact sediments, which range in thickness from ~150 to over 400 m (Fig. 2). Recent numerical modeling by Collins and Wünnemann (2005) illustrates the temporal overlap and transitional nature of these processes and stages.

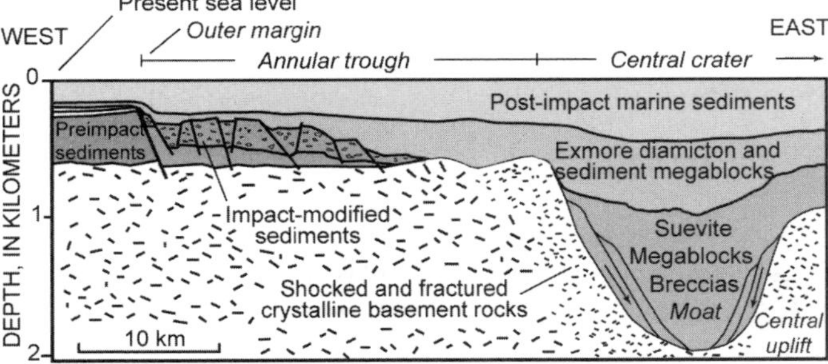

Figure 2. Interpretive radial cross section, western half of Chesapeake Bay impact structure along line shown in Figure 1, ×10 vertical exaggeration. Modified from Figure A7 of Horton et al. (2005d).

IMPACTITES OF THE CHESAPEAKE BAY IMPACT STRUCTURE

The main classes of impactites are shocked rock, impact-melt rock, and impact breccia (Stöffler and Grieve, 1994, 2007; French, 1998; Evans et al., 2005). The term "shocked rock" is used for nonbrecciated rock that has shock-induced features. Impact-melt rock is a crystalline, semihyaline, or hyaline rock that solidified from impact melt. Impact breccia (breccia around, inside, or below impact structures) includes monomict cataclastic breccia (having fragments of uniform composition and free of impact-melt particles), lithic breccia (free of impact-melt particles and generally polymict), and suevite (generally polymict breccia having particles of cogenetic impact-melt rock in addition to lithic and mineral clasts).

The recognition of three categories of impactites in the CBIS (Horton et al., 2005c, 2006a, 2006b) provides insights into different aspects of the crater-forming process. These categories are (1) crystalline-clast suevite and associated crystalline-rock megablocks (monomict cataclastic breccia) in the central crater, (2) impact-modified autochthonous to parautochthonous sediments in the annular trough, and (3) a polymict diamicton of allogenic sediment-clast breccia (Exmore beds in the restrictive sense of Gohn et al., 2005) that blankets the entire structure and nearby areas. Using the proposed classification of impactites by Stöffler and Grieve (1994, 2007), the Exmore beds and the impact-modified sediments are best classified as allogenic lithic breccia, and monomict cataclastic breccia, respectively. This provisional classification is not optimal, however, for distinguishing the characteristics of "wet" impactites formed from low-strength target materials such as water-saturated, unconsolidated sediments. Consequently, this paper uses more explicit sedimentary descriptions and terms where necessary. The list of impactite types discussed in this report is not comprehensive. Breccias that fill the deeper part of the central crater are not explicitly discussed.

Suevite and Megablocks of the Central Crater

The first scientific test hole drilled into the central crater (Cape Charles; Figs. 1 and 3) is located on the central uplift ~1 km from the center of the impact structure (Horton et al., 2004, 2005b; Sanford et al., 2004; Sanford, 2005; Gohn et al., 2007). This test hole primarily produced cuttings, with cored

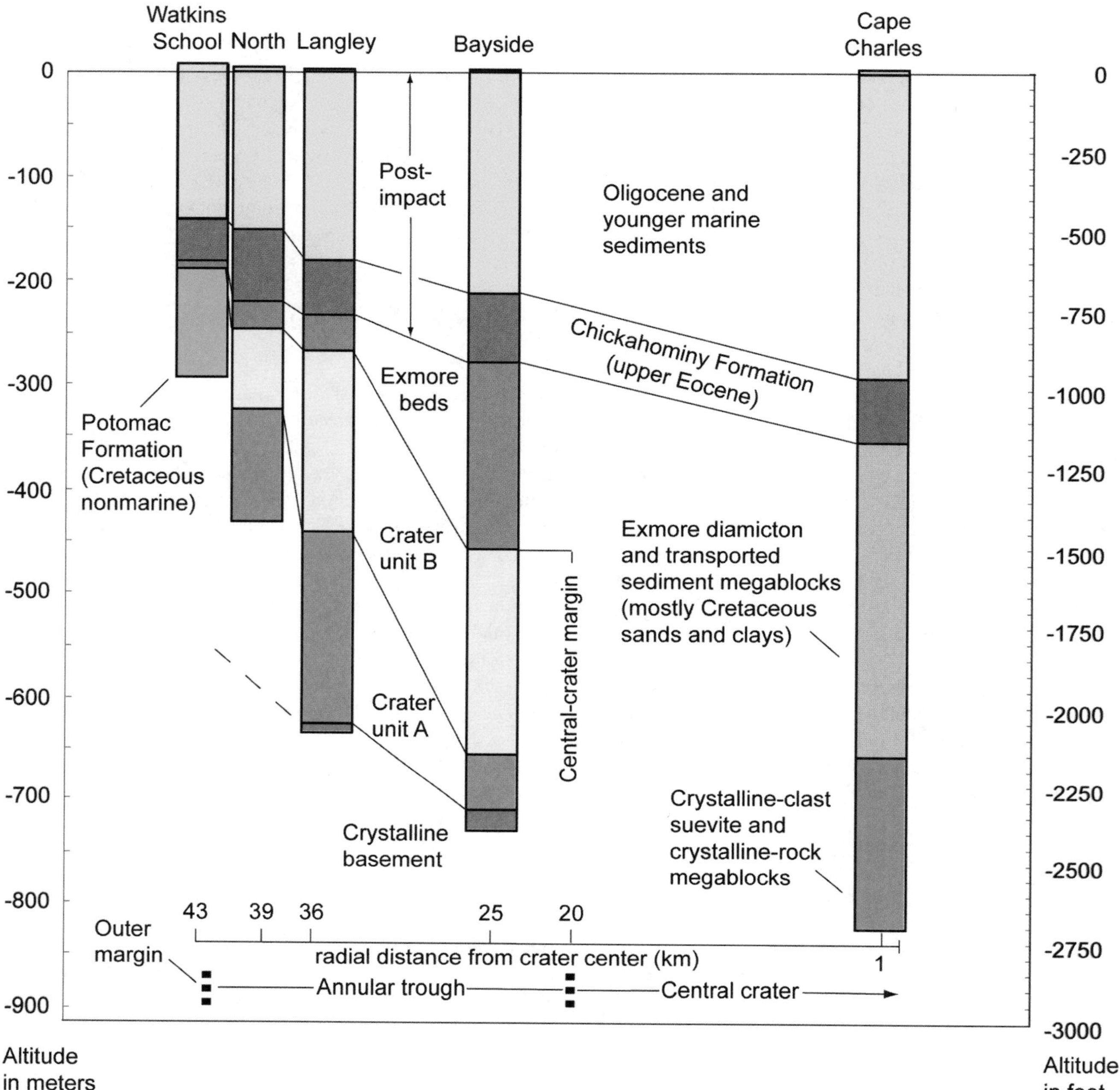

Figure 3. Generalized lithologic sections of four core holes and one partially cored test hole in the Chesapeake Bay impact structure. From left to right, these are the USGS Watkins School core hole, the USGS North core hole, the USGS-NASA Langley core hole, the USGS Bayside core hole, and the partially cored USGS Cape Charles test hole. See map locations (Fig. 1) and details in Appendix. Geologic units are discussed in text.

intervals at 427–433 m and at 744–823 m depths. The deeper cored interval contains a distinctive crystalline-clast suevite that had not been encountered by previous drilling in the annular trough. The suevite underlies the Exmore beds and transported sediment blocks (discussed below), as shown in Figure 3.

The crystalline-clast suevite contains lithic clasts of cataclastically deformed, basement-derived metamorphic and igneous rocks ranging in size from sand to boulders, and less abundant particles of impact-melt rock up to pebble size (Horton et al., 2004, 2005b). Quartzofeldspathic gneiss is the most common type of lithic fragment, followed in abundance by slate and slaty metasiltstone. This suevite is heterogeneous, poorly sorted, and unstratified (Fig. 4A). The matrix is composed of smaller particles identical to the visible clasts.

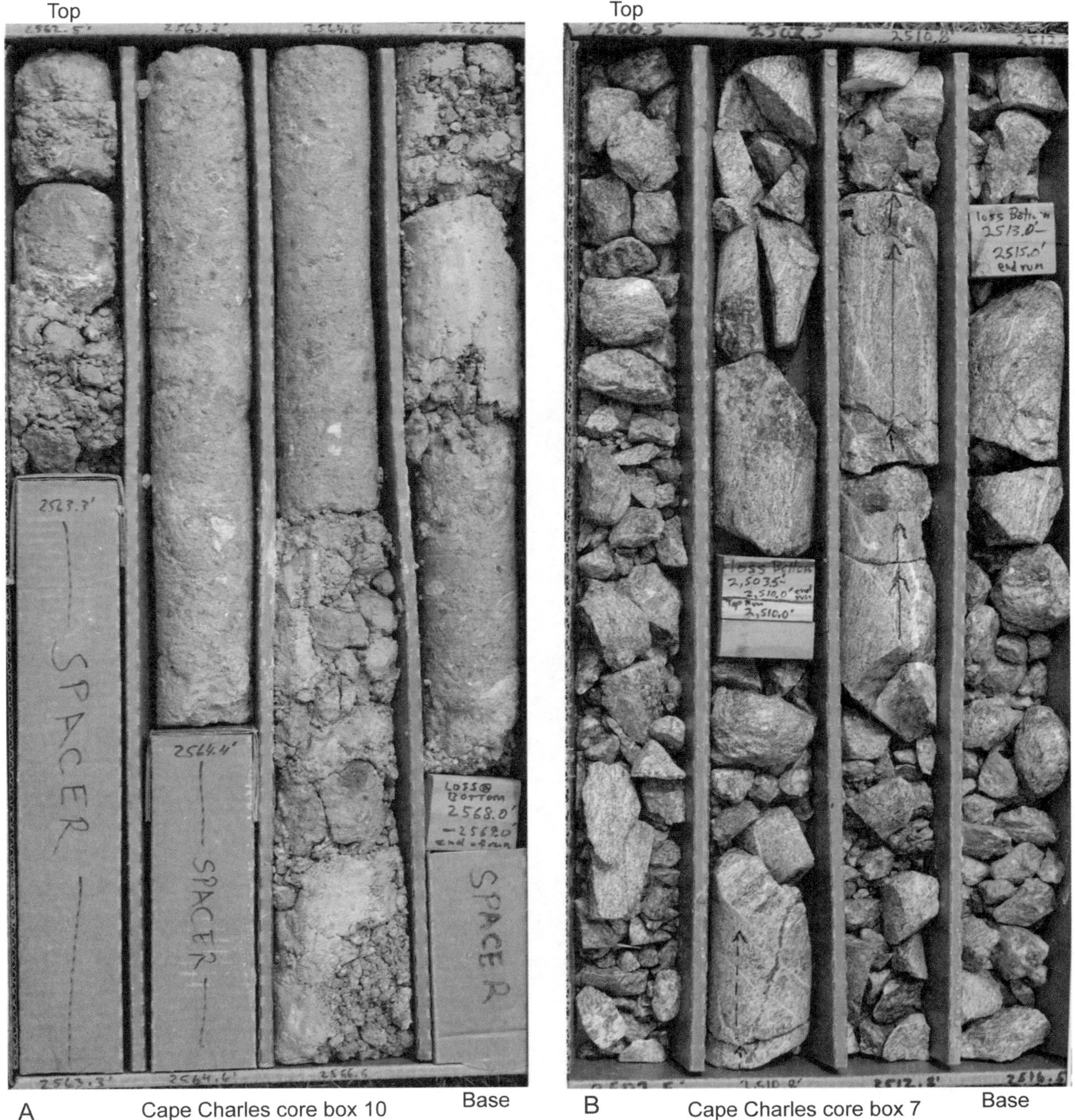

Figure 4. Photographs of suevite and cataclastic gneiss from the central crater, USGS Cape Charles test hole. A: Crystalline-clast suevite in core box 10 from 781.1 to 791.3 m (2562.5–2569.0 ft) depth. B: Monomict cataclastic gneiss from part of megablock >10 m in length within the crystalline-clast suevite in core box 7 from 762.2 to 767.0 m (2500.5–2516.5 ft) depth. Cores of crystalline-clast suevite at this locality contain variable amounts of impact-melt rock, and suevite is present above and below the gneissic megablock shown in B. In core boxes, depth increases from left to right and top to bottom. Nominal core diameter is 6.1 cm, and internal length of core boxes is 61 cm. Drilling depths were recorded in feet, as handwritten on the boxes, and converted to meters for this paper.

Cataclastic fabrics pervade many of the lithic clasts without penetrating the matrix that surrounds these clasts. These fabrics are commonly associated with shocked quartz, suggesting that they formed early in the cratering process. Quartz with multiple intersecting sets of shock-induced planar deformation features (PDFs) is common as grains in the matrix, in gneiss clasts within the suevite, and in associated megablocks of cataclastically deformed gneiss. The PDFs typically are decorated with tiny fluid inclusions, which may be a consequence of annealing under hydrothermal conditions.

The suevite contains clasts of partly glassy impact-melt rock in which flow laminations were warped while they were still plastic. The ragged edges of some melt-rock clasts and intercalated matrix were flattened together, indicating hot compaction (Horton et al., 2005b). Osmium isotopic and platinum-group element analyses confirm that some of the impact-melt clasts contain trace amounts of extraterrestrial meteoritic material (Lee et al., 2005, 2006).

Some megablocks of cataclastically deformed gneiss (Fig. 4B) within or bounded by the suevite exceed 10 m in apparent thickness (Horton et al., 2005b). Two radiometrically dated gneiss megablocks have identical SHRIMP (sensitive high-resolution ion microprobe) U-Pb zircon ages of 612 ± 8 Ma, indicating that a Neoproterozoic basement terrane was excavated by the impact (Horton et al., 2005b). These ages are similar to the U-Pb zircon ages of granites recovered from core holes in the annular trough at Langley and Bayside (Horton et al., 2005e).

Cataclastic fabrics superimposed on the gneiss, such as microfaults, fracture zones, broken and internally deformed mineral grains, and planar fractures in quartz, are variably distributed. Planar fractures associated with planar deformation features (PDFs) in shocked quartz may be related to the initial shock compression and rarefaction, whereas fractures associated with late-stage dilational veins may have formed or reactivated during the central-uplift rebound and collapse. These suevites and megablocks of cataclastic gneiss on the central uplift show pervasive, conspicuous hydrothermal alteration (chloritization and albitization) of metamorphic and igneous mineral assemblages inherited from the preimpact basement target rocks (Horton et al., 2004, 2005b, 2006c, 2006d).

Impact-Modified Sediments of the Annular Trough

The sediment layer of the impact target consisted of (1) a thick, widespread sequence of Lower Cretaceous and basal Upper Cretaceous nonmarine sediments, (2) thin local sequences of Upper Cretaceous marine sediments, and (3) a thin, widespread, seaward-thickening sequence of lower Tertiary marine sediments (Powars and Bruce, 1999; Poag et al., 2004; Gohn et al., 2005). These sediments are not intact in the central crater, where remnants are found only as transported grains, clasts, and slump blocks. However, autochthonous to parautochthonous sections of impact-modified Cretaceous nonmarine sediments are a major, characteristic component of the annular trough.

Impact modifications of the target sediments in the annular trough (Figs. 1–3) include block faulting, fracturing of clays, liquefaction and fluidization of sands, rotation of beds, soft-sediment folds, disaggregation of the near-surface marine sediments, and downward emplacement of the disaggregated marine sediments into the underlying nonmarine sediments (Frederiksen et al., 2005; Gohn et al., 2005; Horton et al., 2005a). With the exception of stratigraphically confined fracturing of oxidized silty clays, none of these features were observed in Cretaceous nonmarine sediments from the Watkins School core, located just outside the outer margin of the annular trough. Elsewhere outside the impact structure, syndepositional sedimentary structures resembling the impact-related structures in the annular trough are rare in the Cretaceous nonmarine sediments compared to the magnitude of impact disruption of the Cretaceous sediments within the annular trough.

Impact-modified sediments of the annular trough were sampled in the Langley, North, and Bayside core holes (Figs. 1 and 3). The Langley and Bayside cores provide complete sections of the impact-modified sediments and samples of the underlying crystalline basement, which consists of Neoproterozoic granite at those locations (Horton et al., 2005a, 2005e).

The impact-modified sedimentary sequence within the annular trough consists of two informally defined units (Fig. 3): (1) A basal crater unit A consists of autochthonous to parautochthonous block-faulted Lower Cretaceous nonmarine sediments, which Gohn et al. (2005) divide into a nonliquefied lower part and a variably liquefied upper part. (2) An overlying, parautochthonous crater unit B consists of blocks of Lower Cretaceous and basal Upper Cretaceous nonmarine sediment disrupted by zones of extensive liquefaction and numerous intercalations of mixed sediments containing mixed-age (Cretaceous and Tertiary) fossils and mineral grains (Frederiksen et al., 2005; Gohn et al., 2002, 2005). The base of crater unit B is defined by the deepest occurrence of glauconite, which marks the lower limit of mixed-sediment intercalations that include material from the top of the target sedimentary section.

Core holes and seismic surveys of the annular trough show the following: (1) The magnitude of impact deformation systematically increases upward within crater units A and B to the base of the overlying Exmore beds, where shocked quartz provides unequivocal evidence of impact-derived material (Catchings et al., 2005b; Gohn et al., 2005; Horton and Izett, 2005; Horton et al., 2005a). (2) The deformed interval generally thickens and its base deepens from the outer margin of the structure inward across the annular trough toward the crater center (Powars and Bruce, 1999; Powars et al., 2003, 2004; Poag et al., 2004) as illustrated in Figures 2 and 3. (3) The deformed interval deepens almost to the top of the crystalline basement in localized extensional structures, including an outer-margin ring graben (Poag et al., 2004) and proposed rings of concentric collapse structures in the annular trough (Powars et al., 2003, 2004). (4) The lower part of the impact-modified sediments (primarily crater unit A) appears on the seismic images as intervals that contain relatively continuous subhorizontal reflections; these are not present across the innermost 8–10 km of the annular trough (Powars et al., 2004). (5) Within this innermost part of the annular trough (e.g., Bayside core), the sedimentary sequence above the crystalline basement and beneath the Exmore beds consists of a relatively thick crater unit B over a relatively thin crater unit A (Fig. 3).

The upward increase in the intensity of impact disruption is most thoroughly documented in the Langley core (Catchings et al., 2005b; Gohn et al., 2005; Horton et al., 2005d). The Neoproterozoic granite at the bottom of the Langley core is not shocked, and no thermal signature of the impact was detected in the granite by fission-track analyses of zircon and apatite (Horton et al., 2005e).

In Cretaceous nonmarine sediments directly above the granite in the Langley core, bedding contacts are horizontal, cross-beds appear undisrupted, and no liquefied sands are present. Gohn et al. (2005) interpreted these sediments as relatively pristine fluvial and deltaic deposits and defined them as the lower part of crater unit A. The upper part of crater unit A contains sections of structureless, unstratified, well-sorted sands with dispersed gravel-sized clasts that are interpreted to be liquefied beds, and clays interbedded with these sands are moderately fractured. Shocked quartz, and the exotic mixed-sediment intercalations characteristic of crater unit B, are absent throughout crater unit A.

Crater unit B at Langley and North consists of blocks and megablocks of Cretaceous nonmarine sediments that range up to tens of meters in size (Gohn et al., 2005). These blocks and megablocks are fractured, variably rotated, and locally exhibit features characteristic of liquefaction (Fig. 5); bedding is inclined locally at dip angles steeper than 60°. Sediment intercalations between

Figure 5. Photographs of impact-modified sediments in the USGS North core (crater unit B). Numbered features 1 and 2 are disrupted, steeply inclined to vertical beds that rotated during liquefaction and fluidization of the underlying sands. Numbered feature 3 consists of massive or irregularly laminated sands indicative of liquefaction. Features 4 and 5 are ball-and-pillow structures characteristic of liquefaction. Feature 6 is a clastic dike of sand that crosscuts bedding in clay, and feature 7 is a block of fractured clay. Includes (left to right) North core box 99 from 282.2 to 284.8 m (925.8–934.5 ft) depth, box 100 from 284.8 to 287.2 m (934.5–942.4 ft) depth, and box 101 from 287.2 to 292.7 m (942.4–960.2 ft) depth. Nominal core diameter is 6.1 cm, and internal length of core boxes is 61 cm (depths handwritten on boxes are in feet).

the blocks and megablocks consist of locally derived Cretaceous clay fragments and quartz-feldspar sand, granules, and pebbles mixed with muddy quartz-glauconite sand (Fig. 6). The glauconitic component in these mixed-sediment intercalations clearly represents disaggregated Upper Cretaceous and lower Tertiary marine sediments from the upper part of the target sedimentary sequence, as confirmed by the rare occurrence of marine microfossils in the matrix (Frederiksen et al., 2005). Glauconite is very rare in the preimpact Lower Cretaceous and basal Upper Cretaceous nonmarine beds. There are no recognizable blocks of Upper Cretaceous or lower Tertiary marine sediments in crater unit B, despite the abundance of sand-sized glauconite grains in the mixed-sediment intercalations. The only shocked quartz that has been found in these impact-modified sediments is in a single rock fragment from muddy glauconitic sand that was emplaced between blocks near the top of crater unit B in the Langley core (Horton and Izett, 2005).

The mixed-sediment intercalations locally penetrate nearly the entire thickness of the sub-Exmore sedimentary section in the Bayside core, located 25 km from the center of the crater (Fig. 3). Crater unit A at Bayside lies directly on granite basement and is no thicker than 54 m. An overlying section of crater unit B containing the localized mixed-sediment intercalations is ≥200 m in thickness. As in the Langley core, the Neoproterozoic granite at the bottom of the Bayside core does not show any discernible thermal or shock-metamorphic effects of the impact (Horton et al., 2005a).

Outside the impact structure, the lower Tertiary marine sediments are consistently found at the top of the preimpact sedimentary sequence in updip and downdip areas, and the Upper Cretaceous marine sediments are consistently found in the downdip area. However, the marine beds are missing from their normal stratigraphic position in all cores within the annular trough.

In summary, impact-modified sediments in the outer annular trough assigned to crater unit A include autochthonous to parautochthonous, block-faulted Lower Cretaceous nonmarine sediments that were locally liquefied only in their upper part. These beds are overlain by crater unit B, which consists of parautochthonous blocks of collapsed Lower and Upper Cretaceous nonmarine sediment disrupted by faults, liquefied sands, fractured clays, and mixed-sediment intercalations that include particles of disaggregated Upper Cretaceous and lower Tertiary marine sediments. Deformation features generally increase in intensity and scale upward and toward the center of the structure, with the local exceptions of an outer ring graben (Poag et al., 2004) and rings of concentric collapse structures (Powars et al., 2003).

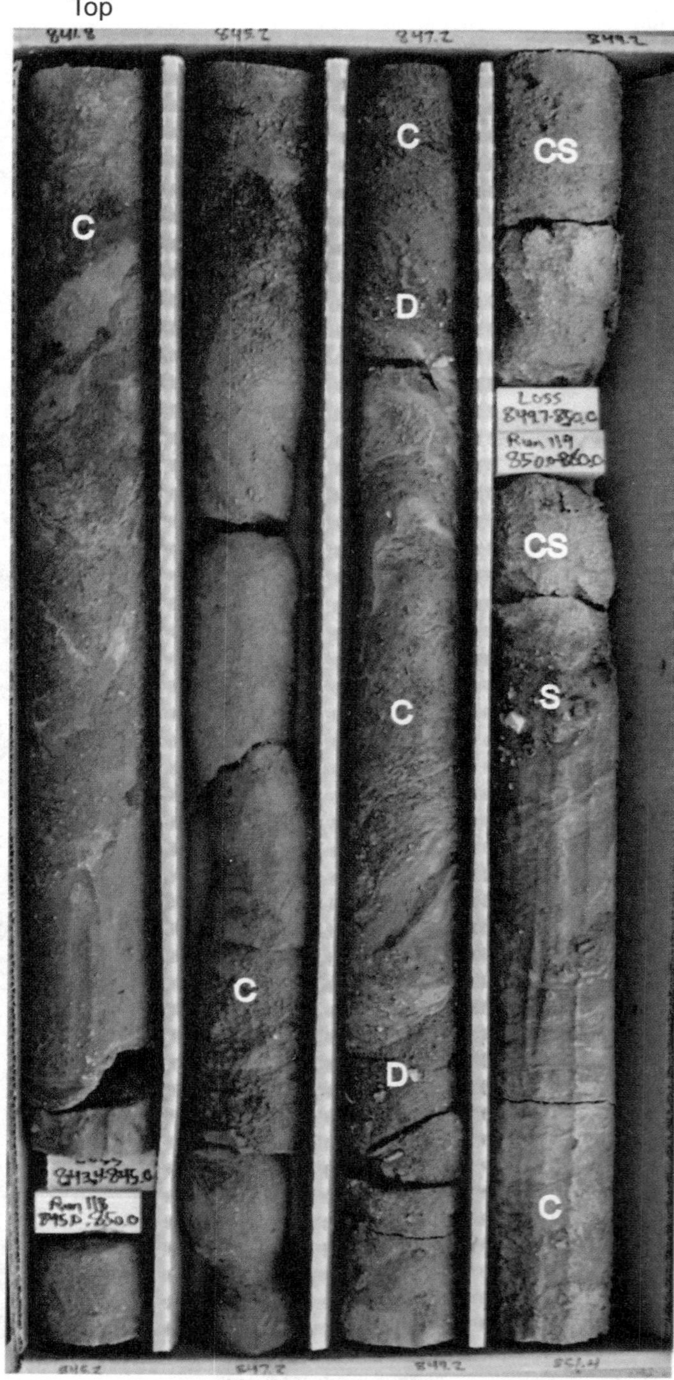

Figure 6. Photograph of impact-modified sediments from the USGS North core, containing injected clastic dikes (D) and a sill (S) of darker muddy quartz-glauconite sand between blocks of Cretaceous nonglauconitic clay (C) and clayey sand (CS). Forceful injection of the dikes is indicated by sharp discordant contacts, sidewall drag folds, apophyses, and plucked sidewall material. North core box 90 from 256.6 to 259.5 m (841.8–851.4 ft) depth. Nominal core diameter is 6.1 cm, and internal length of core box is 61 cm (depths handwritten on box are in feet).

Polymict Diamicton and Upper Stratified Sediments of the Exmore Beds

The informally named "Exmore beds" (Powars et al., 1992) consist primarily of a distinctive sedimentary diamicton. Other informal stratigraphic names previously applied to the Exmore beds include the "Exmore boulder bed" (Poag et al., 1992), the

"Exmore breccia" (Poag, 1996, 1997; Poag et al., 1994, 2004; Poag and Norris, 2005), the "Exmore tsunami-breccia" (Powars and Bruce, 1999; Powars, 2000), "unit C" (Powars et al., 2001), and the "Exmore diamicton" (Horton et al., 2002).

The Exmore diamicton ranges in thickness from 8 to ~200 m and overlies other impactites and preimpact sediments within and near the CBIS. A thin interval of stratified silts and sands (~1–2 m) overlies the diamicton and records the transition from impact-induced to postimpact geologic processes (Poag, 2002; Poag et al., 2004; Gohn et al., 2005; Poag and Norris, 2005). In this report, we define these two units as informal stratigraphic members of the Exmore: (1) a *lower diamicton member* that constitutes most of the Exmore beds, and (2) a thin (less than ~2 m) *upper stratified member* of fine to locally medium sands and laminated clayey silts that overlie the diamicton. The stratified member is equivalent to the "transition sediments" of Gohn et al. (2005) and includes the "fallout layer" and overlying "dead zone" of Poag (2002), Poag et al. (2004), and Poag and Norris (2005).

The Exmore beds are the most widespread impactite unit in the CBIS. They extend slightly beyond the outer margin of the structure and are present in all drill holes of sufficient depth inside the outer margin. A complete, well-preserved section of the Exmore beds within the annular trough at the Langley core hole (Figs. 1 and 3) has been studied in detail by Gohn et al. (2005) and Frederiksen et al. (2005). The other cores through the Exmore beds have not been studied in as much detail, and most of this section in the Cape Charles test hole is represented only by drill cuttings and geophysical logs.

This report uses the informal name, "Exmore beds," as revised by Gohn et al. (2005) on the basis of detailed studies of the Langley core. Correlation diagrams in Horton et al. (2005d, Fig. A6 therein) and Gohn et al. (2005, Fig. C5 therein) compare previous informal usages of the names "Exmore beds" and "Exmore breccia." The name "Exmore breccia" as used by Poag et al. (2004) and Poag and Norris (2005) combines impact-generated sediments of the diamicton, here assigned to the Exmore beds, and underlying impact-modified target sediments of the annular trough, here assigned primarily to crater unit B. We consider the distinction between the products of impact-induced sedimentation and the products of impact modification of preexisting sediments to be important for understanding marine crater materials and processes.

The *lower diamicton member* of the Exmore beds consists of granule- to boulder-sized sediment and rock clasts suspended in a matrix of muddy quartz-glauconite sand and granules (Powars and Bruce, 1999; Powars, 2000; Poag et al., 2004; Gohn et al., 2005; Horton et al., 2005a) (Figs. 7 and 8). The abundant sediment clasts represent all known preimpact Cretaceous, Paleocene, and Eocene sedimentary formations in the impact target area, as confirmed by biostratigraphic ages determined for individual clasts (Frederiksen et al., 2005). Typical sediment-clast lithologic types include (1) oxidized, noncalcareous clays, silts, and quartz-feldspar sands, (2) gray and grayish-green, calcareous and noncalcareous, clays, silts, quartz sands, and quartz-glauconite sands, and (3) minor glauconitic, calcite-cemented sandstones and shelly limestones. Unshocked rounded quartz, feldspar, chert, phosphate, and lithic pebbles derived from the target sediments also are dispersed within the matrix (Gohn et al., 2005; Horton and Izett, 2005). Some large sediment clasts have highly irregular shapes and centimeter-thick alteration rinds.

The lower diamicton member contains sparse clasts of shocked crystalline ejecta from the crystalline basement of the target area. This population of clasts includes rhyolitic felsite, felsic to mafic plutonic rocks, and complex monomict and polymict cataclastic breccias (Horton et al., 2005a; Horton and Izett, 2005). Shock-induced planar deformation features (PDFs) in quartz and feldspar are present in these clasts and provide unambiguous evidence of impact deformation. Shocked quartz also is present as sparse individual sand grains within the diamicton matrix (Koeberl et al., 1996; Horton and Izett, 2005; Horton et al., 2005a).

The diamicton matrix is unsorted and unstratified. It consists of shelly, microfossiliferous, feldspathic, muddy quartz-glauconite sand and granules derived from the preimpact Lower Cretaceous and basal Upper Cretaceous nonmarine sediments, and from the glauconitic Upper Cretaceous and lower Tertiary marine sediments, of the target area. Palynomorphs and calcareous microfossils of Late Cretaceous, Paleocene, and Eocene age are mixed within the matrix (Poag and Aubry, 1995; Edwards and Powars, 2003; Self-Trail, 2003; Poag et al., 2004; Frederiksen et al., 2005). The lower diamicton member has a typically, but not exclusively, matrix-supported fabric. Local concentrations of unoriented cobbles and boulders, particularly near the base of the unit, have clast-supported fabrics.

Normal "coarse-tail" grading (i.e., fining-upward size sorting of only the larger clasts) is one of the few observed sedimentary features in the diamicton, exclusive of structures within individual clasts (Gohn et al., 2005). The presence of two cycles of normal coarse-tail grading in the Exmore section of the Langley core suggests the presence of two depositional cycles; the lower cycle has a thickness of 25.4 m and the upper one has a thickness of 8.4 m (Gohn et al., 2005). Late Cretaceous calcareous nannofossils are present only in the upper cycle, whereas early Tertiary calcareous nannofossils are present in both graded cycles in the Langley Exmore section, suggesting that the two depositional cycles contain material of different provenance (Frederiksen et al., 2005).

In addition to the broad pattern of coarse-tail grading, meter-scale variations in the concentration of clasts are present. However, no stratification, depositional contacts, erosional contacts, or dewatering structures have been observed within the diamicton.

The character of the Exmore lower contact varies with location and with the type of underlying material. It can be difficult to place this contact accurately, owing to the mixing of sediment along this contact and the limited perspective provided by even the best core samples. In areas outside the outer crater margin, where deformation typical of crater unit B is absent, the Exmore lower contact is readily placed at the top of essentially

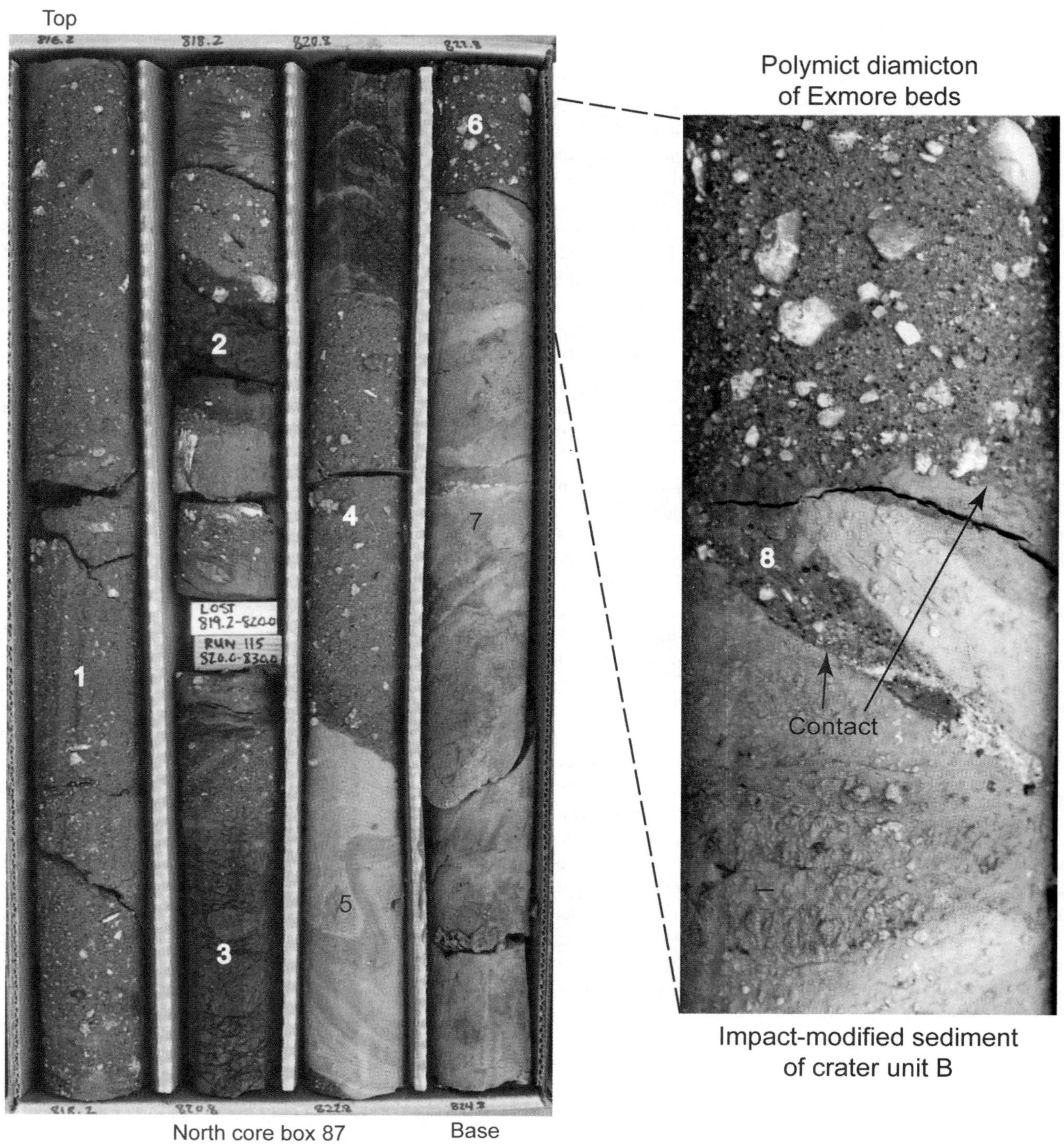

Figure 7. Photograph and close-up image showing the contact between the Exmore beds (polymict diamicton of allogenic sediment-clast breccia) and underlying impact-modified sediments as exposed in the USGS North core. Numbered features of the Exmore beds: 1, 4, 6—muddy quartz-glauconite sand matrix in which local variations are not systematic; 2—clast of dark-greenish-gray, glauconite-rich clay, probably lower Tertiary marine; 3—clast of reddish-brown silty clay, probably Lower Cretaceous nonmarine; 5—probable rip-up clast from below of sandy clay with convolute bedding and soft-sediment slump folds, probably Lower Cretaceous nonmarine. Feature 7 is impact-modified sediment (Lower Cretaceous nonmarine clay) below the Exmore beds. A close-up view of the sharp contact on the right shows downward penetration (8) of the Exmore muddy quartz-glauconite sand matrix into the underlying impact-modified sediment; elongate grains indicate flow parallel to the sides of the feature and are consistent with either injection or infiltration. North core box 87 from 248.8 to 251.4 m (816.2–824.8 ft) depth. Nominal core diameter is 6.1 cm, and internal length of core box is 61 cm (depths handwritten on box are in feet).

Figure 8. Photograph of a section of the Exmore beds (lower diamicton member) in the USGS North core, showing features that characterize this polymict diamicton of sediment-clast breccia. 1—muddy quartz-glauconite sand matrix (dark greenish gray, nonbedded); 2—clast consisting of an altered, vesicular glass bomb of reworked impact-melt rock (dusky yellow, irregular shape); 3—clast of white indurated sandstone, containing glauconite and shell hash, probably lower Tertiary marine; 4—clast of black glauconitic sand, probably lower Tertiary marine; 5, 7, 10—clay clasts (dark greenish-gray clay); 6, 8—muddy quartz-glauconite sand matrix (dark greenish gray, nonbedded) containing pebbles of quartz and feldspar; 9—clast of leucocratic granite containing shocked quartz and calcite veinlets. Includes (left to right) North core box 80 from 227.0 to 229.5 m (744.9–752.8 ft) depth, box 81 from 229.5 to 231.6 m (752.8–760.0 ft) depth, and box 82 from 231.6 to 234.1 m (760.0–767.9 ft) depth. Nominal core diameter is 6.1 cm, and internal length of core boxes is 61 cm (depths handwritten on boxes are in feet).

undeformed, autochthonous Cretaceous sediments. For example, in the Watkins School core (Figs. 1 and 3), the Exmore diamicton unconformably overlies undeformed Cretaceous nonmarine sediments along a sharp erosional contact. Lower Tertiary marine beds, which are present regionally above the Cretaceous section, and some unknown thickness of the Cretaceous sediments, are missing along this unconformity.

In the annular trough, the Exmore beds overlie crater unit B (e.g., Langley and North cores). These two units are superficially similar; both consist of blocks (or clasts) in a finer-grained matrix. However, qualitative differences exist between the block fractions and the matrix fractions in these two units. Large blocks in crater unit B consist only of parautochthonous pieces of disturbed Cretaceous nonmarine sands, silts, and clays. The mixed-sediment intercalations between these blocks consist of fluidized Cretaceous nonmarine sands, quartz-glauconite sands derived from Upper Cretaceous and Tertiary marine sediments, and small blocks of locally derived Cretaceous nonmarine sediments. Clasts of glauconitic Upper Cretaceous and lower Tertiary marine sediments are absent, and pieces of shocked crystalline ejecta are extremely sparse. In contrast, shocked crystalline ejecta, Cretaceous and Tertiary sediments that contain marine microfossils and glauconite, and Cretaceous nonmarine sediments are present as large and small clasts in the Exmore diamicton member

(Frederiksen et al., 2005; Gohn et al., 2005). The matrix between clasts in the Exmore is typically microfossiliferous and shelly, in contrast to sparingly calcareous mixed-sediment intercalations in crater unit B.

In the Bayside core, sediment of the Exmore diamicton closely resembles thick intercalations of mixed sediment observed in crater unit B. The indicated location of the base of the Exmore in this core (Fig. 3) is provisional, owing to this similarity and the inherent limitations of drill core.

In the central crater, the Exmore beds are interpreted to overlie megablocks of allochthonous Cretaceous nonmarine sands and clays on the basis of geophysical logs and cuttings samples from the Cape Charles test hole and the presence of sediment megablocks at this stratigraphic position in the Eyreville core (Gohn et al., 2006a, 2006b, 2006c) (Figs. 1 and 3). The Cape Charles test hole produced spot cores from the Exmore beds (Sanford et al., 2004; Gohn et al., 2007) but not from the allochthonous sediment megablock section; these units at Cape Charles are combined in Figure 3 due to the limitations of drill cuttings. Lithologic contrasts between the Exmore and the underlying transported sediment megablocks probably are very similar to the contrasts between the Exmore and crater unit B in the annular trough.

Clasts are consistently small, typically granules and small pebbles, and sparse in the uppermost 1–2 m of the lower diamicton member. This uppermost muddy sand and fine gravel grades upward into fine to very fine sand and silt of the upper stratified member, typically across a few centimeters.

The *upper stratified member* of the Exmore beds consists of laminated clayey silts and fine sands. In the North core (Figs. 1 and 3), it includes about one meter of cross-laminated fine to medium sands overlain by laminated fine sands and silts (Fig. 9). In the Watkins School core, the stratified member exhibits soft-sediment deformation structures characteristic of dewatering.

The gradational upper contact of the stratified member with clayey marine sediments of the postimpact Chickahominy Formation is difficult to delineate and has been placed differently by different investigators (e.g., Poag et al., 2004; Gohn et al., 2005; Horton et al., 2005d; Poag and Norris, 2005). The principal criteria for identifying this contact are the dominance of thin laminae in the Exmore upper stratified member, and the dominance of burrowing in the Chickahominy. However, in some cored sections across the contact (e.g., the Watkins School core), one generation of subvertical burrows crosses the contact, resulting in the upward and downward transport of sediment that tends to blur the contact. Postimpact slumps within the impact structure and/or erosion outside the structure continued to contribute minor amounts of reworked Exmore sediments, as indicated by reworked microfossils, during deposition of the Chickahominy Formation and younger Tertiary units (Edwards and Powars, 2003; Edwards et al., 2004).

In summary, the lower diamicton member of the Exmore beds consists of mixed-age sediment clasts and minor crystalline-rock clasts in a muddy quartz-glauconite sand and granule matrix. It contains minor amounts of reworked ejecta-derived material, including shocked quartz grains and rock fragments that contain shocked quartz (Koeberl et al., 1996; Horton and Izett, 2005). This sequence of unstratified polymict sedimentary breccias ranges in thickness from 8 to 200 m, and it overlies the other impactites and preimpact deposits. Sedimentary breccia of the diamicton member grades upward into a relatively thin (<2 m) unit of fine sand and silt discussed here as the upper stratified member of the Exmore beds, which grades conformably into postimpact marine sediments.

IMPLICATIONS FOR IMPACT-RELATED PROCESSES

Table 1 summarizes the three classes of impactites discussed in this paper as a basis for interpreting the processes of shock metamorphism, suevite formation and central-uplift rise and collapse, impact modification of target sediments in the annular trough, gravitational collapse and lateral spreading, ocean-resurge erosion and sedimentation, and the transition from synimpact to postimpact sedimentation in a large marine impact structure.

Shock Metamorphism

The impact-modified sediments, the suevite and associated gneiss megablocks, and the Exmore beds have different populations of shock-metamorphosed grains and ejecta-derived lithic clasts that constrain interpretations of their origin and emplacement. The impact-modified sediments and underlying bedrock of the annular trough show no evidence of shock metamorphism, except in the one clast in the matrix between blocks near the top of crater unit B (Horton and Izett, 2005). However, gneiss blocks in the suevite and associated gneiss megablocks within the central crater contain abundant shocked quartz and feldspar (Horton et al., 2005b) and minor amounts of a rare, shock-induced polymorph of anatase and rutile (TiO_2 II; Jackson et al., 2005, 2006). In samples of suevite and associated megablocks from the Cape Charles test hole, the PDFs in shocked quartz typically are decorated with tiny fluid inclusions, probably due to annealing under hydrothermal conditions. Survival of the shock-induced polymorph of TiO_2 in the crystalline-clast suevite and associated blocks suggests that the postshock temperatures did not exceed ~500–550 °C (Horton et al., 2006c, 2006d).

Shocked quartz has been found as grains and in crystalline-rock clasts from the Exmore beds at eight localities (Koeberl et al., 1996; Poag et al., 2004; Horton et al., 2005a; Horton and Izett, 2005); observed variations in shock levels suggest mixing of clastic components from different parts of the evolving crater (Reimold et al., 2002; Poag et al., 2004). The ratio of shocked to unshocked quartz grains in the muddy sand matrix of the Exmore is extremely low, no more than one in several thousand (Horton and Izett, 2005), indicating that the sparse shocked grains are diluted by a vastly larger volume of unshocked matrix material. Shocked feldspar and/or quartz are found in most of the large pebbles, cobbles, and boulders of crystalline rock in the Exmore beds, but not in the sediment clasts that have been studied.

Figure 9. Sedimentary record of the synimpact to postimpact transition in the USGS North core. The section in this core is unusually complete in that over a meter of hummocky (?) cross-laminated medium to fine sand and laminated silt (Es, upper stratified member of the Exmore beds) is present between the allogenic sediment-clast breccia (Ed, lower diamicton member of the Exmore beds) and the marine silty clay of the postimpact Chickahominy Formation (C). Silt laminae (≤1 mm thick) are abundant up to a depth of ~225.08 m (738.5 ft; arrow tip), and sparse shell fragments (<1.5 cm across; not visible in photo) appear above this depth. The exact placement of contacts will require additional study. The upper stratified member is interpreted as a waning stage of resurge deposition and settling of water-suspended sediment. There has been no subsequent erosion of crater-fill deposits. Includes (left to right) North core box 78 from 222.3 to 224.7 m (729.4–737.2 ft) depth, and box 79 from 224.7 to 227.0 m (737.2–744.9 ft) depth. Nominal core diameter is 6.1 cm, and internal length of core boxes is 61 cm (depths handwritten on boxes are in feet).

TABLE 1. SUMMARY OF STUDIED IMPACTITES IN THE CHESAPEAKE BAY IMPACT STRUCTURE

	Crystalline-clast suevite and associated megablocks	Impact-modified sediments	Polymict sediment-clast diamicton and stratified sediments
Informal name(s)	Suevite and associated megablocks at Cape Charles	Crater units A and B	Exmore beds, lower diamicton member, and upper stratified member
Location	Only in central crater	Throughout annular trough	Covers entire structure and nearby areas
Structural position	Allochthonous; underlies polymict sediment-clast diamicton and sediment megablocks	Parautochthonous to autochthonous; underlies polymict sediment-clast diamicton	Allogenic, overlies all other impactites
Major components	Crystalline-rock clasts abundant, impact-melt rock clasts less abundant	Mostly nonmarine (fluvial and deltaic) Cretaceous sediments, poorly consolidated	Muddy quartz-glauconite sand matrix; mixed-age sediment clasts; minor crystalline-rock clasts include reworked shocked lithic ejecta
Structures	Basement-derived gneissic megablocks >10 m	Block-faulted collapse structures, rotated beds, liquefied and fluidized sands, fractured clays; deformation increases upward	Diamicton unstratified and unsorted except for normal coarse-tail grading of large clasts
Shocked quartz	Common	Rare in upper part of crater unit B	Diluted but widespread
Microfossils	No data	Mixed-sediment intercalations contain exotic particles, including mixed-age and impact-damaged microfossils	Mixed ages; impact-damaged microfossils present
Other characteristics	Hydrothermally altered over central uplift	Deepest glauconite in mixed-sediment intercalations defines base of crater unit B and top of crater unit A	Lower diamicton member fines upward into fine sand and silt of upper stratified member
Processes	(1) Impact-melt formation and dispersal; (2) Central-uplift collapse and fragmentation; (3) Mixing of suevite and basement-derived megablocks at the top of the central uplift	(1) Rarefaction-induced fragmentation and sediment injections; (2) Vibration-induced sand liquefaction and fluidization; (3) Gravitational collapse and lateral spreading	(1) Ejection and fallback of shocked target materials into water column; (2) Dispersal and mixing of sediments in the water column; (3) Ocean-resurge erosion and sedimentation with depositional cycles that may represent stages such as anti-resurge and tsunami backwash; (4) Settling of water-suspended sediment

Suevite Formation and Central-Uplift Rise and Collapse

Suevites are defined as breccias that contain clasts of impact-melt rock (Stöffler and Grieve, 1994, 2007). Kieffer and Simonds (1980) and Kieffer (2005) argue that suevites form where the targets contained water or other volatiles; however, water is available in most terrestrial environments, and the occurrence of suevite is not limited to marine impacts. In the CBIS, the crystalline-clast suevite contains melt-rock particles and blocks of shocked cataclastic gneiss and other types of crystalline rocks. This suevite also shows evidence of hydrothermal alteration (Horton et al., 2004, 2005b, 2006c, 2006d). The suevite is found only in the central crater, where it is overlain by transported megablocks of target sediments that are overlain by resurge sediments of the Exmore beds, as interpreted from the Cape Charles test hole (Sanford et al., 2004; Gohn et al., 2007). Based on these superpositional relations, we conclude that the suevite accumulated in the central crater before the arrival of slumped sediments from collapse of the transient crater margin, and before ocean-resurge sedimentation.

Most lithic clasts and grains in the suevite at Cape Charles are similar in composition to the associated gneiss megablocks, although other crystalline-rock clasts (e.g., slate) also are present. This suite of lithic clasts likely represents basement rocks that were transported to their present position by rebound of the central uplift. However, clasts and grains derived from the sedimentary units higher in the target sequence are rare, if present at all, in the suevite.

We initially hypothesized a fallback origin for the suevite (Horton et al., 2006a, 2006b). However, a fallback origin, in the sense of material that was ejected above and fell back into the crater, is here considered unlikely for the suevite at Cape Charles, because the clast population includes an incomplete suite of the known target materials. In addition, there is no specific evidence that impact-melt particles and crystalline-rock debris in the Cape Charles suevite ever left the central crater. Therefore, the Cape Charles suevite is analogous to the "crater suevite," but unlike the "fallout suevite," in the Ries crater (Germany) (Engelhardt and Graup, 1984; Engelhardt, 1997).

The drill core from the deepest 34 m of the Cape Charles test hole consists of variably brecciated megablocks of shocked gneiss (Horton et al., 2005b). Seismic imaging indicates that relatively intact basement rocks of the central uplift are at significantly greater depths (Catchings et al., 2005a, 2006). The largest gneiss megablocks are difficult to reconcile with an ejection and fallback origin. They may have been detached from the basement and entrained with suevitic material (but not ejected beyond the crater) by the excavation flow, and/or detached from the basement and mixed with the suevite during the central-uplift rebound and collapse (Horton et al., 2005b).

Suevite recently was reported from a second core hole in the central crater at Eyreville Farm north of Cape Charles (Gohn et al., 2006a, 2006b, 2006c). Future studies and comparisons of these suevites may provide insights into their origins and associated cratering processes.

Early Impact Modification of Target Sediments in the Annular Trough

Preimpact Cretaceous sediments in the annular trough outside the transient crater were modified initially during the contact-compression and excavation stages of the impact event. These early modifications probably resulted from a combination of processes, including rarefaction-induced fragmentation and clastic-sediment injections, and sand liquefaction and fluidization in response to acoustic-wave vibrations. There was a differential response to these processes in the sand layers versus the more competent clay layers at shallow depths within the parautochthonous sediments (Gohn et al., 2002, 2005). Subsequent modification of these sediments by late-stage gravitational collapse and lateral spreading is discussed in a later section.

Rarefaction-Induced Fragmentation and Injection of Exotic Sediments

Preimpact Upper Cretaceous and Tertiary marine beds are missing from their normal stratigraphic position above impact-modified Cretaceous nonmarine sediments in core holes throughout the annular trough, yet disaggregated particles of glauconite and microfossils of various target ages derived from these beds are found as particles in the mixed-sediment intercalations between blocks of impact-modified sediment in crater unit B. Hence, Gohn et al. (2005) proposed that the near-surface Cretaceous and Tertiary marine sediments, and possibly sediments in the upper part of the Cretaceous nonmarine section, were disaggregated by the downward passage of a high-energy rarefaction wave through the low-tensile-strength sediments and injected downward into underpressured, dilational zones of weakness between more competent layers and blocks. This rarefaction was produced by the downward reflection at the seafloor sediment-water interface of the remnant normal acoustic wave derived from the diminished shock wave (Melosh, 1989; French, 1998).

In crater unit B, the intercalations of glauconite-bearing sediment that contain mixed-age microfossils within and between the blocks of nonmarine sand and clay are consistent with two modes of emplacement. These modes of emplacement are (1) gravitational infiltration of wet sediment into cracks and voids, and (2) injection of fluidized sediment from a domain of higher pressure to one of lower pressure. The processes of infiltration and injection are not mutually exclusive and could have operated simultaneously with liquefaction and fluidization, block faulting, and slumping to enhance the overall mixing of material.

Distinction between the products of infiltration and injection can be difficult or impossible, in part owing to the inherent limitations of drill core. Observed evidence for forceful injection of clastic dikes includes sharp discordant contacts, sidewall drag folds, apophyses, and plucked sidewall material as shown in Figure 6. Other features indicative of clastic injections in crater unit B include center-to-edge size sorting of clasts within sand dikes, and centimeter-scale dikes and sills of exotic matrix that penetrate sediment clasts. The observed downward transport of exotic material for more than a hundred meters also favors injection over infiltration where the exotic material is not significantly denser than the host material.

Infiltration of sediment is a likely mechanism where the vertical transport distance is small, as in the upper part of crater unit B. This process would explain why the only known occurrence of shocked quartz in crater unit B at Langley is in a single rock fragment in muddy glauconitic sand that was emplaced between sediment blocks within 6.4 m of the top of the unit (Horton and Izett, 2005).

Sturkell and Ormö (1997) used the concept of rarefaction-induced fragmentation and injection to explain clastic dikes and sills in strata adjacent to the Ordovician Lockne marine-target crater in Sweden. Figure 10 illustrates an analogous conceptual model for impact modification of sediments in the CBIS annular trough. This model envisions an upward increase in rarefaction-induced extensional fracturing, faulting, and fragmentation of sediments along weak zones, and an increase in dissociation of sediment particles into the water column, as the overburden confining pressure decreased upward; the resulting underpressured zones of dilation along weak layers and fractures in parautochthonous sediments provided the contrast in confining pressure necessary for injection of muddy sand and water slurries containing exotic particles from above.

To summarize the conceptual model: (1) The rarefaction wave caused the entire sediment column to expand upward into the water column. (2) The lower sediments remained coherent and broke into competent clay blocks with liquefied sand interbeds. (3) The upper sediments were completely disaggregated and fluidized. (4) The lifting and pulling apart of competent beds produced underpressured zones of dilation into which some of the fluidized sediment was injected. Extensional fracturing and gravitational filling of open fractures by the fluidized mix was most likely in the upper part of crater unit B.

This conceptual model is consistent with drill-core data and provides a mechanism for the shallow excavation of weak sediments at relatively low (nonshock) strain rates. Such strain rates are consistent with the lack of mineralogical evidence for shock metamorphism in nonmarine Cretaceous parautochthonous sediments and megablocks of crater units A and B in the annular

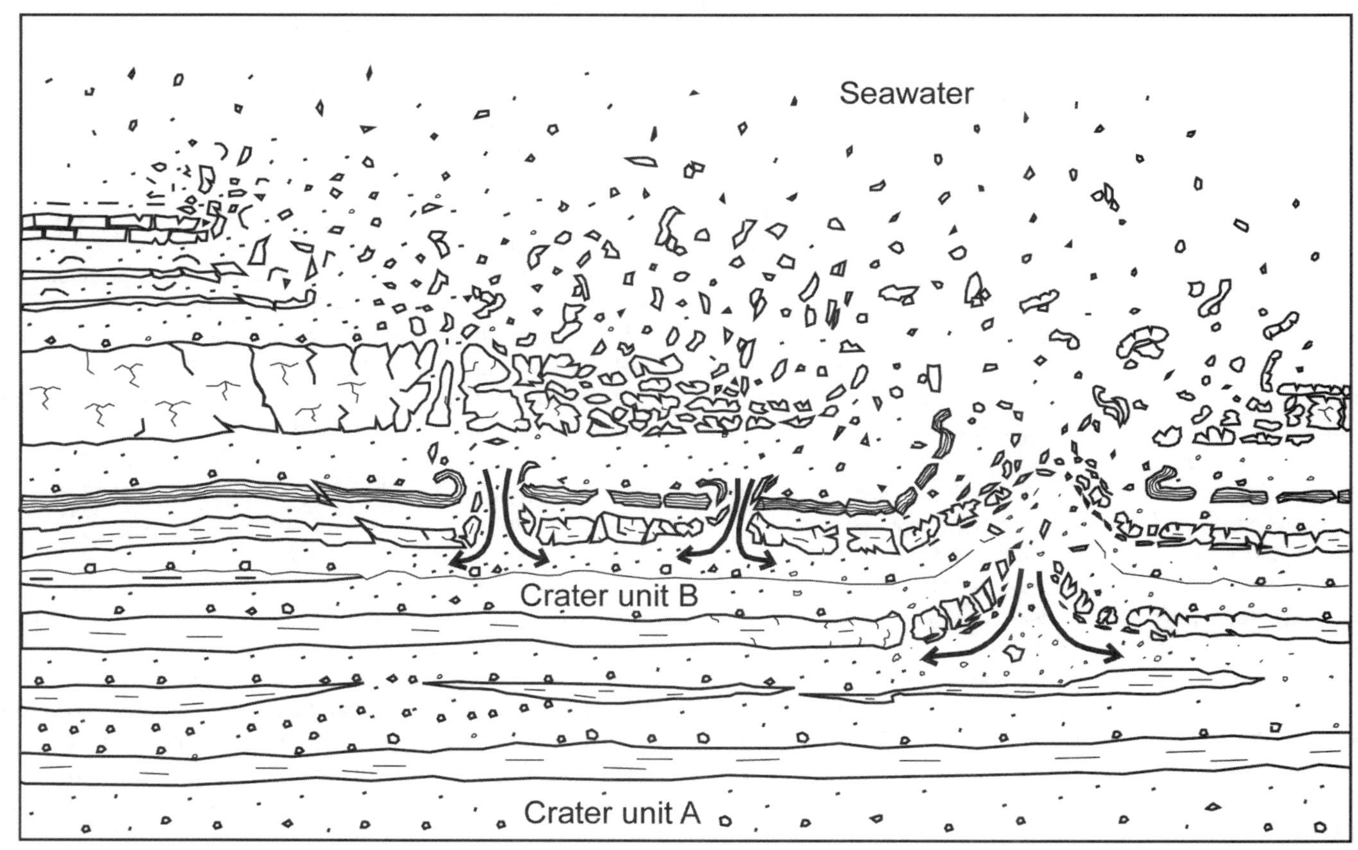

Figure 10. Schematic cross section illustrating conceptual model of rarefaction-induced fragmentation and clastic injections in sediments in the annular trough outside the transient crater. Stippled—sand; small polygons—gravel; dashed—clay; closely spaced line pattern—finely laminated organic-rich clay; irregular T—clay from Cretaceous paleosol; brick pattern—minor calcite-cemented sandstone or limestone. Based on analogy with the marine Lockne crater, Sweden, as interpreted by Sturkell and Ormö (1997).

trough. Theoretical models indicate that peak shock-wave pressures decrease exponentially with distance from the point of impact (Melosh, 1989), and that at a distance near the eventual crater rim, the shock wave diminishes into a normal seismic wave propagating at the velocity of sound in the target material (Kieffer and Simonds, 1980; French, 1998).

The lower depth limit of sediment dissociation and injection (i.e., rarefaction effects) is a yield horizon or zone controlled in part by the confining pressure (Gohn et al., 2005; Horton et al., 2005a). This horizon appears to rise stratigraphically away from the crater center. It defines the boundary of the "strength crater" (Croft, 1981), which marks the transition from thoroughly impact-dissociated materials to materials that were only ruptured or elastically deformed by the impact. In the outer part of the annular trough, this horizon occurs within the lower part of the preimpact Cretaceous sediments rather than at the top of crystalline basement (Horton et al., 2005a). The outer margin of the "strength crater" in this sense is not necessarily the outer limit of faulting produced by late-stage gravitational collapse (discussed below).

Liquefaction and Fluidization of Sands

In the case of impacts into marine targets consisting of wet noncohesive sediments, it is important to distinguish the liquefaction and fluidization of granular sediments that are saturated with pore water from the concept of acoustic fluidization of solid bedrock that is widely used to explain the rheological behavior of materials during impacts (Melosh, 1979, 1989; Melosh and Ivanov, 1999; Collins and Wünnemann, 2005).

Liquefaction is the transformation of a saturated granular material from a solid to a liquefied state as a consequence of increased pore-water pressure; it can be induced by gravity loading, seismic shaking, nonseismic vibrations, or wave-induced shear stresses (Obermeier, 1996; Obermeier et al., 2001; Davies et al., 2004). It occurs in sediments that lack cohesion, most commonly in sands of uniform grain size, where shear strain causes an increase in pore-water pressure and a temporary transfer of grain support to the pore fluid (Obermeier et al., 2001; Davies et al., 2004). Sediments can persist in a liquefied state only during the active disturbance.

Fluidization of sediment is the result of upward fluid flow, and sediment can remain fluidized as long as external fluid is pumped into it (Allen, 1984; Davies et al., 2004). It can be caused by processes such as liquefaction or dewatering due to compaction of underlying layers. Soft-sediment plastic deformation features of muds and cohesionless sands such as convolute bedding, ball-and-pillow structure, and load casts are typical of liquefaction, whereas clastic dikes and sills are most typical of fluidization (Obermeier et al., 2001; Davies et al., 2004).

Sand liquefaction structures observed in impact-modified sediments of the annular trough are similar to those formed by liquefaction of water-saturated unconsolidated sands during earthquakes (e.g., Obermeier et al., 2001; Olson et al., 2003), as noted by Horton et al. (2005a). This observation indicates that vibration-induced liquefaction of water-saturated sand layers occurred in near-surface sediments outside the transient crater.

Sand liquefaction and fluidization occurred at deeper levels and was more pervasive 25 km from the crater center in the Bayside core than it was 36 km from the center in the Langley core. No evidence of sand liquefaction or fluidization was observed in the preimpact section just beyond the outer margin of the annular trough in the Watkins School core (Edwards et al., 2004; Horton et al., 2005a). This progressive downward and outward decrease in fluidization is consistent with a decay of impact energy with distance from the transient crater. Near the outer crater margin, the overburden confining pressure may have been sufficient to suppress the viscous flow of sand except near the surface (Gohn et al., 2002). The liquefaction of sands facilitated lateral spreading of impact-modified sediments in the annular trough as discussed below.

Late-Stage Gravitational Collapse and Lateral Spreading

Late-stage gravitational collapse is the driving force that produces the final crater diameter and morphology (Melosh, 1989; Melosh and Ivanov, 1999). Vertical and lateral variations in the target strength also influence the final morphology and the types and locations of collapse structures (Turtle et al., 2005). The structures produced by late-stage gravitational collapse are superimposed on those formed during the earlier stages of crater formation, potentially affecting the character of impactites as well as their distribution.

The impactite sections described in this report are associated with three major collapse features: (1) the collapse of the transient crater wall that was ~8 km in depth and 28 km in diameter (Collins and Wünnemann, 2005; Crawford and Barnouin-Jha, 2004), (2) the collapse of the central-uplift rebound, and (3) the collapse and inward lateral movement of several hundred meters of unconsolidated sediments within the annular trough surrounding the excavation cavity out to ~43 km from the crater center.

Thick impactite sections produced in the first two collapse features constitute the rock and sediment slump blocks and debris that core holes and geophysical data suggest fills the deeper parts of the 40 km wide final central crater (Horton et al., 2004, 2005c; Sanford et al., 2004; Powars et al., 2005, 2006a, 2006b; Shah et al., 2005; Gohn et al., 2006c, 2007). The third collapse zone caused sediment slumping and movement across the width of the annular trough.

Evidence of gravitational collapse, block faulting, and inward slumping of sediments is conspicuous on seismic profile images across segments of the annular trough (Poag et al., 1999, 2004; Powars and Bruce, 1999). One seismically defined megaslump block is ~1 km high and 2 km long, and it appears to have detached, dropped, and rotated downward at the edge of the central crater (Powars and Bruce, 1999). Additional examples of collapse structures in the annular trough include concentric collapse rings interpreted from seismic data (Fig. 11) (Powars et al., 2003; Catchings et al., 2005b) and a ring graben located at the outer periphery of the annular trough (Poag et al., 2004). The outer edge of the outer ring graben represents the outer margin of the crater and has highly variable width, relief, and morphology that is interpreted to be a product of gravity-driven slumping, ocean-resurge erosion and sedimentation, and variability in the competency of the sediments (Powars and Bruce, 1999).

The liquefaction features and extensional structures observed in drill cores (Figs. 5 and 6), land-based seismic images (Catchings et al., 2005b) (Fig. 11), and marine seismic images (Poag, 1996; Powars and Bruce, 1999; Poag et al., 2004) resemble those of lateral spreads triggered by rapid ground motion during earthquakes, where mass movement on gentle to horizontal slopes is caused by liquefaction of saturated, cohesionless sediments, and the dominant movement is lateral extension accompanied by shear or tensile fractures (Varnes, 1978; Bartlett and Youd, 1995; Rauch and Martin, 2000). Lateral spreading is also consistent with the lack of coherent seismic reflections from sediments in the inner 8–10 km of the annular trough (Powars et al., 2004). The inward deepening of abundant liquefaction features suggests an inward increase in the thickness of mobilized material, and the upward increase in deformation is consistent with an upward increase in lateral displacement. Modeling by Collins and Wünnemann (2005) reproduced the CBIS "inverted sombrero" morphology by gravitational collapse and inward spreading of weak sediments across the annular trough and slumping of the sediments into the collapsed transient crater. Inward translation of weak sediments across the annular trough into the central crater provides the necessary accommodation space for the outer ring graben and rings of concentric collapse structures as well as reduction of sediment volume required for the annular trough to be deeper than the region surrounding the impact structure. Such a mechanism is also required to transport the large sediment blocks found above suevites in the previously excavated central crater (Gohn et al., 2006c, 2007).

The collective body of evidence from drill cores, geophysics, and modeling implies that significant accommodation space was created by lateral movement of weak sediments in the annular trough toward the collapsing transient crater. Numerical simulation of the CBIS suggests that most of the unconsolidated preimpact target sediments in the annular trough within about 10 km of the central crater are displaced laterally for significant distances (Collins and Wünnemann, 2005) and are consistent with marine seismic data that characterize the inner part of the annular trough (Powars and Bruce, 1999; Poag et al., 2004). The numerical models also indicate that the transition from crater excavation to crater wall collapse can overlap in time (Melosh, 1989; Crawford and Barnouin-Jha, 2004; Collins and Wünnemann, 2005; Turtle et al., 2005). Therefore, discussions of these processes and stages as separate entities can be misleading.

Ocean-Resurge Erosion and Sedimentation

The polymict diamicton of the Exmore beds is interpreted as the product of late-stage catastrophic resurge of seawater into the collapsing transient crater (Powars and Bruce, 1999; Poag et al., 2004; Gohn et al., 2005; Horton et al., 2005a). In the central crater, the Exmore beds and transported megablocks of Cretaceous sands and clays overlie the suevite (Figs. 2 and 3), requiring that the suevite was formed in the central crater before the ocean resurge (Horton et al., 2004, 2005b, 2005c). The resurge deposits extend well beyond the location of the original transient crater across the annular trough where they overlie impact-modified parautochthonous sediments.

Just outside the impact structure, thinner resurge deposits erosionally overlie undeformed Cretaceous sediments. Erosive scouring at the base of the outgoing megatsunami and ingoing resurge sediment-gravity flows likely contributed to the observed absence of Late Cretaceous and Tertiary glauconitic marine beds at the top of the preimpact sequence in the annular trough and adjacent areas outside the crater, and to the observed absence of a proximal ejecta field in these areas. Particles derived from these uppermost target marine beds were mixed into the resurge flow and incorporated as grains and clasts into the Exmore diamicton.

Shocked quartz grains and rock fragments that contain shocked quartz were ejected into the water column outside the transient crater, probably as fallback particles from the ejecta curtain, prior to being incorporated into resurge sediments of the Exmore diamicton; sparse shocked material is mixed into and diluted by much larger volumes of unshocked clastic sediment throughout the Exmore section (Horton and Izett, 2005; Horton et al., 2005a).

The principal sedimentary characteristics of the Exmore diamicton (i.e., poor sorting, clasts suspended in matrix, absence of stratification) suggest deposition by high-concentration sediment-gravity flows (Mulder and Alexander, 2001; Gani, 2004). Local concentrations of pebbles and cobbles within the diamicton suggest temporal and spatial variations of flow conditions (e.g., sediment-support mechanism) during sedimentation. These uncommonly large sediment-gravity flows probably were generated by sediment bulking (scouring and entrainment) in the seawater resurge and by the downslope evolution of slump blocks into submarine debris flows.

Despite the apparent absence of internal depositional and erosional contacts, the diamicton probably consists of multiple depositional units that reflect stages of a single ocean-resurge event. High-resolution seismic images suggest the presence of four resurge depositional units in the Langley vicinity (Catchings et al., 2005b, Fig. I9 therein). The coarse-tail grading of large clasts and variations in microfossil content in the diamicton at the Langley core hole suggest the presence of two main resurge cycles at this specific location on the landward side of the annular trough (Gohn et al., 2005). Both of these cycles contain early Tertiary calcareous nannofossils. However, Late Cretaceous calcareous nannofossils, thought to be from more seaward sources to the east and north, are found only in the upper cycle (Frederiksen et al., 2005). The cycles may represent a proximal resurge flow in the annular trough outside the transient crater, followed by the arrival of resurge flows containing particles from

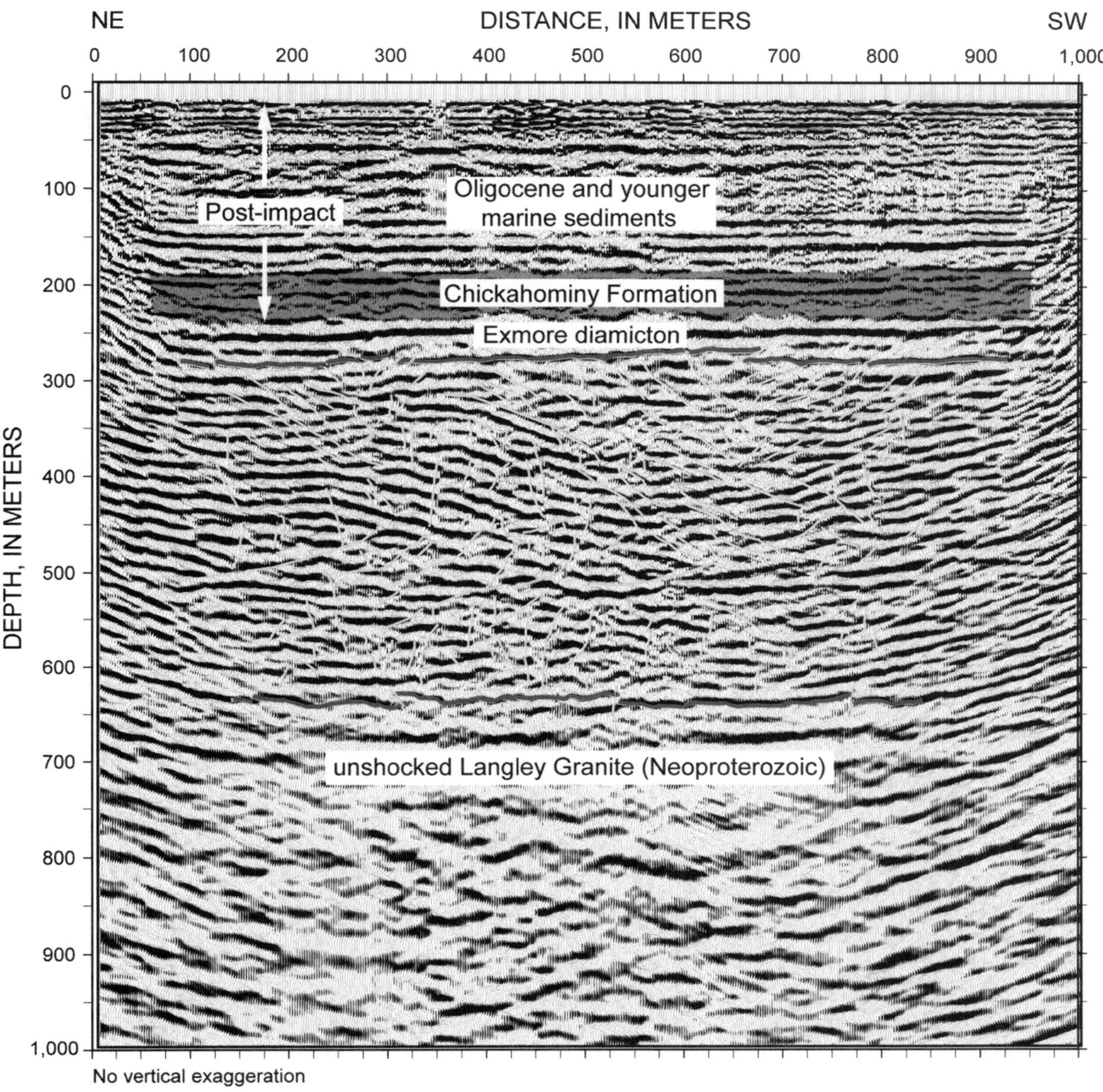

Figure 11. Collapse structure within the annular trough as shown on a high-resolution seismic-reflection profile across the USGS-NASA Langley corehole site. Displacements within the impact-modified sediments (between the red lines) are distributed through numerous small-displacement normal faults (short yellow lines) and bedding-parallel thrust faults, and through fluidized sand layers at different levels within the impact-modified sediments above the crystalline basement (granite) and below the Exmore beds, where their abundance increases upward (Catchings et al., 2005b; Gohn et al., 2005). The 550-m-wide collapse structure in this example is not bounded by major normal faults, and the Langley core shows no evidence for a major décollement at the top of basement at this location (Horton et al., 2005e). Upward dewatering through the fault network may have provided accommodation space in the lower part of the structure that increased upward as a result of lateral spreading. Migrated depth image and interpretation from Catchings et al. (2005b, Figs. I9 and I10 therein).

more distant parts of the structure. Alternatively, the two cycles may represent an inward-directed resurge flow (lower cycle) followed by an outward-directed anti-resurge flow (upper cycle) (e.g., Ormö et al., 2007). Similar resurge cycles in the Lockne marine-target crater and implications for resurge dynamics are discussed in a later section.

Synimpact to Postimpact Transition

The uppermost part of the Exmore diamicton member fines upward into fine sand and clayey silt of the upper stratified member of the Exmore beds. These fine-grained, stratified sediments represent a gradual decrease in the energy of resurge currents and impact-related wave oscillations followed by the settling of impact-generated, suspended sediments from the water column (Poag, 2002; Poag and Norris, 2005; Gohn et al., 2005). This thin stratified layer contains an interpreted biologic "dead zone" and a possible microspherule layer, as described by Poag (2002), Poag et al. (2004), and Poag and Norris (2005). Settling of voluminous suspended particles through the seawater column may have continued for weeks, months, or longer after the resurge flow subsided. During that time, tides, storm waves, currents, and slumps could have periodically resuspended sediments of the upper stratified member. Although the water depth probably was not uniform, the benthic foraminiferal faunas in the Chickahominy Formation suggest postimpact seawater paleodepths of ~300 m (Poag et al., 2004; Poag and Norris, 2005). This value is about twice as deep as the normal storm-wave base, suggesting that any postimpact resuspension of sediments resulted primarily from slumping at the outer margin of the annular trough.

The upper Eocene Chickahominy Formation conformably overlies the upper stratified member of the Exmore (Powars, 2000; Poag and Norris, 2005). The Chickahominy consists of massive to bioturbated to thin-bedded, olive-gray, very compact, clayey silt to silty clay, which contains variable amounts of glauconite, shells, and iron sulfides (Powars and Bruce, 1999; Powars, 2000).

On the New Jersey continental margin, the late Eocene, microspherule-bearing subunit I of McHugh et al. (1998) from ODP Site 904 consists of laminated fine sediments with soft-sediment deformation features and resembles the upper stratified member of the Exmore beds. These authors interpreted subunit I to represent rapid sedimentation of a large volume of suspended sediment during minor current activity.

DISCUSSION

Studies of the Chesapeake Bay impact structure reveal that the central and outer parts of the structure contain different impactite sections. Suevite at the top of the central uplift formed through the accumulation of melt clasts in the transient crater during and after the rise and collapse of the central uplift and before deposition of the overlying slumped sediment blocks and resurge deposits. The impact-modified sediments in the annular trough formed through a combination of processes, including vibration-induced sand liquefaction, rarefaction-induced fragmentation and clastic injections, and gravitational collapse and lateral spreading. An allogenic sediment-clast diamicton blankets the other impactites and preimpact deposits. It formed by the late-stage resurge of seawater carrying eroded continental-shelf sediments and proximal ejecta into the collapsing crater.

The character and distribution of impactites in the CBIS imply that the inverted sombrero shape was influenced by the contrast in strength between poorly consolidated continental-shelf sediments and underlying crystalline bedrock. A recent numerical model that incorporates the observed contrast in strength was able to reproduce this morphology, and it suggests a transient crater diameter of ~28 km (Collins and Wünnemann, 2005). The transient crater expanded by rim collapse along bedrock-rooted detachments to the present (final) central-crater diameter of ~40 km. Impact modification and collapse of weak sediments extended the brim of the structure to the 85 km diameter outer margin of the annular trough (Figs. 1 and 2).

Significance for Impacts on Oceanic "Wet" Targets

The buried Chesapeake Bay impact structure serves as a pristine natural laboratory for studies of physical processes of a large hypervelocity impact into wet, unconsolidated sediments covered by water. These processes are poorly understood relative to those of impacts into consolidated, well-cemented, and crystalline targets (Ormö and Lindström, 2000; Dypvik and Jansa, 2003; French, 2004). Several features distinguish the CBIS from impact craters on Earth that formed on land in rheologically homogeneous targets. These features include (1) the inverted sombrero morphology of a deep central crater and a wide shallow brim formed in weak sediments, (2) a structural domain of collapsed, slumped, fluidized sediments that surrounds the central crater, (3) evidence for rarefaction-induced dissociation and removal of the uppermost weak sediments that surrounded the transient crater, (4) evidence for the erosive nature of the outbound megatsunami linked to the water ejecta curtain and resurge erosion by debris-laden seawater that flowed into the transient crater, (5) high-energy ocean-resurge deposits that resemble sediment-gravity flows, and (6) final settling of late synimpact suspended sediments into thin stratified layers that cap the resurge sediments, indicating little if any erosion of the impactites.

Other terrestrial impact craters that formed in shallow marine environments, including the Lockne crater in central Sweden (Sturkell, 1998), the Manson crater in Iowa, USA (Keiswetter et al., 1996), and the Mjølnir crater in the Barents Sea (Tsikalas et al., 1998), share the inverted sombrero morphology. However, inverted sombrero crater morphologies on Earth and other planetary bodies can have different causes. For example, in layered targets on Mars, some concentric craters appear to be the result of a shallow excavation flow that removed parts of a weak layer surrounding a deeper crater in the basement, whereas others appear to be the result of extensive slumping of an upper weak layer during or after the modification stage (Horton et al., 2006b).

A key relation in numerical models of "wet" impacts is whether seawater (or seawater + weak sediment) depth (H) is greater or less than the impactor diameter (d). Where d < H, the transient water crater is much wider than the underlying transient basement crater, producing a wide zone of shallow excavation; where d > H, the transient water crater is an extension of the transient basement crater, so shallow excavation of the water crater and sediments is not substantially wider or decoupled from excavation of the basement crater (Shuvalov and Artemieva, 2001; Shuvalov and Trubestkaya, 2002). At Chesapeake Bay, the preimpact water depth and sediment thickness (Powars and Bruce, 1999; Horton et al., 2005d) were substantially less than the modeled impactor diameter of ~3 km (Collins and Wünnemann, 2005), so modeling does not support a wide transient water crater or related shallow excavation. However, this does not prevent rarefaction-induced spalling and dissociation of material at the sediment-water interface outside the transient crater.

Explosion-induced liquefaction of water-saturated sediments is well documented in the KOA and OAK nuclear explosion craters on Enewetak Atoll, where the soupy flow produced craters that are far broader and shallower than expected on a dry target (Wardlaw et al., 1991; Melosh, 2001). Numerical simulations of the Chesapeake Bay impact (Collins and Wünnemann, 2005) confirm that fluid-like behavior of waterlogged soft sediments in the upper layer allows these sediments to undergo strong deformation outside the central crater with minimal disturbance of the underlying bedrock.

Impact-modified sediments of the annular trough have features in common with those of other wet-target craters that formed in shallow marine environments. For example, collapse structures dominated by small-offset faults also characterize the outer terrace terrane of the Manson crater, which is underlain by numerous small structural blocks (<200 m across) that are bounded by normal faults (Keiswetter et al., 1996). Injections of fluidized sediments similar to those of crater unit B have been described in the Lockne crater (Sturkell and Ormö, 1997).

The removal of preimpact Upper Cretaceous and Tertiary marine beds throughout the annular trough of the CBIS is attributed, at least in part, to scouring and erosion by powerful resurge currents. Evidence of resurge erosion in other marine impacts includes gullies in the Lockne crater (von Dalwigk and Ormö, 2001), incised rims in the Montagnais crater on the Scotian Shelf near Canada (Jansa et al., 1989; Dypvik and Jansa, 2003), and flat-topped central highs attributed to scouring and reworking in the Mjølnir and Montagnais craters (Dypvik et al., 1996; Dypvik and Jansa, 2003).

Resurge sediments analogous to the lower diamicton member of the Exmore beds include the Lockne Breccia in the Lockne marine impact structure (Lindström et al., 2005) and deposits of submarine debris flows, turbidites, and slumps in and near the Montagnais, Mjølnir, and other marine impact structures. Two main resurge pulses, as recognized in the annular trough of the CBIS, also occurred in the Lockne structure. At Lockne, a combination of evidence from sedimentology and numerical modeling indicates that the initial stage of inward resurge (resurge proper) resulted in the formation of a central water plume that subsequently collapsed into an outward "anti-resurge" (Ormö et al., 2007). The lack of an extensive Exmore-type resurge section at Chicxulub may have been caused by differences in the response of coherent carbonate rocks, rather than unconsolidated sands and muds, in the upper part of the "wet" shallow marine target, and/or by a difference in target water depth. Postimpact suspension deposits similar to the upper stratified member of the Exmore beds have been described in the Mjølnir crater (Dypvik and Jansa, 2003) and as the Loftarstone in the Lockne crater (Lindström et al., 2005).

Implications for the Classification of Marine Impactites

The provisional IUGS (International Union of Geological Sciences) classification of impactites (Stöffler and Grieve, 1994, 2007) recognizes three main classes: shocked rock, impact-melt rock, and impact breccia, as noted above. The impact breccias are grouped into three subclasses as suevite, lithic breccia, and monomict (cataclastic) breccia. Using this classification, the lower diamicton member of the Exmore beds is a polymict, allogenic lithic breccia, although this term is not optimal for distinguishing the characteristics of a fundamentally sedimentary deposit. "Diamicton" is a more informative descriptive term for the unit. Autochthonous to parautochthonous impact-modified sediments of the annular trough would be classified as monomict (cataclastic) breccias using the provisional IUGS classification. However, they are not cataclastic (e.g., Sibson, 1977; Snoke and Tullis, 1998) except locally along faults, and most of the little-disturbed lower part of this unit is inconsistent with the common meaning of "breccia." The term "monomict cataclastic breccia" does not clearly reflect their character or distinguish them from the Exmore beds. Also, the standard classification would require use of the term "monomict" in this report for sediment and rock units that are fundamentally authigenic but are not lithologically homogeneous. The provisional classification of impactites might benefit from refinements to address the descriptive properties of materials formed by impacts into "wet" sedimentary targets, although hierarchical classification is not a substitute for detailed description.

ACKNOWLEDGMENTS

This study was supported by the U.S. Geological Survey (USGS) National Cooperative Geologic Mapping Program through the Chesapeake Bay Impact Crater Project. USGS investigations of the Chesapeake Bay impact structure are conducted in

APPENDIX. RECENT CORE HOLES AND TEST HOLE IN THE CHESAPEAKE BAY IMPACT STRUCTURE

Core hole or test hole (USGS well number)*	Year	Distance from center of impact (km)	Location— city or county (7.5' quadrangle)	Latitude	Longitude	Ground surface elevation	Total depth
Watkins School[†] (59E 32)	2002	43	Newport News, Virginia (Newport News North)	37°04'31.92" N	76°27'30.65" W	27 ft, 8.2 m[§§]	985.3 ft, 300.3 m
North[§] (59H 3)	2001	39	Mathews Co., Virginia (Ware Neck)	37°26'41.27" N	76°23'54.80" W	15 ft, 4.6 m[§§]	1427.5 ft, 435.1 m
Langley[#] (59E 31)	2000	36	City of Hampton, Virginia (Newport News North)	37°05'44.28" N	76°23'08.96" W	7.9 ft, 2.4 m[##]	2083.8 ft, 635.1 m
Bayside #1** (60G 5)	2001	25	Mathews Co., Virginia (New Point Comfort)	37°19'30.79" N	76°17'33.10" W	4 ft, 1.2 m[§§]	1160 ft, 353.6 m
Bayside #2** (60G 6)	2001	25	Mathews Co., Virginia (New Point Comfort)	37°19'30.57" N	76°17'33.25" W	4 ft, 1.2 m[§§]	2390.2 ft, 728.5 m
Bayside #3** (60G 8)	2001	25	Mathews Co., Virginia (New Point Comfort)	37°19'30.86" N	76°17'33.66" W	4 ft, 1.2 m[§§]	164.0 ft, 50.0 m
Cape Charles test hole[††] (62G 24 and 62G 25)	2004	1	Town of Cape Charles, Virginia (Cape Charles)	37°15'33.17" N	76°01'04.83" W	7 ft, 2.1 m[§§]	2699.0 ft, 822.7 m

*USGS water-quality database online at http://waterdata.usgs.gov/va/nwis/qw.

[†]Official site name: USGS Dorothy R. Watkins Elementary School core hole.

[§]Official site name: USGS North core hole.

[#]Official site name: USGS-NASA Langley core hole.

**Official site name: USGS Bayside core hole.

[††]Official site name: USGS Sustainable Technology Park test hole #2.

[§§]Elevation (±1 ft at best) estimated from 1:24,000-scale topographic map.

[##]From Global Positioning System using North American Vertical Datum of 1988 (NAVD88).

cooperation with the Hampton Roads Planning District Commission, the Virginia Department of Environmental Quality, and the National Aeronautics and Space Administration (NASA) Langley Research Center. The Hampton Roads Planning District Commission and the USGS provided funds for drilling the USGS-NASA Langley, North, Bayside, and Watkins School core holes. The Virginia Department of Environmental Quality also provided funds for drilling the North core hole and provided extensive operational support at the other sites. The USGS provided funds for drilling the Cape Charles test hole. We thank Olivier Barnouin-Jha, Scott Harris, Jared Morrow, Jean Self-Trail, and Joseph Smoot for constructive reviews that significantly improved the paper.

REFERENCES CITED

Allen, J.R.L., 1984, Sedimentary structures—Their character and physical basis (unabridged one-volume edition): Amsterdam, Elsevier Press, 1256 p.

Bartlett, S.F., and Youd, T.L., 1995, Empirical prediction of liquefaction-induced lateral spread: Journal of Geotechnical Engineering, v. 121, no. 4, p. 316–329, doi: 10.1061/(ASCE)0733-9410(1995)121:4(316).

Catchings, R.D., Powars, D.S., Gohn, G.S., Goldman, M.R., Horton, J.W., Jr., and Hole, J.A., 2005a, Chesapeake Bay impact crater structure from crustal seismic imaging [abs.]: American Association for the Advancement of Science, Annual Meeting, Washington, D.C., February 17–21, 2005.

Catchings, R.D., Powars, D.S., Gohn, G.S., and Goldman, M.R., 2005b, High-resolution seismic reflection image of the Chesapeake Bay impact structure, NASA-Langley Research Center, Hampton, Virginia, *in* Horton, J.W., Jr., et al., eds., Studies of the Chesapeake Bay impact structure—The USGS-NASA Langley corehole, Hampton, Virginia, and related coreholes and geophysical surveys: U.S. Geological Survey Professional Paper 1688, p. I1–I21.

Catchings, R.D., Powars, D.S., Gohn, G.S., Horton, J.W., Goldman, M.R., and Daniels, D.L., 2006, Anatomy of the Chesapeake Bay impact structure from seismic, geophysical, and borehole data along the Delmarva Peninsula, Virginia, USA: Eos (Transactions, American Geophysical Union), v. 87, no. 52, Fall Meeting Supplement, Abstract S43D-06.

Collins, G.S., and Wünnemann, K., 2005, How big was the Chesapeake Bay impact? Insight from numerical modeling: Geology, v. 33, p. 925–928, doi: 10.1130/G21854.1.

Crawford, D.A., and Barnouin-Jha, O.S., 2004, Computational investigations of the Chesapeake Bay impact structure [abs.]: Houston, Texas, Lunar and Planetary Science Conference XXXV, abstract 1757.

Croft, S.K., 1981, The excavation stage of basin formation: A qualitative model, *in* Proceedings, Conference on Multi-ring Basins: Formation and Evolution, Lunar and Planetary Institute, Houston, November 10–12, 1980, Volume 12: New York, Pergamon Press, p. 207–225.

Davies, N.D., Turner, P., and Sansom, I.J., 2004, Soft-sediment deformation structures in the Late Silurian Stubtal Formation: The result of seismic triggering: Norwegian Journal of Geology, v. 85, p. 233–243.

Dypvik, H., and Jansa, L.F., 2003, Sedimentary signatures and processes during marine impacts: A review: Sedimentary Geology, v. 161, p. 309–337, doi: 10.1016/S0037-0738(03)00135-0.

Dypvik, H., Gudlaugsson, S.T., Tsikalas, F., Attrep, M., Jr., Ferrel, R.E., Krinsley, D.H., Mørk, A., Faleide, J.I., and Nagy, J., 1996, Mjølnir structure, An impact crater in the Barents Sea: Geology, v. 24, p. 779–782, doi: 10.1130/0091-7613(1996)024<0779:MLSAIC>2.3.CO;2.

Edwards, L.E., and Powars, D.S., 2003, Impact damage to dinocysts from the late Eocene Chesapeake Bay event: Palaios, v. 18, p. 275–285.

Edwards, L.E., Powars, D.S., Gohn, D.S., Horton, J.W., Jr., Litwin, R.J., and Self-Trail, J.M., 2004, Inside the crater, outside the crater: Geological Society of America Abstracts with Programs, v. 36, no. 5, p. 266.

Engelhardt, W. von, 1997, Suevite breccia of the Ries impact crater, Germany: Petrography, chemistry and shock metamorphism of crystalline rock clasts: Meteoritics and Planetary Science, v. 32, p. 545–554.

Engelhardt, W. von, and Graup, G., 1984, Suevite of the Ries crater: Geologische Rundschau, v. 73, no. 2, p. 447–481, doi: 10.1007/BF01824968.

Evans, K.R., Horton, J.W., Jr., Thompson, M.F., and Warme, J.E., 2005, The sedimentary record of meteorite impacts: An SEPM research conference: The Sedimentary Record, v. 3, no. 1, p. 4–9, http://www.sepm.org/sedrecord/sedrecord3.1.pdf (September 2007).

Frederiksen, N.O., Edwards, L.E., Self-Trail, J.M., Bybell, L.M., and Cronin, T.M., 2005, Paleontology of the impact-modified and impact-generated sediments in the USGS-NASA Langley core, Hampton, Virginia, *in* Horton, J.W., Jr., et al., eds., Studies of the Chesapeake Bay impact structure—The USGS-NASA Langley corehole, Hampton, Virginia, and related coreholes and geophysical surveys: U.S. Geological Survey Professional Paper 1688, p. D1–D37.

French, B.M., 1998, Traces of catastrophe: A handbook of shock-metamorphic effects in terrestrial meteorite impact structures: Houston, Texas, Lunar and Planetary Institute Contribution 954, 120 p.

French, B.M., 2004, The importance of being cratered: The new role of meteorite impact as a normal geological process: Meteoritics and Planetary Science, v. 39, p. 169–197.

Gani, M.R., 2004, From turbid to lucid: A straightforward approach to sediment gravity flows and their deposits: The Sedimentary Record, v. 2, no. 4, p. 4–8.

Gohn, G.S., Powars, D.S., Quick, J.E., Horton, J.W., Jr., and Catchings, R.D., 2002, Variation of impact response with depth and lithology, outer annular trough of the Chesapeake Bay impact structure, Virginia Coastal Plain [abs.]: Geological Society of America Abstracts with Programs, v. 34, no. 6, p. 465, http://gsa.confex.com/gsa/2002AM/finalprogram/abstract_45483.htm (September 2007).

Gohn, G.S., Powars, D.S., Bruce, T.S., and Self-Trail, J.M., 2005, Physical geology of the impact-modified and impact-generated sediments in the USGS-NASA Langley core, Hampton, Virginia, *in* Horton, J.W., Jr., et al., eds., Studies of the Chesapeake Bay impact structure—The USGS-NASA Langley corehole, Hampton, Virginia, and related coreholes and geophysical surveys: U.S. Geological Survey Professional Paper 1688, p. C1–C38.

Gohn, G.S., Koeberl, C., Miller, K.G., Reimold, W.U., Browning, J.V., Cockell, C.S., Dypvik, H., Edwards, L.E., Horton, J.W., Jr., McLaughlin, P.P., Ormö, J., Plescia, J.B., Powars, D.S., Sanford, W.E., Self-Trail, J.M., and Voytek, M.A., 2006a, Preliminary site report for the 2005 ICDP-USGS deep corehole in the Chesapeake Bay impact crater [abs.]: Lunar and Planetary Science Conference XXXVII, abstract 1713, 2 p., http://www.lpi.usra.edu/meetings/lpsc2006/pdf/1713.pdf (September 2007).

Gohn, G.S., Koeberl, C., Miller, K., Sr., Reimold, W.U., Cockell, C., Dypvik, H., Edwards, L.E., Horton, J.W., Jr., Powars, D.S., and Sanford, W.E., 2006b, ICDP-USGS deep drilling program in the Chesapeake Bay impact structure, Virginia, USA: Geological Society of America Abstracts with Programs, v. 38, no. 7, p. 119.

Gohn, G.S., Koeberl, C., Miller, K.G., Reimold, W.U., Cockell, C.S., Horton, J.W., Jr., Sanford, W.E., and Voytek, M.A., 2006c, Chesapeake Bay impact structure drilled: Eos (Transactions, American Geophysical Union), v. 87, p. 349–355.

Gohn, G.S., Sanford, W.E., Powars, D.S., Horton, J.W., Jr., Edwards, L.E., Morin, R.H., and Self-Trail, J.M., 2007, Site report for USGS test holes drilled at Cape Charles, Northampton County, Virginia, in 2004: U.S. Geological Survey Open-File Report 2007-1074, 22 p., http://pubs.usgs.gov/of/2007/1094/ (October 2007).

Horton, J.W., Jr., and Izett, G.A., 2005, Crystalline-rock ejecta and shocked minerals of the Chesapeake Bay impact structure, USGS-NASA Langley core, Hampton, Virginia, with supplemental constraints on the age of impact, *in* Horton, J.W., Jr., et al., eds., Studies of the Chesapeake Bay impact structure—The USGS-NASA Langley corehole, Hampton, Virginia, and related coreholes and geophysical surveys: U.S. Geological Survey Professional Paper 1688, p. E1–E30.

Horton, J.W., Jr., Aleinikoff, J.N., Izett, G.A., Naeser, N.D., Naeser, C.W., and Kunk, M.J., 2002, Crystalline basement and impact-derived clasts from three coreholes in the Chesapeake Bay impact structure, southeastern Virginia [abs.]: Eos (Transactions, American Geophysical Union), v. 83, Spring Meeting Supplement, abstract T21A-03, p. S351.

Horton, J.W., Jr., Gohn, G.S., Powars, D.S., Jackson, J.C., Self-Trail, J.M., Edwards, L.E., and Sanford, W.E., 2004, Impact breccias of the central uplift, Chesapeake Bay impact structure: Initial results of a test hole at Cape Charles, Virginia: Geological Society of America Abstracts with Programs, v. 36, no. 5, p. 266.

Horton, J.W., Jr., Aleinikoff, J.N., Kunk, M.J., Gohn, G.S., Edwards, L.E., Self-Trail, J.M., Powars, D.S., and Izett, G.A., 2005a, Recent research on the

Chesapeake Bay impact structure, USA—Impact debris and reworked ejecta, *in* Kenkmann, T., et al., eds., Large meteorite impacts, III: Geological Society of America Special Paper 384, p. 147–170.

Horton, J.W., Jr., Gohn, G.S., Jackson, J.C., Aleinikoff, J.N., Sanford, W.E., Edwards, L.E., and Powars, D.S., 2005b, Results from a scientific test hole in the central uplift, Chesapeake Bay impact structure, Virginia, USA [abs.]: Lunar and Planetary Science Conference XXXVI, abstract 2003, 2 p., http://www.lpi.usra.edu/meetings/lpsc2005/pdf/2003.pdf (September 2007).

Horton, J.W., Jr., Gohn, G.S., Powars, D.E., and Edwards, L.E., 2005c, Origin and emplacement of breccias in the Chesapeake Bay impact structure, Virginia, USA [abs.], *in* Evans, K.R., et al., eds., SEPM Research Conference, The Sedimentary Record of Meteorite Impacts, Springfield, Missouri, May 21–23, 2005, Abstracts with Program, p. 19–20.

Horton, J.W., Jr., Powars, D.S., and Gohn, G.S., 2005d, Studies of the Chesapeake Bay impact structure—Introduction and discussion, *in* Horton, J.W., Jr., et al., eds., Studies of the Chesapeake Bay impact structure—The USGS-NASA Langley corehole, Hampton, Virginia, and related coreholes and geophysical surveys: U.S. Geological Survey Professional Paper 1688, p. A1–A24.

Horton, J.W., Jr., Aleinikoff, J.N., Kunk, M.J., Naeser, C.W., and Naeser, N.D., 2005e, Petrography, structure, age, and thermal history of granitic Coastal Plain basement in the Chesapeake Bay impact structure, USGS-NASA Langley core, Hampton, Virginia, *in* Horton, J.W., Jr., et al., eds., Studies of the Chesapeake Bay impact structure—The USGS-NASA Langley corehole, Hampton, Virginia, and related coreholes and geophysical surveys: U.S. Geological Survey Professional Paper 1688, p. B1–B29.

Horton, J.W., Jr., Gohn, G.S., Powars, D.S., and Edwards, L.E., 2006a, Chesapeake Bay impact structure—Influence of a layered, shallow marine target on crater excavation and modification [abs.], *in* Ormö, J., and Bergman, H., eds., Impact craters as indicators for planetary environmental evolution and astrobiology, Abstract volume and program, Östersund (Sweden), June 8–14, 2006, 2 p., http://www.geo.su.se/lockne2006.

Horton, J.W., Jr., Ormö, J., Powars, D.S., and Gohn, G.S., 2006b, Chesapeake Bay impact structure: Morphology, crater fill, and relevance for impact structures on Mars: Meteoritics and Planetary Science, v. 41, p. 1613–1624.

Horton, J.W., Jr., Vanko, D.A., Naeser, C.W., Naeser, N.D., Larsen, D., Jackson, J.C., and Belkin, H.E., 2006c, Postimpact hydrothermal conditions at the central uplift, Chesapeake Bay impact structure, Virginia, USA [abs.]: Lunar and Planetary Science Conference XXXVII, abstract 1842, 2 p., http://www.lpi.usra.edu/meetings/lpsc2006/pdf/1842.pdf (September 2007).

Horton, J.W., Jr., Vanko, D.A., Naeser, C.W., Naeser, N.D., Larsen, D., Jackson, J.C., and Belkin, H.E., 2006d, Hydrothermal alteration of breccias at the central uplift, Chesapeake Bay impact structure: Geological Society of America Abstracts with Programs, v. 38, no. 7, p. 59.

Jackson, J.C., Horton, J.W., Jr., Chou, I.-M., and Belkin, H.E., 2005, A shock-induced polymorph of anatase and rutile from the Chesapeake Bay impact structure, Virginia, USA [abs.], *in* Evans, K.R., et al., eds., SEPM Research Conference, The Sedimentary Record of Meteorite Impacts, Springfield, Missouri, May 21–23, 2005, Abstracts with Program, p. 20.

Jackson, J.C., Horton, J.W., Jr., Chou, I., and Belkin, H.E., 2006, A shock-induced polymorph of anatase and rutile from the Chesapeake Bay impact structure, Virginia, USA: American Mineralogist, v. 91, p. 604–608, doi: 10.2138/am.2006.2061.

Jansa, L.F., Pe-Piper, G., Robertson, B., and Friedenreich, O., 1989, Montagnais: A submarine impact structure on the Scotian shelf, eastern Canada: Geological Society of America Bulletin, v. 101, p. 450–463, doi: 10.1130/0016-7606(1989)101<0450:MASISO>2.3.CO;2.

Keiswetter, D., Black, R., and Steeples, D., 1996, Structure of the terrace terrane, Manson impact structure, Iowa, interpreted from high-resolution seismic reflection data, *in* Koeberl, C., and Anderson, R.R., eds., The Manson impact structure, Iowa: Anatomy of an impact crater: Geological Society of America Special Paper 302, p. 105–113.

Kieffer, S.W., 2005, The role of volatiles in impacts: Implications for the Chesapeake Bay impact [abs.]: American Association for the Advancement of Science Annual Meeting, 17–21 February 2005, Washington, D.C., v. 171, p. A43, CD-ROM, http://php.aaas.org/meetings/Archive_2005/abstracts.php?xabs=834

Kieffer, S.W., and Simonds, C.H., 1980, The role of volatiles and lithology in the impact cratering process: Reviews of Geophysics and Space Physics, v. 18, p. 143–181.

Koeberl, C., Poag, C.W., Reimold, W.U., and Brandt, D., 1996, Impact origin of the Chesapeake Bay structure and the source of the North American tektites: Science, v. 271, p. 1263–1266, doi: 10.1126/science.271.5253.1263.

Lee, S.R., Horton, J.W., Jr., and Walker, R.J., 2005, Osmium-isotope and platinum-group-element systematics of impact-melt rocks, Chesapeake Bay impact structure, Virginia, USA [abs.]: Lunar and Planetary Science Conference XXXVI, abstract 1700, 2 p., http://www.lpi.usra.edu/meetings/lpsc2005/pdf/1700.pdf (September 2007).

Lee, S.R., Horton, J.W., Jr., and Walker, R.W., 2006, Confirmation of a meteoritic component in impact-melt rocks of the Chesapeake Bay impact structure, Virginia, USA—Evidence from osmium isotopic and PGE systematics: Meteoritics and Planetary Science, v. 41, p. 819–833.

Lindström, M., Ormö, J., Sturkell, E., and von Dalwigk, I., 2005, The Lockne crater: Revision and reassessment of structure and impact stratigraphy, *in* Koeberl, C., and Henkel, H., eds., Impact tectonics: New York, Springer, p. 357–377.

McHugh, C.M.G., Snyder, S.W., and Miller, K.G., 1998, Upper Eocene ejecta of the New Jersey continental margin reveal dynamics of Chesapeake Bay impact: Earth and Planetary Science Letters, v. 160, p. 353–367, doi: 10.1016/S0012-821X(98)00096-X.

Melosh, H.J., 1979, Acoustic fluidization: A new geologic process?: Journal of Geophysical Research, v. 84, p. 7513–7520.

Melosh, H.J., 1989, Impact cratering—A geologic process: New York, Oxford University Press, 245 p.

Melosh, H.J., 2001, The physics of crater collapse in saturated media [abs.], *in* Morgan, M.L., and Warme, J.E., eds., Bolide impacts on wet targets: Boulder, Colorado, Geological Society of America Field Forum, Nevada and Utah, April 22–28, 2001, Abstracts, 1 p.

Melosh, H.J., and Ivanov, B.A., 1999, Impact crater collapse: Annual Review of Earth and Planetary Sciences, v. 27, p. 385–415, doi: 10.1146/annurev.earth.27.1.385.

Mulder, T., and Alexander, J., 2001, The physical character of subaqueous sedimentary density flows and their deposits: Sedimentology, v. 48, p. 269–299, doi: 10.1046/j.1365-3091.2001.00360.x.

Obermeier, S.F., 1996, Use of liquefaction-induced features for paleoseismic analysis—An overview of how seismic liquefaction features can be distinguished from other features and how their regional distribution and properties can be used to infer the location and strength of Holocene paleo-earthquakes: Engineering Geology, v. 44, p. 1–76, doi: 10.1016/S0013-7952(96)00040-3.

Obermeier, S.F., Pond, E.C., and Olson, S.M., 2001, Paleoliquefaction studies in continental settings: Geologic and geotechnical factors in interpretations and back-analysis: U.S. Geological Survey Open-File Report 01-029, 75 p., http://pubs.usgs.gov/openfile/of01-029 (September 2007).

Olson, S.M., Green, R.A., and Obermeier, S.F., 2003, Geotechnical analysis of paleoseismic shaking using liquefaction features, part I: Major updating of analysis techniques: U.S. Geological Survey Open-File Report 03-307, 33 p., http://pubs.usgs.gov/of/2003/of03-307/of03-307.pdf (September 2007).

Ormö, J., and Lindström, M., 2000, When a cosmic impact strikes the seabed: Geological Magazine, v. 137, no. 1, p. 67–80, doi: 10.1017/S0016756800003538.

Ormö, J., Sturkell, E., Lindström, M., and Lepinette, A., 2007, Resurge dynamics at the Lockne and Tvären marine-target impact craters analysed with sedimentological and numerical methods [abs.]: Lunar and Planetary Science Conference XXXVIII, abstract 1540, 2 p., http://www.lpi.usra.edu/meetings/lpsc2007/pdf/1540.pdf (September 2007).

Poag, C.W., 1996, Structural outer rim of Chesapeake Bay impact crater—Seismic and borehole evidence: Meteoritics and Planetary Science, v. 31, p. 218–226.

Poag, C.W., 1997, The Chesapeake Bay bolide impact: A convulsive event in the Atlantic Coastal Plain evolution: Sedimentary Geology, v. 108, p. 45–90, doi: 10.1016/S0037-0738(96)00048-6.

Poag, C.W., 2002, Synimpact-postimpact transition inside Chesapeake Bay crater: Geology, v. 30, p. 995–998, doi: 10.1130/0091-7613(2002)030<0995:SPTICB>2.0.CO;2.

Poag, C.W., and Aubry, M.-P., 1995, Upper Eocene impactites of the U.S. East Coast: Depositional origins, biostratigraphic framework, and correlation: Palaios, v. 10, p. 16–43, doi: 10.2307/3515005.

Poag, C.W., and Norris, R.D., 2005, Stratigraphy and paleoenvironments of early postimpact deposits—USGS-NASA Langley core, Chesapeake Bay impact crater, *in* Horton, J.W., Jr., et al., eds., Studies of the Chesapeake Bay impact structure—The USGS-NASA Langley corehole, Hampton, Virginia, and related coreholes and geophysical surveys: U.S. Geological Survey Professional Paper 1688, p. F1–F52.

Poag, C.W., Powars, D.W., Poppe, L.J., Mixon, R.B., Edwards, L.E., Folger, D.W., and Bruce, S., 1992, Deep Sea Drilling Project Site 612 bolide event—New evidence of a late Eocene impact-wave deposit and a possible impact site, U.S. East Coast: Geology, v. 20, p. 771–774, doi: 10.1130/0091-7613(1992)020<0771:DSDPSB>2.3.CO;2.

Poag, C.W., Powars, D.S., Poppe, L.J., and Mixon, R.B., 1994, Meteoroid mayhem in Ole Virginny—Source of the North American tektite strewn field: Geology, v. 22, p. 691–694, doi: 10.1130/0091-7613(1994)022<0691:MMIOVS>2.3.CO;2.

Poag, C.W., Hutchinson, D.R., Colman, S.M., and Lee, M.W., 1999, Seismic expression of the Chesapeake Bay impact crater: Structural and morphologic refinements based on new seismic data, *in* Dressler, B.O., and Sharpton, V.L., eds., Large meteorite impacts and planetary evolution, II: Geological Society of America Special Paper 339, p. 149–164.

Poag, C.W., Koeberl, C., and Reimold, W.U., 2004, The Chesapeake Bay impact crater—Geology and geophysics of a late Eocene submarine impact structure: New York, Springer-Verlag, 522 p. plus CD-ROM.

Powars, D.S., 2000, The effects of the Chesapeake Bay impact crater on the geologic framework and the correlation of hydrogeologic units of southeastern Virginia, south of the James River: U.S. Geological Survey Professional Paper 1622, 53 p., http://pubs.usgs.gov/prof/p1622/ (September 2007).

Powars, D.S., and Bruce, T.S., 1999, The effects of the Chesapeake Bay impact crater on the geological framework and correlation of hydrogeologic units of the lower York-James Peninsula, Virginia: U.S. Geological Survey Professional Paper 1612, 82 p., http://pubs.usgs.gov/prof/p1612/ (September 2007).

Powars, D.S., Mixon, R.B., and Bruce, T.S., 1992, Uppermost Mesozoic and Cenozoic geologic cross section, outer coastal plain of Virginia, *in* Gohn, G.S., ed., Proceedings of the 1988 U.S. Geological Survey Workshop on the Geology and Geohydrology of the Atlantic Coastal Plain: U.S. Geological Survey Circular 1059, p. 85–101.

Powars, D.S., Bruce, T.S., Bybell, L.M., Cronin, T.M., Edwards, L.E., Frederiksen, N.O., Gohn, G.S., Horton, J.W., Jr., Izett, G.A., Johnson, G.H., Levine, J.S., McFarland, E.R., Poag, C.W., Quick, J.E., Schindler, J.S., Self-Trail, J.M., Smith, M.J., Stamm, R.G., and Weems, R.E., 2001, Preliminary geologic summary for the USGS-NASA Langley corehole, Hampton, Virginia: U.S. Geological Survey Open-File Report 01-87-B, 20 p., http://pubs.usgs.gov/of/2001/of01-087/ (October 2007).

Powars, D.S., Gohn, G.S., Catchings, R.D., Horton, J.W., Jr., and Edwards, L.E., 2003, Recent research in the Chesapeake Bay impact crater, USA, part 1: Structure of the western annular trough and interpretation of multiple collapse structures [abs.]: International Conference on Large Meteorite Impacts, 3rd, Noerdlingen, Germany, August 5–7, 2003, abstract 4053, http://www.lpi.usra.edu/meetings/largeimpacts2003/pdf/4053.pdf (September 2007).

Powars, D.S., Gohn, G.S., Horton, J.W., Jr., and Catchings, R.D., 2004, Progress report and continuing proposal for collaborative research on lithostratigraphy, seismostratigraphy, and structure of the Chesapeake Bay impact structure, *in* Edwards, L.E., et al., eds., ICDP-USGS Workshop on Deep Drilling in the Central Crater of the Chesapeake Bay Impact Structure, Virginia, USA, September 22–24, 2003, Proceedings Volume: U.S. Geological Survey Open-File Report 2004-1016, CD-ROM, p. 76–77.

Powars, D.S., Catchings, R.D., Daniels, R.D., Pierce, H.A., Gohn, G.A., Horton, J.W., Jr., and Edwards, L.E., 2005, Anatomy of the central crater of the Chesapeake Bay impact structure [abs.], *in* Evans, K.R., et al., eds., SEPM Research Conference, The Sedimentary Record of Meteorite Impacts, Springfield, Missouri, May 21–23, 2005, Abstracts with Program, p. 28–29.

Powars, D.S., Catchings, R.D., Pierce, H.A., Gohn, G.S., Horton, J.W., Jr., Edwards, L.E., and Daniels, D.L., 2006a, Geophysics reveals anatomy of the Chesapeake Bay impact structure's inner and outer crater [abs.], *in* Ormö, J., and Bergman, H., eds., Impact craters as indicators for planetary environmental evolution and astrobiology, Abstract volume and program, Östersund (Sweden), June 8–14, 2006, 2 p., http://www.geo.su.se/Lockne2006.

Powars, D.S., Catchings, R.D., Gohn, G.S., Horton, J.W., Jr., Edwards, L.E., Daniels, D.L., and Pierce, H.A., 2006b, Insights into the structurally complex inner crater of the Chesapeake Bay impact structure from geophysics and deep test holes: Geological Society of America Abstracts with Programs, v. 38, no. 7, p. 119–120.

Rauch, A.F., and Martin, J.R., III, 2000, EPOLLS model for predicting average displacements on lateral spreads: Journal of Geotechnical and Geoenvironmental Engineering, v. 126, no. 4, p. 360–371, doi: 10.1061/(ASCE)1090-0241(2000)126:4(360).

Reimold, W.U., Koeberl, C., and Poag, C.W., 2002, Chesapeake Bay impact crater: Petrographic and geochemical investigations of the impact breccia fill [abs.]: Geological Society of America Abstracts with Programs, v. 34, no. 6, p. 466, http://gsa.confex.com/gsa/2002AM/finalprogram/abstract_40802.htm (September 2007).

Sanford, W.E., 2005, A simulation of the hydrothermal response to the Chesapeake Bay bolide impact: Geofluids, v. 5, p. 185–201, doi: 10.1111/j.1468-8123.2005.00110.x.

Sanford, W.E., Gohn, G.S., Powars, D.S., Horton, J.W., Jr., Edwards, L.E., Self-Trail, J.M., and Morin, R.H., 2004, Drilling the central crater of the Chesapeake Bay impact structure: A first look: Eos (Transactions, American Geophysical Union), v. 85, p. 369–377.

Self-Trail, J.M., 2003, Shock-wave-induced fracturing of calcareous nannofossils from the Chesapeake Bay impact crater: Geology, v. 31, p. 697–700, doi: 10.1130/G19678.1.

Shah, A.K., Brozena, J., Vogt, P., Daniels, D., and Plescia, J., 2005, New surveys of the Chesapeake Bay impact structure suggest melt pockets and target-structure effects: Geology, v. 33, p. 417–420, doi: 10.1130/G21213.1.

Shuvalov, V.V., and Artemieva, N.A., 2001, Numerical modeling of impact cratering at shallow sea [abs.]: Lunar and Planetary Science Conference XXXII, abstract 1122, 2 p.

Shuvalov, V.V., and Trubestkaya, I.A., 2002, Numerical modeling of marine target impacts: Solar System Research, v. 36, no. 5, p. 417–430, doi: 10.1023/A:1020467522340.

Sibson, R.H., 1977, Fault rocks and fault mechanisms: Geological Society [London] Journal, v. 133, p. 191–231.

Snoke, A.W., and Tullis, J., 1998, An overview of fault rocks, *in* Snoke, A.W., et al., eds., Fault-related rocks: A photographic atlas: Princeton, New Jersey, Princeton University Press, p. 3–18.

Stöffler, D., and Grieve, R.A.F., 1994, Classification and nomenclature of impact metamorphic rocks: A proposal to the IUGS Subcommission on the Systematic of Metamorphic Rocks, *in* Montanari, A., and Smit, J., eds., Post-Östersund newsletter: Strasbourg, European Science Foundation Scientific Network on Impact Cratering and Evolution of Planet Earth, p. 9–15.

Stöffler, D., and Grieve, R.A.F., 2007, A systematic nomenclature for metamorphic rocks: 11. Impactites. Recommendations by the IUGS Subcommission on the Systematics of Metamorphic Rocks. Recommendations, web version of 01.02.07, 15 p. http://www.bgs.ac.uk/SCMR/docs/papers/paper_11.pdf (October 2007).

Sturkell, E.F.F., 1998, The marine Lockne impact structure, Jamtland, Sweden: A review: Geologische Rundschau, v. 87, p. 253–267, doi: 10.1007/s005310050208.

Sturkell, E.F.F., and Ormö, J., 1997, Impact-related clastic injections in the marine Ordovician Lockne impact structure, central Sweden: Sedimentology, v. 44, p. 793–804, doi: 10.1046/j.1365-3091.1997.d01-54.x.

Tsikalas, F., Gudlaugsson, S.T., and Faleide, J.I., 1998, Collapse, infilling, and postimpact deformation at the Mjølnir impact structure, Barents Sea: Geological Society of America Bulletin, v. 110, p. 537–552, doi: 10.1130/0016-7606(1998)110<0537:CIAPDA>2.3.CO;2.

Turtle, E.P., Pierazzo, E., Collins, G.S., Osinski, G.R., Melosh, H.J., Morgan, J.V., and Reimold, W.U., 2005, Impact structures: What does crater diameter mean?, *in* Kenkmann, T., et al., eds., Large meteorite impacts, III: Geological Society of America Special Paper 384, p. 1–24.

Varnes, D.J., 1978, Slope movement types and processes, *in* Schuster, R.L., and Krizek, R.J., eds., Landslides—Analysis and control: Washington, D.C., National Research Council, Transportation Research Board Special Report 176, p. 11–33.

von Dalwigk, I., and Ormö, J., 2001, Formation of resurge gullies at impacts at sea: The Lockne crater, Sweden: Meteoritics and Planetary Science, v. 36, p. 359–369.

Wardlaw, B.R., Quinn, T.M., and Martin, W.E., 1991, Sediment facies of Enewetak Atoll lagoon: U.S. Geological Survey Professional Paper 1513-B, p. B1–B60.

Manuscript Accepted by the Society 10 July 2007

Printed in the USA

The Geological Society of America
Special Paper 437
2008

Alamo Event, Nevada: Crater stratigraphy and impact breccia Realms

Jesús A. Pinto*
INTEVEP-PDVSA, Building South 1, Floor 2, Office 202, 1201, Los Teques, Venezuela

John E. Warme*
Department of Geology and Geological Engineering, Colorado School of Mines, Golden, Colorado 80401, USA

ABSTRACT

Based on evaluation of past results and new research, we have partitioned the distribution of the Alamo Breccia in southeastern Nevada and western Utah into six genetic Realms that provide a working model for the marine Late Devonian Alamo Impact Event. Each Realm exhibits discrete impact processes and stratigraphic products that are enumerated here. The first five form roughly concentric semicircular bands across the Devonian shallow-water carbonate platform. These are: (1) Rim Realm, where a newly defined impact stratigraphy includes both autogenic and allogenic breccias associated with the crater rim; (2) Ring Realm, where breccias are now interpreted to have formed sequentially by seismic shock, passage of the ejecta curtain, tsunami waves or surge, and runoff that accumulated over tilted terrace(s) bounded by syn-Event, ring-forming, listric faults; (3) Runup Realm, where graded breccias were stranded by tsunami surge or waves; (4) Runoff Realm, where sheet-floods carried traces of impact debris across the distal platform beds and channels filled with impact debris; (5) Seismite Realm, where near-surface beds far across the platform were uniquely deformed; and (6) Runout/Resurge Realm, where offshore channels of thick off-platform Alamo Breccia, together with large-scale olistolith(s), signal contemporaneous massive collapse of the platform margin, possibly into the central crater.

Five breccia Units characterize the newly interpreted Rim Realm, in ascending order: (1) deformed target rocks, (2) injected dikes and sills, (3) chaotic fallback, (4) smeared fallback, and (5) resurge. This succession is covered by deepwater limestones deposited inside the crater rim, or across a new slope created after platform margin collapse. Unit 1 exhibits shatter-cone-like structures interpreted as impact products. Newly discovered Ordovician and probable older meter-scale clasts in Unit 3 confirm a minimum excavation depth of 1.5 km. Microscopic components in Units 3 and 4 indicate high pressures (>10 GPa), probable quenched carbonate melt, and accreted

*Pinto: pintojb@PDVSA.com; corresponding author: Warme: jwarme@mines.edu

Pinto, J.A., and Warme, J.E., 2008, Alamo Event, Nevada: Crater stratigraphy and impact breccia Realms, *in* Evans, K.R., Horton, J.W., Jr., King, D.T., Jr., and Morrow, J.R., eds., The Sedimentary Record of Meteorite Impacts: Geological Society of America Special Paper 437, p. 99–137, doi: 10.1130/2008.2437(07). For permission to copy, contact editing@geosociety.org.

particles that may be new kinds of impact products. Postimpact tectonics and other factors obscure the full panorama, including the location and character of the missing central crater, but the assemblage of Realms offers a working model to compare with expected impact paradigms.

Keywords: impact, Devonian, Nevada, Alamo Breccia, crater stratigraphy.

INTRODUCTION

The Late Devonian Alamo Breccia is one of the largest impactogenic stratigraphic intervals discovered on Earth. The Breccia represents diverse and complex large-scale impact-deformed bedrock, ejecta, and reworked debris exposed in 25 or more mountain ranges across present-day southern Nevada and western Utah (Fig. 1). Various facies of Alamo Breccia are conservatively calculated to occupy an area of ~28,000 km^2 and a volume of ~1300 km^3 in southern Nevada alone (Fig. 2).

The Breccia provides rare opportunities to understand impact events. Widespread, tectonically tilted and eroded ranges exhibit expansive outcrops that are mostly bare owing to desert climate and sparse vegetation. They reveal the base, top, internal character, and stratigraphic context of the Breccia. Drilling, seismic acquisition, or other indirect data procurements are not required. Although the Breccia formed at ~382 Ma, it is precisely dated within a narrow fossil zone that is correlated throughout the study area and provides the potential to correlate it to far-field effects elsewhere.

The Alamo Event occurred in a marine setting. The target was mainly a thick interval of Paleozoic carbonate rocks that were redistributed as complex Alamo Breccia facies. These facies occur singly or combined in six genetic "Realms" that are introduced and explained herein (Fig. 2). Thus far only one extensive outcrop along Tempiute Mountain (TEM, TMS, Figs. 1 and 2) contains unambiguous Alamo near-crater signatures, elaborated here for the first time. This location is interpreted as part of the crater rim, is illustrated and interpreted on Figures 2–13, and is hereafter referred to as the "Rim Realm," or TEM.

An array of different Alamo impact breccias (Warme, 2004) are distributed away from the Rim Realm in an eastward-extended, apparently semicircular pattern across the Guilmette Formation, which is the local name for the shallow-water carbonate platform that bordered western North America in Late Devonian time. These breccias are interpreted as roughly proximal to distal away from the Rim Realm at TEM, and occur exclusively upon the platform environment in what we interpret as Ring, Runup, Runoff, and Seismite Realms (Figs. 2, 14–19). Figure 20 is a summary diagram showing our view of the genesis of crater stratigraphy at TEM and its relationship to the adjacent Realms on the platform. West of TEM the concentric pattern breaks down, and the Alamo Breccia appears in channels of the Runout/Resurge Realm (Fig. 2) that overlie a regional deepwater unconformity and are associated with one or more kilometer-scale likely olistoliths.

The purpose of this contribution is to merge the results presented in previous reports with new data and analyses to present a three-dimensional genetic reconstruction of the products of the Alamo Impact Event. Our new contributions are mainly outcomes of J. Pinto's (2006) dissertation research, which include documentation and interpretation of the crater rim stratigraphy at TEM, Alamo Breccia variability in the Ring Realm, and potential far-field Alamo seismites. Pinto's field and laboratory methods are outlined in his dissertation, including automated methods for quantitative textural analyses and the results that are presented herein.

We intend our outcomes to help reveal the as yet undiscovered location and precise rock facies at the point of impact, the size and composition of the projectile, and the original size and geometry of the crater before postimpact tectonic pulses complicated the geology of the study area. At many localities the sedimentological details about the Breccia and their interpretation as impact phenomena remain superficial compared to the opportunities they present. Satisfactory assembly of the full impact scenario is still lacking, fraught with the Enigmas listed under following headings. Most notable is the well-documented compelling but indirect evidence for deep crater excavation, but unknown deep crater location and character.

PAST WORK AND NEW PERSPECTIVES

The Alamo Breccia was recognized and informally named in 1990 as a convulsive deposit captured in the Guilmette Formation (Warme et al., 1991). The age, distribution, stratigraphic context, and sedimentological aspects were treated in several papers and theses. Major reviews combined with new findings are the reports of Warme and Sandberg (1995, 1996); Sandberg et al. (1997, 2002); Warme and Kuehner (1998); Morrow et al. (1998, 2005); and Warme et al. (2002). Discovery, naming, and early research results on the Breccia were reviewed by Warme and Kuehner (1998). Additional reports and theses include Warme (1991, 2004); Warme et al. (1991, 1993); Ackman (1991); Yarmanto (1992); Leroux et al. (1995); Chamberlain and Warme (1996); Kuehner (1997); Morrow (1997, 2000, 2004); Sandberg et al. (2003, 2005, 2006); Chamberlain (1999); Morrow and Sandberg (2003a, 2003b, 2006, 2007); Koeberl et al. (2003); Warme and Pinto (2005); Pinto and Warme (2004, 2006); Pinto (2006); Anderson and Tapanila (2007); and Tapanila and Anderson (2007).

Age and Member Status

The age of the Alamo Event was dated as early Frasnian (*Palmatolepis punctata* conodont Zone) by Sandberg (Sandberg and Warme, 1993; Warme and Sandberg, 1995, 1996),

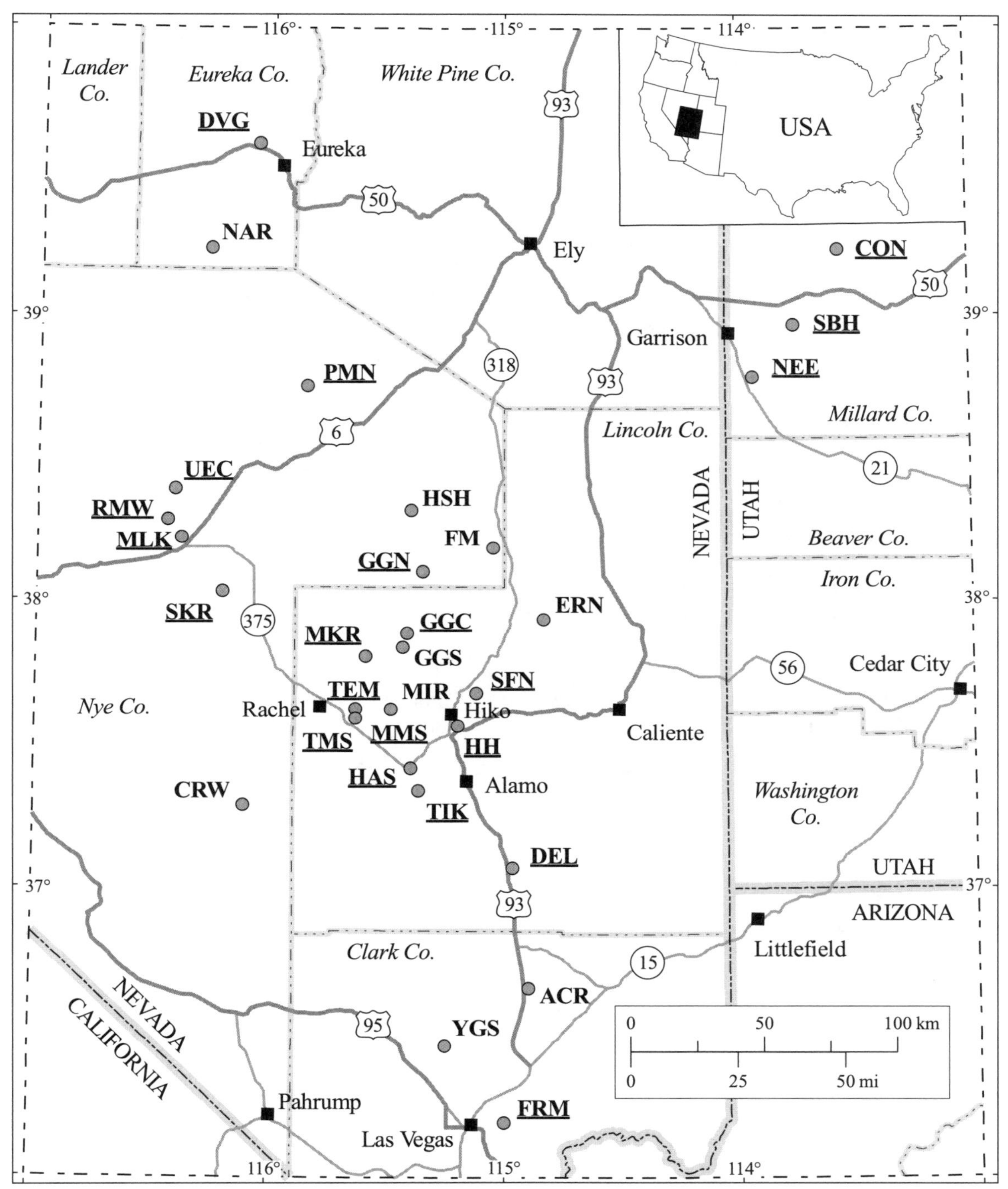

Figure 1. Map showing most of the documented Alamo Breccia localities in mountain ranges across Nevada and Utah. Three-letter codes are locations where the Breccia has been studied and placed on maps by Morrow et al. (1998, 2001, 2005), Pinto and Warme (2006), Morrow and Sandberg (2006), and Sandberg et al. (2006). Underlined localities are discussed in text. See Figure 2.

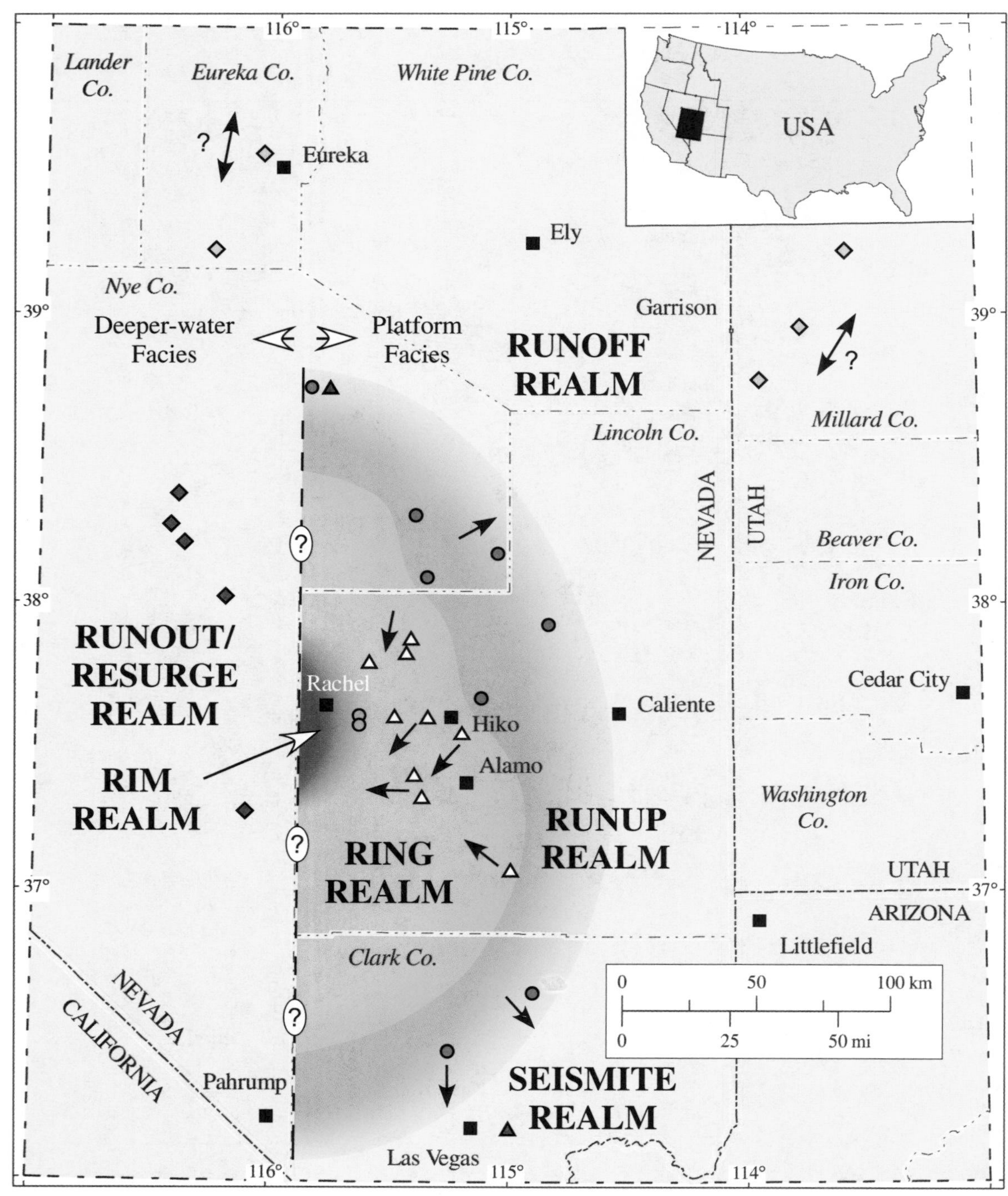

Figure 2. Area of Figure 1 showing distribution of Alamo Breccia Realms, unrestored. Queried north-south line separates the roughly concentric Realms on the Late Devonian eastern shallow-water carbonate platform of the Guilmette Formation from western Alamo Breccia outcrops in deeper-water off-platform environments. Double orange circles in the Rim Realm are part of a continuous outcrop on Tempiute Mountain from TEM to TMS, and referred to as TEM in text. Other symbols represent locations within Realms: white triangles—Ring Realm; red circles—Runup Realm; yellow diamonds—Runoff Realm; green triangles—Seismite Realm; blue diamonds—Runout/Resurge Realm. Arrows represent paleocurrent directions in Alamo Breccia.

who estimated the absolute age as ~367 Ma interpolated from radiometric dates far younger and older than the *punctata* Zone. Based on newer dates given by Kaufmann (2006), the age was adjusted to ~382 Ma (Morrow and Sandberg, 2005), with a possible error of ~± 4 m.y. Late Devonian conodont biofacies assemblages indicated a target water depth of ~300 m, and reworked older species provided evidence for crater excavation to Lower Ordovician levels (Sandberg and Warme, 1993; Warme and Sandberg, 1995, 1996) and perhaps Upper Cambrian (Morrow et al., 1998, 2001, 2005; Morrow and Sandberg, 2003a, 2006).

Reso (1963) chose the top of a persistent cliff-forming, stromatoporoid-rich interval for the boundary between lower and upper members of the Guilmette Formation. That interval is now the middle Alamo Breccia Member, formally proposed by Sandberg et al. (1997), that separates the as yet informal lower and upper members. Within the Runout/Resurge Realm west of the platform, the Alamo Breccia Member takes the form of channels and lenses and associated olistoliths that unconformably intercalate with deeper-water formations (Sandberg et al., 2006) rather than the Guilmette.

Morrow et al. (1998) offered possible regional and global effects of the Alamo Event, and concluded that it was "subcritical" as a cause of mass mortality; the closest critical event occurred at the Frasnian-Famennian boundary ~6 m.y. later (Kaufmann, 2006).

Impact Evidence

Morrow et al. (2005) compiled the known attributes that document the impact genesis of the Alamo Breccia that include (1) character and distribution of the megabreccia, (2) dikes and sills of dolomite breccia with quartzose matrix injected in target strata at TEM, (3) carbonate accretionary lapilli and bombs, (4) shock-metamorphosed quartz grains, (5) iridium anomalies, (6) shock-induced lateral delamination and deformation of platform strata, (7) ejected and redeposited conodonts as old as Late Cambrian, and (8) channelized Alamo debris in offshore settings. The crater stratigraphy, documented on Tempiute Mountain herein, provides important new evidence for impact.

Shocked Quartz

Quartz grains with unusual parallel microstructures in the Breccia were first noted by Ackman (1991) and illustrated by Warme and Sandberg (1995, 1996). These features are mostly planar fractures (PFs) or annealed, highly decorated planar deformation features (PDFs). Leroux et al. (1995), using transmission electron microscopy, identified up to six sets of micrometer-scale PDFs in Alamo quartz grains, as well as other crystal defects such as inclusion trails, rotated domains, and crystal mosaicism. Morrow and Sandberg (2001, 2006) and Morrow et al. (2005) quantified the spectrum of quartz shock effects, expanded the distribution of the grains in the Breccia, and discussed the ubiquitous embedded hematite crystals that uniquely characterize the grains wherever the Breccia occurs. Harris and Morrow (2007) proposed that the crystals are diagenetically altered micro-impactors originally embedded while the grains were in the hypervelocity ejecta curtain.

Carbonate Lapilli

Carbonate accretionary lapilli and lapillistone beds associated with the Alamo Breccia in several localities within the Ring Realm were first documented by Kuehner (1997) and Warme and Kuehner (1998) as impact "spherules," followed by detailed description and interpretation of their genesis as accretionary lapilli by Warme et al. (2002). The lapilli and the host lapillistone matrix both contain very fine-grained shocked quartz crystals, and together may be the most important evidence for the impact origin of the Alamo Event (Warme, 2004). One large-scale pure limestone bomb, discussed under the *Ring Realm* heading, mimics volcanic bombs in form and structure (Fig. 16).

Geochemical Signatures

Warme and Sandberg (1995, 1996) reported a minor iridium (Ir) anomaly with the maximum content of 133 parts per trillion (ppt) ~1.5 m below the top of Unit A (see *Ring Realm*) at Meeker Peak, Worthington Mountains (MKR, Fig. 1). Newer samples from several stratigraphic levels at five localities yielded a maximum value of only 51 ppt ~27 m from the base of Unit A at Hiko Hills (HH, Fig. 1), but contained the highest chromium and nickel content (Koeberl et al., 2003). Background Ir concentrations of ~2–25 ppt in adjacent carbonate rocks are slightly lower, but regarded as significant, compared to Alamo Breccia samples. Morrow and Sandberg (2006) reported moderate enrichment of siderophile and lithophile elements in Alamo-related channel deposits at Devils Gate, within the Runoff Realm in central Nevada (DVG, Fig. 1). A sample from the basal channel fill yielded an Ir content of ~183 ppt, which is the maximum Breccia Ir value detected so far. Warme and Sandberg (1995, 1996) ascribed the overall poor enrichment of siderophile elements in the breccia to rapid rates of deposition and dilution within the Breccia by preexisting disturbed and ejected rock, which constitutes probably >99% of the Breccia in the study area. Koeberl et al. (2003) suggested that the low values resulted from an achondritic projectile or that the breccia failed to catch or preserve the signature. These alternatives are not mutually exclusive.

Alamo Impact Models and Excavation Depth

Warme and Sandberg (1995, 1996) proposed that seismic waves caused the margin of the Guilmette carbonate platform to collapse westward into deeper water, and giant tsunamis generated by the impact and/or by underwater slide(s) caused damaging uprush and backwash of marine water that sorted the debris across the platform. They implied that the debris in their Zone 1 (see *Rim Realm*) could represent part of the crater.

Kuehner (1997) and Warme and Kuehner (1998) adapted the genetic model of Oberbeck et al. (1993), originally proposed for a shallow terrigenous shelf target, to the Alamo Breccia on the carbonate platform. The scenario explained many features of the Alamo Breccia that included stacked graded beds generated by a succession of tsunami-scale waves and currents. However, Morrow et al. (1998) concluded, on the basis of ejected Ordovician and younger conodonts recovered from the Breccia, that the impact excavated to lower Paleozoic formations and occurred in an off-platform setting. They calculated a crater depth of ~1.5 km and diameter of ~38–100 km, and proposed a location centered in Sand Spring Valley ~30 km northwest of TEM (Figs. 1 and 2). Subsequent discoveries in the western part of the Runout/Resurge Realm documented Alamo Breccia channels and a probable kilometer-scale olistolith in deeper-water settings (Morrow et al., 2001, 2005; Sandberg et al., 2006). Based on paleocurrent directions, the proposed deep crater was relocated, but still within the broad eastern half of the Runout/Resurge Realm where crucial outcrops are missing (e.g., Morrow et al., 2001, 2005).

In an amazing discovery, Morrow et al. (2005) recovered Cambrian to Devonian conodonts from lapillistone clasts in the Breccia in the West Pahranagat Range (TIK, Fig. 1). They also reported polycrystalline quartz grains from a Runoff Realm channel (DVG, Fig. 1), possibly derived from the Lower Cambrian Prospect Mountain Quartzite or even deeper pre-Paleozoic formations. Based on these findings, Morrow et al. (2005) presented a new estimated penetration depth of 1.7–2.5 km, final crater diameter of 44–65 km, and again chose a possible crater location between the Hot Creek Range (RMW, Fig. 1) and TEM.

Pinto and Warme (2004) interpreted the carbonate breccias and associated bedrock deformation in Devonian strata exposed at TEM (Figs. 1–5) and proposed five types of deformational units, described herein, as part of the Alamo crater. Rare meter-scale Eureka Quartzite clasts, recently found in the fallback interval at TEM (Unit 3, see below), confirm minimum excavation down to at least Ordovician levels.

ALAMO ENIGMAS

Current knowledge of the Alamo Event leaves several general problems that are unsolved or concluded by only circumstantial evidence. These issues are listed here so that the reader understands what is known versus what needs resolution in each of the Realms described and interpreted in the following section. Some solutions are proposed in *Discussion of Enigmas* at the end of this report.

Breccia Asymmetry

The basic semicircular distribution of impact facies in the Realms on the eastern carbonate platform is not matched by impact facies in the western deepwater Runout/Resurge Realm (Fig. 2). The plan geometry does not conform to standard notions of circular impact craters.

Crater Depth

The crater fragment on TEM exhibits impact stratigraphy only as deep as the top of the Lower Devonian Sevy Dolostone or equivalent, whereas the overlying fallback interval (Unit 3 in Rim Realm) contains Middle Ordovician or older sandstone fragments, and the widespread Alamo Breccia in more distant Realms contains ejected early Paleozoic conodonts and mineral grains of perhaps older provenance.

Crater Depth versus Breccia Area and Volume

The diminutive apparent depth of excavation at TEM fails to account for the immense area and volume of the Breccia outside the Rim Realm.

Deep Crater Location

The location of the required deeply penetrated crater has not been discovered.

Target Water Depth

The crater at TEM is excavated into shallow- to moderate-depth platform limestones and dolostones, whereas the paleoecology of contemporaneous Late Devonian and older ejected conodont assemblages recovered from the Alamo Breccia imply an off-platform target bathymetry.

Rim Realm to Runout/Resurge Realm Disconnect

There is no discovered transition between the eastern shallower carbonate platform at TEM and the western off-platform siliciclastic deepwater Runout/Resurge Realm containing Alamo Breccia channels. Crucial outcrops are lacking.

Rim Realm to Ring Realm Disconnect

TEM is separated from the nearest Ring Realm outcrops (MMS, Fig. 1) by only ~12 km on the surface today (Figs. 1 and 2), with no transitional outcrops between the two Realms. Strata below, within, and above the Alamo Breccia at TEM abruptly change to much different characteristics in the Ring Realm. In contrast, the stratigraphic succession is similar and predictable across thousands of square kilometers of the Ring Realm.

Ring Realm Impact Structure

The vast ~13,000 km^2 Ring Realm is puzzling for several reasons. If the area represents rings, the final crater is very large. However, the ring portion is covered by a comparatively thin 50–100 m Alamo Breccia that shows no directional thickness trends, and everywhere sharply overlies undamaged bedrock of the Upper Devonian shallow-water carbonate platform.

Ring Realm to Runup Realm Transition

Breccia thicknesses change abruptly from ~50–100 m in the Ring Realm to <10 m in the closely adjacent Runup Realm (Fig. 2). No intermediate thicknesses have been found.

Source of Quartz Grains

The outcropping crater fragment at Tempiute Mountain does not deeply pierce sandstone, leaving unexplained the pathway of the copious loose shocked quartz grains in the matrix of the carbonate Breccia.

Role of Lapilli and Lapillistone in Event Scenario

Alamo lapilli have been discovered in only two locations. Individual rare lapilli were recently discovered in the matrix of the fallback breccia on Tempiute Mountain (Unit 3, Figs. 2, 5, 9–11). In contrast, layered beds of lapilli occur across the Ring Realm, but only in broken lapillistone clasts that represent <<1 vol% of the Breccia (Warme et al., 2002). These two occurrences must be explained within the context of the phases of the Alamo Event.

The following descriptions and interpretations of the *Lateral Realms* bear on many of the problems listed above, and are applied to their solution under *Discussion of Enigmas*.

LATERAL REALMS

When its wide extent and thickness variation across the carbonate platform were realized, the Alamo Breccia was divided into three eastward-directed concentric semicircular Zones (Warme and Sandberg, 1995, 1996). Zone 1 was, and still is, represented only by Tempiute Mountain (TEM) outcrops, labeled herein as the Rim Realm. Zone 2 is now labeled Ring Realm, and Zone 3 now the Runup Realm. Studies by Kuehner (1997) and Warme and Kuehner (1998) on the sedimentology of the Alamo Breccia at six localities provided evidence for the broadly predictable but internally complex organization of the Breccia in each Zone. They presented a cross-section that showed how the Breccia cuts across progressively older beds of the lower Guilmette Formation and thickens from the Runup Realm (their Zone 3) westward across the Ring Realm (Zone 2) down to Middle Devonian beds of the Rim Realm at TEM. New discoveries stretch the Breccia farther east into western Utah (Runoff Realm) and west of TEM into central Nevada (Runout/Resurge Realm) (Fig. 2).

Rim Realm

Rock exposures on Tempiute Mountain (TEM, Fig. 1) represent the Rim Realm (Fig. 2) where the mapped Breccia (Fig. 3) crops out for ~7 km along strike. In former reports the Breccia was described as being ~100–130 m in thickness, greater than at any other locality. Puzzlingly, the largest clasts in the basal ~30 m were only a few tens of meters across, compared to the much larger basal clasts in the Ring Realm. The 100 m thick remainder of the interval was composed of two massive graded beds that were relatively finer-grained and better sorted than the graded beds of the Ring and Runup Realms. At TEM we now recognize a thicker ~430 m package of impact breccias divided into five Units. Unit 1 represents ~300 m of broken and sheared parautochthonous autogenic dolostones injected by allogenic breccia dikes and sills of Unit 2 (Figs. 4–6, 8). Units 1 and 2 are beneath the previously labeled Alamo Breccia at TEM. The overlying allogenic heterolithic Alamo Breccia interval is reinterpreted as Unit 3 fallback breccia, Unit 4 partially melted fallback breccia, and Unit 5 graded resurge beds (Figs. 9–13).

Pre-impact Stratigraphy in the Rim Realm

Pre-impact Paleozoic strata of Tempiute Mountain include dolomites, limestones, and subordinate sandstones deposited along the shallow marine continental margin of western North America, described and analyzed by many workers (e.g., Stewart, 1980; Poole et al., 1992; Kendall et al., 1997; Cook and Corboy, 2004). The formations that have been mapped or otherwise recognized on Tempiute Mountain (Tschanz and Pampeyan, 1970; Sandberg et al., 1997; Chamberlain, 1999; Morrow et al., 2005) are shown on the reconnaissance geologic map of Figure 3. We use the Sentinel Mountain Dolostone for the dolostone interval overlying the Oxyoke Canyon Sandstone at TEM, but recognize that it may include some of the overlying Bay State Dolostone as shown by Sandberg et al. (1997, 2005). The interval is more like the peritidal Sentinel Mountain at its type area (Nolan et al., 1956; Merriam, 1963) than the equivalent but more intertidal Simonson Dolostone typical of the Ring Realm east of Tempiute Mountain.

Figure 3 shows newly traced boundaries of some of the formations identified on the west face of Tempiute Mountain, including the location of the stratigraphic profile shown in Figure 4 and the stratigraphic column of Figure 5. The Oxyoke Canyon and Sentinel Mountain were certainly deformed by the Alamo Impact, but older formations in the area are included because they may have undergone deformation on scales too large to see, even on the extensive outcrops of the mountain face.

Rim Realm Impact Stratigraphy

The stratigraphic column of Figure 5 shows the package of five impactogenic Units and their upper and lower bounding formations at TEM.

Unit 1: Deformed target rocks. Unit 1 contains the lowest and thickest impactogenic interval recognized at TEM (Figs. 3–8). It is composed of ~300 m of disturbed Oxyoke Canyon Sandstone and Sentinel Mountain Dolostone that overlie the Lower Devonian Sevy Dolostone. The upper part of the Sevy is silicified in irregular patches, tens of meters across, that may represent an effect of the Alamo Event. The Sevy contact with Unit 1 is either erosional (Fig. 5) or marked by an irregular slip surface that is interpreted as a fundamental detachment at the base of the impact breccia at TEM (Figs. 3–6). The upper boundary of Unit 1 (Fig. 5)

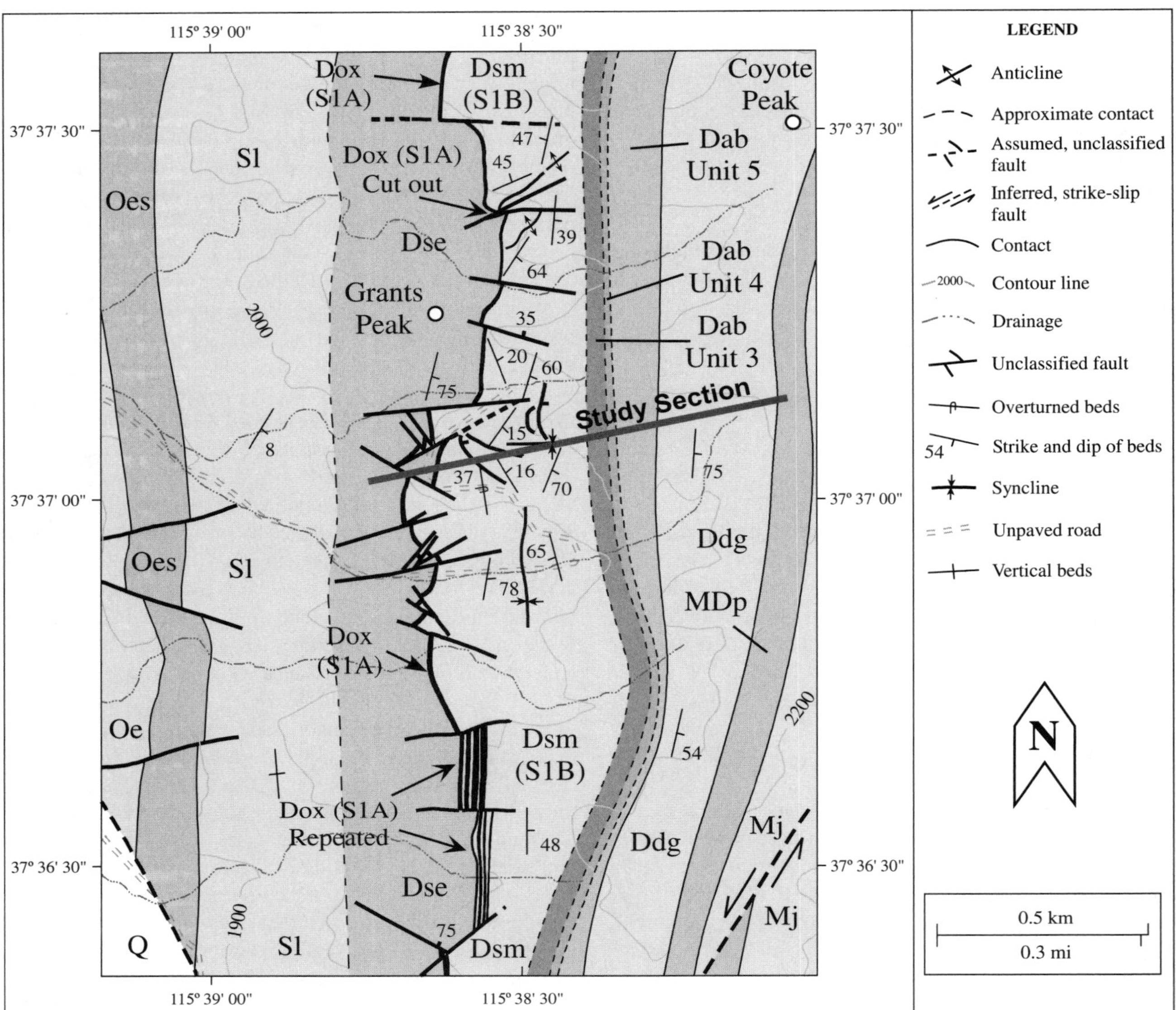

Figure 3. Reconnaissance map showing a portion of the west face of Tempiute Mountain (TEM, TMS, Fig. 1), and location of the study section shown on Figures 4 and 5. Contours in meters. Paleozoic formations and ages are Ordovician Eureka Quartzite (Oe) and Ely Springs Dolostone (Oes), Silurian Laketown Dolostone (Sl), Lower Devonian Sevy Dolostone (Dse), Lower to Middle Devonian Oxyoke Canyon Sandstone (Dox, bold lines labeled S1A for Subunit 1A), Middle Devonian Sentinel Mountain Dolostone (Dsm, labeled S1B for Subunit 1B), Upper Devonian Alamo Breccia (Dab) Units 3–5, Upper Devonian Devils Gate Limestone (Ddg), Mississippian-Devonian Pilot Shale (MDp), and Mississippian Joana Limestone (Mj). Dab 1 to Dab 5 are newly named and described impactogenic Units 1–5 at Tempiute Mountain. Unit 1 combines the Oxyoke Canyon Sandstone and Sentinel Mountain Dolostone, which were damaged and interpreted as autogenic impact breccias. The Oxyoke is offset, cut out, and repeated by numerous faults that also affect portions of the underlying Sevy Dolostone and of the overlying Sentinel Mountain Dolostone. The faults are interpreted as planes where damaged formations along the crater rim adjusted during later phases of the Alamo Event. The adjustment was completed by the time allogenic Units 3–5 were deposited as Alamo Breccia.

is marked by an abrupt shift to Unit 3 that includes changes from autogenic monomict to allogenic polymict clasts, from exclusively dolostone to predominantly limestone, and a transition to smaller clasts. Addition of the Oxyoke Canyon and the Sentinel Mountain adjusts the thickness of the impactogenic interval at TEM from ~100–130 m of previous reports to ~430 m (Fig. 5).

The Oxyoke Canyon and Sentinel Mountain are combined into a single Unit because they both are pervasively and sharply intruded by the polymict breccias of Unit 2. For discussion, Unit 1 is divided into Subunit 1A (S1A, Oxyoke Canyon Sandstone) and Subunit 1B (S1B, Sentinel Mountain Dolostone) (Pinto and Warme, 2006) (Figs. 3–5). Where preserved, the original contact

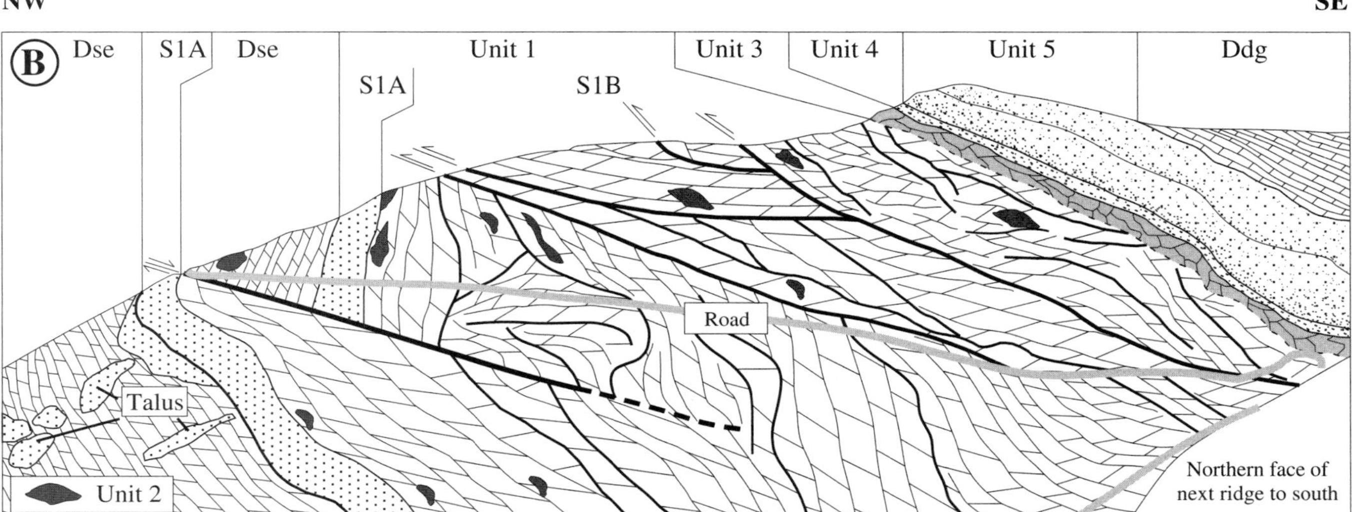

Figure 4. A: View looking northeast at ridge and south-facing slope, where traverse of study section (Fig. 3) and columnar section (Fig. 5) are located. Distance along ridge between faulted base of Sevy Dolostone and top of Unit 5 is ~0.9 km. Note abandoned road (see B) and scattered light-hued mine tailings. B: Ridge contains Sevy Dolostone (Dse) overlain by impact-damaged Oxyoke Canyon Sandstone (S1A) and Sentinel Mountain Dolostone (S1B) that are injected by dike-and-sill system of Unit 2, some of which is shown as dark patches of the dominant irregular sills. The Sentinel Mountain Dolostone is partitioned by major faults that bound packages of smaller fault- and fracture-bounded blocks that create a thick autogenic monomict breccia. Units 3 (fallback breccia), 4 (partial melt breccia), and 5 (thick graded resurge beds) are allogenic polymict breccias that complete the exposed crater succession, and are overlain by post-Event Devils Gate Limestone (Ddg).

between these formations is gradational. However, in most places on Tempiute Mountain the approximate contact is sharp, irregular, separated by the injections of Unit 2, or marked by glide surfaces and breccias that are interpreted as impact phenomena.

Subunit 1A: Oxyoke Canyon Sandstone. Subunit 1A has a maximum preserved thickness of ~15 m (Figs. 4–6). It consists of cross-bedded quartz sandstones intercalated with quartzose dolostones and dolomitic quartzose sandstones (Fig. 5). S1A displays degrees of brittle and ductile damage along strike where it was fractured, faulted, folded, and partially fluidized (Figs. 3–6). It is locally cut out, or repeated as many as four times over a stratigraphic distance of ~60 m, as mapped along strike (Fig. 3).

The quartz grains are monocrystalline, rarely polycrystalline, set in a uniform dolomicrite. Most quartz grains are heavily stressed, as indicated by undulose extinction and unevenly spaced planar and curved fractures. Some grains display as many as three sets of evenly spaced parallel and planar lamellae that are interpreted as planar fractures (PFs) (cf. French, 1998). Inclusions are common and typically exhibit subhedral to euhedral crystals of dolomite and less common hematite. Morrow et al. (2005) reported more than three sets of planar deformation features (PDFs) in a few quartz grains from S1A. Some are studded with the common subhedral to euhedral hematite crystals.

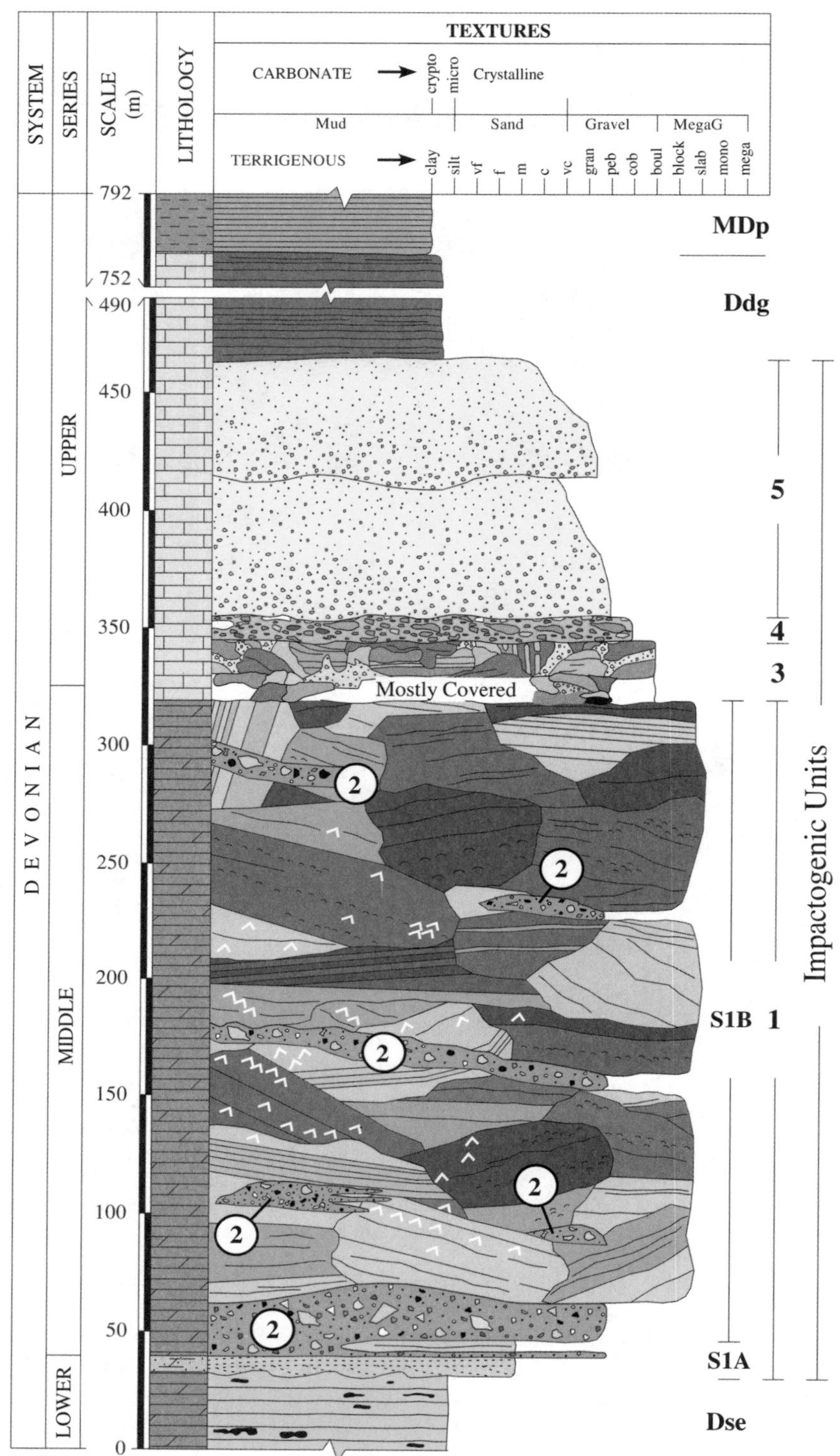

Figure 5. Columnar section, located on Figures 3 and 4, showing the thickness and character of impactogenic Units 1–5, the underlying upper part of the Sevy Dolostone (Dse), the overlying deepwater Devils Gate Limestone (Ddg), and the Pilot Shale (MDp). Autogenic impact breccias of Unit 1 are the Oxyoke Canyon Sandstone (S1A) and Sentinel Mountain Dolostone (S1B). Unit 2 is injected polymict breccia (Fig. 8) that tends to be thickest where it swells, pinches, and separates S1A and S1B (Fig. 6). S1B contains shatter-cone-like structures (cone symbols), interpreted as impact-induced (Fig. 7). Units 3 and 4 are composed of allogenic limestone clasts (Figs. 9, 10, 12) with varied matrices that bear shocked quartz, lapilli, and other fallback particles interpreted as shock and melt indicators (Fig. 11). Unit 5 is two thick resurge breccias (Fig. 13). "MegaG" terms are clast categories of megagravels that extend the Udden-wentworth particle size scale beyond "boulder" (Blair and MacPherson, 1999).

Subunit 1B: Sentinel Mountain Dolostone. Subunit 1B is ~285 m of gray to black Sentinel Mountain Dolostone. Beds range in thickness from a few centimeters to ~1 m, are well laminated to internally massive, and are composed of variable crystal sizes and sedimentary structures. Abundant fossils in the darker, bioturbated dolostones include brachiopods, gastropods, rugose corals, and crinoid columnals that all indicate oxygenated subtidal marine water. Concentrations of stromatoporoid heads and *Amphipora*-like forms that can tolerate more restricted conditions occur at many levels. Fossils are scarce to absent in the lighter-colored, thinner-bedded dolostones; laminated intervals represent the high intertidal to supratidal portions of shallowing-upward carbonate platform cycles.

Bedding attitudes change abruptly and erratically over short distances, caused by rotation of fault-bounded blocks (Figs. 4 and 5). Irregular folds and normal and reverse faults permeate the section on scales of meters to tens of meters (Figs. 3 and 4). The middle half of the Subunit is pervasively shattered. It contains unusual shatter-cone-like features that we interpret as impact phenomena (see below).

Monomict breccia in pods and lenses up to 4 m wide and centimeter-scale branching pipes are common within and between the blocks and are interpreted as impact-induced rock fluidization and mobilization. They are filled with locally sourced clasts 1–3 cm across, subangular to subrounded, set in a calcareous sandy to silty matrix. They show no internal organization; breccia and host rock compositions are similar, and the contact between them can be either transitional or sharp.

Interpretation. Unit 1 is interpreted as disturbed target rock near the crater rim based on (1) pervasive brittle deformation, (2) abundant monomict breccias, (3) sheared clast margins, (4) planar fractures in quartz grains of S1A, (5) shatter-cone-like structures, and (6) intrusion of dikes and sills of polymict breccia containing variably shocked quartz grains (see *Unit 2*). These attributes match products produced by the passage of impact shock waves through target rocks (Collins et al., 2004; Dressler

Figure 6. Contact where Sevy Dolostone (Dse) is cut by detachment surface (arrows) and overlain by Subunit 1A, Oxyoke Canyon Sandstone (S1A), which is ~15 m in thickness. The heavily quartzitic S1A is overlain by Unit 2 sills and low-angle dikes that swell and pinch along strike and form thick pods and lenses in Sentinel Mountain Dolostone (S1B).

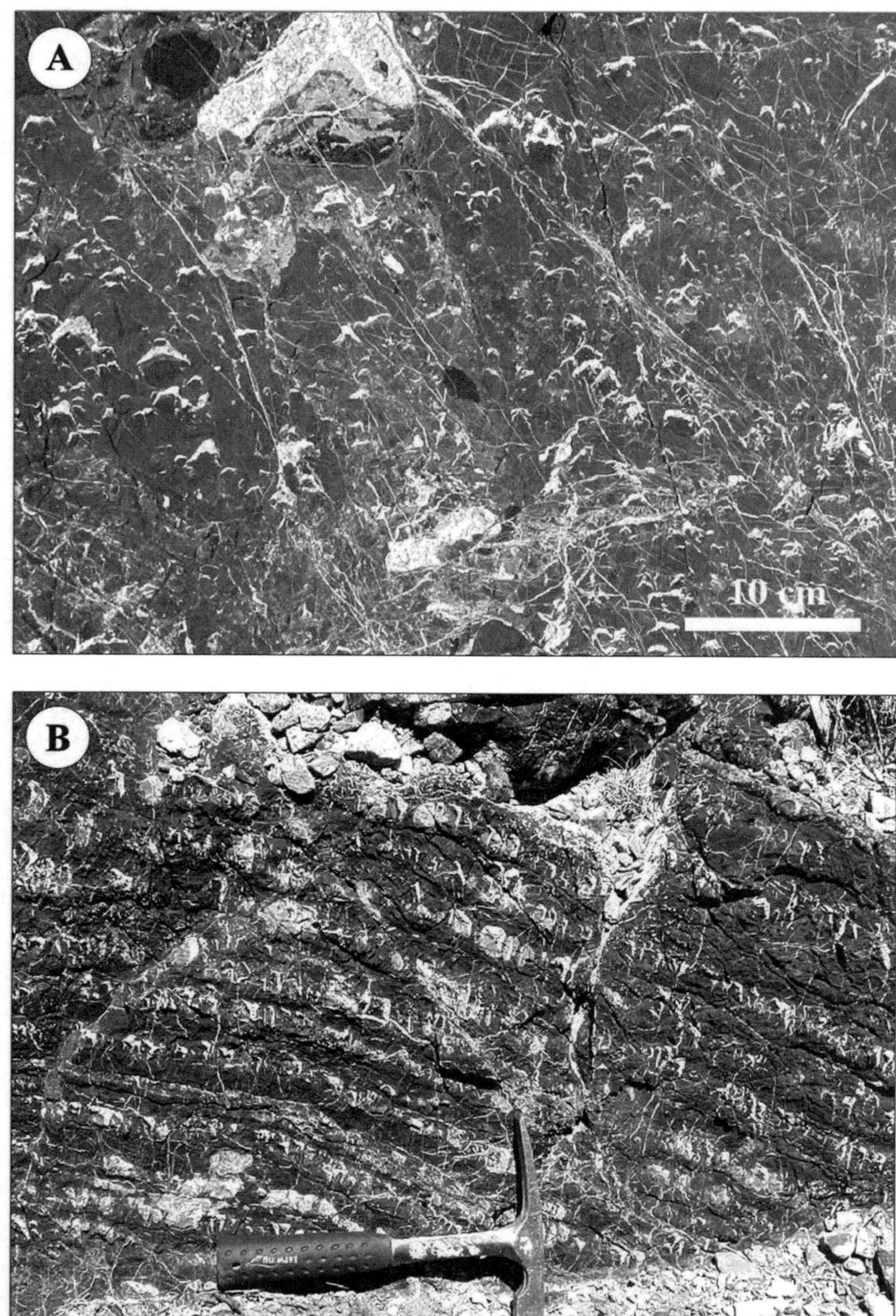

Figure 7. Arrays of shatter-cone-like and related structures in dolostones (Subunit 1B) interpreted as impact phenomena. A: Dispersed cone-shaped fractures ranging from 10 cm (upper left) to ~1 mm in cross section. White chevrons are cross sections of cone-shaped fractures that opened and only partially closed, then filled with cement that is now dolomite. B: Unusual bed-parallel rows of roughly cone-shaped structures. Hammer is 28 cm in length.

and Reimold, 2004). The absence of high-pressure indicators such as planar deformation features (PDFs) in target quartz grains suggests, however, that shock pressures must have been <8 GPa (French, 1998, and references therein). Given that peak shock waves attenuate exponentially from ground zero (Dence et al., 1977; Melosh, 1989), the degree of shock damage close to the crater rim is expected to be relatively low (French, 1998).

We conclude that features in Unit 1 at Tempiute Mountain represent deformation away from the crater center, but within the crater rim. This interpretation is further supported by the occurrence of Ordovician, and probably older, target strata in the overlying Unit 3. Deep-seated target material mostly occurs inside the tectonic rim of large impact structures (Hörz et al., 1983). Similar disturbed target strata were documented below the floors or along walls of craters at Slate Islands in Canada (Dressler and Sharpton, 1997), Sierra Madera in Texas (Howard et al., 1972), and Ries in Germany (Skála, 2002). Discussion at the end of this report and Figure 20 depict our estimate of the original position of TEM with respect to the central crater.

Shock-induced conical fractures and associated structures. Centimeter- to millimeter-scale fractures of conical and other geometries in S1B are interpreted as the result of moderate shock deformation. They occur mainly in the black facies of the middle part of the Subunit (Figs. 7A and 7B). They are not cone-in-cone, stromatactis, large-scale fenestrae, or zebra-like features, and are not specifically associated with fractures or faults. Abundant examples were created when portions of the dark beds were displaced downward along cone-shaped fractures, leaving open space that was later filled with white cement that created sedimentary structures with distinct light-over-dark segments. The spaces may have filled with instantly fluidized rock that served to keep them open. Within individual beds, the distribution of fractures is irregular or uniform (compare Figs. 7A and 7B), isolated or grouped, or preferentially developed along thin-bedded to laminated intervals. A spectrum of other geometries was created by the same process. They permeate some clasts as light-over-dark patches as small as one millimeter across; examples were illustrated by Warme and Kuehner (1998) but were not closely studied or interpreted by them. Thin-section examination reveals that these fractures occur where the host rock is pervasively brecciated on a millimeter scale, with only slight dilation and rotation of the tiny fragments. Contact between the fractures and the host rock is sharp. Axes of the conical fractures are oriented, with apices pointing toward bed tops (Pinto and Warme, 2004). Apical angles are commonly near 90°, but range up to 120° (Figs. 7A and 7B).

Except for the lack of longitudinal striations, which may never have developed or are obscured by diagenesis, these structures show the characteristics of shatter cones described at other impact sites. Alternatively, they may constitute a newly recognized response of carbonate target rocks to impact shock.

Shatter cone formation has been attributed to interactions between rock heterogeneities and shock wave front (Johnson and Talbot, 1964; Baratoux and Melosh, 2003), interference between wave front and reflected waves (Gash, 1971), or tensile rock fracturing along wave fronts (Sagy et al., 2004). Johnson and Talbot (1964) and Baratoux and Melosh (2003) suggested that hoop tensional stress is generated around particles in heterogeneous rocks that reduce shock wave speed and propagate conical fractures. Based on preliminary study, this model explains the conical shape, cavity formation, and displaced material inside the structures at TEM. Field and petrographic observations show heterogeneities in S1B to be mineral grains, fossil fragments, interstices, or bedding surfaces (Fig. 7B); the latter could cause the variety of nonconical light-over-dark structures that permeate patches of S1B. Milton (1977, Fig. 4 therein) illustrated shatter cones similar to those in Figure 7B, preferentially developed along thin rippled sandstone beds in the Gosses Bluff crater in Australia.

Unit 2: Polymict breccia dikes and sills with quartz-rich matrix. Multiple dikes and sills penetrate Unit 1 (Figs. 4–6, 8). They contain a distinctive polymict breccia with a quartz-rich dolomicrite matrix (Fig. 8). We have identified 14 different kinds of clasts, most of which seem exotic and have not been correlated with the formations on Tempiute Mountain. The clasts are dispersed and not fitted. Breccia bodies within the fissures range up to ~12 m across and ~3 m thick; the thickest ones lie along the top of Subunit 1A, where they pinch, swell, and abruptly change direction at generally low angles (Fig. 6). Most dikes are dome- or lens-shaped, and branch, pinch out, or bifurcate complexly over short distances. Sills are more common, range from ~40 to 80 cm in thickness, and extend for a few to tens of meters along outcrops. A distinctive, irregular, 2 to 5 cm thick finer-grained monomict breccia commonly separates wall rock from the central polymict breccia (Fig. 8A). It apparently developed as a buffer and contains tabular and wedge-shaped fragments of wall rock preserved in the process of being torn away.

Framework accounts for ~20%–75% of Unit 2 breccia, and includes diverse fragments of dolostone, sandstone, invertebrate fossils, and rare preexisting breccia of unknown origin. Clasts range from a few millimeters to ~2 m, commonly ~3 cm. They are angular to subangular, poorly to fairly sorted, and either display a chaotic organization or grade from coarse to fine from the center to the walls. They have well-defined edges and sharp contacts with Unit 1 rocks (Fig. 8A). Invertebrate fossils occur in clasts or loose, and include domal stromatoporoid heads, gastropods, and rugose corals. Larger clasts occur mainly in the larger bodies at and near the base of S1B, where they were clearly sourced from S1A. Several maintain remnants of Oxyoke Canyon Sandstone bedding but are recumbently folded and/or twisted (Fig. 8B).

Matrix accounts for ~25%–80% of the breccia, and consists of 5%–20% quartz grains embedded in a uniform dolomicrite. They are subangular to rounded, poorly sorted, fine- to medium-grained, and very similar to those in S1A Oxyoke Canyon Sandstone. Most quartz grains are monocrystalline; <10% are polycrystalline. Many exhibit strong undulose extinction; others display unevenly spaced, curved fractures. Some grains contain up to three sets of evenly spaced, planar parallel cleavage interpreted as planar fractures (PFs).

Figure 8. Unit 2 injections. A: Discordant contact, with dike (above) sheared across beds of Subunit 1B Sentinel Mountain Dolostone (below). Note thin light-hued transitional monomict breccia at margin of dike. Hammer head is 16 cm in length. B: Doubly folded and partly fluidized clast of Subunit 1A (Oxyoke Canyon Sandstone) carried along wide Unit 2 dike with otherwise homogeneous fill. Marking pen in upper center is 13 cm in length.

Interpretation. We interpret Unit 2 breccia as the result of forceful injection during cratering. The general crosscutting relationships and contrasting compositions between the breccia and country rock (Fig. 8A), ductile and brittle deformation of large clasts (Fig. 8B), and exotic quartz grains all suggest powerful and instantaneous injection such as attained during hypervelocity collisions (French, 1998). The association of Unit 2 breccia with the fractured, shattered, and fluidized dolostones of S1B supports this concept.

Kuehner (1997) and Warme and Kuehner (1998) interpreted Unit 2 as part of a dike-and-sill system intruded into the host rocks by ascendant flow during the Alamo Impact Event. However, our geometric analyses indicate that Unit 2 breccia was mainly injected laterally from the impact point as shown in the summary diagram of Figure 20. Dike and sill emplacement most likely occurred along fractures or dilation paths (Figs. 6, 8, 20) during the excavation stage of cratering (Melosh, 1989; Spray, 1998). Injection pathways probably consisted of fissures or zones of weakness newly developed in response to shock disturbance (Collins et al., 2004).

Comparable polymict breccia dikes have been documented in other impact structures, including Chicxulub in Mexico (Dressler et al., 2003), Popigai in Russia (Kettrup et al., 2003), Ries in Germany (Stöffler et al., 1977), Slate Islands in Canada (Dressler and Sharpton, 1997; Dressler et al., 1999) and Lockne in Sweden (Sturkell and Ormö, 1997). Plastically deformed clasts and loosened quartz grains derived from S1A (Fig. 8B) suggest that bedrock was weakened and moved away. A similar brecciation process is expected below the crater floor and/or wall as a result of acoustic fluidization (Melosh, 1989). Oscillatory reworking and/or repeated injection probably occurred during the modification stage if the mass slumped into the crater and adjusted to the new location (Dressler and Reimold, 2004).

Unit 2 breccias were most likely injected radially away from the impact point and preserved beyond the transient crater, as depicted in Figure 20. Variably shocked quartz grains and the diversity of exotic clasts support this interpretation. The restriction of Unit 2 to TEM suggests that rocks there are the closest known to the missing central crater.

Unit 3: Polymict chaotic fallback breccia. Unit 3 is a continuous, chaotic, allogenic breccia pile mapped across the top of S1B (Figs. 3–5). The huge exclusively dolostone slabs at the top of Unit 1 are replaced by the meter-scale, predominantly limestone clasts of Unit 3. Unit 3 is 15–30 m in thickness and composed of 22 or more lithologically different kinds of clasts (Figs. 9, 10) and a diverse matrix (Figs. 11A–11F). Larger fragments are partially enclosed and injected by the finer-grained matrix that contains recently found shocked quartz grains (Fig. 11F) and rare accretionary lapilli (Figs. 11B and 11C). Except for a few identified dolostone and sandstone blocks that indicate penetration as deep as Ordovician formations, most lithologies are unfamiliar; they probably represent the maximum diversity of components in the Alamo Breccia.

Framework accounts for 75%–80% of the breccia and includes a variety of chaotically mixed exotic limestone and subordinate dolostone and sandstone clasts (Figs. 9, 10). The separate allogenic clasts range from ~20 cm to >20 m across. The largest ones are mainly along an uneven surface at the top of Unit 1; most clasts in the upper part range from ~2 to 5 m across and tend to be sorted by size along irregular layers. Clasts are block- to slab-shaped, angular to subangular, rarely rounded, and tightly packed (Fig. 9). Most exhibit diffused and deformed margins (Fig. 10), and some deeply interpenetrate. Limestone clasts constitute >85% of the framework and vary from light gray to black, massive to well bedded or laminated, and are generally devoid of obvious macrofossils. Dolostone clasts account for <10% and consist of light gray, thin-bedded to massive facies similar to those of the Sevy and Laketown Dolostones. Sandstone clasts represent <5% of the framework, and are of two kinds. A quartzose sandstone is petrographically similar to the Eureka Quartzite. A finer-grained or cherty sandstone was not recognized in the stratigraphic section at TEM and is tentatively correlated with the Cambrian Prospect Mountain Quartzite. A lens-shaped clast of unusual globular microtexture (Fig. 11A) occurs in the middle part of the Unit, and is described under the following heading.

The matrix is also a breccia (Fig. 10), accounts for <25% of the Unit, and is squeezed between clasts; no clast >20 cm is entirely enclosed by matrix. The matrix is a mixture of limestone, dolostone, sandstone, and subordinate invertebrate fossil fragments set in a gray, finer-grained, quartz-bearing calcareous groundmass. Invertebrate fragments include stromatoporoids, rugose corals, brachiopods, calcispheres, probable sponge spicules, and undetermined forms. The margin of some clasts was preserved in the process of transforming into matrix. The matrix locally exhibits flame-like structures produced by interlayered black bands of limy micrite and contains carbonate accretionary lapilli, probable carbonate spherules, and rare carbonate clasts enveloped by the black bands (Figs. 11B–11E).

Microscopic components in Unit 3. Petrographic examination of Unit 3 rock thin-sections reveals sedimentary particles from preexisting formations mixed with unusual microscopic components that were generated during the Alamo Event. Their analyses and interpretation have only begun, but these components seem undescribed in other impact studies. They include several kinds of accretionary lapilli, shocked quartz grains, unusual carbonate crystals, mineral aggregates, probable spherules, and a variety of possible melt or partial melt products.

Figure 11 illustrates selected examples from Units 3 and 4. Figure 11A shows the globular texture of a lens-shaped clast from the middle of Unit 3. The millimeter-scale rounded to oval globules are composed of calcite enclosed by a blocky to fibrous silica rim that was perhaps cement. Smaller examples have a micritic texture; larger ones typically have bladed calcite crystals that form a feathery texture. Most calcite crystals seem randomly oriented, but some radiate from local centers. This clast contains many voids that represent dissolved globules, surrounded by the tiny silica crystals. Except for their limy composition and feathery texture, the globules resemble dolomite spheroids described by Ocampo et al. (1996) from the K-T

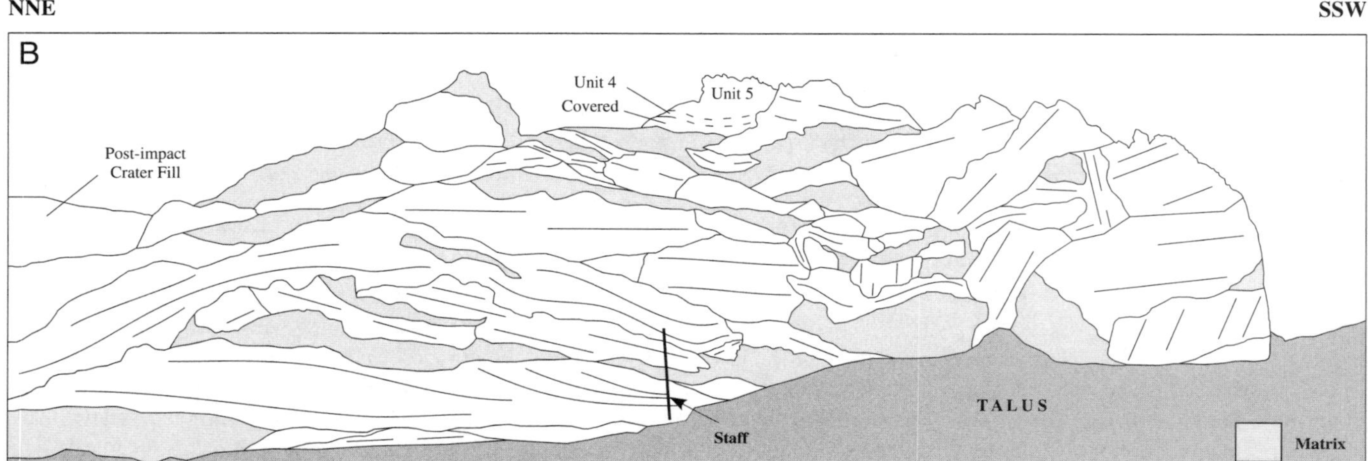

Figure 9. View eastward of upper half of Unit 3 outcrop on ridge of study section (Figs. 3–5). A: Photo of heterogeneous clasts and outline of larger clast boundaries. B: Diagram showing chaotic distribution of clasts and matrix. Shaded areas are patches of matrix containing only smaller clasts. View is ~25 m wide; staff is 1.5 m long.

impact ejecta blanket on Albion Island, Belize. The spheroids were inferred to represent fallout from the Chicxulub Impact vapor plume, possibly as liquid carbonate.

Figures 11B and 11C are two kinds of accretionary lapilli from the matrix of Unit 3. Figure 11B illustrates an oval lapillus ~2.3 mm across composed of nucleus, mantle, and crust (terms recommended by Warme et al., 2002). The nucleus contains blocky calcite crystals rimmed by a smooth, invaginated wall. A small carbonate fragment in the center may be an inner nucleus. The mantle is ~0.1–0.5 mm thick, graded fining outward, and consists of one layer of very fine- to fine-crystalline calcite and minor rounded, unaltered and shocked tiny quartz grains and fine-grained carbonate clasts. The crust is a dark, <0.1 mm thick, wavy boundary against adjoining components, and consists of thinly laminated hematite-rich micrite. This example is similar to some lapilli documented in the Ring Realm, but the invaginated boundary of the nucleus is unusual. The nucleus may represent a recrystallized particle, or an internal gas- or liquid-filled vesicle that was deformed upon impact or by early burial, then later filled by crystallization of the liquid or by burial cement. The hematite-rich crust may represent a chemical reaction during the last phase of ballistic transportation or shortly after deposition. Volcanic analogues precipitate several kilometers away from pyroclastic vents (Schumacher and Schmincke, 1995).

Figure 11C shows a drop-shaped accretionary lapillus composed of a nucleus, compound mantle, and thin crust. The nucleus is a medium-sized, rounded quartz grain barely resolvable in the photo. The mantle is ~1.2–1.5 cm across and contains three concentric layers of very fine- to fine-crystalline calcite, minor fine-grained and rounded quartz grains, opaque minerals, and fine-grained carbonate clasts. The inner layer is homogeneous, very fine-crystalline calcite, and resembles some Unit 3 matrix. The middle and outer layers grade outward. The crust is partially rimmed by an ~0.1 mm thick, thinly laminated, hematite-rich micrite, which penetrates the matrix. Multiple concentric layers are rare in the better-known lapillistone of the Ring Realm, but

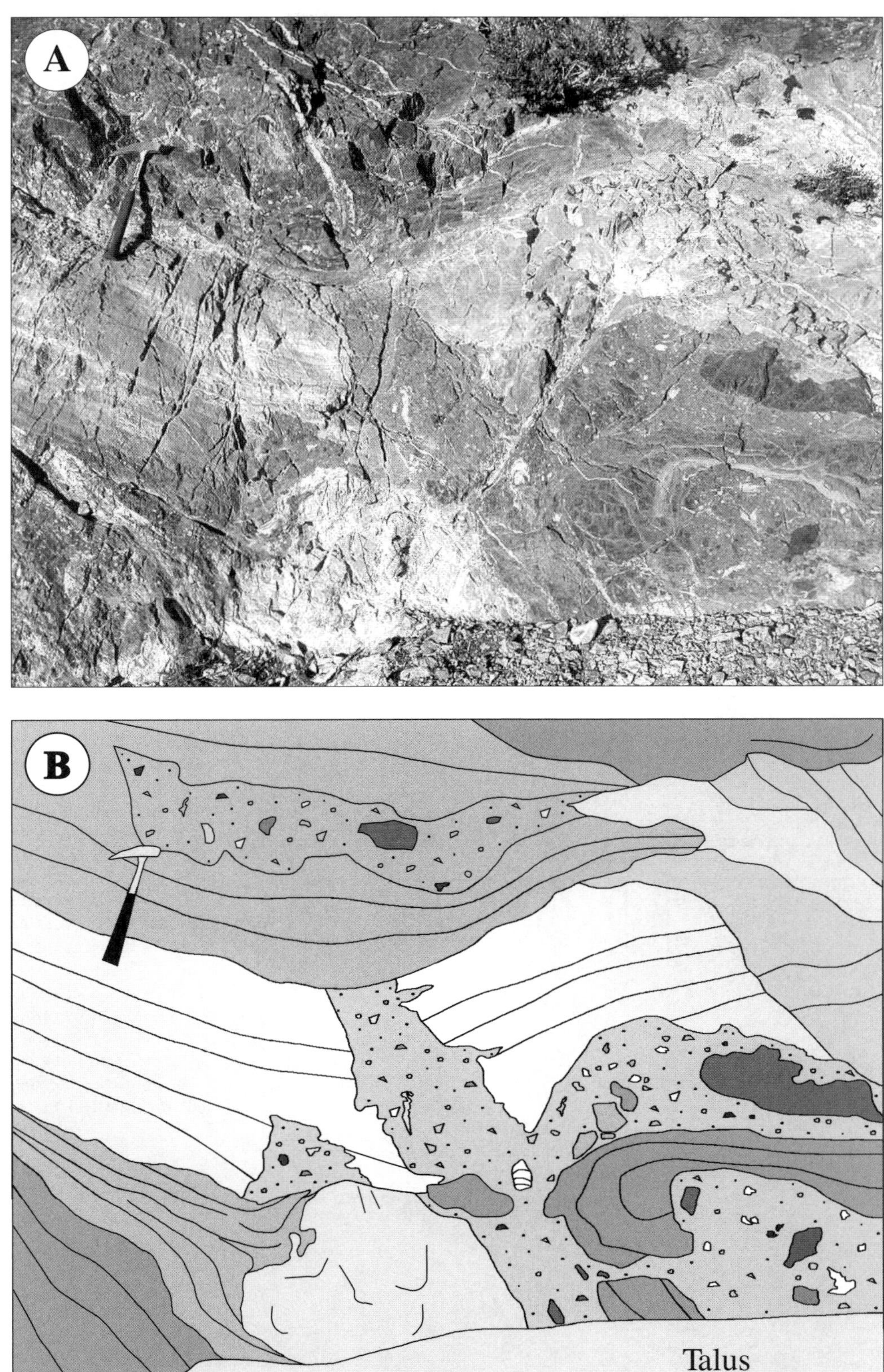

Figure 10. Photograph and drawing of Unit 3 matrix and clast relationships. Clasts are cracked and broken, sheared and deformed at margins, and injected by matrix that is both imported and locally generated. Hammer is 28 cm in length.

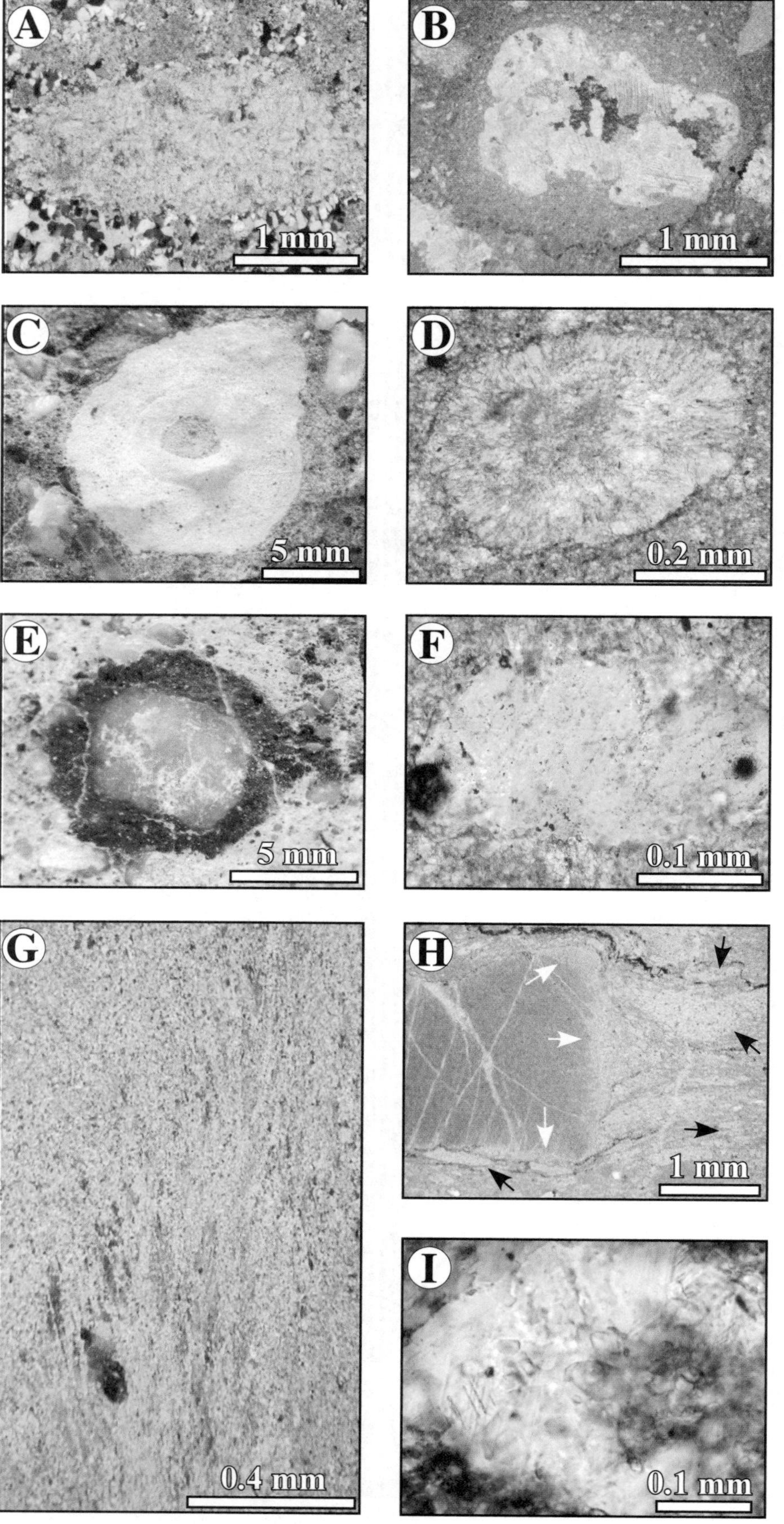

Figure 11. Thin-section photomicrographs and surface photos (C, E) of microscopic and small-scale features of Unit 3 (A–F) and Unit 4 (G–I). All photomicrographs in cross-polarized light. A: Globular calcite particle from homogeneous clast, showing microscopic feathery crystals that are potentially quenched melt, encased in rim of silica crystals. B: Accretionary lapillus from matrix with invaginated nucleus that may represent originally gas- or liquid-filled cavity later filled with cement. C: Lapillus from matrix showing history of multiple accretion. Nucleus is tiny quartz crystal (barely resolvable in photo) surrounded by three cycles of mantle accretion and coated with thin hematite-bearing crust. D: Calcite body with irregular nucleus, mantle, and thin crust. Mantle exhibits bladed calcite that locally displays a feathery texture that implies crystallization from melt. E: Spherical particle of calcite and amorphous silica enclosed by black band of hematite-rich microcrystalline calcite that originated by flowage from right side. F: Quartz grain with three sets of decorated planar deformation features (PDFs) (not focused) and embedded euhedral hematite crystals. G: Unit 4 matrix with concentration of feathery calcite crystals that radiate at acute angles, together with other bladed and fibrous calcite. H: Small clast with altered edges (white arrows) enveloped by flow structure with augen-like patches (black arrows). I: Quartz grain with as many as four sets of poorly preserved PDFs, mottled extinction, and small, embedded hematite crystals.

similar examples were reported proximal to volcanic vents by Schumacher and Schmincke (1995), who suggested that multiple accretion cycles occurred during flight in turbulent clouds of particles. Our example may represent three cycles of rapid accretion in the expansion phase of the turbulent vapor cloud and early rapid precipitation near the crater, which fits our genetic scenario for Unit 3 (see *Summary: Alamo Impact Structure*).

The oval calcite body in Figure 11D is from Unit 3 matrix, is ~0.5 mm across, and has a three-layer structure somewhat similar to accretionary lapilli. The nucleus is ~0.15–0.2 mm across and consists of microcrystalline calcite. Its boundaries are indistinct, probably recrystallized. The mantle is ~0.06–0.16 mm thick and composed of radiating crystals of fibrous or bladed calcite, which locally display a feathery texture; compare with Figure 11G. These crystals radiate from the nucleus both inward and outward. Minor opaque minerals are present in both the nucleus and mantle. The crust is <0.05 mm thick, irregular, and consists of hematite-bearing microcrystalline calcite, slightly finer-crystalline than the Unit 3 matrix. This particle resembles spherules described from ejecta beds in Precambrian successions in Africa and Australia (e.g., Lowe et al., 2003; Simonson, 2003) and ejecta deposits associated with the Chicxulub Event in Mexico (Schulte and Kontny, 2005). The radiating calcite crystals suggest nucleation cores along the outer and inner boundaries of a liquid carbonate mantle. Alternatively, the body may represent a biogenic or other sedimentary particle, but the lack of organized concentric or radial growth layers, the eccentric nucleus, and the fan-shaped, bladed calcite crystals of the mantle that radiate both outward and inward are inconsistent with such interpretation (Simonson, 2003).

Figure 11E shows a spherical body, ~1 cm across, from the Unit 3 matrix that is composed of a nucleus and a mantle that mimics accretionary lapilli. The nucleus is an ~6 mm diameter, rounded carbonate clast of white, milky, microcrystalline calcite. The patchy texture results from inclusions of amorphous silica. The clast is enclosed by a black band of homogeneous fine-crystalline calcite. The band composition appears similar to that of the flame-like structures in Unit 3 matrix. In this example, the black calcite apparently fed from the right side of the photo and encased the nucleus. Similar-looking bodies occur in Chicxulub ejecta (Schulte and Kontny, 2005). They were interpreted as lapillus-like spherules based on the concentric accretion around the nucleus. In our example, however, the black mantle lacks concentric structure, was clearly fed through the matrix, and apparently has no equivalent in other impact deposits that have been described. The closest comparison comes from clasts wrapped by fiamme in pyroclastic deposits from the Canary Islands, interpreted as interstitial flowage of hot, fluid volcanic ejecta that then cooled within the hot volcanic pile (Kobberger and Schmincke, 1999). Unit 3 black bands may represent hot or partially melted carbonate that flowed during degassing, dehydration, or initial compaction of the deposit (Warme et al., 2002).

Figure 11F shows a medium-grained, subangular quartz grain from Unit 3 matrix that contains as many as three sets of poorly exhibited PDFs. The fractures are decorated, indicate shock pressures >10 GPa (French, 1998), and have inclusions of the subhedral to euhedral hematite crystals (black) that are uniquely associated with quartz grains in the Alamo Breccia wherever it occurs (Warme and Sandberg, 1995, 1996; Morrow et al., 2005; Morrow and Sandberg, 2006), and to our knowledge not reported from other impact cites. They were identified as pseudomorphs after magnetite or pyrite (Warme et al., 2002; Morrow et al., 2005). Such crystals grew and commonly displaced quartz in grain interiors and around peripheries to yield a studded appearance (Fig. 11F). The volume of original quartz in some studded grains was reduced by replacement to less than 50%. The crystals also occur as tiny internal inclusions, along planar microstructures of quartz grains, in various sizes scattered in breccia matrices, and in the flame-like structures in lapillistone (Warme et al., 2002). They are so abundant in the various products of the Alamo Event that they represent a large total volume of iron. We speculate that a volatized iron meteorite or iron-rich comet could provide sufficient free iron to source the original magnetite and pyrite crystals.

The largest clasts in Unit 3 were previously equated with the Unit C megaclasts in the Ring Realm (Fig. 2). However, Unit 3 clasts reach a maximum size of only a few tens of meters and do not match the underlying Unit 1 shattered dolostone bedrock, whereas the Ring Realm clasts are as long as 500 m and are delaminated from similar but undisturbed bedrock beneath them.

Interpretation. We reinterpret Unit 3 as a fallback breccia. It contains a spectrum of variously sized exotic clasts that include Ordovician Eureka Quartzite, a complex matrix with carbonate accretionary lapilli, and a stratigraphic position above the highly disturbed autogenic dolostones of Unit 1. The identification of the clasts is unfinished; most of them lack obvious macrofossils. However, one yielded conodonts that suggest a late Middle Devonian or slightly younger age (C.A. Sandberg via J.R. Morrow, 2006, personal commun.), approximately equivalent to the Fox Mountain Formation that underlies the base of the early Late Devonian Guilmette Formation. Unit 3 likely rests upon some level of the Sentinel Mountain Dolostone (or Bay State Dolostone; Sandberg et al., 1997); thus some or all of the overlying Fox Mountain was ejected and expected to contribute clasts to the fallback breccia, and to debris cascading from the inner crater rim.

Unit 3 was most likely deposited just beyond the transient cavity and preserved in the adjusted crater rim. As mentioned by Hörz et al. (1983), deep-seated ejected target rocks are also expected to accumulate at the inner crater rim of large impact structures, where they represent late-stage excavation when big clasts were no longer jetted beyond the crater rim (Nolan et al., 1996). The dominance of unfamiliar clasts in Unit 3 suggests that they represent early Paleozoic and perhaps older formations. In earlier stages similar clasts were probably ejected far outside the rim and occur as unidentified components of the Ring Realm and beyond. Some of the fractured younger strata along the crater rim likely slumped into the crater during the modification stage. Similar fallback deposits were documented above disturbed target rock in other impact structures, for example the megabreccia of Popigai, Russia (Masaitis et al., 1999; Vishnevsky and Montanari, 1999), and the Bunte Breccia of Ries, Germany (Pohl et al., 1977; Engelhardt, 1990). The summary Figure 20 depicts the stratigraphic column at TEM to represent breccias formed during the evolution of the transient crater, then preserved during crater adjustment.

Unit 4: Polymict smeared fallback breccia. Unit 4 is a continuous band of distinctive Alamo Breccia above Unit 3 (Figs. 3–5), but varies laterally in thickness, composition, and texture. It is 10–15 m in thickness, and to date contains 15 documented different kinds of clasts that are much smaller and better sorted than in Unit 3, set in a quartz-bearing calcareous matrix (Fig. 12). The lower contact with Unit 3 is sharp, appears to be controlled by the pre-existing large-clast topography along the top of Unit 3, and was emplaced during a later phase of the Alamo Event. Across this boundary, dolostone and sandstone clasts increase at the expense of limestones. The most striking feature is local expression of a marbly, fluidal fabric, which is evident in both outcrop and thin sections (Figs. 11 and 12). Evidence for partial melting is feathery or bladed calcite crystals (Fig. 11G) and augen-like flow structures (Fig. 11H). Quartz grains show low- to moderate-stage shock metamorphism (Fig. 11I).

Framework clasts occupy 70%–80% of Unit 4 breccia, composed mainly of comparable amounts of limestone and dolostone and less quartzose siltstone and sandstone. Clasts are generally elongated, subangular to rounded, rarely angular, and tightly packed. Sizes range downward from 15 cm in the longest dimension; most are less than 5 cm. The largest clasts are limestone, commonly broken stromatoporoids. Figure 12 shows the response of different kinds of clasts to deformation and alteration according to their lithology and/or competence, and how some deformed to give the smeared appearance. The matrix accounts for 20%–30% of Unit 4 breccia. It is micro- to fine-crystalline calcite (>87%) that contains limestone fragments, broken invertebrate fossils (<10%), quartz grains (<3%), and bladed calcite crystals of the same feathery structure that occurs in Unit 3.

Microscopic components in Unit 4. Figures 11G and 11H illustrate microscopic carbonate components in the matrix of Unit 4. The feathery texture in portions of the matrix is shown

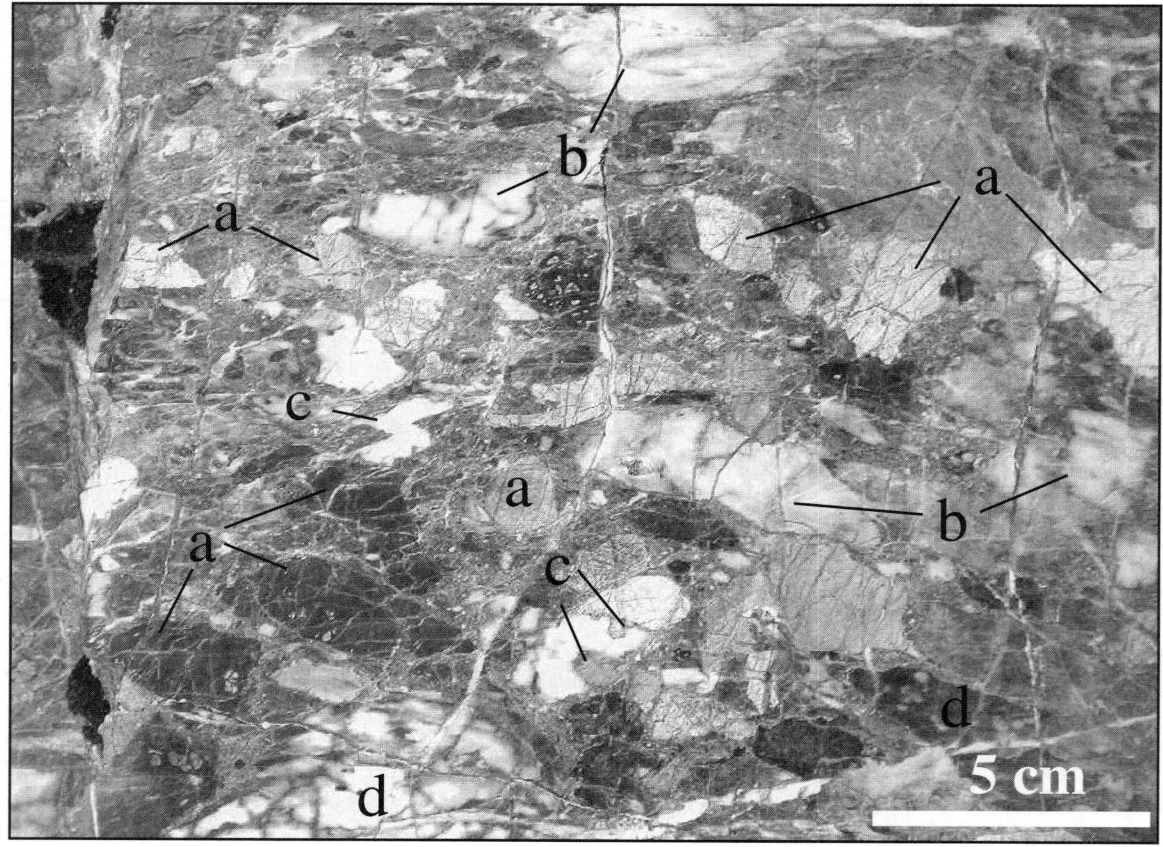

Figure 12. Unit 4 sorted and smeared clasts. Unit may have been hot when deposited. Labels indicate how different lithologies responded to shear and/or compaction as evidenced by aligned clasts: a—fractured brittle limestones and dolostones; b—smeared, blurred limestones with smoothed edges; c—ductile, plastic limestones commonly interpenetrated by dolostones; d—fractured, dilated, and partially smeared clast. Smaller clasts appear structureless or undeformed.

in Figure 11G. Bladed and minor fibrous calcite crystals are arranged at acute angles. This texture is considered a signature for crystallized carbonate melt, for example in the Chicxulub structure (Claeys et al., 2003; Deutsch et al., 2003). If this texture signifies quenched carbonate melt, then other unusual co-occurring components may represent new classes of melt and quench products.

Figure 11H displays a microscopic deformation fabric in the matrix of Unit 4. Stretched patches of fine-crystalline calcite (black arrows) form eye-shaped lenses ("augens") in a microcrystalline calcite groundmass. The augens vary in size from less than 0.1 mm to 1 mm and are elongated along fluid pathways, in this case around a small grain whose margins were altered and eroded (white arrows). Fractures in the clast are truncated at the margins. Irregular, dark, hematite-rich seams in the matrix crosscut formation of the augens. These augen-like structures are similar to metamorphic fabrics developed under high strain rates in tectonic settings (Borradaile et al., 1982). The fabric is also similar to that documented in water-saturated pyroclastic deposits (e.g., Kobberger and Schmincke, 1999), but lack observed welded particles. Flowage may be caused by late-stage thermal cracking and/or fluid reactions during cooling, a process also reported in volcanic ejecta (e.g., Wolff and Wright, 1981; Allen, 2004), or simply by burial effects. Similar structures occur in hydrothermally altered carbonate rock, suggested as decalcification during emplacement of hot fluids (Spirakis and Hey, 1988).

We suggest that Unit 4 fabric results from still hot secondary flow, as proposed for the origin of the black fiamme-like structures in the matrix of Unit 3 (Fig. 11E) and the lapillistone of the Ring Realm. Given the likely hot matrix of Unit 3, such clast alteration may result during flowage and cooling of the matrix grains and/or related fluids.

The medium-sized, angular quartz grain of Figure 11I from the Unit 4 matrix displays the common small subhedral hematite inclusions (black dots) and as many as four sets of PDFs that are not well shown in this view. This grain indicates shock pressures >10 GPa (French, 1998).

Interpretation. Unit 4 has an apparent bulk composition similar to Units 3 and 5, which suggests the same provenance, but differs in clast size, fabric, texture, and modal composition. We interpret Unit 4 as a heated, partially melted fallback breccia that had a complex depositional history. The feathery calcite structure in Figure 11G is interpreted as quenched melt, as suggested in other impact structures (Claeys et al., 2003; Deutsch et al., 2003). The broad variety of clasts, shocked quartz grains, plastically deformed clast edges, probable carbonate crystallization from melt, and preferred orientation of clasts all suggest fallout deposition of heated debris and redeposition of the combined slurry over the fallout breccia of Unit 3, most likely inside the crater rim (Fig. 20).

Combined processes could include (1) fallback deposition of hot, viscous, vaporous ejecta that was sorted in flight during a late phase of transient crater evolution, (2) collapse down the inner slope of the cavity, which could also sort the clasts, or ponding in depressions near the rim, (3) friction, strain, and possibly melting produced by shearing, and (4) flattening while in a weakened state during flow or early compaction from overlying resurge debris (Unit 5). Fluidization by shearing and partial melting during secondary flow was discussed by Spray (1998).

Unit 5: Polymict resurge Alamo Breccia. Unit 5 is a remarkably thick and homogeneous deposit of resedimented carbonate clasts that strongly contrasts with the thinner but more complex clast-and-matrix fabric of underlying Units 3 and 4 (Fig. 5). This uppermost Alamo Breccia interval at TEM forms a conspicuous wide outcrop band across the whole mountain (Figs. 3–5). It ranges from ~100 to 120 m in thickness and is composed of only two graded beds. It contains 15 or more clast types as well as typical Alamo Breccia shocked and studded quartz grains. These clast-supported, clean-washed, and well-sorted beds (Fig. 13) resemble those of Unit A in the Ring Realm (described below) but are much thicker, better sorted, and finer-grained. The lower contacts can be gradational but are commonly erosional and irregular with relief as much as ~1 m. Clast sizes in the lower bed range from ~30 cm near the base to very coarse sand at the truncated top, and from granules upward to carbonate sand and mud in the upper bed. These beds generally lack traction current structures, except for subtle wavy cross-bedding and thin parallel laminations in the topmost meter of the upper bed. Local concentrations of slightly coarser material form bands and lenses as thick as 10 cm that we interpret as depositional pulses and possible local mixing from loading and dewatering.

Framework accounts for ~80%–90% of the breccia and comprises clasts of limestones, dolostones, stromatoporoids, and subordinate quartzose sandstones and siltstones (Fig. 13). Near the top of the lower bed and throughout the upper bed the framework is clast-supported and generally very well sorted. Clasts are generally equant, subangular to well rounded, and rarely angular. They are generally loosely packed, except at the base of the lower bed where they are compact and display preferred orientation and imbrication that indicate a southwest paleoflow.

The matrix accounts for ~10%–20% of the breccia and consists of a gray to dark-gray micrite, probably mostly cement, that contains quartz grains (<1%) and whole or fragmented invertebrate fossils (~5%). The fossil assemblage includes brachiopods, calcispheres, sponges, stromatoporoids, corals, and probable sponge spicules and ostracodes.

Quartz grains are studded with the subhedral to euhedral hematite crystals typical of those recovered in all Alamo Breccia locations. The grains represent much less than 1% of the volume of Unit 5, and are concentrated in the upper part. They are fine- to medium-grained, subrounded to well rounded, and moderately sorted. Grains are largely monocrystalline; less than 10% are polycrystalline. Most exhibit undulatory extinction, and ~20% contain up to three sets of PFs and PDFs that indicate moderate- to high-stage shock metamorphism. Morrow et al. (2005) reported that ~15% of grains recovered in insoluble residues from conodont samples contain more than three sets of decorated PDFs.

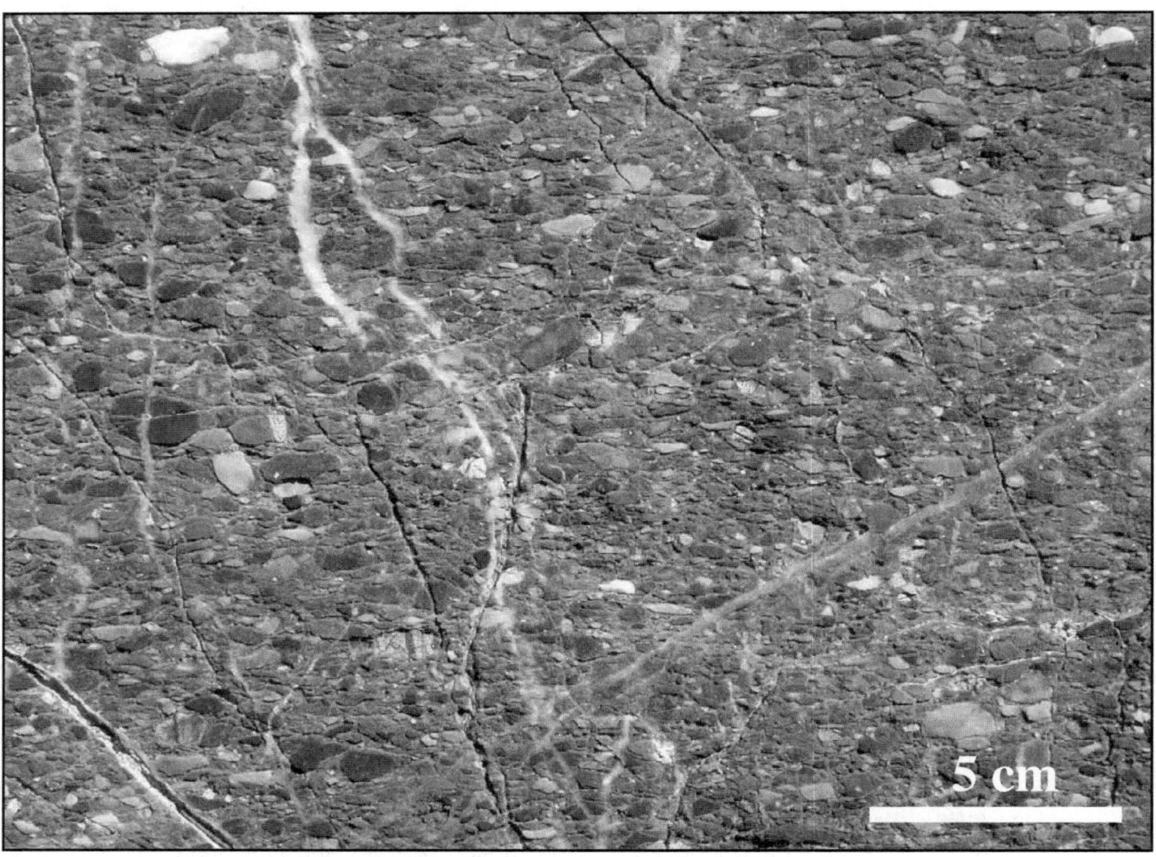

Figure 13. Unit 5: Well-sorted, clast-supported, clean-washed resurge breccia typical of the middle part of the lower graded bed.

Interpretation. Units 3–5 were measured as ~100–135 m in thickness on three different ridges on Tempiute Mountain (Chamberlain and Warme, 1996; Warme and Kuehner, 1998; Pinto, 2006; this report, Fig. 5), and designated as part of the Alamo Breccia (e.g., Warme and Sandberg, 1995, 1996; Warme and Kuehner, 1998) based on the graded beds and other superficial likenesses with the Breccia in the Ring Realm. Units 3 and 4 are now differentiated by their exotic clasts and unusual matrices that imply heat and rapid, high-energy transportational and depositional processes. The thicker Unit 5 was probably deposited cool, has a very different clean-washed fabric, and is interpreted as presorted, resedimented resurge into the crater. However, we still regard all three Units as allogenic Alamo Breccia. Alamo Breccia in the Ring Realm similarly is interpreted to contain both primary and reworked ejecta as well as reworked carbonate platform debris.

Various thicknesses of the Units 3–5 interval may reflect filling of irregular topography along the crater wall as the cavity adjusted, or of slope scars where parts of the crater slid away during late phases of the Alamo Event. Evidence for such adjustment is the repeated and omitted Oxyoke Canyon Sandstone at the base of the Unit 1 breccia pile (Fig. 3). Unit 5 compares with other impact-related resurge breccias at Chesapeake Bay in the eastern United States (Exmore Breccia: Poag, 1997; Horton et al., 2005), Kaluga in Russia (Masaitis, 2002), Lockne in Sweden (von Dalwigk and Ormö, 2001), and Mjølnir in the Barents Sea (Dypvik et al., 2004).

Post–Alamo Event beds. The post-Event interval rests conformably over Alamo Breccia Unit 5 (Figs. 3 to 5). The diffuse graded bed at the top of the Breccia imperceptibly merges with the overlying similar fine-grained, dark gray limestones at the base of the post-Event deepwater marine beds. This younger interval is ~235 m thick and was described, dated and identified as the Frasnian and Famennian Devils Gate Limestone by Sandberg et al. (1997). The lower few tens of meters are absent of benthic shelled invertebrates except for very rare, monospecific, disarticulated and resedimented brachiopod shells. Beds are laminated, and the few ichnofossils are small-scale, lack diversity, mostly resemble *Chondrites*, and are restricted to intervals only a few centimeters thick. However, pelagic fossils are present that include calcispheres, stylolinids, conodonts, ostracodes, and other forms of possible planktonic or nektonic habit (Sandberg et al., 1997; Morrow, 2000; Morrow and Sandberg, 2003b).

Interpretation. Units 3–5 separate the underlying autogenic outer platform-margin Sentinel Mountain breccias of Unit 1 from an overlying deepwater facies of the Alamo Breccia that accumulated well below wave base. The early post-Event beds were deposited under anaerobic or dysaerobic benthic conditions. Low diversity and scarcity of shelly benthos and ichnofauna indicate that the beds were completely isolated from the contemporaneous shallow Guilmette platform of the more eastern Alamo Breccia Realms, but possible planktonic and nektonic forms indicate connection of overlying surface waters with an open marine environment (Warme and Kuehner, 1998). This restricted benthic habitat could represent resumption of slope facies of the Devils Gate Limestone that was assumed to already exist at TEM at the time of the Alamo Event (Sandberg et al., 1997), deposition across a new slope created by massive sloughing of the continental margin as part of the Alamo Event, or accumulation inside or across the adjusted rimmed or partially rimmed crater that was preserved.

On the basis of conodont ages and facies resemblances, Sandberg et al. (1997) and Morrow (2000) compared the post-Event interval at TEM with the deepwater upper member of the Devils Gate Limestone at its type locality in central Nevada (DVG, Fig. 1). They labeled the TEM beds as Devils Gate, and implied that an eastward-directed salient of slope facies rather than platform facies existed at TEM at the time of the Alamo Event. However, at DVG the Alamo Breccia occurs only as a meter-scale channel fill within the Runoff Realm. The channel is intercalated with cyclic shallow-water carbonate platform facies of the lower member of the Devils Gate, which persisted there as ~100 m of platform facies after Alamo Event time (Sandberg et al., 1997, 2003, 2005), in contrast with the post-Event, deeper-water, restricted limestones of the same age over the Alamo Breccia at TEM. Drowning of the platform and the shift to deeper-water facies that characterize the upper member at Devils Gate occurred during the Early *rhenana* conodont Zone (Sandberg et al., 1997, 2003), which was approximately three million years later than the *punctata* Zone age of the Alamo Breccia (Kaufmann, 2006). At TEM, the Early *rhenana* Zone occurs above ~60 m of deepwater limestones, containing three conodont zones, over the top of the Alamo Breccia (Morrow, 2000). If the much later transgression at DVG was relatively quick and regional, then the surficial target rocks at TEM may have also still been shallow carbonate platform or platform margin facies that were blasted away during impact, and may help account for the abundant shallow-facies domal stromatoporoids that characterize the Alamo Breccia of the Rim, Ring, and Runout/Resurge Realms. These relationships suggest that instantaneous deepening occurred at TEM, as a result of the Alamo Event that was as much as three million years before transgression from platform to slope facies at Devils Gate. Although the initiation of somewhat similar deepwater facies is diachronous, we have used Devils Gate Limestone to label the post-Event beds between the Alamo Breccia and the Pilot Shale at TEM (Figs. 3–5). Because the Alamo Breccia also contains independent paleoecological evidence for penetration of contemporary deepwater facies (e.g., Warme and Sandberg, 1995, 1996), we conclude that ejected surface rocks ranged from shallow to deep facies, so that the target zone was along the platform margin.

Significance of Units 1–5

A persistent caveat for all our interpretations is that Tempiute Mountain is likely within a thrust slice that moved a significant distance eastward toward the Ring Realm and probably also moved away from the central crater (see *Discussion of Enigmas*). However, damaged bedrock (Units 1 and 2) and early-Event impact deposits (Units 3–5) demand that TEM rocks were originally not far from the central crater. Evidence presented above suggests that Units 3–5 at TEM were likely deposited on the inner slope of the crater rim, as shown in the synthesis diagram of Figure 20 at the end of this report. Figure 20 depicts TEM at the adjusted inner rim of an intact crater, but the same Unit 3–5 stratigraphy could develop on a slide scar over Units 1 and 2. The latter solution requires that soughing must have occurred as part of the Event, not later, because Units 3 and 4 have properties that require accumulation during transient crater evolution. In either case, most of the crater may have disintegrated and is preserved only as the channeled debris and olistoliths discovered in the Runout/Resurge Realm, described below.

Ring Realm

The Ring Realm, called "Zone 2" by Warme and Sandberg (1995, 1996) and in subsequent papers by them and various coauthors, is an expanse of Alamo Breccia that covers ≥13,000 km^2 (Fig. 2) and contrasts with the Rim Realm in important ways. The Breccia crops out in ten or more mountain ranges and is usually expressed as part or all of a prominent cliff within the lower part of the shallow-water carbonate platform facies represented by the Guilmette Formation. It encompasses the type locality of the middle Alamo Breccia Member of the Guilmette (Sandberg et al., 1997) (HAS, Fig. 1). The following descriptions are used to interpret the former Zone 2 as a terrain of crater rings that formed by the action of seismic waves and creation of listric faults, quickly followed by the ejecta curtain and later by tsunami uprush, backwash, and platform drainage all coincident with listric fault adjustments (see *Discussion of Enigmas*).

Ring Realm Vertical Facies

Sandberg and Warme (1993), Warme and Sandberg (1995, 1996), and subsequent combinations of authors adopted and illustrated a fourfold vertical classification of facies for the Alamo Breccia, Units D to A in ascending order, that best represent the Breccia within the Ring Realm. These Units are described below and illustrated in three different locations (Figs. 14, 15, 17).

Unit D. Unit D is an unusual <3 m thick monomict limestone breccia that defines the base of the Alamo Breccia only in the Ring Realm. It overlies ~125 m of the lower member of the Guilmette Formation, which contains ~25 typical shorter

Milankovitch-scale (tens of thousands of years), ~1–5 m thick, upward-shallowing carbonate depositional cycles, and underlies the megaclasts of Unit C (Figs. 14, 15, 17). It was described and illustrated in previous publications (Warme and Sandberg, 1995, 1996; Kuehner, 1997; Sandberg et al., 1997; Warme and Kuehner, 1998). It is a horizontal to subhorizontal interval that extends for tens to hundreds of meters beneath the length of megaclasts that define Unit C.

This thin interval is interpreted as dilated and fluidized bedrock that was preserved along a detachment surface between intact platform carbonate beds. It developed at a shallow 50–100 m depth below the contemporary surface of the platform, probably where the influence of seismic waves traveling laterally through the platform was compensated for by overburden pressure. The underlying platform is undamaged. Monomict breccia similar to Unit D occurs at the shredded edges of some Unit C megaclasts, around some larger Unit B clasts, and in decimeter- to meter-sized pockets within the matrix of Unit B. All these are interpreted as places where rock was fluidized.

Unit C. Unit C represents an interval of discontinuous megaclasts that range up to ~500 m long and ~80 m thick. The clasts were delaminated and mobilized at or near the base of the Breccia where they captured and preserved Unit D facies (Figs. 14, 15, 17). Most are parallel to the underlying bedrock, but some are tilted a few degrees; others collided, peeled up, and were preserved as spectacular intraformational folds (Fig. 14) (Warme and Sandberg, 1995, 1996). The delaminated clasts of the platform were misinterpreted by prior petroleum, mining, and academic geologists who measured numerous stratigraphic sections of the Guilmette Formation, mainly because the megaclasts are generally parallel to bedding, contain the expected Guilmette sedimentary cycles, and the stromatoporoid-rich matrix of Unit B between and over them was regarded as reef talus (Reso, 1963; Dunn, 1979) or channel or storm deposits. Other workers mistook the matrix for fault-, karst-, or solution-collapse breccias that are widespread in the carbonate rocks of Nevada.

Unit B. Unit B is composed of chaotic, polymict, generally matrix-supported breccia that superficially resembles the products of debris flows. It overlies Unit C clasts or extends to the base of the Breccia where they have parted (Figs. 14 and 15) and Unit D was unprotected and mixed with Unit B. It is interpreted as various proportions of ejecta and platform debris that were mixed by the processes of ballistic bombardment, settling, fluid (water, steam) evasion, and downslope adjustment and/or massive evacuation of the platform.

Unit A. Unit A is a normally graded, clean-washed, clast-supported, polymict breccia, interpreted as gravity-flow deposits with sedimentary structures similar to turbidites. It is the most organized, universally present, and easily recognized Unit in the Ring and Runup Realms (Fig. 2), and similar to the two much thicker, clean-washed resurge beds of Unit 5 in the Rim Realm (Figs. 4, 5, 13). Units D to B are much more complex and variable. Unit A normally contains two or more stacked graded beds that are each thinner and finer-grained upward. Kuehner (1997) and Warme and Kuehner (1998) showed that Units B and A combined contain as many as five graded intervals. The lower ones in Unit B were detected only by statistically measuring clast lengths, which have large size ranges and are more chaotically dispersed; the better-organized and finer-grained upper graded beds are bounded by load and flame structures and are easily discriminated as Unit A. At most localities the uppermost graded interval is commonly <1 m in thickness, grades from carbonate gravel and/or sand to mud, and the upper part is burrowed, indicating a return to marine habitat after the Alamo Event (Tapanila and Ekdale, 2004). Less common are Unit A tops that are gravel, perhaps the stranded crests of very large-scale subaqueous dunes that formed during the final pulse of platform drainage.

Unit A in the Ring Realm is interpreted as a result of sediment-surface leveling when waters from tsunamis and/or voluminous condensation from the wet impact drained across the platform. Imbrication of smaller clasts in the upper part of the Alamo Breccia in the Ring Realm indicate westward paleoflow directions, toward the general area of the Rim Realm (Warme and Kuehner, 1998) and the platform margin, as do large-scale rip-ups and the orientation of folded megaclasts (Fig. 14).

A complex series of tsunamis were likely generated by two processes. The first follows the model of Oberbeck et al. (1993) in which the first of multiple tsunamis are created by the collapse of the wet-impact slurry rim of the transient crater. Collapse outward creates an initial wave, and collapse inward creates a central-crater spout; collapse of the spout initiates a second tsunami. As each tsunami returns to the crater, a new spout forms and fails, creating more waves until they completely damp out. This mechanism was used by Warme and Kuehner (1998) to explain the stacked graded beds in the Ring Realm that become thinner and finer-grained upward. The second process is slumping off of the platform margin (Warme and Sandberg, 1995, 1996).

Ring Realm Examples

Outcrop belts of the Breccia within each range of the Ring Realm show rather uniform characteristics, such as thickness, character of upper and lower boundaries, and gross internal structures. However, these properties differ significantly between ranges. The following three examples illustrate only some of the variety of interpreted processes and resulting products that formed during the Alamo Event, but they serve to help envision the diverse character of the Ring Realm.

West Pahranagat Range: Intraformational folds and large-scale imbrication. Figure 14 is a composite diagram showing some of the features exposed in a 1.5 km long cliff face in the West Pahranagat Range, ~6 km north of the type section (HAS, Fig. 1). The Alamo Breccia is 60 m thick and shows discontinuous Unit C megaclasts over Unit D monomict breccia. Where the megaclasts parted, Unit D is erased and Unit B chaotic breccia extends down to the base of the Breccia interval.

As shown along this outcrop, two Unit C megaclasts collided. One was wedged under the other and created a spectacular intraformational fold ~30 m high, diagrammed in Figure 14 and

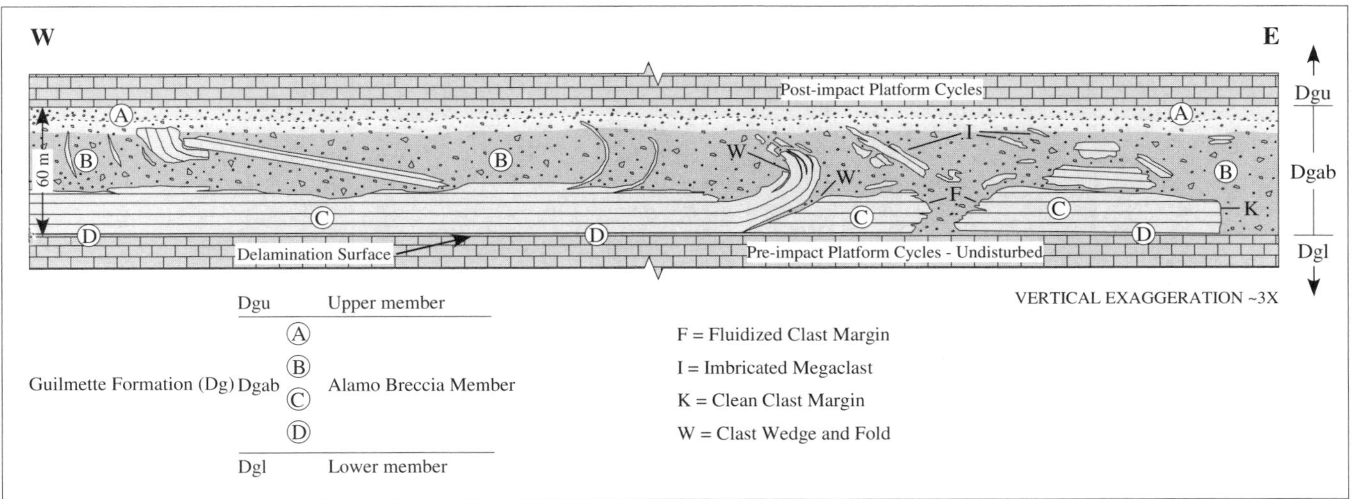

Figure 14. Composite diagram showing characteristics of peel structure and other large-scale features along the ~1.5 km length of Alamo Breccia outcrop ~6 km north of the type locality, West Pahranagat Range (HAS, Fig. 1). Unit D is the monomict breccia that separates underlying bedrock from Unit C megaclasts, some of which were preserved in the process of peeling up and disintegrating. The discontinuous megaclasts are separated and overlain by Unit B chaotic breccia. Unit A graded beds cap the Breccia interval. Sense of breccia flow is east to west. Note scale: Alamo Breccia here is 60 m in thickness.

illustrated by photos in Warme and Sandberg (1995, 1996). The top of the fold was truncated by high-velocity currents flowing westward (right to left in Fig. 14); the exposed bed ends were splayed open. Two thinner intervals are peeled back in the same manner. Similar peel structures on variable scales were discovered near Meeker Peak in the Worthington Mountains (MKR, Fig. 1; Warme and Sandberg, 1995), ~40 km to the northwest. The upper half of Unit B contains a spectrum of clast sizes and shapes, some as long as 50 m and imbricated with an eastward dip, indicating westward paleoflow. Some elongate fragments are clast-supported; others are perched within finer matrix but have the same dip orientation. Lapillistone clasts as deep as 30 m from the top indicate that they were reworked by high-velocity currents that occurred during a late phase of the Alamo Event when tsunami backwash, resurge, or voluminous condensed water drained the carbonate platform or rushed into the new crater. Near the base of Unit B some decimeter- to meter-scale Breccia fragments are delicately folded, fractured, parted, injected, have fluidized margins, and otherwise appear to have been heated or weakened by seismic shaking, acoustic fluidization (Melosh, 1989; Melosh and Ivanov, 1999), passage of the ejecta curtain, or other impact processes during an early phase of the Event, and escaped the reworking by currents that occurred in the upper half of Unit B.

Hiko Hills: Platform evacuation structure. Figure 15 shows a south-facing outcrop at the southern end of the Hiko Hills (HH, Fig. 1) where the Alamo Breccia is ~90 m thick. Figure 15A shows the predictable shallow-water carbonate platform cycles of the Guilmette Formation under the Alamo Breccia, and similar cycles preserved within an 80 m thick Unit C megaclast on the west (left) side; this is the thickest Alamo Breccia and intact Unit C clast that we have discovered in the Ring Realm. The normal platform cycles were reestablished over the breccia in the center and right side of the photo, and a discontinuous coral-sponge buildup, as much as ~40 m in height, was established over the breccia on the left side (R, Fig. 15B). The so-called "Reso's Reef," 50 m high, also developed directly over the Alamo Breccia ~14 km northwest of this locality, near Mount Irish (MIR, Fig. 1) (Reso, 1963; Dunn, 1979; Chamberlain and Warme, 1996; Sandberg et al., 1997; Chamberlain, 1999). Figure 15B shows the locations of lapillistone clasts (L). Some represent early precipitation during the Alamo Event, and are mixed into Unit B as much as 50 m below the top of the Breccia. Others trail along an interval only ~3 m from the top of the well-sorted Unit A Breccia (black circles), suggesting that they precipitated there during a lull in the more violent processes and then were broken up by the next pulse of wave uprush, backwash, or platform drainage.

A much larger limestone bomb from the upper part of the Alamo Breccia at Hiko Hills duplicates the morphology of volcanic bombs, is 16 cm in length, and is fusiform with twisted ends (Fig. 16). It evolved in flight when already-accreted lapilli and carbonate mud in the impact plume amalgamated and formed concentric shells that culminated as a sphere 9 cm in diameter. Some of the larger lapilli on the interior accreted around nuclei of brachiopod shell fragments. The sphere was partially cemented, but the outer layers were ductile, deformed, and the mass became streamlined as it spiraled in flight. It must have evolved from ductile to fully cemented while airborne because it was not damaged when it landed or while energetically reworked in the clean-washed upper part of Unit A. Ballistic cementation of Alamo carbonate lapilli was discussed by Warme et al. (2002). The

WSW ENE

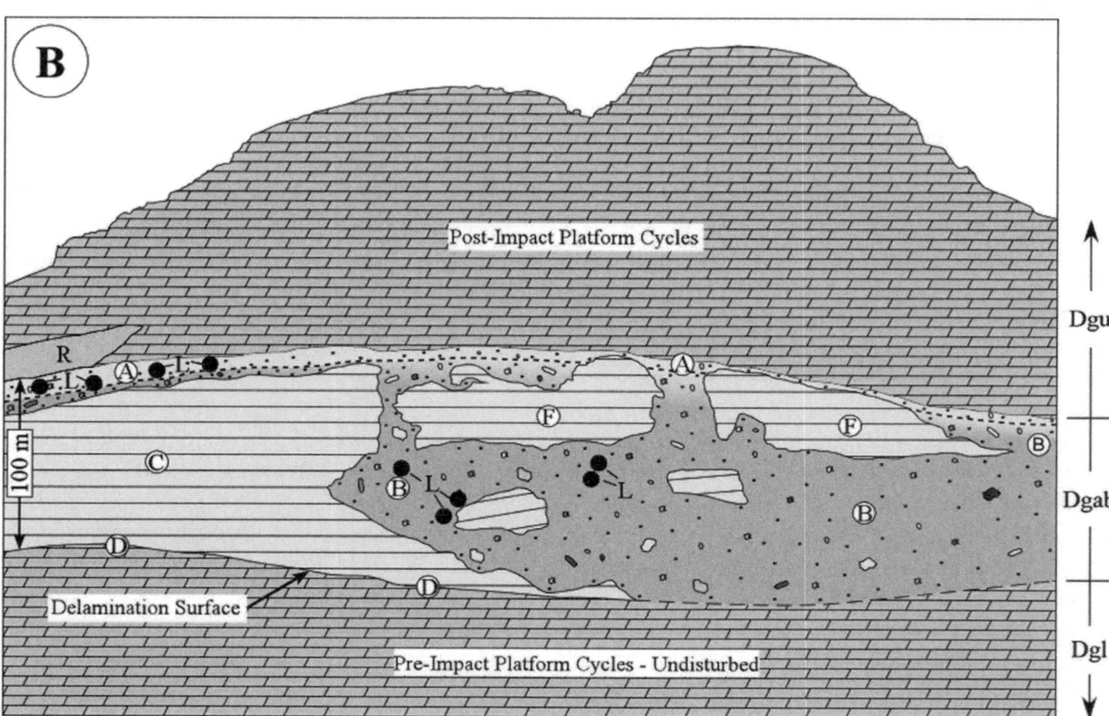

Figure 15. Photograph and diagram of Alamo Breccia evacuation structure at the south end of Hiko Hills (HH, Fig. 1). Breccia interval is 100 m in thickness. A: View to northward showing characteristic shallow-water carbonate platform cyclic bedding of the Guilmette Formation above and below Breccia. B: Left: unusually thick (~80 m) preserved Unit C megaclast with thin (~10 m) Unit A graded beds over top. Center and right: Cyclic carbonate beds equivalent to the megaclast disintegrated down to the level of detachment (Unit D), except for an upper finger (F) of beds that extended over newly formed Unit B chaotic breccia. The finger broke into pieces as some of the underlying Unit B was evacuated. Lapillistone clasts (L) as much as 50 m under the finger represent early-precipitated beds that were broken and mixed with Unit B. One or more later lapilli beds precipitated and were preserved as a discontinuous trail of broken and smeared lapillistone within in a graded bed of Unit A (black circles, upper left), which formed across the whole area in a late phase of the Alamo Event after the finger collapsed. Post-Event deposits in this area include sponge-coral reefs (R) up to ~40 m high.

bomb provides independent evidence that lapilli were cemented in flight, landed, and reworked quickly before the final phases of the Event across the Ring Realm. Other classes of Alamo bombs were described in the Runout/Resurge Realm by Morrow et al. (2005) and Sandberg et al. (2006).

Unit B chaotic breccia abuts the right side of the intact Unit C interval (Fig. 15). A preserved finger of the upper beds extends over Unit B debris. The finger was eroded across the top, tapered and bent downward at the end, and eventually broken and parted in two places that allowed overlying already well-sorted Unit A Breccia to cascade downward while the underlying Unit B breccia was evacuated. Mass movement of Unit B and collapse of the finger at Hiko Hills occurred after passage of the ejecta blanket, formation of some lapillistone beds, and/or tsunami runup and backwash that sorted the debris that formed the lower part of Unit A. However, the upper Unit A graded beds that cap the breccia succession are undisturbed, indicating that they did not participate in the adjustment, and the finger collapsed before terminal drainage of the platform at Hiko Hills. Tsunami backwash and/or precipitation runoff may have occurred for days, but after adjustment that mixed early-formed lapillistone clasts throughout much of Unit B.

We propose that Unit B adjusted as part of crater ring formation, outlined under *Discussion of Enigmas*. Alternatively, the Ring Realm may have connected to the Rim Realm or platform margin by large-scale sheets of debris flows, or networks of resurge gullies, which funneled debris seaward from far into the carbonate platform. If headward erosion or gully migration undercut the Unit C finger shown in Figure 15, it happened early in the Alamo Event before accumulation of the undisturbed upper beds of Unit A.

Delamar Mountains: Lifted and stacked tabular megaclasts. The Alamo Breccia at the southern end of the Delamar Mountains (DEL, Figs. 1, 2, 17) exhibits a third expression of the Alamo Breccia in the Ring Realm. Although this location is ~85 km southeast of the Rim Realm, near the apparent outer edge of the Ring Realm, its ~90 m thickness is greater than the Breccia in most Ring Realm localities and much greater than in the nearby Runup Realm where the Breccia is <10 m in thickness. At this locality ~80% of the Breccia is composed of complexly stacked, large-scale, tabular megaclasts that are commonly 100 m or more in length and 5–10 m in thickness. Only ~20% is matrix (Fig. 17).

An interval of the carbonate platform appears to have been detached and separated into sheet-like megaclasts that sheared across one another. They are gently folded, locally buckled, and each is about the same thickness, not graded, from the base to the top of the Breccia (Fig. 17). In places beneath the Breccia a subtle, incipient Unit D indicates where an interval of overlying beds were preserved in the process of being dilated, detached, and forming the Unit C megaclasts. However, at DEL a similar layer of incipient Unit D is preserved *within* several of the lower stacked megaclasts, as if the clasts were detached a few meters lower than the initial Unit D interval during a secondary process.

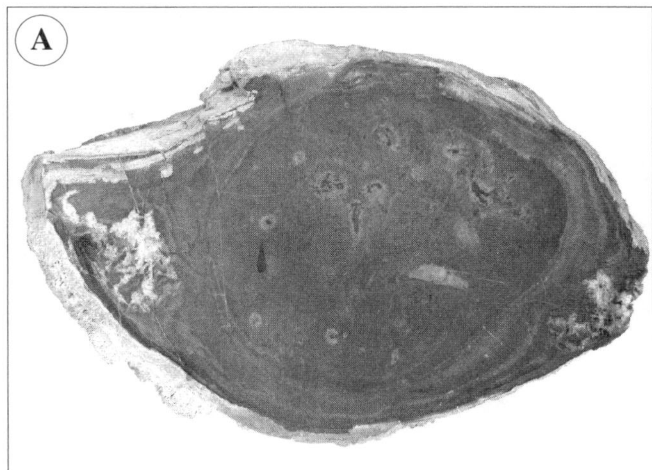

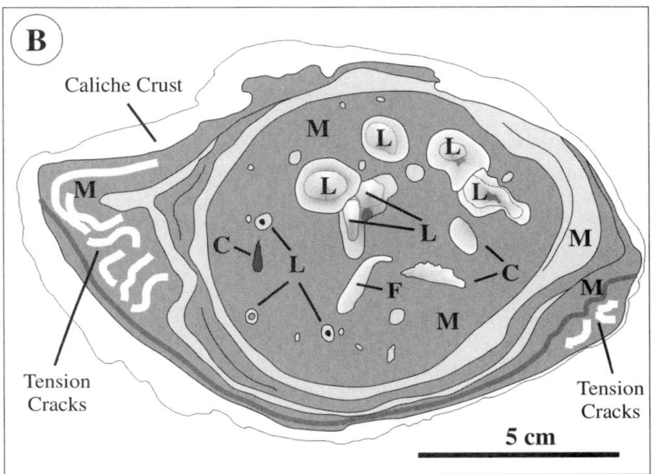

Figure 16. Large, fusiform, pure limestone bomb from the upper part of Alamo Breccia at Hiko Hills. Note scale. A: Polished surface showing internal components. B: Matrix of fine-grained particles (M), small clasts (C), fossil fragments (F), earlier-formed lapilli as much as 1 cm in diameter (L). The accreted concentric spheres started to cement and harden. While the outer shells were still ductile, the mass started to spin, formed a long axis, and opened tension cracks at each end that later filled with calcite cement. The caliche crust formed during near-surface weathering.

Ends or buckled folds of clasts penetrate the top of the Breccia, where the fine-grained graded interval of Unit A partially engulfs them. The matrix resembles well-sorted Unit A Breccia, between the tabular clasts deep into the Breccia interval. This large-scale fabric of megaclasts and pre-sorted matrix is unique for the Breccia, and suggests that the tabular megaclasts were detached and stacked in a second movement. They were mixed with matrix that was already sorted by currents. Some matrix was still in motion, or remobilized, and deposited as the graded and leveled interval of Unit A that caps the Breccia.

Directions of apparent shear and abruptly abutted Unit C clasts (Fig. 17) indicate a problematic en masse eastward movement of

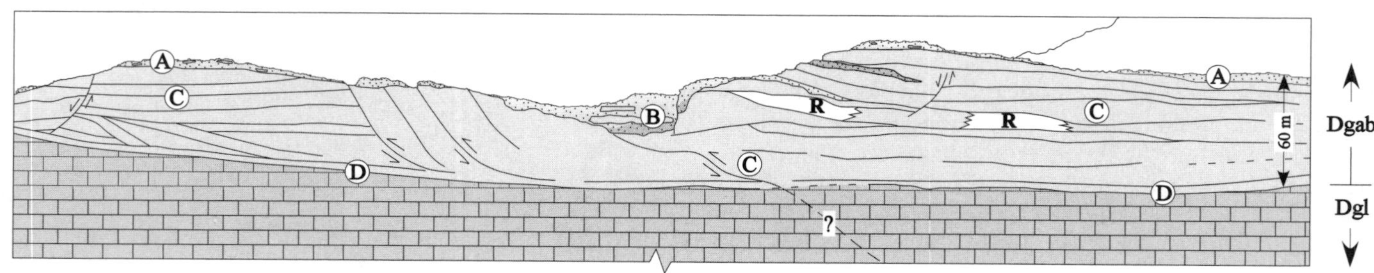

Figure 17. View toward south-southwest of Alamo Breccia outcrop that caps the dip slope at the southern end of the Delamar Mountains (DEL, Fig. 1). Image is tilted to restore horizontal, which is indicated by trend of ridge of the Sheep Range in distance at right in top image and slanted line in bottom image. Top: Photo of northern slope of one of several east-west ridges that display Alamo Breccia with unusual tabular megaclasts and very little matrix. Bottom: Note 60 m scale of total Breccia. Units of the Breccia include the detachment surface (D), jumbled megaclasts that continue to the top of the Breccia in most places (C), limited internal chaotic Breccia matrix (B), and topmost graded beds that are now in the process of eroding away (A). R—stromatoporoid sponge reefs within displaced clasts. Relative movement of the clasts is eastward, as discussed in text.

the Breccia at DEL, away from the presumed westward direction of the platform margin and the present position of the Rim Realm, and opposed to the northwest direction given by imbrication of the small clasts in Unit A that were the last to be reworked (Warme and Kuehner, 1998). (See *Discussion of Enigmas*.)

Runup Realm

The Runup Realm on Figure 2, labeled "Zone 3" by Warme and Sandberg (1995, 1996) and later reports, is a more distal semicircular band where the Alamo Breccia is <10 m thick, and commonly only 1–1.5 m. The distance from the northern to the southern known exposures is ~260 km, and the present known area of this Realm is ~12,500 km^2. Where the Breccia is several meters thick, it matches Units B and A of the Ring Realm; where only ~1 m thick it is a single graded bed like Unit A. Each locality has a different expression and appears to contain much locally reworked platform debris. Imbricated clasts suggest paleoflow away from the Ring Realm; thus these beds are interpreted as stranded runup of one or more tsunami waves or surges across the platform (Warme and Kuehner, 1998). Both matrix and clasts of the Breccia in the Runup Realm are dolomitized and generally lighter gray than in the Ring Realm. At some localities the upper surface contains desiccation cracks or karst features that are overlain and preserved by the first post-Event Milankovitch-scale cyclic sea-level rise. These attributes indicate that the Breccia was stranded across the inner platform above contemporaneous sea level, and that the Alamo Event probably occurred during the lowstand portion of a short-term Milankovitch cycle.

Runoff Realm

Beyond the Runup Realm, four anomalous channels, all most likely of *punctata* Zone age, were discovered and described by Sandberg et al. (2003), Morrow et al. (2005), and Morrow and Sandberg (2006). Channel fills are commonly ~1 m thick, and contain the distinctive hematite-studded quartz grains that signify the Alamo Event. All four localities were nominated as Alamo Event tsunami uprush and/or backwash deposits that represented the distal effect of a megatsunami wave-train that reached ~300 km from the general impact site in Nevada. One channel is at Devils Gate in central Nevada (DVG, Figs. 1 and 2), where the channel fill is intercalated with the carbonate platform facies of the lower member of the Devils Gate Formation. The fill is roughly normally graded, contains stromatoporoid debris and clumps of individuals that apparently lived for some time along the base, is more terrigenous and thus more recessive-weathering than the enclosing carbonate beds, and contains as much as 183 ± 27 ppt Ir (Morrow and Sandberg, 2006), which is the highest Ir value measured thus far for beds of the Alamo Event.

Three other localities are in western Utah (NEE, SBH, and CON, Figs. 1 and 2). Except for the 10 cm thick slab pictured by Morrow and Sandberg (2006), we have not seen the complex heterolithic bed in the Confusion Range (CON), which is reported to total ~6 m in thickness (Morrow and Sandberg, 2006). The Needle Range East (NEE) and South Burbank Hills (SBH) localities were reported to contain an ~1-m-thick, quartz rich, shallowly channeled bed that was correlated between the ranges and assumed to be widespread. Upon learning of this discovery, we interpreted the channels to have been eroded by runoff after massive condensation and precipitation of the marine Alamo impact, and rapidly filled by fallout from the debris-rich suspended load.

Subsequently it was discovered that two similar beds are present at NEE and SBH (Sandberg and Morrow, 2007), only a few meters apart, but separated by more normal carbonate platform facies that probably represent millennia of accumulation. Our preliminary field study indicates that both beds are seemingly continuous, are shallowly channeled so that their thicknesses change along strike, are roughly graded with ripup clasts concentrated near the bases, and exhibit climbing current- and wave-ripples modified by soft-sediment deformation in the upper parts. These features are consistent with flashy sheetwash erosion and rapid deposition of sediment-choked currents that crossed the platform. The currents may have been caused by runoff from catastrophic rainfalls, backwash from seismic-, tsunami-, or storm-induced seiches, or other rare but plausible events. Sandberg and Morrow (2007) proposed that two separate, short intervals of voluminous rainfall and runoff, well after the Alamo Event, tapped into a surface reservoir of quartz sand grains, which most likely resided in streams and beaches at the periphery of the carbonate platform. They report feldspar and quartz grains possibly sourced from Granite Mountain, north of the Dugway Range, Utah. All evidence indicates a widespread Runoff Realm as labeled on Figure 2.

Seismite Realm

Very unusual deformed beds occur in Devonian beds on Frenchman Mountain (FRM, Figs. 1 and 2) that we interpret as Alamo seismites. Frenchman Mountain is within the Las Vegas Valley shear zone. Rocks of the Mountain may have been laterally displaced in the shear zone, ~60 km from the east (Fryxell and Duebendorfer, 2005), perhaps >250 km away from the present Rim Realm. The stratigraphic section contains a reduced section of Paleozoic formations. The Sultan Formation is the only Devonian formation and is divided into three Members: Ironside, Valentine, and Crystal Pass. The Ironside and Valentine are undifferentiated on Frenchman Mountain (Matti et al., 1993; Castor et al., 2000), but are considered to be Frasnian in age, and the Crystal Pass to be Frasnian to Famennian (Bereskin, 1982). Thus the Sultan is of similar age to the Guilmette Formation in our main study area to the north, and the Ironside and Valentine Members likely encompass the moment of the Alamo Event.

The uppermost 10 m interval of the Ironside/Valentine contains sedimentary structures that are unique for platform carbonates, based on our experience. They occur in a discrete interval that was mapped for almost 5 km along the west face of Frenchman Mountain and traced another 4 km northward onto adjacent outcrops (Pinto, 2006). The interval encompasses four persistent bed-parallel units that each contain a different suite of fractured, sheared, rotated, and injected clasts, some of which float in what we interpret as locally fluidized-rock matrix (Figs. 18 and 19). The layers are separated by less-altered country rock. They show no evidence of cavities from karst or tepee formation, other penecontemporaneous processes, and no system of later crosscutting tectonic fractures or faults that are focused on the interval. Each unit is interpreted to represent somewhat different responses to seismic waves from the Alamo Event. On the whole they compare to many examples from elsewhere that are interpreted as seismites (Ettensohn et al., 2002). For example, deformed beds near the Dead Sea were explained as seismites by Marco and Agnon (1995); their Figure 3 shows an interval where beds were detached, lifted, injected, and increasingly dispersed from the base upward, culminating in a matrix-supported breccia that closely matches Unit 4 on Figures 18 and 19A.

The Ironside/Valentine interval is directly overlain by very different, apparently shallower facies of the Crystal Pass Member, which suggests that the sedimentary environment across the carbonate platform was abruptly transformed by the influence of the Alamo Event. The unusual intervals on Frenchman Mountain are the subject of continued study.

Seismic deformation was presumably extensive all across the brittle carbonate platform, but the undetached near-surface interval beyond the rings would most likely preserve the seismites. The Seismite Realm labeled near Frenchman Mountain on Figure 2 might extend to Portuguese Mountain in the far northern part of the Runup Realm (PMN, Fig. 1). Sandberg et al. (2002) proposed that a 20 cm deep wedge-shaped fissure, filled with overlying Alamo Breccia, was a product of seismic action. Larger disruptions have been discovered in the same area (J. Morrow, 2007, personal commun.). The present distance between PMN and FRM is almost 300 km, and could have been greater if Frenchman Mountain rocks were carried westward in post-Devonian time (Fryxell and Duebendorfer, 2005). If the features at PMN are seismites, then that locality exhibits properties of both the Runup and Seismite Realms.

Runout/Resurge Realm

Discoveries of deepwater channels in south-central Nevada (UEC, RMW, MLK, and SKR, Fig. 1) (Morrow et al., 2001, 2005; Morrow and Sandberg, 2003a; Sandberg et al., 2005, 2006), correlative with the Alamo Breccia on the carbonate platform, show that large volumes of rock were moved and/or ejected from shallower to deeper ocean. The channels may be discontinuous, but some are thicker than 40 m (Morrow et al., 2005). They contain abundant fossils sourced from platform and slope environments,

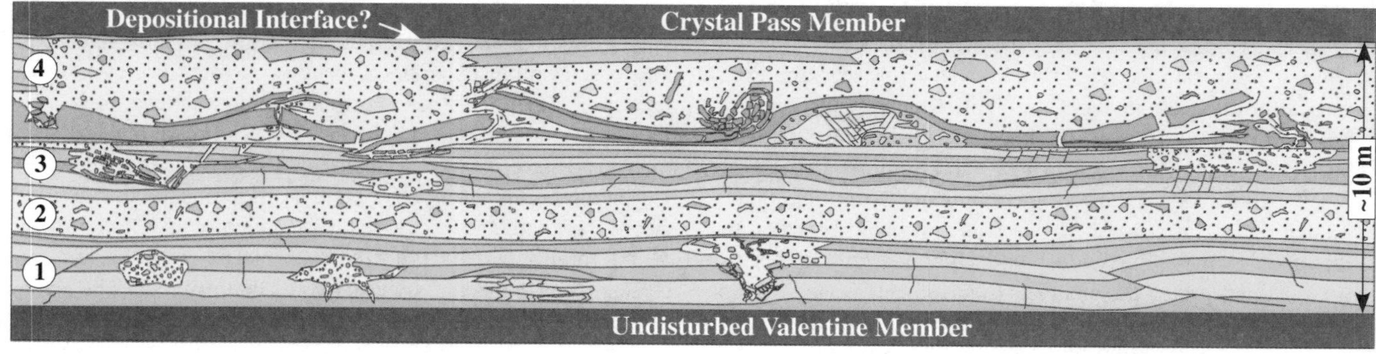

Figure 18. Composite diagram showing Units 1–4 (circled) and variety of deformation structures (interpreted as seismites) in the upper 10 m of the Valentine Member, Upper Devonian Sultan Formation, Frenchman Mountain (FRM, Fig. 1). Degree and style of deformation varies between Units, but is laterally consistent within them. Unit 1: relatively undisturbed interval with irregular pockets that contain dilated, fractured, folded, faulted, and brecciated rock fabric. Unit 2: pervasively brecciated interval confined between two thin, deformed, ductile layers that separate it from Units 1 and 3. Unit 3: dilated, fractured, faulted, and brecciated interval with unusual load-like structures and internal families of high-angle faults. Unit 4: thickest and most disturbed interval; lower contact with Unit 3 is generally sharp, but overlain by an interval that was fluidized, fractured, micro-folded and -faulted, peeled up, and brecciated. See Figure 19A. Units 3 and 4 are locally reorganized into complex sag structures. The upper part of Unit 4 is a breccia of dispersed clasts, capped by a 0.1–0.6-m-thick quartzose dolomite sand of unknown origin. It was injected from below, and probably expelled at the surface as sand fountains. Note 10 m scale on right side of diagram.

deepwater rip-up clasts, shocked quartz grains, accretionary bombs, and glassy material. Moreover, Sandberg et al. (2006) described a kilometer-scale olistolith of Devonian slope facies, overlain by channelized pockets of breccia, in the area of deep-water Alamo Channels in the southern Hot Creek Range (MLK, Figs. 1 and 2). These occurrences show that a massive volume of Devonian rock was ejected from the crater, swept from the platform, and/or displaced by platform margin collapse, thus the "Runout" term for this Realm.

Morrow et al. (2001) determined paleocurrent directions from the deep-water channels, and concluded that resurge currents from the Alamo Event converged toward a point that was proposed to be the central crater, thus the additional "Resurge" term for this Realm. The location is in the gap between the known deepwater deposits and TEM (Fig. 2). No direct evidence exists for a crater in this gap. Sandberg and Morrow (2007) most recently offered that the crater was "tectonically dismembered and buried, most likely beneath the Roberts Mountain Allochthon," which is in central Nevada.

DISCUSSION OF ENIGMAS

The preceding summaries, new descriptions, and interpretations of the Alamo Breccia Realms help resolve many of the long-standing Alamo Enigmas.

Breccia Asymmetry

The early Late Devonian carbonate platform contains facies of the Alamo Breccia that are preserved in roughly concentric semicircular eastward-extending bands that are interpreted as proximal to distal impactogenic units classified into the genetic Realms shown on Figure 2. The western half of the expected circular impact pattern is missing and replaced by an area west of the north-south questioned line on Figure 2 that is apparently devoid of Alamo Breccia outcrops, and by a more distant zone of channels and at least one olistolith that defines the Runout/Resurge Realm. A circular crater would encompass these western zones. Several hypotheses are erected to explain this asymmetry; these include the following: (1) The missing portion exists but is still buried, or was uplifted and eroded away, and not available for outcrop study. (2) The western half of the crater exists in the area of no Breccia outcrop, but is covered by Cenozoic volcanic flows and/or valley fill. (3) Post-Devonian thrust faults displaced the western half eastward, and some of it may actually be unrecognized as significantly allochthonous in ranges of the Ring Realm, which could contain both halves of the concentric Realms, compressed and probably laterally mixed. (4) Post-Devonian extensional tectonics moved the missing portion westward with respect to TEM, where it has not been recognized, is buried, or is eroded away. (5) Post-Event Paleozoic and Mesozoic thrusts, later extension, and perhaps lineament shear thoroughly dismembered the crater so that reconstruction will continuously evolve along with new discoveries and tectonic analyses. Hypotheses 3–5 imply that the central parts of the original crater could be almost anywhere with respect to present-day Breccia outcrops in or near the study area of Figure 1. (6) The western half detached and disintegrated during the Alamo Event, and the target rocks are represented by resedimented debrites, turbidites, and olistostromes in a widespread Runout/Resurge Realm.

These options are not independent; significant crater modifications during the Event could be masked by post-Event tectonics up to the present, requiring a multistep reassembly.

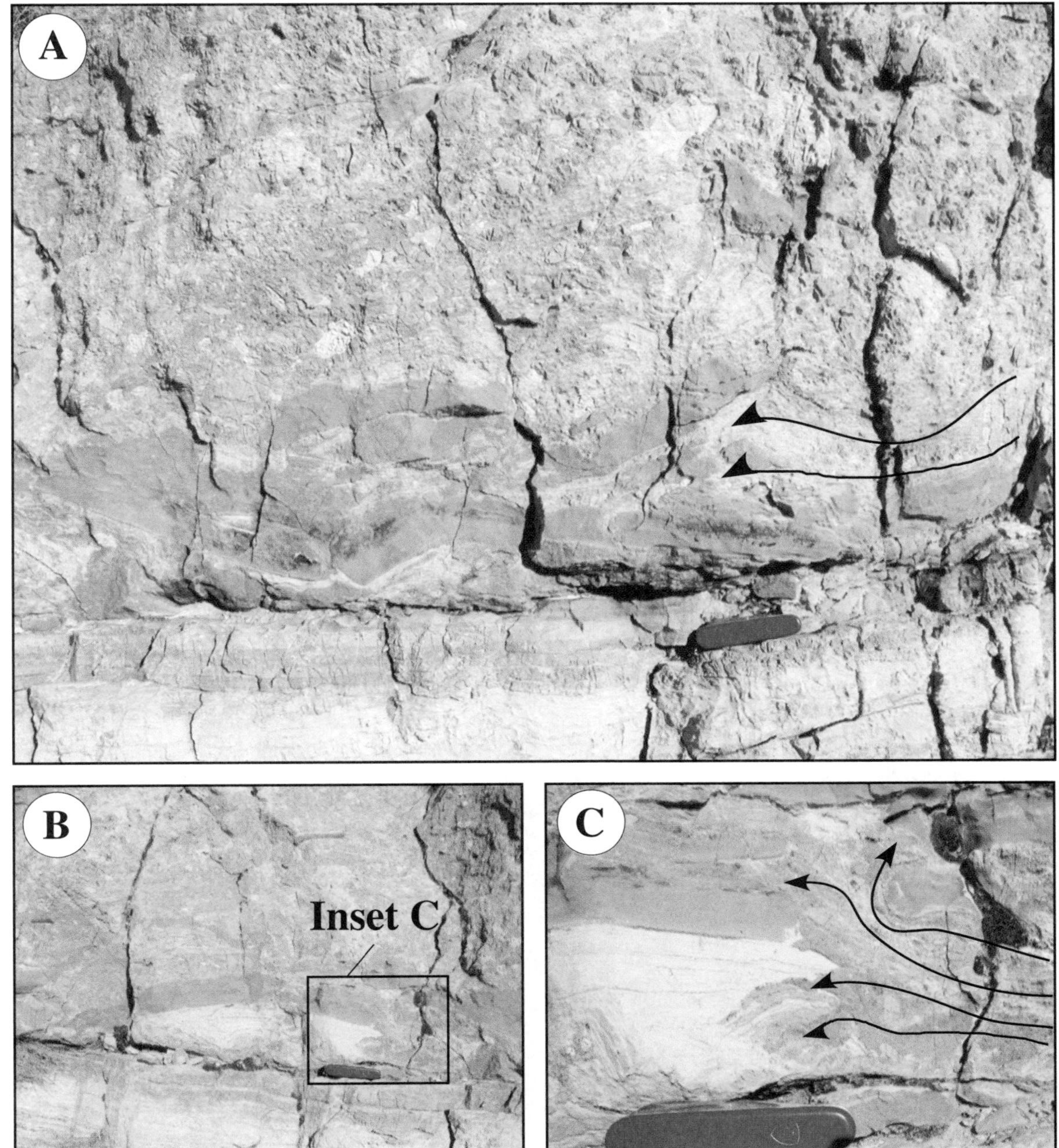

Figure 19. Upper part of seismite interval on Frenchman Mountain. A: Lower part of Unit 4 (Fig. 18), showing intact bed at base, lifted and injected beds across center, and homogenized, dispersed clasts above. B: Intact bed below, light-hued clast above that was injected and overlain by fluidized matrix. C: Detail of B showing flow lines. Knife is 8 cm long.

However, hypothesis 6 is enforced by new discoveries of the 1 km long olistolith and associated deepwater channels in the Runout/Resurge Realm (Sandberg et al., 2006), which is a major step toward solving the asymmetry problem. We anticipate that more voluminous platform and slope debris, and kilometer- to even mountain-scale detached blocks, will be recognized in central Nevada and even farther away as part of the Alamo Event.

Crater Depth

The crater rim breccia succession at TEM (Fig. 5) rests over intact Lower Devonian and older beds, which accounts neither for the excavation of lower Paleozoic rocks represented by the clasts of the Unit 3 fallback interval there, nor for the lower Paleozoic conodonts and mineral fragments scattered throughout the Breccia Realms. We conclude that the succession at TEM represents the margin of a much deeper and larger crater. Rocks of Tempiute Mountain may have been displaced on one or more thrusts away from the main crater, and/or impact shock waves weakened adjacent central target rocks and triggered unusually massive slides that displaced the central parts of the crater into deeper water. Even under normal conditions, carbonate platform margins worldwide are prone to detach and slump seaward, as documented along the Nevada platform margin throughout the Paleozoic (e.g., Cook and Corboy, 2004; Stevens and Stone, 2006).

Crater Depth versus Breccia Area and Volume

The ~1300 km^3 volume of the semicircular ~28,000 km^2 area shown on Figure 2, the uncalculated area and volume of Breccia in the Runoff and Runout/Resurge Realms, plus unknown values for lost ejecta and distant fallout, total an immense volume of rock that was displaced during the Alamo Event. The limited crater stratigraphy at TEM must represent the margin of a large, deep crater.

Deep Crater Location

A crater deep enough to expel lower Paleozoic rocks into the Devonian Alamo Breccia has not been discovered. Standing options are that the central crater is still buried, was uplifted and eroded away, or was sloughed away into the Runout/Resurge Realm and dismembered.

Target Water Depth

Paleoecologic analyses of reworked but contemporaneous Late Devonian conodonts captured in the Alamo Breccia indicate that they were expelled from off-platform facies, whereas the Lower and Middle Devonian rocks on Tempiute Mountain are carbonate platform facies. A large crater at the continental margin could have encompassed both shallow platform and deeper slope environments. Alternatively, two or more deep-water craters formed simultaneously, for which no independent evidence exists.

Rim Realm to Runout/Resurge Realm Disconnect

The platform to off-platform facies transition could be missing for the same reasons that the central crater and western half of the concentric rings are missing: they may be still buried, uplifted and eroded away, or sloughed away during or after the Alamo Event. The prospect remains that the crater fragment at TEM and/or the deepwater channels and olistolith that represent the Runout/Resurge Realm are not in their original positions with respect to one another, owing to post-Event tectonic rearrangement of crustal blocks across the study area.

Rim Realm to Ring Realm Disconnect

The disparity is remarkable between the properties of the impactogenic breccias at TEM and the numerous sites of the Alamo Breccia within the Ring Realm. The TEM Breccia package overlies a Middle Devonian formation, exhibits damaged bedrock (Unit 1), injected breccias (Unit 2), allogenic fallback breccias with probable melt indicators (Units 3 and 4), and thick but relatively fine-grained resurge beds (Unit 5); it underlies deep-water limestones (Fig. 5). Across the vast Ring Realm the Breccia overlies, and was delaminated from, undamaged bedrock of the Upper Devonian shallow-water, cyclic carbonate platform facies of the lower Guilmette Formation, is mainly composed of locally derived disintegrated Guilmette, and is overlain by resumed shallow-water carbonate cycles of the upper Guilmette. We interpret this disparity as additional evidence for tectonic shortening on one or more thrust faults mapped between the ranges of the two Realms (Tschanz and Pampeyan, 1970; Chamberlain, 1999; Taylor et al., 2000).

Ring Realm Impact Structure

In the Ring Realm the relationship between the Alamo Breccia and underlying bedrock is puzzling. The Ring Realm Breccia is clearly an impact product, covers an estimated 13,000 km^2, and includes locally derived bed-parallel megaclasts as long as 500 m (Figs. 14, 15, 17) that demonstrate energetic processes across a large area. However, the thickness of the Breccia ranges between only 50 and 100 m, and floats on undamaged underlying bedrock.

In the following scenario, impact-induced seismic ground waves detached, segmented, and severely damaged the Unit C megaclasts over the level of Unit D. Field evidence shows that the cyclic platform carbonates were well stratified, fully cemented, brittle, and connected by subtle facies changes to the adjoining Rim Realm target-zone carbonates, which were also cemented, brittle, and probably responded similarly to the intense seismic waves of the event. Sediments of meter-scale carbonate platform depositional cycles worldwide become cemented early, commonly directly on the seafloor, before initiation of the next transgressive/regressive cycle. The desiccation cracks, karst breccias, and remnant soils that cap such individual, fully cemented cycles of the Guilmette platform indicate that early cementation occurred frequently, followed by exposure. These observations are consistent with the concept that seismic surface waves laterally delaminated brittle platform beds and formed Unit D no more than 100 m below the contemporary platform surface. Compressive Love waves and following orbital Rayleigh waves could detach and fracture the interval into segments represented by Unit C megaclasts. Margins of the megaclasts were damaged when they oscillated horizontally against each other, created large-scale folds where some were peeled back over others, and formed gaps between them (Fig. 14). The waves shook, broke, mixed and fluidized the detached bedrock, perhaps well beyond the Ring Realm (see *Seismite Realm*). The subsequent ejecta curtain added volume to the platform debris, proven by the Unit B content of accretionary lapilli clasts and exotic lower Paleozoic conodonts and mineral grains.

Following the seismic waves and ejecta curtain, adjustment of the transient crater included a family of distant, concentric, listric ring faults and intervening broad terraces. The rings were probably offset along radial accommodation faults. The listric faults soled out at or near the level of Unit D, and rotated to transform the terraces into a set of outward-tilted half graben (Fig. 20). The result is annular asymmetric troughs separated by raised ridges across the Ring Realm. Tilted terraces between listric faults created variable accommodation space that may explain why the Breccia does not progressively thin away from the Rim Realm. For example, the thickness of the Alamo Breccia at Hiko Hills (HH, Figs. 1 and 15) is ~100 m, at the West Pahranagat Range (HAS) ~20 km to the west it is ~60 m (Figs. 1 and 14), and at Monte Mountain South (MMS, Fig. 1), which is located between these two sections, it is only ~50 m. Accommodation space would increase down the slope of each tilted block. All the new space was filled and leveled during the runup and runoff phases of the Event. A tilted graben could explain the secondary adjustment of Unit B Breccia at Hiko Hills (Fig. 15). Gently tilted platform beds under the Breccia could explain the secondary eastward movement of the tabular clasts at the Delamar Mountains (Fig. 17), where they may have slid and stacked against the outermost ring-fault footwall. The outermost ring would mark the distal limit of the Ring Realm, beyond which only thin Runup Realm Alamo Breccia occurs.

The above scenario must have played out within a day. The wet impact spawned ensuing waves, currents, and runoff that then reworked, sorted, and washed broken rock to form the clast-supported graded beds of Unit A that cover the entire Ring Realm.

Alternative processes that could detach an interval of upper platform beds include regional depression and rapid rebound of the upper crust outside the transient crater. Rebound could cause temporary tilting and detachment that was responsible for the outward movement and stacking of clasts at the Delamar Mountains, as opposed to accumulation by gravitational sliding down back-tilted terraces, or accumulation as megatsunami wrack.

Ring Realm to Runup Realm Transition

As shown on Figure 2, Ring Realm Alamo Breccia as thick as 100 m (e.g., HH, GGC) is positioned close to Runup Realm outcrops where the Breccia is <10 m (e. g., SFN, GGN). Runup Realm Breccia is commonly capped by desiccation cracks and karst features, created after impact debris from tsunami uprush was stranded above sea level. These observations suggest that sea level across the platform at the exact time of impact was at a low phase within one of the shorter Milankovitch-scale depositional cycles. In contrast, all Ring Realm Breccia is capped by bioturbated carbonate muds, sands, or gravels, by carbonate buildups as much as 50 m in height, and by deeper conodont biofacies. These relationships support the idea that the Ring Realm segment of the platform was lowered during the Alamo Event, possibly by rotation of listric faults (Fig. 20).

Source of Quartz Grains

The shocked and studded Alamo Breccia quartz grains (Warme and Sandberg, 1995, 1996; Morrow et al., 2005) have not been discovered in sandstone clasts outside of the Rim Realm. They occur only as loose grains. Silt-sized to very fine sand-sized grains from the Breccia are commonly fractured, but larger intact grains are well rounded and appear to be sourced from the Oxyoke Canyon Sandstone. The Oxyoke Canyon at TEM is brecciated, fluidized, and carried along in the Unit 2 dike-and-sill network, but was not fully ejected and therefore not a local source of the quartz grains that occur across the whole expanse of the Breccia. The central crater elsewhere fully penetrated the Oxyoke Canyon Sandstone, as well as the much deeper Eureka Quartzite (Fig. 20), and liberated the quartz grains.

Role of Lapilli and Lapillistone in Event Scenario

The distribution of Alamo lapilli and lapillistone can be used to refine the sequence of activity and duration of the Alamo Event. Whether or not precipitation of lapilli was somewhat continuous, to our knowledge they were preserved under only two circumstances: as individual lapilli mixed into the matrix of the Unit 3 fallback on Tempiute Mountain (Figs. 11B and 11C), and as clasts of broken lapillistone scattered in the Breccia Units B and A of the Ring Realm. As explained by Warme et al. (2002), carbonate lapilli can harden in flight, and precipitated lapillistone beds can continue to cement and completely harden in as little as a few minutes or hours.

Tempiute Mountain lapilli are rare, scattered, generally coarser-grained, and most are irregularly shaped compared to the Ring Realm examples. They are interpreted to have formed early during the Event, perhaps in a base surge associated with near-target ejecta fallout. In contrast, the distant lapilli in the Ring Realm accumulated as almost pure lapillistone beds, as thick as 1.2 m. A few hundred of the clasts have been found, but they represent <<1% of the clasts in the Ring Realm Breccia (Warme et al., 2002). They exhibit well-differentiated nuclei, accreted mantles, and crusts, which probably represent long flight times. Because lapillistone clasts from the same Ring Realm outcrops contain a range of lapilli sizes, internal stratigraphy, and cementation histories, we conclude that specific beds accumulated during pause(s) between more violent processes that are signified by the hiatuses that separate each graded bed in the Ring Realm. Thin lapillistone beds accumulated during the pauses; each bed was then reworked into the thick breccias that accumulated between pauses. The lapillistone cemented to varying degrees. Well-cemented beds were hard, brittle, and easily fractured and parted as they were transported and buried by newly overlain Breccia. Others were still plastic and molded between carbonate clasts of the Breccia. Each lapillistone bed was dismembered and largely destroyed by subsequent processes, but before the final graded bed of Unit A accumulated across the Ring Realm. These relationships indicate that the entire Breccia genesis in the Ring Realm was completed within a few hours or days. Much of the deposit was significantly reworked and adjusted during the short time between initial accumulation of chaotic Unit B and final deposition of the sorted beds of Unit A, as exemplified at the HH locality (Fig. 15).

SUMMARY: ALAMO IMPACT STRUCTURE

Figure 20 depicts a working theoretical model for the immediate results of the Alamo Impact. The model is based on new field observations at Tempiute Mountain (Figs. 1–5), previous investigations on the Alamo Breccia throughout south-central Nevada and western Utah (reviewed above), estimates of crater dimensions (Morrow et al., 1998, 2005), and the results of recent numerical and other models of impact cratering (Ormö and Miyamoto, 2002; Shuvalov and Trubestkaya, 2002; Gisler et al., 2004; Ivanov, 2004; Ormö and Lindström, 2005). Notions of crater variability based on target rock properties are expanding as more craters are discovered and studied on Earth and in space (e.g., Spudis, 1993; Melosh and Ivanov, 1999; Collins and Turtle, 2003; Lana et al., 2006). Each real crater apparently exhibits unique characteristics so that generalizations about crater size versus properties such as presence or absence of rings, central peaks, and peak rings have significant exceptions. It is agreed that target rock composition and structure are fundamental properties that control the evolution of any crater, as is water depth for marine impacts.

Evidence exists for the water depth on the Devonian carbonate platform and seaward slope or ramp at the time of impact. Displaced contemporary *punctata* Zone conodont assemblages recovered from the Breccia indicate that the depth of the seaward part of the target was 300 m or more (Sandberg and Warme, 1993; Warme and Sandberg, 1995, 1996). However, we do not know the depth to which the platform was flooded, or where the shoreline was located. The interval of platform carbonates in the Guilmette Formation under the Alamo Breccia, and preserved within Unit C megaclasts of the Ring Realm, exhibits ~30 short-term Milankovitch-scale (tens of thousands of years) cycles of sea-level change, when marine water repeatedly flooded and drained the platform. Approximately 20 additional younger cycles are preserved under the thinner Breccia in the Runup Realm (Warme and Kuehner, 1998). The Event could have occurred at any stage of the contemporaneous cycle, but desiccated and karsted tops of the Alamo Breccia at locations in the Runup Realm suggest that it occurred during the drawdown segment.

Our new data from TEM confirm a minimum crater excavation down to the Ordovician Eureka Quartzite, and are used to construct Figure 20. They do not conflict with the latest estimates by Morrow et al. (2005) for an impact crater that ranges in depth from ~1.5 to 2.5 km and in diameter from ~40 to 60 km.

Figure 20A illustrates the following scenario for the Alamo Event. A symmetrical crater was excavated near the margins of the Devonian carbonate platform. A gentle westward tilt is built in but is not obvious on this diagram of small scale and no vertical exaggeration. Mainly carbonate formations were excavated down to the Ordovician Eureka Quartzite. Figure 20B could represent any position around the eastern margin of the crater, or any part of the margin if TEM is significantly displaced by thrust faults. Figure 20C shows how the TEM columnar section of Figure 5 could represent the impact stratigraphy at the inner margin of the crater rim. The diverse autogenic and allogenic breccias, highly disturbed and intruded beds below allogenic breccias, low shock-pressure-induced deformation in parautochthonous minerals and rocks, absence of definite PDFs in the autogenic target rocks (see Kieffer and Simonds, 1980), and apparent shallow crater excavation depth (see *Discussion of Enigmas*) all enforce the conclusion that the impactogenic rocks exposed at TEM were located inside the crater rim, but not near the center. Figure 20D shows our interpretation of the Alamo Breccia across the Ring Realm, calibrated to a representative columnar section measured near Meeker Peak in the Worthington Mountains (MKR, Fig. 1) (Ackman, 1991).

During the excavation stage (Fig. 20B), target carbonate formations were differentially broken, displaced, melted, and vaporized (French, 1998; Ivanov, 2004). Voluminous heated water and variably-sized and -shocked rock particles were ejected and formed a high-velocity curtain and high-altitude plume. Fine-grained ejecta accreted in flight and accumulated as carbonate lapillistone that was deposited as much as 75 km away from the Rim Realm; the rest mixed with the breccias newly generated by seismic surface waves in the Ring Realm and beyond it.

Wet, brecciated, central target rocks were radially injected into the less shock-metamorphosed strata of crater rim area Unit 1 (Figs. 19B and 19C), some of which were probably overturned (e.g., Sturkell and Ormö, 1997), resulting in the polymict breccia of Unit 2 (Figs. 6 and 8). The Oxyoke Canyon Sandstone, near the base of the stratigraphic column (Fig. 5), was deformed and locally mobilized (Figs. 3 and 5). The Oxyoke Canyon must have been fully penetrated nearer the crater center, so that abundant shocked and liberated quartz grains were ejected, and also injected as a component of Unit 2. The injected breccia may have oscillated, pumped, and mixed in the Unit 2 dike-and-sill network as the crater walls slumped, fault blocks rotated, and other adjustments occurred during the succeeding modification stage, as documented in other crater studies (Dressler and Sharpton, 1997; Dressler et al., 1999).

Fallback clasts and fragments in various stages of damage and melt were quickly deposited as Units 3 and 4 (Figs. 9–12) during crater modification. They were later covered by the two normally graded beds of Unit 5 (Fig. 13) that partially filled the newly provided accommodation space. In the crater model of Figure 20, the well-sorted and fine-grained sediment of Unit 5 traveled long distances, and was probably mixed with reworked ejecta, rim-wall debris, and platform rocks brought into the crater through rim channels cut by resurge currents and/or tsunami backwash (Ormö and Miyamoto, 2002; Shuvalov and Trubestkaya, 2002; Gisler et

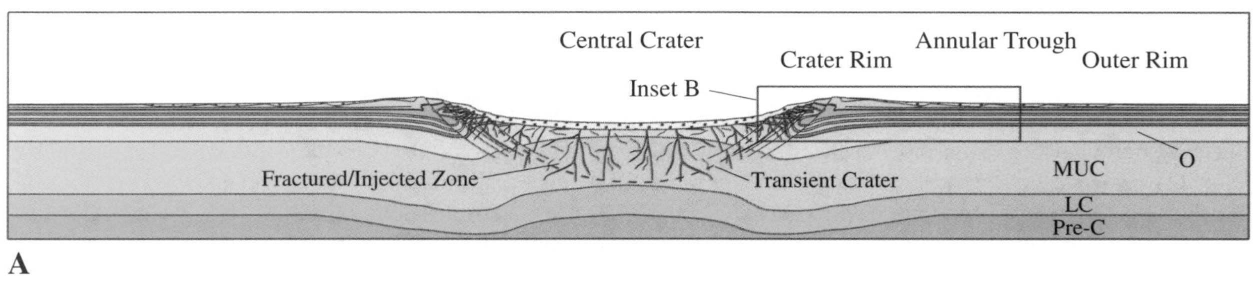

A

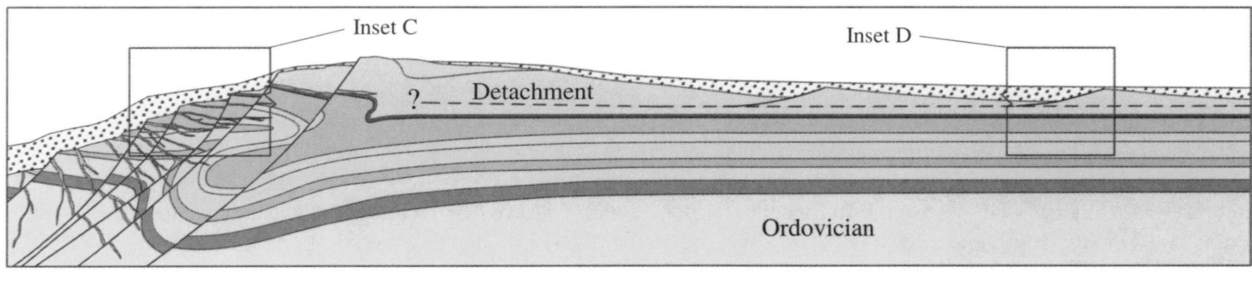

B

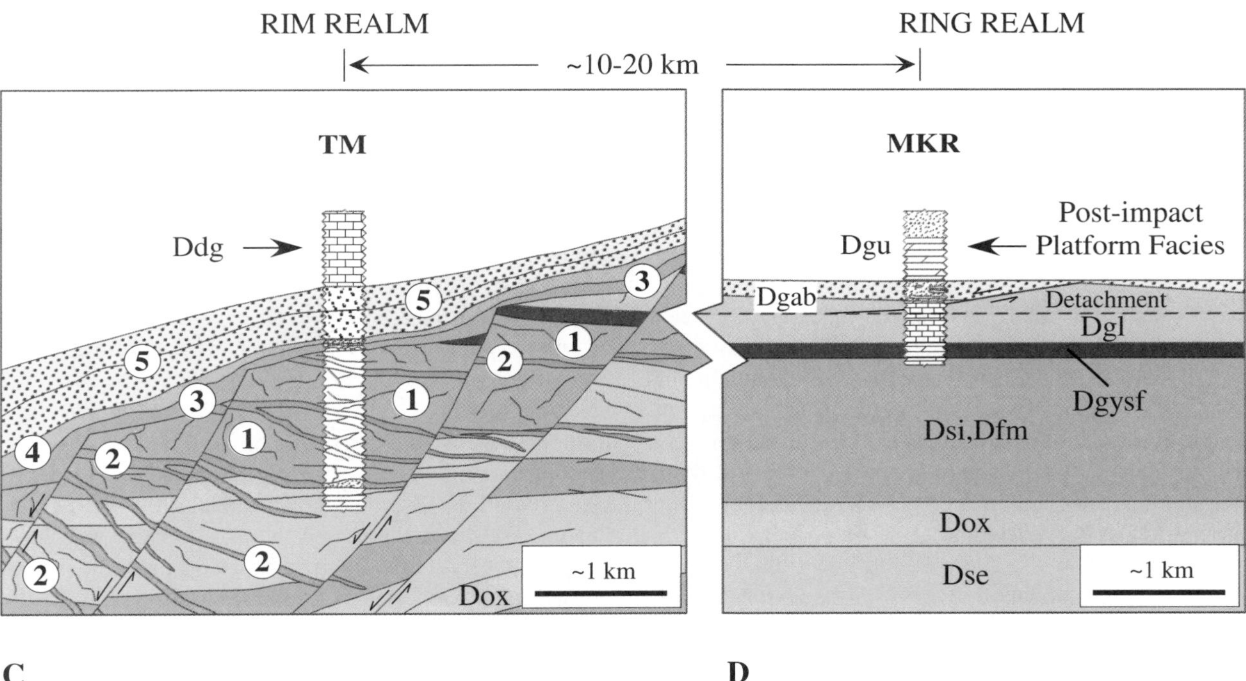

C

D

Figure 20. Working model of Alamo impact structure that accounts for known properties of the allogenic Alamo Breccia in the Rim and Ring Realms and the newly interpreted autogenic impact breccia in the Rim Realm. A: Cross section showing crater center, crater rim, annular trough(s), and outer rim. Pre-C—Precambrian; LC—Lower Cambrian; MUC—Middle and Upper Cambrian; O—Ordovician, overlain by Silurian Laketown Dolostone and Devonian formations. B: Inset from Figure 20A showing segment of crater rim, represented in this report by the Rim Realm at Tempiute Mountain (TEM, Figs. 1 and 2), and annular trough(s), represented by the Ring Realm on the carbonate platform (Fig. 2). C: Inset from Figure 20B showing the Devonian impactogenic breccias that occur at TEM; superimposed stratigraphic column (Fig. 5) shows autogenic breccias of Units 1 and 2 (circled) and allogenic Alamo Breccias of Units 3–5 (circled). Unit 2 is pervasively injected into Unit 1 Oxyoke Canyon Sandstone and Sentinel Mountain Dolostone. D: Inset from Figure 20B showing proposed listric faults that create tilted terraces and form troughs and rings, with superimposed columnar section from Meeker Peak, Worthington Mountains (MKR, Fig. 1). Dse—Sevy Dolostone; Dox—Oxyoke Canyon Sandstone; Dsi—Simonson Dolostone; Dfm—Fox Mountain Formation; Dgysf—"yellow slope-forming interval" of Guilmette Formation; Dgl—lower member Guilmette Formation; Dgab—Alamo Breccia Member of Guilmette Formation; Dgu—upper member Guilmette Formation; Ddg—Devils Gate Limestone.

al., 2004). Following all stages of adjustment, anoxic marine sediments accumulated across the Rim Realm at TEM (Fig. 5). The beds may represent sedimentary fill in the new crater, or cover across the new or previously existing slope there.

Figure 20A depicts a symmetrical crater on a slightly tilted carbonate platform. Our data all converge on the concept that the impact occurred on the carbonate platform margin and adjoining slope or ramp into deep water. Units 3 and 4 accumulated on the inner crater rim, coincident with evolution of the transient crater to the adjustment stage. Alternatively, almost the entire crater quickly slid away from the damaged and injected rocks of Units 1 and 2, Units 3 and 4 accumulated as fallback over the new slope, and Unit 5 blanketed the succession with platform runoff. In the latter scenario TEM holds the near-crater autogenic formations that were damaged by the impact and preserved, but the crater center and crater symmetry were destroyed. In this case, matching impact stratigraphy from other portions of the rim never formed, or disintegrated.

Figures 20B and 20D depict the hypothesized Ring Realm east of the Rim Realm. The brittle carbonate platform responded to seismic surface waves whereby a relatively thin 50–100 m interval of carbonate beds delaminated, broke into long segments, vibrated, and partially disintegrated before being further damaged by the high-velocity ejecta curtain. Families of concentric listric faults, probably offset by radial accommodation faults, created back-tilted terraces that accumulated the impact debris typified there by Units D to A (Figs. 14, 15, 17). We developed this part of the scenario using observed features in the Ring Realm, rather than attempting to match other existing models. Overviews of the existing literature imply that crater ring genesis is not fully understood; the configuration of their vertical third dimension is seldom visible and subject to debate (e.g., Spudis, 1993). It is likely that ring formation and diversity is controlled by physical properties of the target rocks (Melosh and Ivanov, 1999), as well as the size of the event, so that variables such as lateral heterogeneity, vertical stratification, cementation, water content, and porosity all affect whether rings form and what types develop.

Runoff from tsunamis and condensed water filled most of the Ring Ream accommodation space with the abundant loose impact debris. Post-Event marine environments covered all of the Ring Realm. Final-stage runoff may have left giant dunes of kilometer-scale wavelengths and intervening troughs that explain the diversity of depositional facies and faunas over Breccia Unit A (Tapanila and Ekdale, 2004). Water depths may have also been controlled by continued adjustment of the ring faults and terraces, and by dewatering and differential compaction of the impact debris. The morphology of this system may have been similar to the annular troughs of the Popigai impact structure in Russia (Masaitis et al., 1999) and the Chicxulub crater in Mexico (Pope et al., 2005).

Beyond the proposed rings at least some of the Alamo Breccia in the Runup Realm was stranded above sea level, dried out, karsted over millennia, then covered by the next Milankovitch-scale marine transgression over the platform. Even more distal platform rocks preserved channels of the Runoff Realm and disturbed intervals of the Seismite Realm.

Channels of the Runout/Resurge Realm (Fig. 2) contain abundant platform debris, are associated with one or more Middle Devonian limestone olistoliths, and intercalate with Upper Devonian siliciclastic, turbiditic, deepwater facies (Morrow et al., 2005). These deposits may represent the most proximal locations to the central crater, which is still sought.

Abundant evidence for the Alamo Impact Event has been gathered for the past 18 years. Absorbing problems are still left open, such as the exact location and water depth at the target center, the maximum depth of excavation, and the extent of post-Event tectonic redistribution of the original positions of the various outcrops by Paleozoic and Mesozoic tectonic shortening, followed by more recent extension and lineament offset. A major obstacle to better solutions of these problems is the lack of a comprehensive analysis of the tectonic history, which when achieved would facilitate a more complete Late Devonian paleogeographic reconstruction for the entire study area. Conversely, the known properties and regional trends of the Alamo Breccia should be used to test the outcome of proposed models for the needed tectonic reassembly.

ACKNOWLEDGMENTS

New research on the Rim Realm at Tempiute Mountain, new recognition and analysis of disturbed beds on Frenchman Mountain in the Seismite Realm, and new research on the three examples from the Ring Realm were all part of Pinto's (2006) Ph.D. research at the Colorado School of Mines. Much of Pinto's work was supported by Petróleos de Venezuela S.A. (PDVSA), which is gratefully acknowledged. Warme's research is self-funded. We have borrowed heavily from the research results of others who work on the Alamo Breccia, most prominently Jared Morrow at San Diego State University and Charles Sandberg and colleagues at the U.S. Geological Survey. We thank them, and the many others who have contributed observations, ideas, discussions, and skepticism that pushed us onward. Alan Chamberlain discovered the interval we interpret as seismites on Frenchman Mountain, as well as numerous other Alamo Breccia locations that we would not have discovered on our own. The Department of Geology and Geological Engineering at Colorado School of Mines provided research facilities. We thank John Skok at Colorado School of Mines for preparing thin sections and providing other laboratory expertise.

We very sincerely thank Charles Sandberg, Jens Ormö, and Jared Morrow for their expert, knowledgeable, and painstaking reviews of the submitted draft of this manuscript, which we have substantially corrected and revised. We hope this work-in-progress report is a useful contribution toward our understanding of the Alamo Event, and take full responsibility for its content.

REFERENCES CITED

Ackman, B.W., 1991, Stratigraphy of the Guilmette Formation, Worthington Mountains and Schell Creek Range, southeastern Nevada [M.S. thesis]: Golden, Colorado School of Mines, 207 p.

Allen, S.R., 2004, Complex spatter- and pumice-rich pyroclastic deposits from an andesitic caldera-forming eruption: The Siwi pyroclastic sequence, Tanna, Vanuatu: Bulletin of Volcanology, v. 67, p. 27–41, doi: 10.1007/s00445-004-0358-6.

Anderson, J.R., and Tapanila, L., 2007, Reconstructing the aftermath of the Late Devonian Alamo meteor impact in the type section of the Pahranagat Range, SE Nevada: Geological Society of America Annual Meeting, Denver, Abstracts with Programs, v. 39, no. 6, p. 145.

Baratoux, D., and Melosh, H.J., 2003, The formation of shatter cones by shock wave interference during impacting: Earth and Planetary Science Letters, v. 216, p. 43–54, doi: 10.1016/S0012-821X(03)00474-6.

Bereskin, S.R., 1982, Middle and Upper Devonian stratigraphy of portions of southern Nevada and southeastern California, *in* Powers, R.B., ed., Geologic studies of the Cordilleran Thrust Belt, volume 2: Denver, Colorado, Rocky Mountain Association of Geologists, p. 751–764.

Blair, T.C., and MacPherson, J.G., 1999, Grain-size and textural classification of coarse sedimentary particles: Journal of Sedimentary Research, v. 69, p. 6–19.

Borradaile, G.J., Powell, C.M., and Bayly, M.B., 1982, Atlas of deformational and metamorphic rock fabrics: Berlin, New York, Springer-Verlag, 564 p.

Castor, S.B., Faulds, J.E., Rowland, S.M., and dePolo, C.M., 2000, Geologic Map of the Frenchman Mountain Quadrangle, Clark County: Nevada Bureau of Mines and Geology Map 127, scale 1:24,000, 1 sheet, 15 p. text.

Chamberlain, A.K., 1999, Structure and Devonian stratigraphy of the Timpahute Range, Nevada [Ph.D. thesis]: Golden, Colorado School of Mines, 564 p.

Chamberlain, A.K., and Warme, J.E., 1996, Devonian sequences and sequence boundaries, Timpahute Range, Nevada, *in* Longman, M.W., and Sonnenfeld, M.D., eds., Paleozoic Systems of the Rocky Mountain Region: Denver, Colorado, Society for Sedimentary Geology, Rocky Mountain Section, p. 63–84.

Claeys, P., Heuschkel, S., Lounejeva-Baturina, E., Sanchez-Rubio, G., and Stöffler, D., 2003, The suevite of drill hole Yucatán 6 in the Chicxulub impact crater: Meteoritics and Planetary Science, v. 38, p. 1299–1317.

Collins, G.S., and Turtle, E.P., 2003, Modeling complex crater collapse: Lunar and Planetary Institute, Workshop on Impact Cratering, abstract 8037, 2 p.

Collins, G.S., Melosh, H.J., and Ivanov, B.A., 2004, Modeling damage and deformation in impact simulations: Meteoritics and Planetary Science, v. 39, p. 217–231.

Cook, H.E., and Corboy, J.J., 2004, Great Basin Paleozoic carbonate platform: Facies, facies transitions, depositional models, platform architecture, sequence stratigraphy, and predictive mineral host models: U.S. Geological Survey Open-File Report 2004-1078, 129 p.

Dence, M.R., Grieve, R.A.F., and Robertson, P.B., 1977, Terrestrial impact structures: Principal characteristics and energy considerations, *in* Roddy, D.J., et al., eds., Impact and explosion cratering: Planetary and terrestrial implications: New York, Pergamon, p. 247–275.

Deutsch, A., Langenhorst, F., Hornemann, U., and Ivanov, B.A., 2003, On the shock behavior of anhydrite and carbonates: Is post-shock melting the most important effect? Examples from Chicxulub: Lunar and Planetary Institute, International Congress on Large Meteorite Impacts, 3rd, abstract 4080, 2 p.

Dressler, B.O., and Reimold, W.U., 2004, Order or chaos? Origin and mode of emplacement of breccias in floors of large impact structures: Earth-Science Reviews, v. 67, p. 1–54, doi: 10.1016/j.earscirev.2004.01.007.

Dressler, B.O., and Sharpton, V.L., 1997, Breccia formation at a complex impact crater: Slate Islands, Lake Superior, Ontario, Canada: Tectonophysics, v. 275, p. 285–311, doi: 10.1016/S0040-1951(97)00003-6.

Dressler, B.O., Sharpton, V.L., and Copeland, P., 1999, Slate Islands, Lake Superior, Canada: A mid-size, complex impact structure, *in* Dressler, B.O., and Sharpton, V.L., eds., Large meteorite impacts and planetary evolution, II: Geological Society of America Special Paper 339, p. 109–124.

Dressler, B.O., Sharpton, V.L., Morgan, J., Buffler, R., Moran, D., Smit, J., and Urutia, J., 2003, Investigating a 65-Ma-old smoking gun: Deep drilling of the Chicxulub impact structure: Eos (Transactions, American Geophysical Union), v. 84, p. 125–130.

Dunn, M.J., 1979, Depositional history and paleoecology of an Upper Devonian (Frasnian) bioherm, Mount Irish, Nevada [M.S. thesis]: Binghamton, State University of New York, 133 p.

Dypvik, H., Sandbakken, P.L., Postma, G., and Mørk, A., 2004, Early post-impact sedimentation around the central high of the Mjølnir impact crater (Barents Sea, Late Jurassic): Sedimentary Geology, v. 168, p. 227–247, doi: 10.1016/j.sedgeo.2004.03.009.

Engelhardt, W. von, 1990, Distribution, petrography, and shock metamorphism of the ejecta of the Ries Crater in Germany: A review: Tectonophysics, v. 171, p. 259–273, doi: 10.1016/0040-1951(90)90104-G.

Ettensohn, F.R., Rast, N., and Brett, C.E., 2002, Ancient seismites: Geological Society of America Special Paper 359, 200 p.

French, B.M., 1998, Traces of catastrophe: A handbook of shock-metamorphic effects in terrestrial meteorites impact structures: Lunar and Planetary Institute Contribution 954, 120 p.

Fryxell, J.E., and Duebendorfer, E.M., 2005, Origin and trajectory of the Frenchman Mountain block, an extensional allochthon in the Basin and Range province, southern Nevada: Journal of Geology, v. 113, p. 355–371, doi: 10.1086/428810.

Gash, P.J.S., 1971, Dynamic mechanism for the formation of shatter cones: Natural Physical Science, v. 230, p. 32–35.

Gisler, G.R., Weaver, R.P., and Mader, C.L., 2004, Two- and three-dimensional asteroid impact simulation: Computing in Science and Engineering, v. 6, p. 46–55, doi: 10.1109/MCISE.2004.1289308.

Harris, R.S., and Morrow, J.R., 2007, Possible origin of enigmatic hematite inclusions in shocked quartz from the Late Devonian Alamo impact breccia: Geological Society of America Annual Meeting, Denver, Abstracts with Programs, v. 39, no. 6, p. 311.

Horton, J.W., Jr., Aleinikoff, J.N., Kunk, M.J., Gohn, G.S., Edwards, L.E., Self-Trail, J.M., Powars, D.S., and Izett, G.A., 2005, Recent research in the Chesapeake Bay impact structure, USA: Impact debris and reworked ejecta, *in* Kenkmann, T., et al., eds., Large meteorite impacts, III: Geological Society of America Special Paper 384, p. 259–280.

Hörz, F., Ostertag, R., and Rainey, D.A., 1983, Bunte Breccia of the Ries: Continuous deposits of large impact craters: Reviews of Geophysics and Space Physics, v. 21, p. 1667–1725.

Howard, K.A., Offield, T.W., and Wilshire, H.G., 1972, Structure of Sierra Madera, Texas, as a guide to central peaks of lunar craters: Geological Society of America Bulletin, v. 83, p. 2795–2808, doi: 10.1130/0016-7606(1972)83[2795:SOSMTA]2.0.CO;2.

Ivanov, B.A., 2004, Heating of the lithosphere during meteorite cratering: Solar System Research, v. 38, p. 266–278, doi: 10.1023/B:SOLS.0000037462.56729.ba.

Johnson, G.P., and Talbot, R.J., 1964, A theoretical study of the shock wave origin of shatter cones [M.S. thesis]: Dayton, Ohio, Air Force Institute of Technology, 60 p.

Kaufmann, B., 2006, Calibrating the Devonian time scale: A synthesis of U-Pb ID-TIMS ages and conodont stratigraphy: Earth-Science Reviews, v. 76, p. 175–190, doi: 10.1016/j.earscirev.2006.01.001.

Kendall, G.W., Johnson, J.G., Brown, J.O., and Klapper, G., 1997, Stratigraphy and facies across Lower Devonian–Middle Devonian boundary, central Nevada: American Association of Petroleum Geologists Bulletin, v. 67, p. 2199–2207.

Kettrup, B., Deutsch, A., and Masaitis, V.L., 2003, Homogeneous impact melts produced by a heterogeneous target? Sr-Nd isotopic evidence from the Popigai crater, Russia: Geochimica et Cosmochimica Acta, v. 67, p. 733–750, doi: 10.1016/S0016-7037(02)01143-2.

Kieffer, S.W., and Simonds, C.H., 1980, The role of volatiles and the lithology in the impact cratering process: Reviews of Geophysics and Space Physics, v. 18, p. 143–181.

Kobberger, G., and Schmincke, H.-U., 1999, Deposition of rheomorphic ignimbrite D (Mogán Formation), Gran Canaria, Canary Islands, Spain: Bulletin of Volcanology, v. 60, p. 465–485, doi: 10.1007/s004450050246.

Koeberl, C., Huber, H., Morgan, M., and Warme, J.E., 2003, Search for an extraterrestrial component in the Late Devonian Alamo impact Breccia, Nevada: Results of iridium measurements, *in* Koeberl, C., and Martínez-Ruiz, F., eds., Impact markers in the stratigraphic record: Berlin, Springer-Verlag, p. 315–332.

Kuehner, H.-C., 1997, The Late Devonian Alamo impact Breccia, southeastern Nevada [Ph.D. thesis]: Golden, Colorado School of Mines, 327 p.

Lana, C., Romano, R., Reimold, U., and Hippertt, J., 2006, Collapse of large complex impact craters: Implications from the Araguainha impact structure, central Brazil: Geology, v. 34, p. 9–12, doi: 10.1130/G21952.1.

Leroux, H., Warme, J.E., and Doukhan, J.-C., 1995, Shocked quartz in the Alamo Breccia, southern Nevada: Geology, v. 23, p. 1003–1006, doi: 10.1130/0091-7613(1995)023<1003:SQITAB>2.3.CO;2.

Lowe, D.R., Byerly, G.R., Kyte, F.T., Shukolyukov, A., Asaro, F., and Krull, A., 2003, Spherule beds 3.47–3.24 billion years old in the Barberton

Greenstone Belt, South Africa: A record of large meteorite impacts and their influence on early crustal and biological evolution: Astrobiology, v. 3, p. 7–48, doi: 10.1089/153110703321632408.

Marco, S., and Agnon, A., 1995, Prehistoric earthquake deformations near Masada, Dead Sea graben: Geology, v. 23, p. 695–698, doi: 10.1130/0091-7613(1995)023<0695:PEDNMD>2.3.CO;2.

Masaitis, V.L., 2002, The Middle Devonian Kaluga impact crater (Russia): New interpretation of marine setting: Deep Sea Research Part II: Topical Studies in Oceanography, v. 49, p. 1157–1169, doi: 10.1016/S0967-0645(01)00142-4.

Masaitis, V.L., Naumov, M.V., and Mashchak, M.S., 1999, Anatomy of the Popigai impact crater, Russia, *in* Dressler, B.O., and Sharpton, V.L., eds., Large impacts and planetary evolution, II: Geological Society of America Special Paper 339, p. 1–17.

Matti, J.C., Castor, S.B., Bell, J.W., and Rowland, S.M., 1993, Geologic map of Las Vegas NE quadrangle: Nevada Bureau of Mines and Geology Map 3CG, scale 1:24,000, 1 sheet.

Melosh, H.J., 1989, Impact cratering: A geologic process: New York, Oxford University Press, 245 p.

Melosh, H.J., and Ivanov, B.A., 1999, Impact crater collapse: Annual Review of Earth and Planetary Sciences, v. 27, p. 385–415, doi: 10.1146/annurev.earth.27.1.385.

Merriam, C.W., 1963, Paleozoic rocks of Antelope Valley Eureka and Nye Counties Nevada: U.S. Geological Survey Professional Paper 423, 69 p.

Milton, D.J., 1977, Shatter cones: An outstanding problem in shock mechanics, *in* Roddy, D.J., Pepin, R.O., and Merrill, R.B., eds., Impact and explosion cratering: New York, Pergamon Press, p. 703–714.

Morrow, J.R., 1997, Shelf-to-basin event stratigraphy, conodont paleoecology, and geologic history across the Frasnian-Famennian (F-F, mid-Late Devonian) boundary mass extinction, central Great Basin, western U.S. [Ph.D. thesis]: Boulder, University of Colorado, 355 p.

Morrow, J.R., 2000, Shelf-to-basin lithofacies and conodont paleoecology across Frasnian-Famennian (F-F, mid-Late Devonian) boundary, Central Great Basin (western USA): Courier Forschungsinstitut Senckenberg, v. 219, 57 p.

Morrow, J.R., 2004, New evidence on size and marine site of Late Devonian Alamo impact, southern Nevada: Geological Society of America Abstracts with Programs, v. 36, no. 5, p. 265.

Morrow, J.R., and Sandberg, C.A., 2001, Distribution and characteristics of multi-sourced shock-metamorphosed quartz grains, Late Devonian Alamo Impact, Nevada: Lunar and Planetary Institute Contribution 1080, abstract 1233, 2 p.

Morrow, J.R., and Sandberg, C.A., 2003a, Late Devonian Alamo Event, Nevada, USA: Multiple evidence of an off-platform marine impact: Lunar and Planetary Institute Contribution 1167, abstract 4055, 2 p.

Morrow, J.R., and Sandberg, C.A., 2003b, Late Devonian sequence and event stratigraphy across the Frasnian-Famennian (F-F) boundary, Utah and Nevada, *in* Harries, P.J., ed., High-resolution approaches in stratigraphic paleontology: Dordrecht, Kluwer Academic Publishers, p. 351–419.

Morrow, J.R., and Sandberg, C.A., 2005, Revised dating of Alamo and some other Late Devonian impacts in relation to resulting mass extinction, *in* METSOC 2005, Gatlinburg, Tennessee, September 12–16: Meteorics and Planetary Science, v. 40, suppl., p. A106.

Morrow, J.R., and Sandberg, C.A., 2006, Onshore record of marine impact-generated tsunami, 382 Ma Alamo Event, Nevada and Utah, USA: Impact craters as indicators for planetary environmental evolution and astrobiology: Östersund, Sweden, Lockne 2006 meeting, June 8–14, 2006, Abstract Volume, 2 p.

Morrow, J.R., Sandberg, C.A., Warme, J.E., and Kuehner, H.-C., 1998, Regional and possible global effects of sub-critical Late Devonian Alamo impact Event, southern Nevada, USA: Journal of the British Interplanetary Society, v. 51, p. 451–460.

Morrow, J.R., Sandberg, C.A., and Poole, F.G., 2001, New evidence for deeper water site of Late Devonian Alamo impact, Nevada: Lunar and Planetary Institute Contribution 1080, abstract 1018, 2 p.

Morrow, J.R., Sandberg, C.A., and Harris, A.G., 2005, Late Devonian Alamo impact, southern Nevada, USA: Evidence of size, marine site, and widespread effects, *in* Kenkmann, T., et al., eds., Large meteorite impacts, III: Geological Society of America Special Paper 384, p. 259–280.

Nolan, M.C., Asphaug, E., Melosh, H.J., and Greenberg, R., 1996, Impact craters on asteroids: Does gravity or strength control their size?: Icarus, v. 124, p. 359–371, doi: 10.1006/icar.1996.0214.

Nolan, T.B., Merriam, C.W., and Williams, J.S., 1956, The stratigraphic section in the vicinity of Eureka, Nevada: U.S. Geological Survey Professional Paper 276, 77 p.

Oberbeck, V.R., Marshall, J.R., and Aggarwal, H., 1993, Impacts, tillites, and the breakup of Gondwanaland: Journal of Geology, v. 101, p. 1–19.

Ocampo, A.C., Pope, K.O., and Fischer, A.G., 1996, Ejecta blanket deposits of the Chicxulub crater from Albion Island, Belize, *in* Ryder, G., Fastovsky, D., and Gartner, S., eds., The Cretaceous-Tertiary Event and other catastrophes in Earth history: Geological Society of America Special Paper 307, p. 75–88.

Ormö, J., and Lindström, M., 2005, New drill-core data from the Lockne crater, Sweden: The marine excavation and ejection processes, and post-impact environment: 36th Lunar and Planetary Science Conference, abstract 1124, 2 p.

Ormö, J., and Miyamoto, H., 2002, Computer modeling of the water resurge at a marine impact: The Lockne crater, Sweden: Deep Sea Research Part II: Topical Studies in Oceanography, v. 49, p. 983–994, doi: 10.1016/S0967-0645(01)00143-6.

Pinto, J.A., 2006, Alamo impact event, Late Devonian, Nevada: Crater phenomena and distal products [Ph.D. thesis]: Golden, Colorado School of Mines, 191 p.

Pinto, J.A., and Warme, J.E., 2004, Crater stratigraphy and carbonate impact signatures, Late Devonian Alamo event, Nevada: Geological Society of America Abstracts with Programs, v. 36, no. 5, p. 265.

Pinto, J.A., and Warme, J.E., 2006, Alamo event crater and the impact realms of the Alamo Breccia, southern Nevada and environs, USA: Impact craters as indicators for planetary environmental evolution and astrobiology: Östersund, Sweden, Lockne 2006 meeting, June 8–14, 2006, Abstract Volume, 2 p.

Poag, C.W., 1997, The Chesapeake Bay bolide impact: A convulsive event in Atlantic Coastal Plain evolution: Sedimentary Geology, v. 108, p. 45–90, doi: 10.1016/S0037-0738(96)00048-6.

Pohl, J., Stöffler, D., Gall, H., and Ernstson, K., 1977, The Ries impact crater, *in* Roddy, D.J., Pepin, R.O., and Merrill, R.B., eds., Impact and explosion cratering: New York, Pergamon Press, p. 343–404.

Poole, F.G., Stewart, J.H., Palmer, A.R., Sandberg, C.A., Madrid, R.J., Ross, R.J., Jr., Hintze, L.F., Miller, M.M., and Wrucke, C.T., 1992, Latest Precambrian to latest Devonian time: Development of a continental margin, *in* Burchfiel, B.C., Lipman, P.W., and Zoback, M.L., eds., The Cordilleran orogen: Conterminous U.S.: Boulder, Colorado, Geological Society of America, Geology of North America, v. G-3, p. 9–56.

Pope, K.O., Ocampo, A.C., Fischer, A.G., Vega, F.J., Ames, D.E., King, D.T., Jr., Fouke, F.W., and Wachman, R.J., and Kletetschka, G., 2005, Chicxulub impact ejecta deposits in southern Quintana Roo, Mexico, and central Belize, *in* Kenkmann, T., Hörz, F., and Deutsch, A., eds., Large meteorite impacts, III: Geological Society of America Special Paper 384, p. 171–190.

Reso, A., 1963, Composite columnar section of exposed Paleozoic and Cenozoic rocks in the Pahranagat Range, Lincoln County, Nevada: Geological Society of America Bulletin, v. 74, p. 901–918, doi: 10.1130/0016-7606(1963)74[901:CCSOEP]2.0.CO;2.

Sagy, A., Fineberg, J., and Reches, Z., 2004, Shatter cones: Branched rapid fractures formed by shock impact: Journal of Geophysical Research, v. 109, B10209, doi: 10.1029/2004JB003016.

Sandberg, C.A., and Morrow, J.R., 2007, Late Devonian 382 Ma Alamo Impact: proximal and distat effects, Nevada and Utah: Geological Society of America Annual Meeting, Denver, Abstracts with Programs, v. 39, no. 6, p. 372.

Sandberg, C.A., and Warme, J.E., 1993, Conodont dating, biofacies, and catastrophic origin of Late Devonian (early Frasnian) Alamo Breccia, southern Nevada: Geological Society of America Abstracts with Programs, v. 25, no. 3, p. 77.

Sandberg, C.A., Morrow, J.R., and Warme, J.E., 1997, Late Devonian Alamo impact Event, global Kellwasser events, and major eustatic events, eastern Great Basin, Nevada and Utah, *in* Link, P.K., and Kowallis, B.J., eds., Proterozoic to recent stratigraphy, tectonics, and volcanology, Utah, Nevada, southern Idaho and central Mexico: Brigham Young University Geology Studies, v. 42, pt. 1, p. 129–160.

Sandberg, C.A., Morrow, J.R., and Ziegler, W., 2002, Late Devonian sea-level changes, catastrophic events, and mass extinctions, *in* Koeberl, C., and MacLeod, K.G., eds., Catastrophic events and mass extinctions: Geological Society of America Special Paper 356, p. 473–487.

Sandberg, C.A., Morrow, J.R., Poole, F.G., and Ziegler, W., 2003, Middle Devonian to Early Carboniferous event stratigraphy of Devils Gate and northern Antelope Range sections, Nevada, USA: Courier Forschungsinstitut Senckenberg, v. 242, p. 187–207.

Sandberg, C.A., Poole, F.G., and Morrow, J.R., 2005, Milk Spring channels provide further evidence of oceanic, >1.7-km-deep Late Devonian Alamo Crater, southern Nevada: Lunar and Planetary Institute Contribution 1234, abstract 1538, 2 p.

Sandberg, C.A., Poole, F.G., and Morrow, J.R., 2006, Reinterpretation of Milk Spring and other oceanic deposits resulting from the 382 Ma Alamo impact, southern Nevada, USA: Impact craters as indicators for planetary environmental evolution and astrobiology: Östersund, Sweden, Lockne 2006 meeting, June 8–14, 2006, Abstract Volume, 2 p.

Schulte, P., and Kontny, A., 2005, Chicxulub impact ejecta from the Cretaceous-Paleogene (K-P) boundary in northeastern Mexico, *in* Kenkmann, T., Hörz, F., and Deutsch, A., eds., Large meteorite impacts, III: Geological Society of America Special Paper 384, p. 191–221.

Schumacher, R., and Schmincke, H.-U., 1995, Models for the origin of accretionary lapilli: Bulletin of Volcanology, v. 56, p. 626–639.

Shuvalov, V.V., and Trubestkaya, I.A., 2002, Numerical modeling of marine target impacts: Solar System Research, v. 36, p. 417–430, doi: 10.1023/A:1020467522340.

Simonson, B.M., 2003, Petrographic criteria for recognizing certain types of impact spherules in well-preserved Precambrian successions: Astrobiology, v. 3, p. 49–65, doi: 10.1089/153110703321632417.

Skála, R., 2002, Shock-induced phenomena in limestones in the quarry near Ronheim, the Ries Crater, Germany: Czech Geological Survey Bulletin, v. 77, p. 313–320.

Spirakis, C.S., and Hey, A.V., 1988, Possible effects of thermal degradation of organic matter on carbonate paragenesis and fluorite precipitation in Mississippi Valley–type deposits: Geology, v. 16, p. 1117–1120, doi: 10.1130/0091-7613(1988)016<1117:PEOTDO>2.3.CO;2.

Spray, J.G., 1998, Localized shock- and friction-induced melting in response to hypervelocity impact, *in* Grady, M.M., Hutchison, R., McCall, G.J.H., and Rothery, D.A., eds., Meteorites: Flux with time and impact effects: Geological Society [London] Special Publication 140, p. 195–204.

Spudis, P.D., 1993, The geology of multi-ring impact basins: Cambridge, UK, Cambridge University Press, 263 p.

Stevens, C.H., and Stone, P., 2006, Submarine gravity slides on the Paleozoic continental slope at the western edge of the Great Basin, east-central California: A mechanism for development of unconformities in slope environments: Stratigraphy, v. 3, p. 139–149.

Stewart, J.H., 1980, Geology of Nevada: A discussion to accompany the geologic map of Nevada: Nevada Bureau of Mines and Geology Special Publication 4, 136 p.

Stöffler, D., Ewald, U., Ostertag, R., and Reimold, W.U., 1977, Research drilling Nördlingen (1973) (Ries): Composition and texture of polymict impact breccias: Geologica Bavarica, v. 75, p. 163–189.

Sturkell, E.F.F., and Ormö, J., 1997, Impact-related clastic injections in the marine Ordovician Lockne impact structure, central Sweden: Sedimentology, v. 44, p. 793–804, doi: 10.1046/j.1365-3091.1997.d01-54.x.

Tapanila, L.M., and Anderson, J.R., 2007, Survey of ichnological response following the Late Devonian Alamo Impact Event, southeastern Nevada: Geological Society of America Annual Meeting, Denver, Abstracts with Programs, v. 39, no. 6, p. 206.

Tapanila, L.M., and Ekdale, A.A., 2004, Impact of an impact: benthic recovery immediately following the Late Devonian Alamo Event: Geological Society of America Abstracts with Programs, v. 36, no. 5, p. 313.

Taylor, W.J., Bartley, J.M., Martin, M.W., Geissman, J.W., Walker, J.D., Armstrong, P.A., and Fryxell, J.E., 2000, Relations between hinterland and foreland shortening: Sevier orogeny, central North American Cordillera: Tectonics, v. 19, p. 1124–1143.

Tschanz, C.M., and Pampeyan, E.H., 1970, Geology and mineral deposits of Lincoln County, Nevada: Nevada Bureau of Mines and Geology Bulletin, v. 73, 187 p.

Vishnevsky, S., and Montanari, A., 1999, Popigai impact structure (Arctic Siberia, Russia): Geology, petrology, geochemistry, and geochronology of glass-bearing impactites, *in* Dressler, B.O., and Sharpton, V.L., eds., Large impacts and planetary evolution, II: Geological Society of America Special Paper 339, p. 19–59.

von Dalwigk, I., and Ormö, J., 2001, Formation of resurge gullies at impacts at sea: The Lockne crater, Sweden: Meteoritics and Planetary Science, v. 36, p. 359–369.

Warme, J.E., 1991, The Alamo Breccia: Catastrophic Devonian platform deposit in southeastern Nevada, *in* Event markers in Earth history: Program Abstracts, International Union of Geological Sciences, Joint Meeting of IGCP Projects 216, 293, and 303, August 20–30, 1991, Calgary, p. 75.

Warme, J.E., 2004, The many faces of the Alamo impact Breccia: Geotimes, v. 49, p. 28–31.

Warme, J.E., and Kuehner, H.-C., 1998, Anatomy of an anomaly: The Devonian catastrophic Alamo impact breccia of southern Nevada: International Geology Review, v. 40, p. 189–216.

Warme, J.E., and Pinto, J.A., 2005, Alamo enigmas: Too much area and volume, too little crater and symmetry, *in* Evans, K.R., et al., eds., Society for Sedimentary Geology Research Conference: The Sedimentary Record of Meteorite Impacts, Springfield, Missouri, May 21–23, 2005, Abstracts with Program, p. 31–32.

Warme, J.E., and Sandberg, C.A., 1995, The catastrophic Alamo Breccia of southern Nevada: Record of a Late Devonian extraterrestrial impact: Courier Forschungsinstitut Senckenberg, W: Ziegler Commemorative, v. 188, p. 31–57.

Warme, J.E., and Sandberg, C.A., 1996, Alamo megabreccia: GSA Today, v. 6, no. 1, p. 1–7.

Warme, J.E., Chamberlain, A.K., and Ackman, B.W., 1991, The Alamo event: Devonian cataclysmic breccia in southeastern Nevada: Geological Society of America Abstracts with Programs, v. 23, no. 2, p. 108.

Warme, J.E., Ackman, B.W., Yarmanto, and Chamberlain, A.K., 1993, The Alamo Event: Cataclysmic Devonian breccia, southeastern Nevada, *in* Gillespie, C.W., ed., Structural and stratigraphic relationships of Devonian reservoir rocks, east-central Nevada: Reno, Nevada, Petroleum Society of Nevada, 1993 Field Conference, Guidebook, p. 157–170.

Warme, J.E., Morgan, M., and Kuehner, H.-C., 2002, Impact-generated carbonate accretionary lapilli in the Late Devonian Alamo Breccia, *in* Koeberl, C., and MacLeod, K.G., eds., Catastrophic events and mass extinctions: Geological Society of America Special Paper 356, p. 489–504.

Wolff, J.A., and Wright, J.V., 1981, Rheomorphism of welded tuffs: Journal of Volcanology and Geothermal Research, v. 10, p. 13–34, doi: 10.1016/0377-0273(81)90052-4.

Yarmanto, 1992, Sedimentology and stratigraphic setting of a Devonian carbonate breccia, northern Pahranagat Range, Nevada [M.S. thesis]: Golden, Colorado School of Mines, 218 p.

Manuscript Accepted by the Society 10 July 2007

Printed in the USA

The Geological Society of America
Special Paper 437
2008

An examination of the Simpson core test wells suggests an age for the Avak impact feature near Barrow, Alaska

Arthur C. Banet
3050 Flyway Avenue, Anchorage, Alaska 99516, USA

J.P.G. Fenton
Fugro Robertson Ltd., Llandudno North Wales, LL30 ISA, UK

ABSTRACT

Exploration wells and interpretations of two-dimensional seismic surveys delineate a circular feature in the subsurface ~12 km southeast of Barrow, Alaska. This is the Avak impact structure. It is at ~1 km in depth, and it is ~10 km in diameter. Displaced and chaotic stratigraphy, shatter cones, and shocked quartz occur at the Avak #1 well, which tested the central uplift of the Avak feature. To date, the age of the Avak impact has been poorly constrained.

Examinations of core holes from comparatively shallow wells drilled ~50 km east of the Avak impact site show there is an assemblage of exotic rock fragments within a section of marine mud and sand. This is a breccia composed of angular to rounded rock fragments composed of black shale, reddish-brown argillitic siltstone, chert, and quartzite pebbles in sizes up to ~5 cm. These lithologies are representative of the stratigraphic section that was ejected by the Avak event.

These exotic rock fragments occur within the marine mudstone of the Seabee Formation and within poorly sorted, disorganized sands lacking sedimentary structures. Palynology shows that the marine mudstone of the Seabee Formation in this area is Turonian to Coniacian in age. Palynology of the ejecta matrix provides a middle to late Turonian date for the Avak impact event.

Keywords: Alaska, palynology, ejecta, Turonian, Avak.

INTRODUCTION

The Avak structure (71°15.1′N, 156°27.3′W) is the northernmost of 11 Cretaceous-age impacts documented to occur within or immediately peripheral to a Cretaceous-age seaway in western North America (Fig. 1). The Avak impact is located ~12 km southeast of Barrow, Alaska. It is ~10 km in diameter, and it is buried beneath ~1 km of stratigraphically disrupted sedimentary cover. The Avak impact has no surface expression. Drilling and seismic data acquired by the U.S. Navy's exploration (1944–1953 of the National Petroleum Reserve–Alaska, NPR-A [formerly Naval Petroleum Reserve #4]) showed there is a circular, chaotically disturbed zone. Subsequent drilling defined the South and East Barrow gas fields, which flank the impact feature. Drilling

Banet, A.C., and Fenton, J.P.G., 2008, An examination of the Simpson core test wells suggests an age for the Avak impact feature near Barrow, Alaska, *in* Evans, K.R., Horton, J.W., Jr., King, D.T., Jr., and Morrow, J.R., eds., The Sedimentary Record of Meteorite Impacts: Geological Society of America Special Paper 437, p. 139–145, doi: 10.1130/2008.2437(08). For permission to copy, contact editing@geosociety.org.

Figure 1. Western North American Seaway and 11 Cretaceous impacts. Shaded areas represent land superimposed over current outline of North America; modified from http://www.unb.ca/passc/ImpactDatabase/NorthAmerica.html.

in the central uplifted portion of the chaotic zone (Avak #1 well) found an uplifted and fault-repeated sedimentary section with accompanying breccia intervals.

Collins (1961) and Robinson and Brewer (1964) initially described cuttings and cores from the Avak #1 well, the surrounding wells from the South Barrow gas field wells, and the Simpson core hole tests. They subsequently suggested Avak to be a hypervelocity impact feature (Collins and Robinson, 1967). Kirschner et al. (1992) presented a rigorous interpretation of Avak based on seismic interpretations, well analyses, and Bouguer gravity modeling. In cores from the Avak #1 well, which penetrated the central uplift, they also found displaced and cataclastically deformed sediments, uplifted and repeated strata, and polymict breccias. They identified multiple sets of shatter cones and kinked chlorite clays in the uplifted sands and shales. They also identified shocked quartz within breccia near the base of the deformed sediments.

Geophysical data show that the Avak feature is ~8–10 km in diameter. Seismic and gravity interpretations resolve a circular upturned rim surrounding an uplifted central peak. Well and seismic data show that the sedimentary section peripheral to the Avak feature is greatly disturbed. Kirschner et al. (1992) suggest an age of Avak that ranges between Cenomanian and Late Pliocene. Homza (2004) speculates that widespread, soft-sediment slumping and scour events in the Torok Formation (Aptian) found by seismic programs in the Fish Creek area, ~120 km southeast of Avak, may have been triggered by the impact. The slumps and scour are north-trending linear features with displacements toward the east. Gas from some of the wells in the Barrow fields has 2.7%–4% helium content, in contrast to the ~0.1% helium content of other North Slope gases (Moore and Sigler, 1987). The higher helium content may reflect that basement rocks were fractured during the impact, allowing for the migration of deeply seated gas into the Barrow fields (Banet, 2002).

GEOLOGY

Subsurface North Slope geology shows there are two major depositional sequences that overlie the basement rocks (Lerand, 1973). Hubbard et al. (1987) modified this to three depositional sequences. The basement complex in northern Alaska is an antiformal feature called the Barrow Arch, which plunges toward the east-southeast from the Barrow area. Tectonic activity along the arch has resulted in limited deposition and subsequent local erosion of the major depositional sequences along the arch. The basement is ~1 km deep in the Barrow area, ~3.2 km deep in the Prudhoe Bay area, and over 5 km deep where the deepest wells penetrate a carbonate, shale, and schist sequence near the Canning River, which borders ANWR (Arctic National Wildlife Refuge) (Fig. 2). The north flank of the Barrow Arch is fault-bounded with down-to-the-north displacements and localized graben development (Cole et al., 1997; Moore et al., 2004). Away from the Barrow Arch, sedimentary cover increases in thickness regularly to the south. With this geometry, the Barrow Arch is the migrational focus for most of the oil and gas in the northern Alaska.

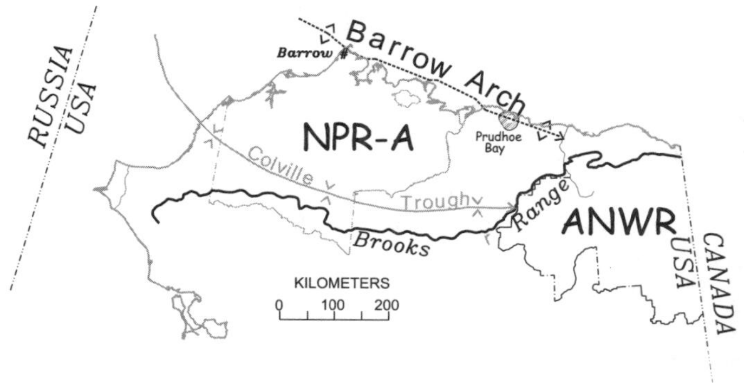

Figure 2. Northern Alaska, showing the major tectonic elements—the Barrow Arch, Colville Trough, and the Brooks Range mountain front—and land management areas NPR-A (National Petroleum Reserve–Alaska) and ANWR (Arctic National Wildlife Refuge).

Exploration wells between the Barrow area and Prudhoe Bay show that basement geology consists of predominantly dark gray to black argillite and subordinate amounts of red argillite around the Barrow gas fields, black graptolitic shale of Silurian or Devonian age around the Prudhoe Bay field (Carter and Laufeld, 1975, and Churkin, 1975), and undated carbonates, shale, and schist near the Canning River. Seismic shows that much of the basement consists of predominantly poorly defined and discontinuous reflectors beneath the sedimentary section. However, there are areas, such as around the Barrow and Cape Simpson areas, where strong and continuous reflectors are steeply dipping.

Hubbard et al. (1987) presented an integrated interpretation of the major depositional episodes across northern Alaska (Fig. 3). They define three major megasequences: the Ellesmerian, Beaufortian, and Brookian. The Ellesmerian sequence consists of trailing continental margin carbonates, shales, and sandstones, which overlie the basement section. The Ellesmerian sediments range in age from Mississippian through Triassic. These sediments have a northern source and are from a petrologically mature provenance. Carbonate lithologies of the Lisburne Group (Upper Mississippian) are the most widespread of North Slope lithologies. The Ivishak Formation (Triassic) of the Sadlerochit Group is the major reservoir of the North Slope's oil and gas accumulations. The Shublik Formation (Triassic) consists of limestone, organic-rich marl, and shale. The Sag River sand overlies the Shublik Formation. Well and seismic data show there are multiple episodes of pronounced thinning and local erosion across the Barrow Arch.

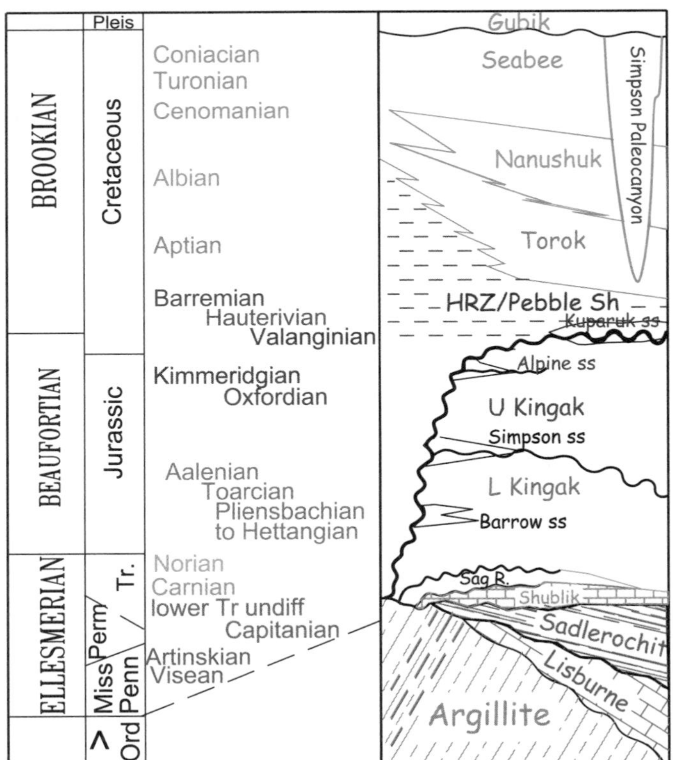

Figure 3. Stratigraphic column for the Avak area and northeast NPR-A showing the major stratigraphic units, their ages, and how the Lower Cretaceous unconformity erodes into the basement (argillite) along the Barrow Arch. HRZ—highly radioactive zone.

The Beaufortian sequence (Jurassic through Lower Cretaceous) is predominantly marine shelf sediments. The Kingak Shale (Jurassic) records marine shelf facies that have sandstones derived from local uplifts along the Barrow Arch. The sands are of local extent and rarely exceed ~30 m in thickness. They are mostly quartzose with conspicuous glauconite and ancillary carbonate cements and minerals. This mineralogy reflects the nature of the eroding Ellesmerian sediments along the Barrow Arch. Within the Kingak, the unconformities and sands along the Barrow Arch record the episodic uplifts associated with the opening of the Arctic Ocean basin. Thick sequences of Beaufortian sediments fill grabens that formed contemporaneously along the north flank of the Barrow Arch. In contrast, the Kingak thickens to the south, and sand content decreases accordingly.

Sands at the regional Lower Cretaceous unconformity record the end of local uplifts along the Barrow Arch. These locally deposited sands and the Pebble Shale/Kalubik Formation (Barremian to lower Aptian; Carman and Hardwick, 1983) overlie the Lower Cretaceous unconformity. The Pebble Shale/Kalubik is black, organic, sooty shale with conspicuous rounded pebbles and floating quartz grains. The overlying highly radioactive zone (HRZ; gamma ray logs > 150 API) consists of black, platy, and highly organic shale. It is a regional marker on logs. Lower Brookian sediments (Albian to Cenomanian) overlie the HRZ.

Brookian clastics prograded east to northeasterly across northern Alaska, overstepping the Barrow Arch and depositing sediment into the Beaufort Sea basin. Well data show that Brookian sedimentation consists of three regionally wide, northeasterly prograding pulses of deep marine through nonmarine facies (Hubbard et al. 1987). These are the Torok/Nanushuk, the Seabee/Tuluvak, and the Sagavanirktok/Canning sequences. Shales and sands of the Torok/Nanushuk (Aptian through Cenomanian) couplet are the basal episode of sedimentation. The Torok is predominantly gray to dark gray, silty shale and mudstone that resolves on seismic as prominent clinoforms/foresets. Turbidite fan and channel sands at the base of the Torok are very fine-grained, lithic-rich, and grade into siltstones. Seismic resolves the basal Torok as hummocky reflectors.

Above the Torok Formation, the Nanushuk Formation comprises paralic through lower deltaic and nonmarine sands, siltstones, and coals. Seismically, these are well-defined, discontinuous, and predominantly flat-lying, topset reflectors. Both seismic and outcrop mapping show that the marine and nonmarine facies interfinger (Mull et al., 2003). The Nanushuk sands are mostly very fine- to fine-grained in the lower paralic facies, and become fine- to coarse-grained, upsection, in the nonmarine facies. In contrast to the Ellesmerian and Beaufortian lithologies, Brookian sands are chert litharenites with conspicuous volcanic and argillitic rock fragments and clays. Sands are typically gray on fresh fracture because of the chert and rock fragments. Nanushuk sands typically weather to tan.

Marine mudstones and silty shales of the Seabee Formation (Cenomanian through Coniacian) disconformably overlie the Nanushuk. This is a major flooding event that delineates the onset of middle Brookian sedimentation. Cuttings show that the Seabee consists of light gray to gray silty mudstone. At outcrops and in wells from south of the Barrow Arch area, the lower Seabee consists of interbedded black platy shale and locally abundant bentonite. The nonmarine facies of the east-northeasterly prograding Seabee/Tuluvak couplet is found mostly east of NPR-A. The Sagavanirktok/Canning sequence is eroded from all but the northeast portion of NPR-A. It does not occur in the Barrow or Simpson areas.

Up to ~33 m of Gubik Formation (Plio-Pleistocene) unconformably overlies the Cretaceous bedrock across northern NPR-A. Near the present coast, the Gubik is predominantly marine, and it is ~20 m thick across the Avak feature on the Barrow Peninsula and at the Simpson Peninsula (Fig. 4). Its lithology consists of unconsolidated and interbedded clay, silt, sand, and cobbles. The sands and cobbles (to ~5 cm) are predominantly black, white, and yellow quartz and chert. There are rare mafic and carbonate rock fragments.

BRECCIA IN THE SIMPSON CORE TEST WELLS

The Simpson Peninsula is 50–70 km southeast of Avak (Fig. 4). Between 1945 and 1953, the U.S. Navy drilled 31 shallow core tests and one deep oil exploration well in the Simpson Peninsula area to delineate the extensive oil seeps. The Simpson cores are ~10 cm in diameter, and they penetrated up to 760 m of Brookian sediments. Core data show that the contact between the Nanushuk sands (lower Brookian) and the overlying marine mudstones of the Seabee Formation (middle Brookian) is markedly uneven across the Simpson Peninsula (Collins and Robinson, 1967; Banet and Fenton, 2005).

In well cuttings, Nanushuk sands are typically tan to buff-colored, very fine- to fine-grained chert litharenites. Bedding is

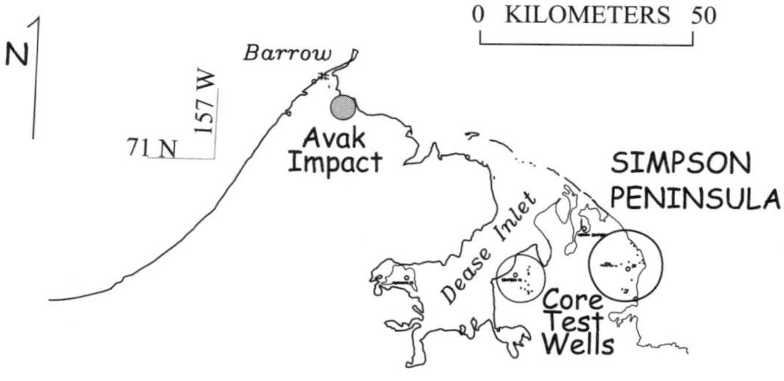

Figure 4. Location and areal distribution of the 31 Simpson core test wells in relation to the Avak impact.

common. Palynology from Simpson core test (SCH) #25, ~70 m beneath the breccia (315.5 m measured depth) shows that the sands of the Nanushuk are late Albian in age. In contrast, the Seabee mudstones are light gray, massive mudstones lacking sedimentary features. Palynology shows that the age of the Seabee is Turonian to Coniacian in the Cape Simpson wells (Table 1). Both the Nanushuk and the Seabee are poorly consolidated in these cores.

Banet (2002) recently studied the Simpson core tests to identify the extent of oil staining for the assessment of oil and gas resources in NPR-A. Robinson and Brewer's (1964) notation of breccia in the marine mudstone of the Seabee section appeared to warrant further study. Examination of five cores from Simpson core tests (#13, #17, #22, #25, and #29; Table 1) along the east side of the Simpson Peninsula shows there are numerous angular, exotic, and polymict clasts that compose the breccia within the Seabee mudstone section. The breccia clasts, which are up to ~2 cm, are most conspicuous in cores from SCH #25 and #29 (Fig. 5). Based on palynological interpretation (sparse flora consisting of *Trilobosporites*), at least one of the large black shale clasts is Lower Cretaceous. The age and lithology are consistent with these clasts correlating to the HRZ or the Pebble Shale.

Additional clasts occur within a section of light to medium gray, poorly sorted, very fine- to coarse-grained and silty sands and gray silt. These sands are predominantly gray and black chert arenites that have no bedding features (Fig. 5). Simpson core tests #22 and #25 have polymict clasts of red to brown and tan, rounded to angular siltstone, gray massive limestone, and laminated limestone. The clasts are up to ~2 cm in size and show no preferred orientation within the poorly sorted matrix. Palynology data suggest that the matrix for these clasts is middle to late Turonian in age, based on the co-occurrence of *Raphidodinium fucatum*, *Stephodinium coronatum*, *Rhyptocorys veligera*, and *Isabelidinium*? *magnum* (Fenton, 2004). The matrix material is comparatively well indurated. Similar red-brown argillitic siltstone and gray limestone clasts occur in the breccia found in the core from Avak #1 well.

French (1998) and Poag (2005a, 2005b) present outcrops and cores of verified impact breccias from multiple locations and ages. Photos of breccias associated with these known impacts resemble the material found in the Simpson core tests. The anomalous size, angularity, and distribution of the polymict clasts (or breccia) and their stratigraphic position found in the Simpson core tests are entirely consistent with an impact origin. The large black shale clasts floating within the mudstone matrix of the Seabee suggest they may represent ballistic sedimentation from the Avak impact. The similar distribution of reddish-brown argillitic siltstone and limestone clasts suggests they may also be from ballistic sedimentary processes. However, the poorly sorted, chert-rich sand matrix in which they occur does not exclude their origin from a base surge. The massive gray mudstone of the Seabee, which overlies these breccia clasts, indicates that relatively quiescent processes returned and persisted after the event (Fig. 5).

TABLE 1. PALYNOLOGY DATA FOR SIMPSON CORE TEST WELLS

Simpson core test (#)	Sample depth (m)	Lithology*	Age	Dating criteria
13	68.6	silty mudstone	Turonian to Coniacian	*Nelsoniella aceras sensu McIntyre, Raphidodinium fucatum, Spongodinium "cristatum" FRL*
13	187.5	silty mudstone	Turonian to Coniacian	*Heterosphaeridium difficle, Nelsoniella aceras sensu McIntyre*
17	144.2	silty mudstone	Turonian to Coniacian	*Nelsoniella aceras sensu McIntyre*
22	243.8 to 246.9	silty mudstone above breccia	Turonian to Coniacian	*Nelsoniella aceras sensu McIntyre, Wallodinium krutzschi, Chatangiella ditissima*
22	247.8	silty mudstone and pebbles	Turonian to Coniacian	*Nelsoniella aceras sensu McIntyre, Chatangiella ditissima*
22	246.9 to 248.4	silty mudstone 0.3 m below pebbles	Turonian to Coniacian	*Nelsoniella aceras sensu McIntyre, Hystrichodinium furcatum, Tubulospina oblongata, Isabelidinium? globosum*
25	238.0	silty mudstone	Turonian to Coniacian	*Nelsoniella aceras sensu McIntyre*
25	239.6	silty mudstone	Turonian to Coniacian	*Nelsoniella aceras sensu McIntyre*
25	244.1 to 247.2	sandy to silty mudstone above breccia	Turonian to Coniacian	*Nelsoniella aceras sensu McIntyre*
25	247.2 to 253.6	black shale breccia clast	Early Cretaceous	*Trilobosporites* spp. (sparse fauna suggests highly radioactive zone of Pebble Shale)
25	250.5 to 253.6	breccia matrix silty mudstone	late to middle Turonian	*Raphidodinium fucatum, Stephodinium coronatum, Rhyptocorys veligera, Isabelidinium magnum,* and specimens of *Chatangiella*
25	250.5 to 253.6	silty sandstone and mudstone	Turonian	*Nelsoniella aceras sensu McIntyre, Hystrichodinium furcatum, Florentinia deanei*
25	257.0 to 260.0	silty mudstone with coaly lamina	Albian	*Psendoceratium securigerum* (no younger than Albian) likely Nanushuk
25	315.5	dark gray mudstone with plant debris	late Albian	*Chichaouadinium* cf. *vestitum, Apteodinium grande, Luxadinium propatulum*
29	120.7 to 124.0	silty mudstone above breccia	Turonian to Coniacian	*Nelsoniella aceras sensu McIntyre*
29	139.9 to 143.0	sand and angular rock fragments	middle–late Turonian	*Wallodinium lunum, Hystrichodinium furcatum, Isabelidinium? globosum, Isabelidinium magnum, Stephodinium coronatum*
29	207.0 to 210.0	silty mudstone >30 m below breccia	late Albian	*Luxadinium propatulum, Epelidosphaeridia spinosa*

*All are light to medium gray.

In order to better preserve the core collection of GMC (Alaska Department of Natural Resources Geologic Materials Center), the megascopic clasts of red-brown argillitic siltstone and gray limestone were not sampled for palynology. However, examination of cores of basement rocks from nearby E. Simpson #2 oil well shows there is present not only the abundant black argillite that is so common through out the Avak #1 well, but also a reddish-brown argillite found in the Simpson core tests and at the Avak #1 well. These oil well cores provide some indirect correlation of the megascopic clasts in the Simpson core tests. The light to medium gray limestone clasts appear to correlate to the Shublik Formation (Triassic).

Seismic mapping identifies north-draining erosional incisions in the subsurface of the Simpson Peninsula. These are the Simpson Paleocanyons. Seismic shows that the Simpson Paleocanyons have eroded deeply and into the Torok Formation at the Simpson Peninsula. J.S. Kelley (2005, personal commun.) suggests that these polymict and megascopic clasts are related to deposition within the Simpson Paleocanyons and are not breccia from the Avak impact. However, the paleocanyons have not penetrated a section that would produce clasts of the HRZ, the Shublik Formation, or any of the basement lithologies (Fig. 3). The nearest outcrops from which these facies could be incorporated into the sedimentary regime are over 300 km to the south. The Simpson Paleocanyons are not known to extend much farther south than the Simpson Peninsula (Kirschner and Rycerski, 1988). In addition, clasts of the sizes noted would not likely be transported from 300 km, nor would they remain subangular.

SCH # 25 247.2 to 253.6 m

SCH # 29 133.5 to 139.9 m.

SCH # 29 136.8 to 139.9 m.

Avak #1 690.3 m.

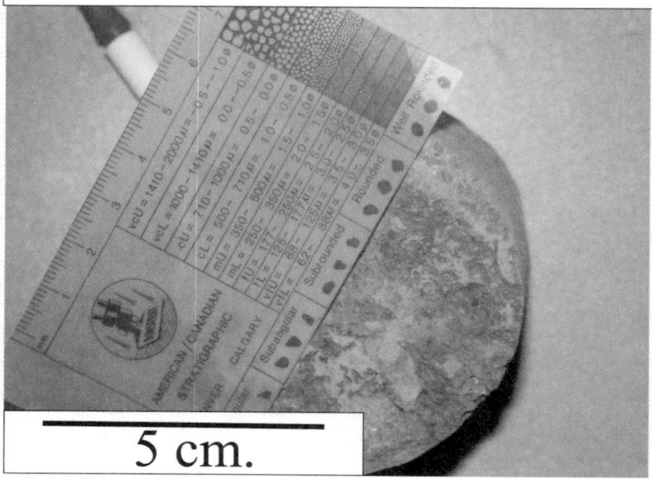

E. Simpson #2 2265 m.

Figure 5. Left column shows black and platy shale clasts from SCH #25 and #29, and red-brown argillitic siltstone and gray limestone clasts from SCH #29. Note that these polymict clasts are comparatively large, angular to subrounded, and unevenly distributed in the matrix. The gray shale matrix is the Seabee, and the chert-rich sand is likely a part of the impact ejecta.

CONCLUSIONS

Polymict and megascopic clasts have been found in shallow cores from the Simpson Peninsula. The clasts are in a matrix of gray mudstones of the Seabee Formation (Turonian to Coniacian). They include angular and black shale clasts from the HRZ, the Pebble Shale (Lower Cretaceous), semirounded, red-brown argillitic siltstones that are likely from the basement rock at Avak, and gray, rounded to angular limestone fragments that may be from the Shublik Formation (Upper Triassic). These angular clasts floating within the unstructured mudstones of the Seabee Formation closely resemble verified impact-related breccias found elsewhere. Their distribution and stratigraphic position suggests that these clasts in the Simpson core tests may represent ballistic fallout sedimentation from the Avak impact near Barrow, Alaska. Poorly sorted clasts lacking preferred orientation within the medium- to coarse-grained chert-litharenite sandstones may represent ballistic or even surge deposition. In addition, the contact between the Seabee and the underlying Nanushuk is markedly uneven across the Simpson Peninsula. The uneven contact and the poorly sorted and exotic, angular to rounded clasts are consistent with ejecta deposition (French, 1998).

The Avak impact near Barrow penetrated, deformed, and exhumed Brookian, Beaufortian, Ellesmerian, and basement rocks. The polymict clasts found in the Simpson core tests reflect some of the lithologies excavated by the Avak event. These exotic clasts are within the Seabee Formation. Palynological data from above the breccia, from the matrix of the breccia, and from immediately below the breccia suggest a middle to late Turonian age for the deposition of the clasts and consequently the age of the Avak impact.

ACKNOWLEDGMENTS

A.C. Banet would like to express his appreciation for Jim Fenton's important palynological contribution, for the assistance of John Reeder of the Geological Material Center in acquiring core samples, and for the support of Dave Buthman of Chevron in pursing impact-related features in Alaska. J.P.G. Fenton publishes with the permission of the directors of Fugro Robertson Limited.

REFERENCES CITED

Banet, A.C., 2002, Weird scenes inside the gold mine: Finding debris from the Avak impact [abs.]: Joint Technical Conference, Pacific Section of American Association of Petroleum Geologists and Western Region, Society of Petroleum Engineers, Anchorage, Alaska, 2002, abstract and poster.

Banet, A.C., and Fenton, J.P.G., 2005, A serendipitous examination of well cores suggests an age for the Avak impact feature near Barrow, Alaska [abs.]: Society of Economic Paleontologists and Mineralogists Research Conference, The Sedimentary Record of Meteorite Impacts, Springfield, Missouri, 2005, May 21–23, 2005.

Carman, G.J., and Hardwick, P., 1983, Geology and regional setting of the Kuparuk oil field: American Association of Petroleum Geologists Bulletin, v. 67, no. 6, p. 1014–1031.

Carter, C., and Laufeld, S., 1975, Ordovician and Silurian fossils in well cores from North Slope of Alaska: American Association of Petroleum Geologists Bulletin, v. 53, no. 3, p. 457–464.

Churkin, M., Jr., 1975, Basement rocks of the Barrow Arch, Alaska and circum-Arctic Paleozoic mobile belt: American Association of Petroleum Geologists Bulletin, v. 59, no. 3, p. 451–456.

Cole, F., Bird, K.J., Toro, J., Roure, F., O'Sullivan, P.B., Pawlewicz, M., and Howell, D.G., 1997, An integrated model for the tectonic development of the frontal Brooks Range and Colville Basin 250 km west of the Trans-Alaska Crustal Transect: Journal of Geophysical Research, v. 102, p. 20,685–20,708, doi: 10.1029/96JB03670.

Collins, F.R., 1961, Core tests and test wells, Barrow area, Alaska: U.S. Geological Survey Professional Paper 305-K, p. 569–644.

Collins, F.R., and Robinson, F.M., 1967, Subsurface, stratigraphic and economic geology, northern Alaska: U.S. Geological Survey Open-File Report 67-64, 252 p., 12 sheets.

Fenton, J.P.G., 2004, Cape Simpson core tests: Palynological analyses of selected samples from core tests 13, 17, 22, 25, and 29: Fugro Robertson Report No. 6449/1b, available at the Geological Materials Center, Eagle River, Alaska.

French, B.M., 1998, Traces of catastrophe: A handbook of shock-metamorphic effects in terrestrial meteorite impact structures: Houston, Lunar and Planetary Institute Contribution 954, 120 p.

Homza, T.X., 2004, A structural interpretation of the Fish Creek Slide (Lower Cretaceous) northern Alaska: American Association of Petroleum Geologists Bulletin, v. 88, no. 3, p. 265–278.

Hubbard, R.J., Pape, J., and Rattey, R.P., 1987, Geologic evolution and hydrocarbon habitat of the "Arctic Alaska Microplate," *in* Tailleur, I.L., and Weimer, P., eds., Pacific Section of Society of Economic Paleontologists and Mineralogists and the Alaska Geological Society, v. 50, p. 797–830.

Kirschner, C.E., and Rycerski, B.A., 1988, Petroleum potential of representative stratigraphic and structural elements in the National Petroleum Reserve in Alaska, *in* Gryc, G., ed., Geology and exploration of the National Petroleum Reserve in Alaska, 1974 to 1982: U.S. Geological Survey Professional Paper 1399, p. 191–208.

Kirschner, C.E., Grantz, A., and Mullen, M.W., 1992, Impact origin of the Avak structure, Arctic Alaska, and genesis of the Barrow gas fields: American Association of Petroleum Geologists Bulletin, v. 76, no. 5, p. 651–679.

Lerand, M., 1973, Beaufort Sea, *in* McCrossan, R.G., ed., The future petroleum provinces of Canada—Their geology and potential: Canadian Society of Petroleum Geologists Memoir 1, p. 315–386.

Moore, B.J., and Sigler, S., 1987, Analyses of natural gases: U.S. Bureau of Mines Report BuMines IC 9129, 101 p.

Moore, T.E., Potter, C.J., and O'Sullivan, P.B., 2004, The Brooks Range Alaska: The consequence of four distinct tectonic events [abs.]: Geological Society of America Abstracts with Programs, v. 36, no. 5, p. 270.

Mull, C.G., Houseknecht, D.W., and Bird, K.J., 2003, Revised Cretaceous and Tertiary stratigraphic nomenclature in the Colville Basin, northern Alaska: U.S. Geological Survey Professional Paper 1673, 59 p.

Poag, C.W., 2005a, The marine sedimentary record of impact paleoenvironmental effects: A field study at Chesapeake Bay [abs.]: SEPM Research Conference, The Sedimentary Record of Meteorite Impacts, Springfield, Missouri, May 21–23, 2005.

Poag, C.W., 2005b, Eastern rim of the Chesapeake Bay impact crater: Morphology, stratigraphy, and structure, *in* Kenkmann, T., et al., eds., Large meteorite impacts, III: Geological Society of America Special Paper 384, p. 117–130.

Robinson, F.M., and Brewer, M.C., 1964, Core tests, Simpson area, Alaska, with a section of temperature measurement studies: U.S. Geological Survey Professional Paper 305-L, p. 645–730.

Manuscript Accepted by the Society 10 July 2007

The Geological Society of America
Special Paper 437
2008

Impact stratigraphy: Old principle, new reality

Gerta Keller*
Department of Geosciences, Princeton University, Princeton, New Jersey 08544, USA

ABSTRACT

Impact stratigraphy is an extremely useful correlation tool that makes use of unique events in Earth's history and places them within spatial and temporal contexts. The K-T boundary is a particularly apt example to test the limits of this method to resolve ongoing controversies over the age of the Chicxulub impact and whether this impact is indeed responsible for the K-T boundary mass extinction. Two impact markers, the Ir anomaly and the Chicxulub impact spherule deposits, are ideal because of their widespread presence. Evaluation of their stratigraphic occurrences reveals the potential and the complexities inherent in using these impact signals. For example, in the most expanded sedimentary sequences: (1) The K-T Ir anomaly never contains Chicxulub impact spherules, whereas the Chicxulub impact spherule layer never contains an Ir anomaly. (2) The separation of up to 9 m between the Ir anomaly and spherule layer cannot be explained by differential settling, tsunamis, or slumps. (3) The presence of multiple spherule layers with the same glass geochemistry as melt rock in the impact breccia of the Chicxulub crater indicates erosion and redeposition of the original spherule ejecta layer. (4) The stratigraphically oldest spherule layer is in undisturbed upper Maastrichtian sediments (zone CF1) in NE Mexico and Texas. (5) From central Mexico to Guatemala, Belize, Haiti, and Cuba, a major K-T hiatus is present and spherule deposits are reworked and redeposited in early Danian (zone P1a) sediments. (6) A second Ir anomaly of cosmic origin is present in the early Danian. This shows that although impact markers represent an instant in time, they are subject to the same geological forces as any other marker horizons—erosion, reworking, and redeposition—and must be used with caution and applied on a regional scale to avoid artifacts of redeposition. For the K-T transition, impact stratigraphy unequivocally indicates that the Chicxulub impact predates the K-T boundary, that the Ir anomaly at the K-T boundary is not related to the Chicxulub impact, and that environmental upheaval continued during the early Danian with possibly another smaller impact and volcanism.

Keywords: stratigraphy, impacts, Chicxulub, K-T boundary.

*gkeller@princeton.edu

Keller, G., 2008, Impact stratigraphy: Old principle, new reality, *in* Evans, K.R., Horton, J.W., Jr., King, D.T., Jr., and Morrow, J.R., eds., The Sedimentary Record of Meteorite Impacts: Geological Society of America Special Paper 437, p. 147–178, doi: 10.1130/2008.2437(09). For permission to copy, contact editing@geosociety.org.

INTRODUCTION

The law of superposition—simply stated, that *in any undisturbed sediment sequence the oldest bed is at the base and the youngest at the top*—has served generations of geologists and led the advance in understanding Earth's history. It has remained the fundamental principle of *Stratigraphy* in relative age dating, permitting the reading of any rock record as an ordered sequence of events from oldest to youngest, regardless of the time period in which it was deposited, or the nature of events that caused their deposition. *Biostratigraphy* is arguably the most reliable and commonly applied age dating technique in stratigraphy. It makes use of the fossil record and its unique evolutionary events, such as species originations and extinctions, unique population turnovers, and individual species blooms, and ties these to the paleomagnetic and radiometric records.

As radiometric dating has steadily improved to error bars of only a few hundred thousand years and scientists have tried to decipher and date the sequence of critical events in Earth's history with ever greater age control, the law of superposition has remained more relevant than ever. This is particularly so for major mass extinctions where a sequence of closely spaced events over a few hundred thousand years may have contributed to the catastrophic demise of life evident in the fossil record. But frequently, radiometric dating cannot decipher the order of such closely spaced events because they may fall within the error limits of the dating methods. In such cases, the only way to determine the time sequence of events leading up to the catastrophe—and hence the potential cause and effect—is the stratigraphic relationship of these events to each other as seen in the sediments and fossils within each rock unit that reveal environmental and biotic effects over time. Stratigraphy and biostratigraphy are thus the handmaidens of geologic history.

During the last decade, the old law of superposition, and stratigraphy and biostratigraphy in particular, has come under attack by what could be called *impact exuberance*. This new reality is driven largely by the belief that large impacts must cause catastrophic mass extinctions and that the telltale evidence of impacts, such as impact breccia, glass spherules, shocked minerals, and iridium anomaly, must therefore be coincident in time with the biotic catastrophe regardless of the sediment layers separating the impact signals from the mass extinction. The most egregious example of this impact exuberance is the K-T boundary mass extinction and the Chicxulub impact. The theory that a large impact caused the mass extinction is widely considered as substantiated by the iridium anomaly at the K-T boundary, and the Chicxulub crater is assumed to be the evidence for this impact. Of course, the association makes sense in theory. But the ground truth—the sedimentary and stratigraphic record—tells another story.

In reality, the most continuous K-T sections show a wide stratigraphic separation between the Chicxulub impact spherule ejecta layer and the K-T mass extinction and iridium anomaly, as for example in NE Mexico, Texas, and even the Chicxulub crater itself (Keller et al., 2003a, 2004a, 2004b, 2007, 2008). Nevertheless, the strong belief in the cause-and-effect scenario, termed the *strong expectation syndrome* by Tsujita (2001), has led some workers to ignore stratigraphic principles. This has resulted in a variety of imaginative interpretations linking Chicxulub impact spherules and the K-T mass extinction and iridium anomaly in a cause-and-effect scenario, including a giant impact-generated megatsunami, backwash and crater infill, large-scale regional slumping due to impact-induced earthquakes, and even injection of layered sediments over wide regions by massive earthquakes (e.g., Smit et al., 1992, 1996, 2004; Smit, 1999; Soria et al., 2001; Lawton et al., 2005; Schulte et al., 2006, 2008). Others found support for a cause-and-effect scenario in a few deep-sea sections of the Caribbean where the Chicxulub spherule ejecta layer and the K-T mass extinction are in close stratigraphic proximity, though associated with disturbed sediments, erosion, and condensed sedimentation (e.g., Sigurdsson et al., 1997; Norris et al., 1999, 2000; Klaus et al., 2000; MacLeod et al., 2006). In effect, interpretations are frequently driven by the *strong expectation syndrome* that any sediments between the K-T boundary and impact ejecta layers must be related to the Chicxulub impact.

This is unfortunate. Impacts are in fact invaluable stratigraphic tools when used within the context of stratigraphy and biostratigraphy. In theory, they provide an instantaneous time horizon in a rock sequence and are correlatable, wherever the particular impact ejecta layer can be found. This study evaluates impact signals in stratigraphy, their application, pitfalls and potential with respect to the K-T boundary and Chicxulub impact events. It addresses the reliability of impact signals as age markers, the problems of reworked impact ejecta, the continuity and completeness of the stratigraphic record, and the necessity of integrating impact signals into the bio-, chemo-, and isotope stratigraphic records. Examples are drawn from all areas with reported impact ejecta, including Texas, Mexico, Haiti, Cuba, Central America, and the deep sea (Fig. 1).

IMPACT STRATIGRAPHY

Impacts occurred throughout Earth's history with some regularity—small and large impacts, isolated or in clusters—and most do not coincide with mass extinctions (Fig. 2). Thus, impacts are not unique by themselves, nor do they generally cause mass extinctions, but they leave unique impact signals that vary from iridium anomalies to shocked minerals, Ni-rich spinels, glass spherules, shatter cones, and suevite breccias. Impacts vary in the signals they leave behind, but most signals are unsuitable as stratigraphic markers because of their sporadic occurrences and limited distributions.

Although frequently ignored, impacts may come in clusters over a period of as little as a couple of hundred thousand years (Napier, 2001, 2006), as notably documented for the late Eocene (Keller et al., 1983; Montanari and Koeberl, 2000; Poag et al., 2002; Harris et al., 2004), K-T boundary (Kelley and Gurov, 2002; Stewart and Allen, 2002; Keller et al., 2003a),

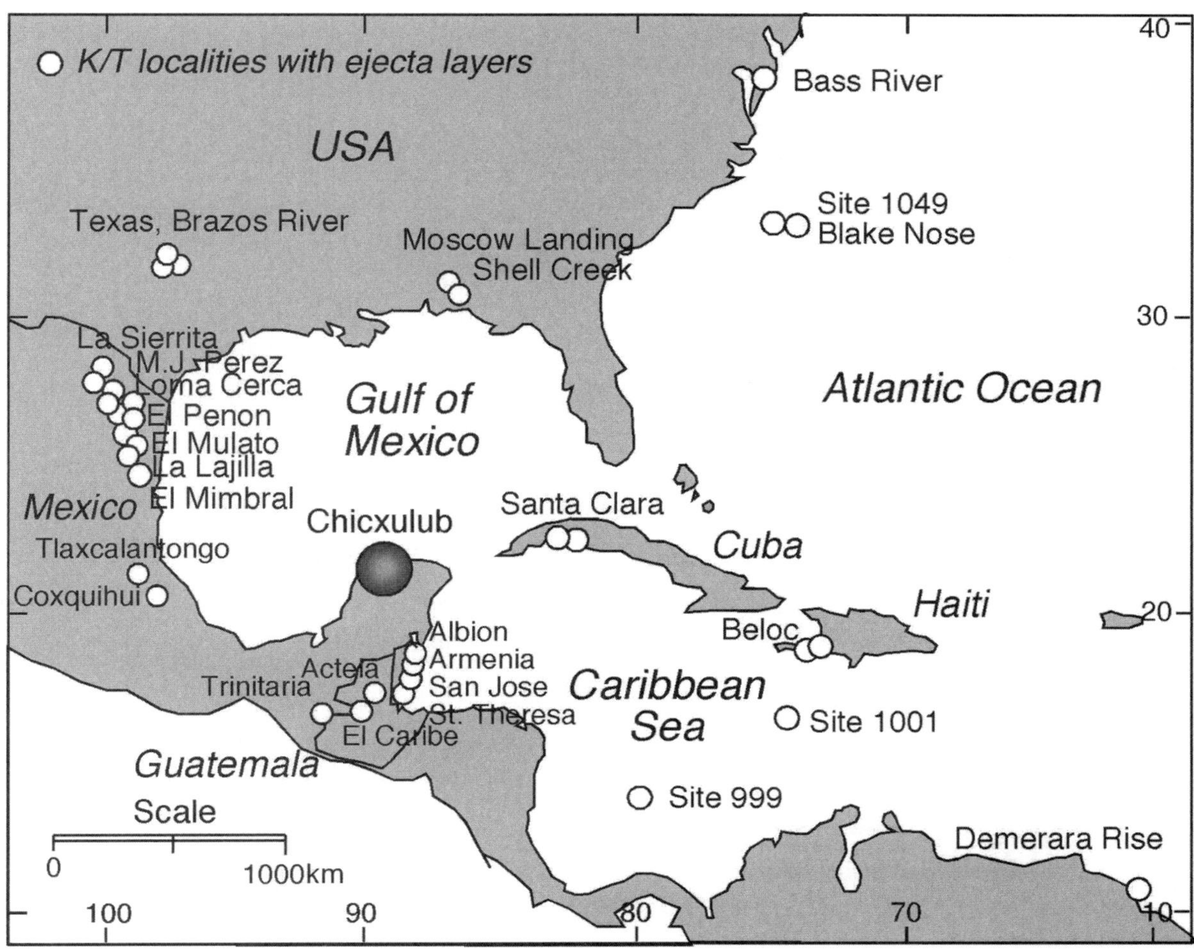

Figure 1. Locations of K-T sequences with Chicxulub impact spherule ejecta in the southern United States, Central America, and Caribbean.

Late Devonian (Glikson and Haines, 2005; Ma and Bai, 2002; Keller, 2005), and Precambrian (Glikson et al., 2004). Therefore, impact signals in close stratigraphic proximity cannot be assumed to represent one and the same impact event. This was first demonstrated by Keller et al. (1983) for the late Eocene and later confirmed by many studies (review in Montanari and Koeberl, 2000). The separation of impact signals in close stratigraphic proximity is therefore an opportunity to decipher the historical record and evaluate biotic consequences. *This is the promise and potential value of impact stratigraphy.*

Impact stratigraphy is part of event stratigraphy that incorporates unique short-term historical events, which leave signals in the sedimentary rocks, into a comprehensive scheme of relative age from oldest to youngest in any given rock sequence. The sedimentary and fossil records before and after the impact signal yield the clues to the environment. Although this seems like a straightforward application, in fact it is complicated by the particular sedimentary environment and the completeness of the sedimentary records. For example, impact signals that are well separated in high-sedimentation marginal shelf environments are frequently juxtaposed in condensed sequences of the deep sea (e.g., Bass River, Blake Nose, Demerara Rise). Hiatuses may have removed sediments containing the impact signals, and erosion and redeposition of impact ejecta into younger sediments results in disparate age relationships (e.g., Mexico, Haiti, Belize, Guatemala). Stratigraphy and biostratigraphy are the primary tools that can unravel the complex postdepositional history of the sedimentary record. But to do so, high-resolution biostratigraphy is necessary.

High-Resolution K-T Biostratigraphy

Planktic foraminifera provide excellent biomarkers for the K-T boundary transition because they suffered the most severe mass extinction of all microfossil groups. All tropical and subtropical specialized large species (2/3 of the assemblage) died out at or shortly before and after the K-T boundary, and their extinction was followed by the rapid evolution and diversification of Danian species beginning within a few centimeters of the boundary clay and Ir anomaly in most sequences. Critical biomarkers

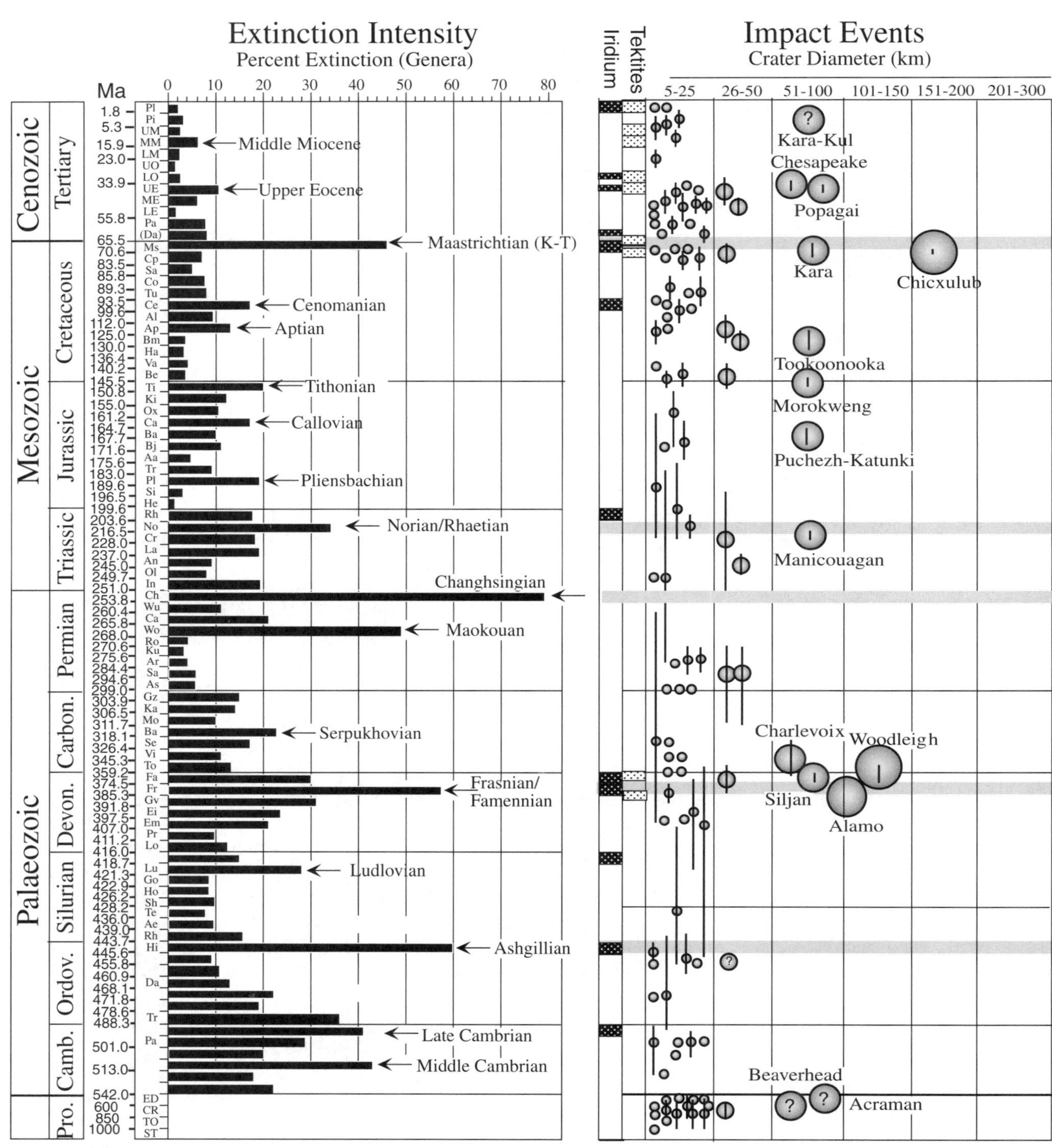

Figure 2. Mass extinctions and impacts during the Phanerozoic. Stratigraphic subdivisions and numerical ages from the 2004 International Commission of Stratigraphy Chart of Gradstein and Ogg (2004). The extinction record is based on genus-level data by Sepkoski (1996) and follows an earlier compilation by MacLeod (1998, 2003), with some modification based on Hallam and Wignall (1997). The number of impact events and the size and age of craters follows largely the Earth Impact Database (Grieve, 2005). Modified after Keller, 2005.

for the K-T boundary clay zone P0 include the first appearances of *Parvularugoglobigerina extensa*, *Globoconusa daubjergensis*, *Eoglobigerina eobulloides*, and *Woodringina hornerstownensis* (Fig. 3). The first appearances of *P. eugubina* and/or *P. longiapertura* mark the base of zone P1a, which spans the range of these species. Subdivision of this zone into P1a(1) and P1a(2) based on the first appearances of *Subbotina triloculinoides* and *Parasubbotina pseudobulloides* provides an important additional age control for the early Danian (Keller et al., 1995, 2002a). Below the K-T boundary, the most important biomarker is the range of *Plummerita hantkeninoides*, which marks zone CF1 and spans the last 300 k.y. of the Maastrichtian (Pardo et al., 1996). Ages for these biozones are estimated based on paleomagnetic stratigraphy, radiometric dates, and extrapolation based on average sediment accumulation rates (Fig. 3). In this study, the stratigraphic evaluation of Chicxulub impact markers uses the Danian zonation of Keller et al. (1995, 2002a) and the Maastrichtian zonation of Li and Keller (1998a) because they provide higher-resolution age control compared with the zonal scheme of Berggren et al. (1995) and Caron (1985).

Defining the K-T Boundary

The El Kef section of Tunisia was officially designated the Cretaceous-Tertiary (K-T) boundary global stratotype section and point (GSSP) at the International Geological Congress in Washington, D.C., in 1989. The official definition of the K-T boundary was never published but was summarized in Keller et al. (1995). The K-T boundary was defined based on (1) a major lithological change from carbonate-rich Maastrichtian sediments to a black organic-rich clay layer, called the "boundary clay," with a thin, 3–4 mm oxidized red layer at the base, (2) an iridium anomaly largely concentrated in the red layer, (3) the mass extinction in marine plankton, particularly the extinction of all tropical and subtropical planktic foraminiferal species nearly coincident with the base of the clay layer, (4) the first appearance of Danian species

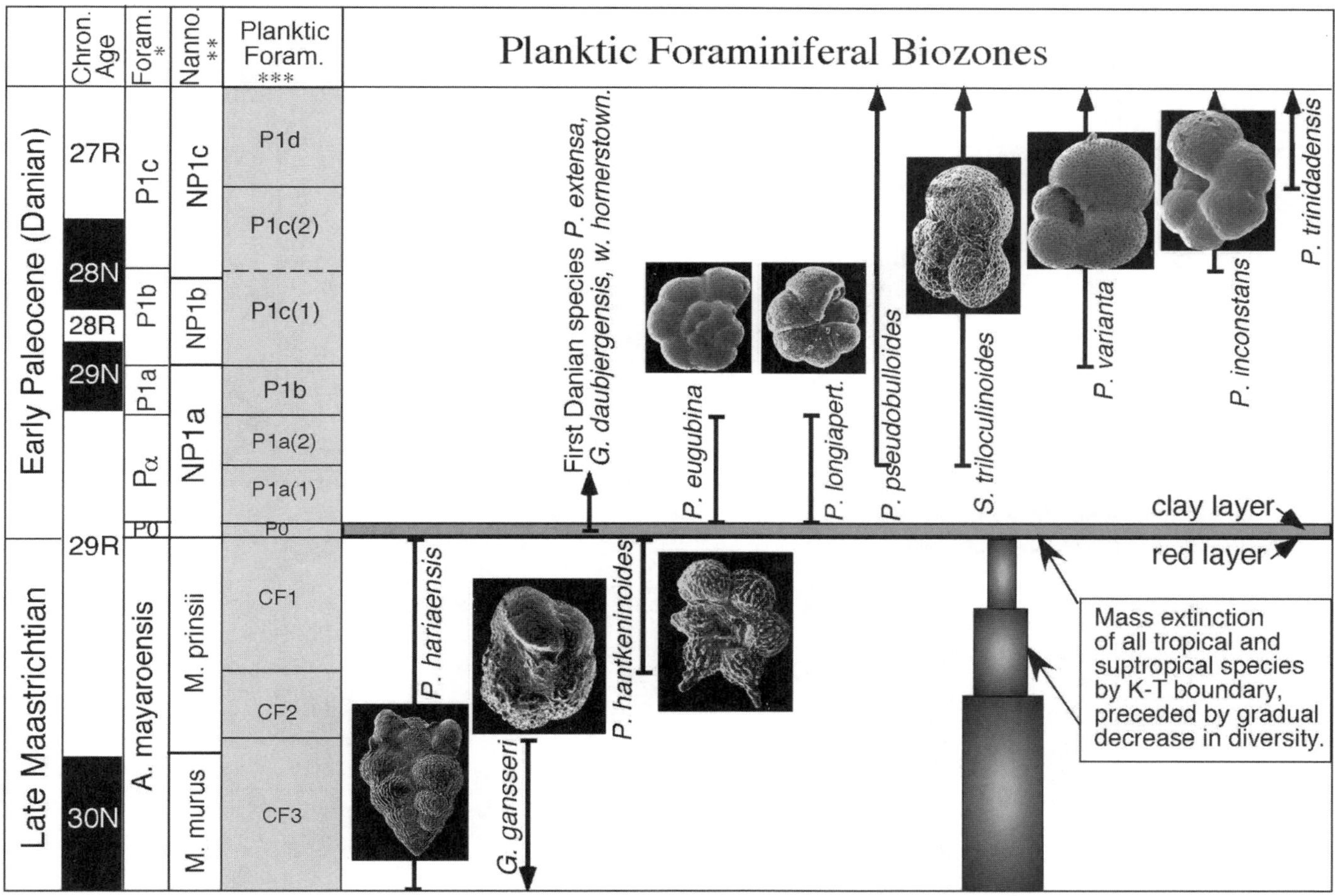

Figure 3. High-resolution planktic foraminiferal biozonation for the Cretaceous-Tertiary transition used in the stratigraphic analysis of impact ejecta deposits. Note that this biozonation significantly refines the resolution for the late Maastrichtian zonal scheme by replacing the upper *A. mayaroensis* zone by three biozones and by subdividing the *P. eugubina* zone P1a into two subzones based on the first appearances of *P. pseudobulloides* and *S. triloculinoides*. Modified after Keller et al., 2003a.

immediately (~1–5 cm) above the extinction horizon, and (5) a 2‰–3‰ negative shift in $\delta^{13}C$ values of marine carbonate (Keller et al., 1995) (Fig. 4). All five have remained remarkably consistent K-T boundary markers in marine sequences worldwide.

Gradstein and Ogg (2004; Molina et al., 2006) recently introduced a K-T boundary definition that reduced these identifying criteria to just the "Ir anomaly associated with a major extinction horizon" (see International Commission on Stratigraphy Web site on GSSPs). This is unfortunate because anomalous Ir concentrations are not unique to the K-T boundary, or may be absent in K-T sediments, whereas the extinction horizon is minor in many microfossils and macrofossils (e.g., palynomorphs, diatoms, radiolarian, benthic foraminifera, ostracods, ammonites, bivalves, see review in Keller, 2001; MacLeod, 1998) and highly reduced in shallow-water environments (e.g., planktic and benthic foraminifera, nannofossils) (Méon, 1990; Keller, 1989; Tantawy, 2003; Culver, 2003). The best results are thus obtained by using all five criteria. But where this is not possible, the shift in $\delta^{13}C$ values of marine carbonate is a unique global oceanographic marker for the K-T boundary and mass extinction horizon, and the first appearance of Tertiary species relative to the $\delta^{13}C$ and the K-T boundary clay appears isochronous globally in shallow or deep environments.

IMPACT MARKERS

Impact signals are the boutique markers of stratigraphy—exquisite but rare to find. Out of a variety of potential impact markers, including iridium and other PGE (platinum group elements) anomalies, impact glass spherules, shocked minerals, Ni-rich spinels, shatter cones, suevite breccia, and impact tsunami deposits, only the Ir anomaly and Chicxulub impact spherules are in practice widespread reliable stratigraphic impact markers. Most other impact signals are at this time unsuitable because of their sporadic occurrences, disputed origins, and limited distributions.

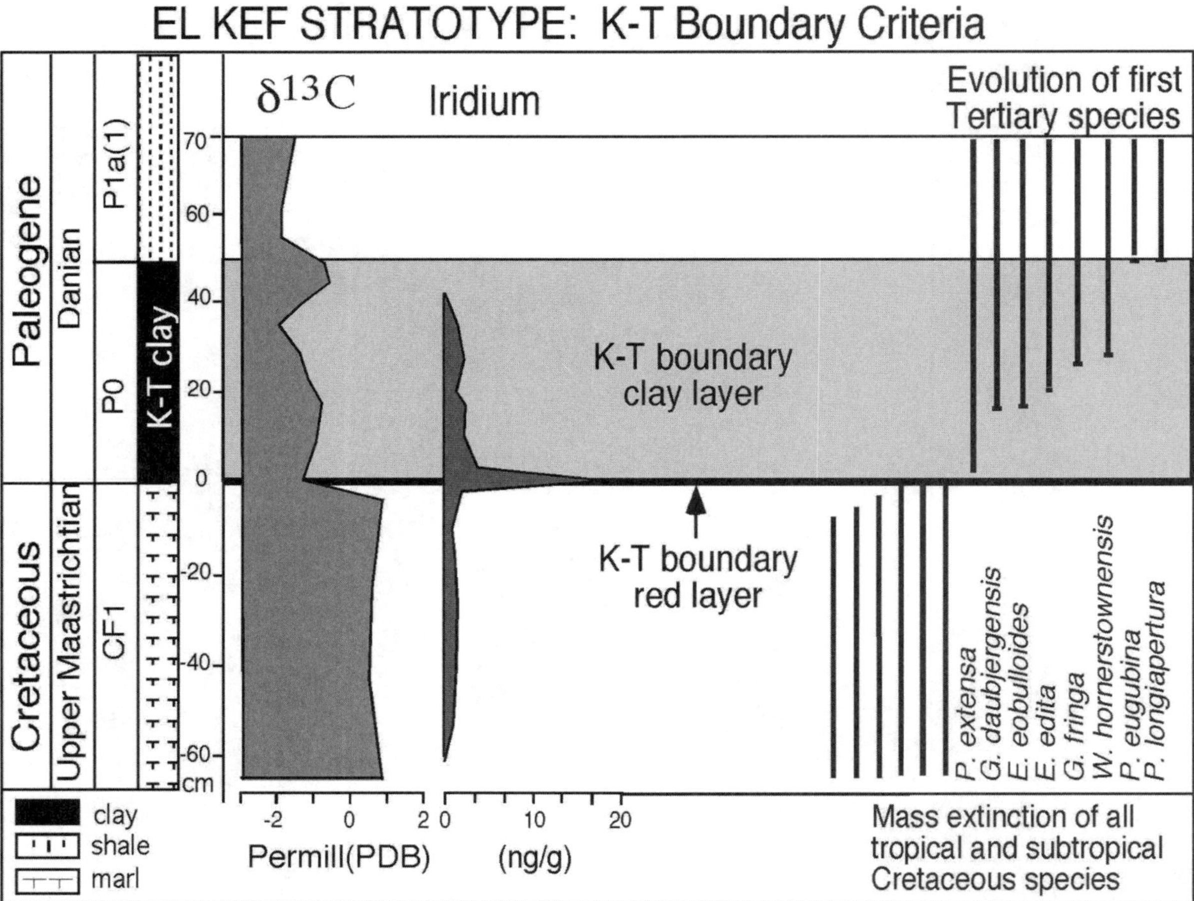

Figure 4. K-T boundary-defining criteria based on the El Kef stratotype. The K-T boundary is defined by (1) the mass extinction of tropical and subtropical planktic foraminifera, (2) the first occurrence of Tertiary species, (3) the 2‰–3‰ drop in $\delta^{13}C$ values marking a global oceanographic change, (4) a lithological change from the carbonate-rich Cretaceous sediments to a dark organic-rich clay with a red oxidized layer at the base, and (5) anomalous concentrations of iridium and other PGEs in the red layer and/or boundary clay. Note that Chicxulub impact spherules are not present at El Kef and are not part of the K-T boundary-defining criteria. Modified after Keller et al., 1995.

Iridium

Stratigraphic Position of Iridium and the K-T Boundary

Anomalous concentrations of iridium were first discovered in the K-T boundary clay in Gubbio, Italy, and linked to an extraterrestrial impact by Alvarez et al. (1980). Since then the Ir anomaly has been identified worldwide in K-T clays, coincident with a 2‰–3‰ negative shift in $\delta^{13}C$ values, the mass extinction of planktic foraminifera, followed by the evolution of the first Danian species. All of these markers therefore have become key identifying criteria for the K-T boundary. As noted above, the K-T boundary is easily recognized in the field by the abrupt lithologic change from carbonate-rich Maastrichtian sediments to dark organic-rich clay that is commonly very thin (1–5 cm) with a thin, 2–4 mm red layer at the base that contains maximum Ir concentrations (Fig. 4). Today, the presence of an Ir anomaly in the K-T clay layer is generally considered as (1) key evidence for a meteorite impact, (2) that it occurred precisely at the K-T boundary, and (3) that the impact either caused or at least substantially contributed to the mass extinction. This rule of thumb has served well. However, some caution is in order because times of multiple impacts may produce multiple Ir anomalies (Fig. 2).

Iridium Anomaly in Early Danian Zone P1a

A little-noticed second Ir anomaly has been observed in the early Danian zone P1a (middle to upper part of the *P. eugubina* zone; Fig. 3). This Ir anomaly has been observed to date in Haiti, central Mexico (Coxquihui), southern Mexico (Bochil), and Guatemala (Keller et al., 2001, 2003a, 2003b; Stinnesbeck et al., 2002; Stüben et al., 2002, 2005), and also in the Indian Ocean ODP Site 752B (Michel et al., 1991). At Site 752B, the K-T Ir anomaly is large (~4 ppb), followed by a second Ir anomaly (2 ppb) 60 cm above it in the *P. eugubina* zone P1a (Keller, 1993). In the Beloc sections of Haiti the Ir anomaly is concentrated in a thin red layer above a 5 cm thick cross-bedded sandstone within subzone P1a(1) and well above the spherule deposit and K-T boundary (Fig. 5). The chondrite-normalized PGE pattern indicates a cosmic origin with higher Ir values compared to Pt and Pd (Stüben et

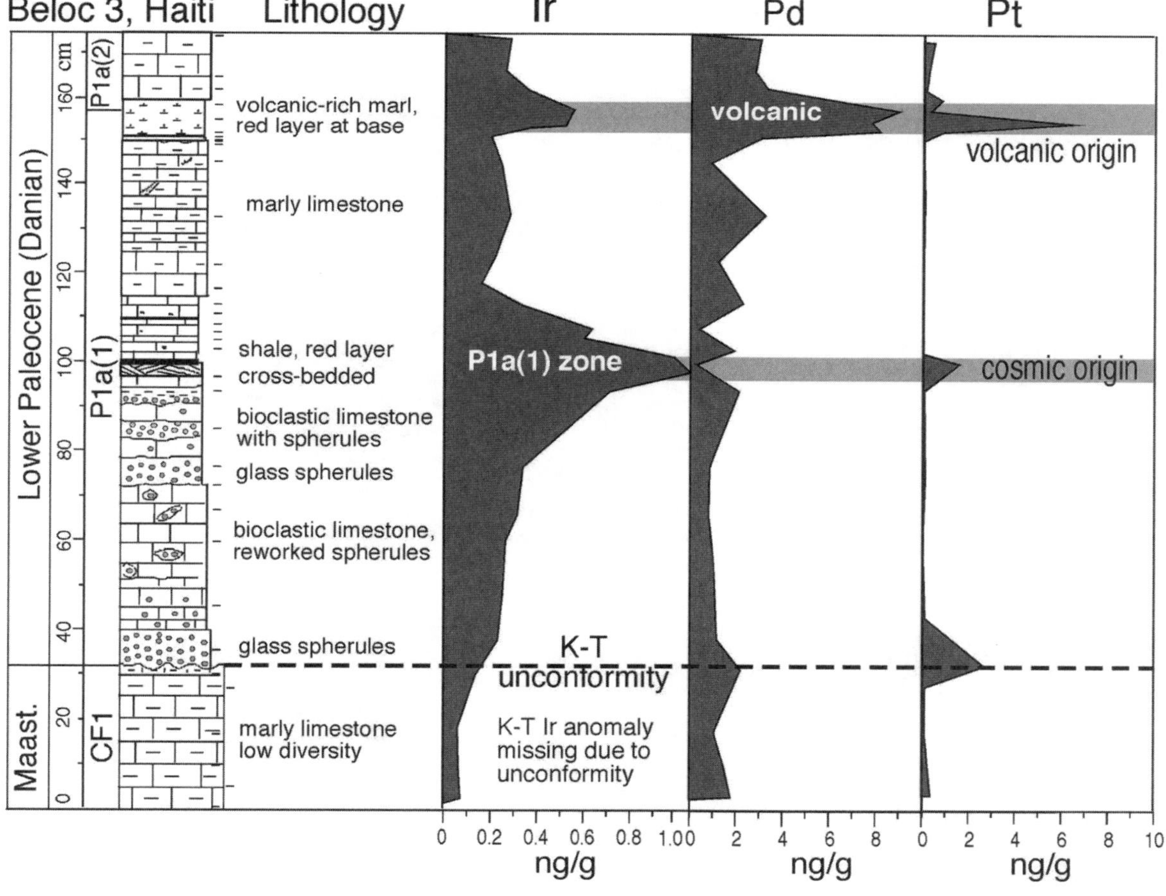

Figure 5. The K-T section of Beloc 3 in Haiti shows a lithological sequence with two PGE anomalies. The lower P1a(1) Ir anomaly is of cosmic origin, the upper small Ir and large Pd and Pt anomalies indicate a volcanic origin. The K-T boundary is missing due to erosion. Note that although the Haiti sections are stratigraphically nearly complete (e.g., all zones present except for the boundary clay), the Chicxulub spherules are reworked within early Danian sediments. Modified after Keller et al., 2001.

al., 2002). In the same section 50 cm higher up is a second PGE anomaly in a volcaniclastic layer with a basalt-like pattern (minor Ir [0.5 ng/g], major Pt and Pd enrichments [7.5 ng/g]) suggesting a volcanic source.

In Bochil, Chiapas (Mexico), the same pattern of cosmic and volcanic anomalies was observed (Keller et al., 2003a; Stüben et al., 2005) (Fig. 6), although the P1a(1) Ir anomaly is stratigraphically closer to the K-T boundary due to erosion of the lowermost Danian. This is indicated by the simultaneous and abundant first appearance of six Danian species (*Parvularugoglobigerina eugubina*, *P. longiapertura*, *P. extensa*, *Globoconusa daubjergensis*, *Woodringina hornerstownensis*, *Chiloguembelina midwayensis*). In Coxquihui, central Mexico, the K-T boundary is marked by a 2 cm spherule layer above a thin silty sandstone that abruptly terminates Maastrichtian marls of zone CF1 (Fig. 7). No Ir anomaly is present. Abundant Danian species indicative of subzone P1a(1) are present in the overlying 20 cm of clayey marl and mark a hiatus, with the boundary clay and lower part of P1a(1) missing (Stinnesbeck et al., 2002). A 60 cm thick spherule layer with abundant reworked Maastrichtian species overlies the clayey marl and indicates that this layer is reworked from Maastrichtian sediments. In the undisturbed marls above is an Ir and Pd anomaly of probable cosmic origin, which marks a Danian impact, similar to Bochil and Beloc.

Thus, multiple Ir and PGE anomalies of both cosmic and volcanic origins may be present in the early Danian, and these could be mistaken for the K-T Ir anomaly in condensed sequences. High-resolution biostratigraphic control is therefore vital to determine the continuity of the sedimentation record and identify the K-T Ir anomaly. A detailed investigation of Ir concentrations and their stratigraphic positions at the K-T boundary, the early Danian, and late Maastrichtian zone CF1 is yet to be done.

Iridium Profiles

Iridium and other PGE anomalies have received much attention as indicators of extraterrestrial sources, such as asteroids, comets, cosmic dust, or impact ejecta. This has been justified on the basis that iridium is highly depleted in Earth's crust but enriched in some asteroids and comets and in Earth's interior, where it is brought to the surface by volcanic activity. Because it is assumed that iridium is extruded over a long time period and therefore likely distributed over broad low plateaus, a sharp peak in iridium concentrations is generally interpreted as a cosmic marker. However, a volcanic origin cannot be excluded because

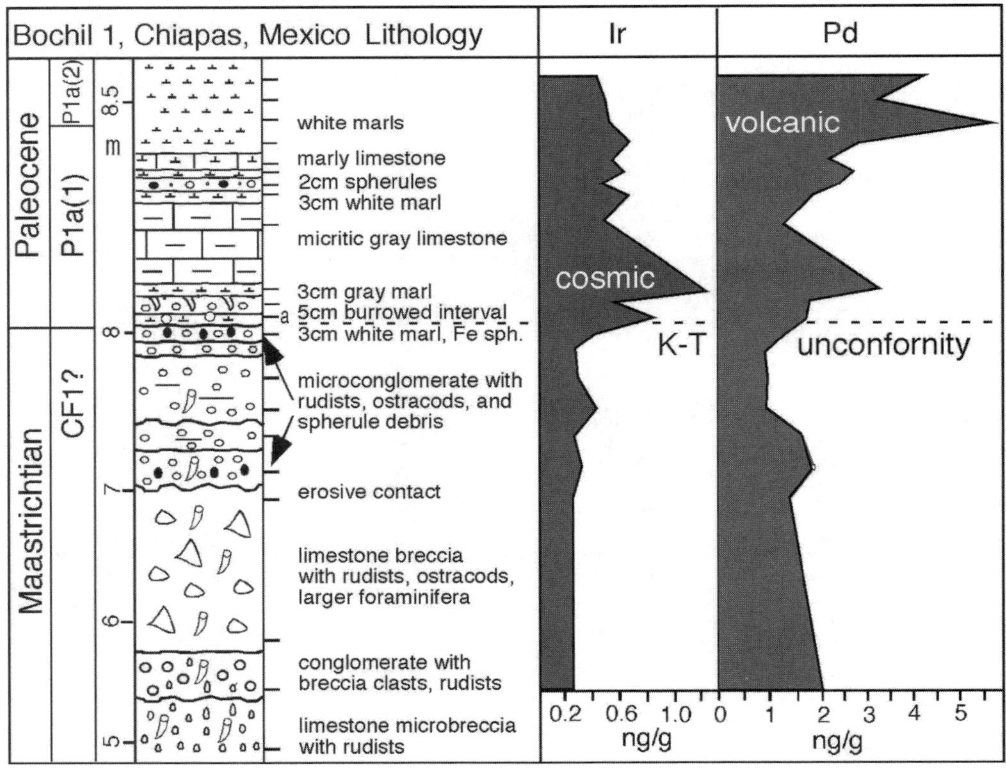

Figure 6. At Bochil 1 in Chiapas, Mexico, altered glass spherules are present at the base and top of a microconglomerate layer of probable latest Maastrichtian age. The first marl layer above this microconglomerate contains early Danian *P. eugubina* subzone P1a(1) planktic foraminifera, as well as Ir and Pd anomalies. A second Pd anomaly near the P1a(1)-P1a(2) boundary appears to be of volcanic origin, similar to Beloc, Haiti. Modified after Keller et al., 2003a.

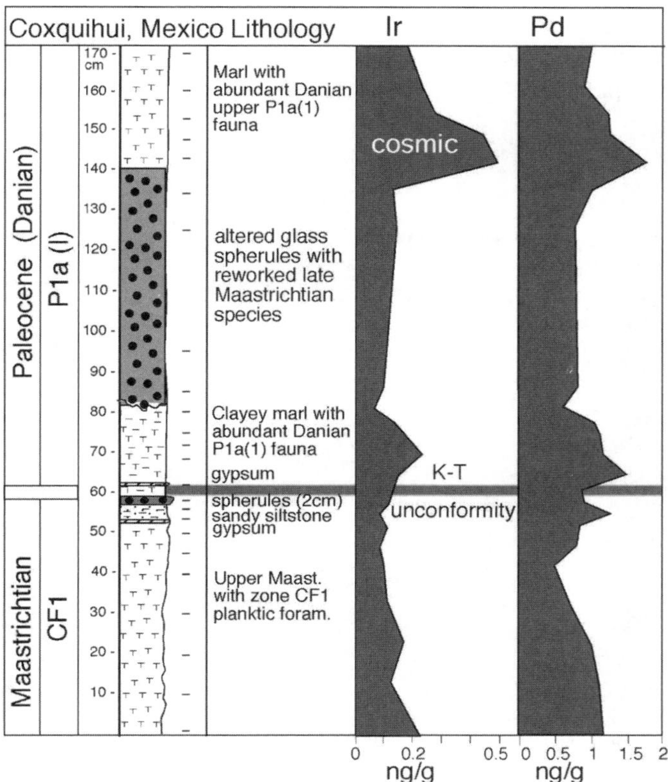

Figure 7. At Coxquihui in central Mexico, the K-T boundary is marked by an unconformity above a 2 cm thick spherule layer, which underlies Danian marls with an abundant P1a(1) assemblage. A 60 cm thick spherule-rich layer with abundant Maastrichtian species indicates redeposition of the Chicxulub spherule layer in the early Danian. The Ir anomaly above it appears to be of cosmic origin and correlates with similar anomalies in Bochil and Beloc. Modified after Stinnesbeck et al., 2002.

recent studies have shown that Deccan eruptions near the K-T boundary occurred rapidly (Chenet et al., 2007). Moreover, sharp peaks may also result from condensed sedimentation, which must be ruled out by normalizing to elements that are constantly present in sediments (e.g., Al or Ti). Ir concentrations at the K-T boundary typically are 1 ppb or more, but smaller 0.5–1.0 ppb concentrations are common. Even smaller concentrations (0.2–0.4 ppb) are generally suspect and should be interpreted with great caution, as they may result from a variety of terrestrial processes.

Sawlowicz (1993) discussed a host of processes that could be responsible for the K-T boundary Ir anomaly and other PGE concentrations. Although the original K-T Ir anomaly is generally considered extraterrestrial, postdepositional oxidation and reprecipitation is a major concern. The K-T red layer consists of thin iron oxide–stained laminae with common iron oxide spherules highly enriched in Ir and noble metals that probably resulted from oxidation of primary iron sulfides (Brooks et al., 1985). Oxidation within this red layer may have caused the degradation of PGE-rich organic compounds and precipitation of native metals and their alloys (Elliott et al., 1989). Such red oxidized layers are commonly found at lithologic boundaries. For example, in the Beloc sections of Haiti, the two early Danian zone P1a(1) Ir and PGE anomalies are each associated with thin red oxidized layers, one above a rippled sandstone and the other below a volcaniclastic layer (Fig. 5). A PGE anomaly may result from the removal of primary components, or in the case of the K-T red layer the addition of mineralizing solutions to a redox boundary. Sawlowicz (1993) concluded that the unusually high Ir concentration at the K-T boundary is probably the result of an extraterrestrial impact and complex terrestrial processes.

Remobilization at Redox Boundaries

Ir anomalies typically occur in organic-rich shales or clays, which serve as low-permeability redox boundaries, and are associated with sandstones or carbonates. In these sediments Ir can move both upward and downward at redox boundaries (Gilmore et al., 1984; Tredoux et al., 1988; Sawlowicz, 1993; Wang et al., 1993). This process may explain the minor (0.2–0.3 ppb) Ir enrichments within the sandstone complex at El Mimbral in NE Mexico and Brazos-1 in Texas. At El Mimbral I, the main Ir anomaly is in the basal Danian clay above the sandstone complex that infills a submarine channel and coincident with the abrupt and abundant first appearance of four Danian zone P1a(1) species (Keller et al., 1994a; Rocchia et al., 1996). Minor Ir enrichment occurs at the top of the underlying rippled calcareous sandstone (Fig. 8A). At El Mimbral II, located outside the submarine channel and ~100 m from El Mimbral I, the sandstone complex is reduced to the topmost 20–25 cm thick rippled sandy limestone. At this section, the Ir anomaly begins in the red layer, coincident with the first appearance of Danian zone P1a(1) species, and reaches maximum concentrations ~7 cm above (Fig. 8B) (Stinnesbeck et al., 1993; Keller et al., 1994a, 1994b). Similar to El Mimbral I, minor (0.2 ppb) Ir concentrations occur in the underlying rippled sandy limestone.

Similarly, at Brazos-1 (Texas) the main Ir anomaly is in a thin red-brown clay and 1 cm sand layer ~17–19 cm above the sandstone complex, as noted in three different studies (Ganapathy et al., 1981; Asaro et al., 1982; Rocchia et al., 1996). Two minor Ir enrichments occur in the sandstone complex below, just above the calcareous claystone horizon (CCH) and in the laminated sandstone layer above the hummocky sandstone (Fig. 9A). These minor enrichments are likely due to postdepositional processes discussed in Sawlowicz (1993). A photo of the Brazos-1 outcrop with the Ir data superimposed (Fig. 9B) illustrates the distinct lithologic separation of the main Ir anomaly and the top of the event deposit.

In Mexico and Texas the minor Ir enrichments in sandstones below the main Ir anomaly have been interpreted as evidence that the sandstone complex was generated by an impact tsunami (Smit et al., 1992, 1996; Hansen et al., 1993; Smit, 1999; Heymann et al., 1998; Schulte et al., 2006). In this scenario the main Ir anomaly is interpreted as fallout after deposition of the sandstone

Figure 8. A: Ir anomaly from Mimbral I, superimposed over outcrop lithology from which it was measured (data from Rocchia et al., 1996). Note that only minor Ir concentrations occur in the topmost rippled sandy limestone of the sandstone complex, and the main anomaly is coincident with the abrupt and abundant appearance of Danian planktic foraminifera in the claystone above (Keller et al., 1994a). B: Ir anomaly from Mimbral II, measured ~100 m from Mimbral I where the sandstone complex consists only of the topmost 20–25 cm thick rippled sandstone. Note that the Ir anomaly begins in the K-T red layer above the rippled sandy limestone and reaches maximum values 7 cm above.

complex by a tsunami generated by the Chicxulub impact, an interpretation that ignores the presence of bioturbated horizons within the sandstone complex that indicates long-term deposition (Keller et al., 2003a, 2007; Gale, 2006). A more likely interpretation is postdepositional remobilization. The complex behavior of iridium and other PGEs in sedimentary environments is still poorly understood. Therefore, such small anomalies must be considered with caution and the sedimentary sequences interpreted based on the integration of geochemical, mineralogical, sedimentologic, and paleontologic data.

Iridium and Chicxulub Impact Spherules?

No iridium anomaly has ever been found in stratigraphic layers associated with Chicxulub impact ejecta, whether in the impact crater or spherule ejecta layers, in Texas, Mexico, Haiti, Belize, or Guatemala (Rocchia et al., 1996; Keller et al., 2003a; Schulte et al., 2003; Harting, 2004; Stüben et al., 2002, 2005). This suggests that the Chicxulub impact was a "dirty snowball"–type comet without iridium enrichment. However, close stratigraphic proximity of an Ir anomaly and Chicxulub impact spherules in some terrestrial and deep-sea sections is frequently cited as unequivocal evidence that the Chicxulub impact is of K-T age (e.g., Norris et al., 1999, 2000; MacLeod et al., 2006; Morgan et al., 2006). These claims are examined below.

IMPACT SPHERULES

K-T Boundary Spherules—Not Chicxulub Origin

Several hundred K-T boundary sequences have been studied to date around the world, and most complete sedimentary records share the same characteristics: an organic-rich boundary clay with a thin red layer at the base, iridium and other PGE anomalies usually concentrated within the red layer, a negative shift in $\delta^{13}C$ (though not in high latitudes; Barrera and Keller, 1994), and the mass extinction of Cretaceous species and evolution of Tertiary species (Fig. 4). In all these sequences only two types of spherules are common: (1) the tiny iron oxide spherules, which mostly occur in encrusting clusters or singly in the red layer or K-T clay; these are highly enriched in Ir and noble metals that may have resulted from oxidation of primary iron sulfides (Brooks et al., 1985); and (2) pyrite framboids, which are very common in the organic-rich clay and are indicative of low-oxygen conditions (Wignall et al., 2005).

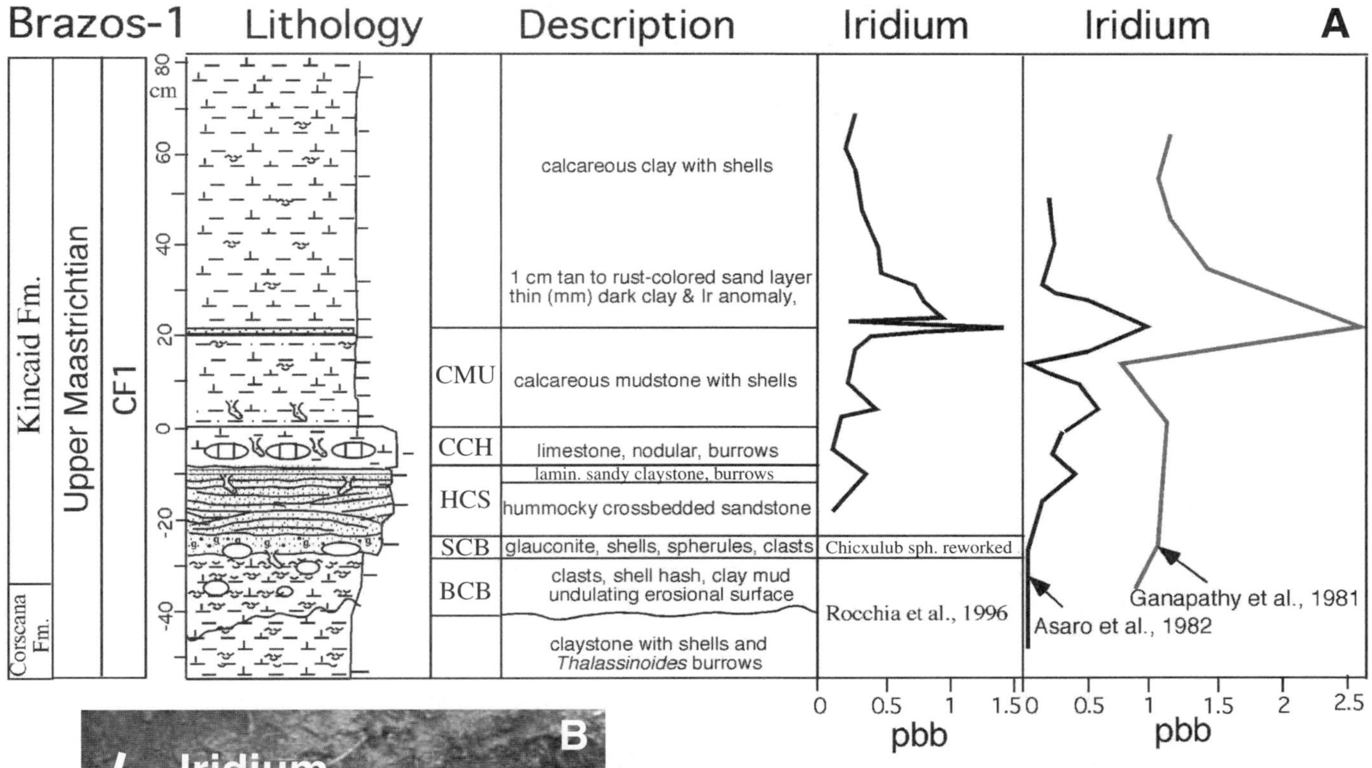

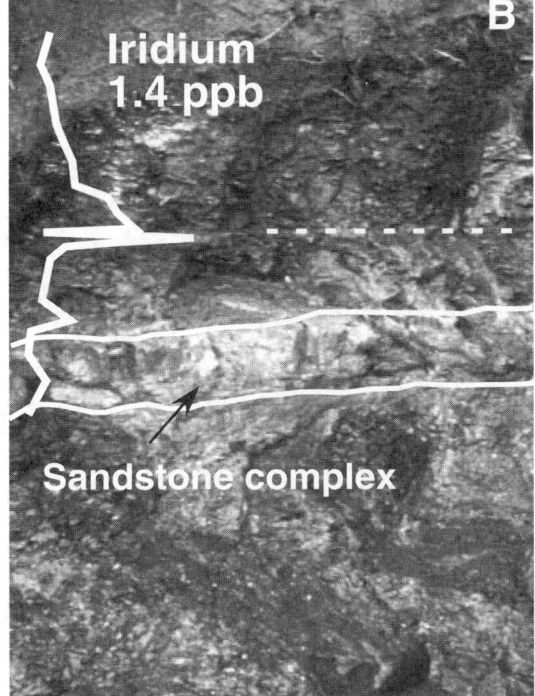

Figure 9. A: At the classical Brazos-1 riverbank section near the Route 413 bridge across the Brazos River, three Ir studies found the anomaly in a thin red-brown layer and 1 cm sandstone 17–19 cm above the top of the calcareous claystone horizon (CCH) that forms the top of the event deposit. The two minor Ir enrichments just above the CCH within a laminated sandy claystone are likely due to postdepositional processes. Unit labels from Yancey (1996). BCB—basal conglomerate bed; SCB—spherulitic conglomerate bed; HCS—hummocky sandstone unit; CCH—calcareous clayey horizon; CMU—calcareous mudstone unit. B: Photo of the K-T boundary and Ir anomaly at Brazos-1. Note that the minor Ir enrichment above the calcareous claystone horizon (CCH) is likely the result of postdepositional processes. Ir data from Rocchia et al. (1996). Photo courtesy of T. Yancey.

A few sections contain rare clay spherules of ~1–2 mm in size, often compressed and iron-coated (e.g., Spain [Agost], Tunisia [El Kef and Elles], Israel [Mishor Rotem]). Smit (1999) interpreted these K-T clay spherules as altered Chicxulub impact glass, although they largely consist of glauconite clay. These spherules are unlike the clay-altered glass spherules characteristic of the Chicxulub impact with their internal calcite-filled air bubbles, and the glass alteration product is a Cheto- or Mg-smectite clay, rather than glauconite (Keller et al., 2003a, 2003b, 2004a, 2004b, 2007, 2008). No relic impact glass has ever been detected in the K-T clay, and the clay content is unlike any altered Chicxulub impact glass. Even in Mexico, no Chicxulub impact spherules occur in the K-T red layer and boundary clay (e.g., La Sierrita, La Parida, El Mimbral)—but they are abundantly present below the sandstone complex, which is well below the K-T boundary, as well as in early Danian sediments above the K-T boundary (e.g., Haiti, Belize, Guatemala). At this time, the rare K-T clay spherules largely appear to be of normal sedimentary origins, though detailed mineralogical and geochemical studies have yet to be done to compare them to the Chicxulub impact glass spherules.

Chicxulub Impact Glass Spherules

Chicxulub impact spherules are almost always spherical or compressed oval, range in size from 1 to 5 mm, and generally contain multiple air cavities as seen in thin sections (Fig. 10). Relic glass, and sometimes well-preserved glass spherules and glass shards can be found in cemented spherule rocks. Spherules that have been eroded and redeposited generally contain a matrix rich in clasts, clastic minerals (Fig. 10A–C), and shallow-water debris (wood, plants, shallow-water foraminifera; Smit et al., 1992; Keller et al., 1994a). In contrast, spherules and glass shards from the original ejecta layer are cemented by a calcite matrix and may show concave-convex contacts that indicate deposition while still malleable (Fig. 10D–F) or postdepositional compression alteration. The spherule glass is mostly altered to green clay (Mg-smectite) with only relic glass preserved. Nevertheless, in outcrops the spherulitic texture is preserved because calcite infills air cavities (Fig. 11). When these samples are washed in the laboratory the green clay frequently dissolves, leaving the white calcite spheres. Throughout Central America, the Caribbean, around the Gulf of Mexico, southern United States, and New Jersey, these spherules have the same physical and chemical characteristics. They are linked to the Chicxulub impact based on their geographic distribution, ^{39}Ar/^{40}Ar ages close to the K-T boundary (Sigurdsson et al., 1991; Swisher et al., 1992; Dalrymple et al., 1993), and chemical similarity to Chicxulub melt rock (Izett et al., 1991; Blum et al., 1993; Koeberl et al., 1994; Chaussidon et al., 1996; Schulte et al., 2003; Harting, 2004).

Stratigraphic Position of Spherule Deposits

Chicxulub impact spherules are excellent stratigraphic markers in Central America, around the Gulf of Mexico, and in Texas (Fig. 1), where they are relatively easy to spot in outcrops by their coarse-grained spherulitic texture and green color (Fig. 11). They have never been observed in the K-T boundary clay, red layer, and Ir anomaly. Most studies of the 1990s describe spherule deposits in NE Mexico ranging from 5 cm to 2 m thick and directly underlying a sandstone complex consisting of 1–3 m of sandstone followed by 1–2 m of alternating sand-silt-marl layers (Figs. 8A and 8B). These sandstone complexes are commonly interpreted as impact tsunami deposits because of their stratigraphic proximity

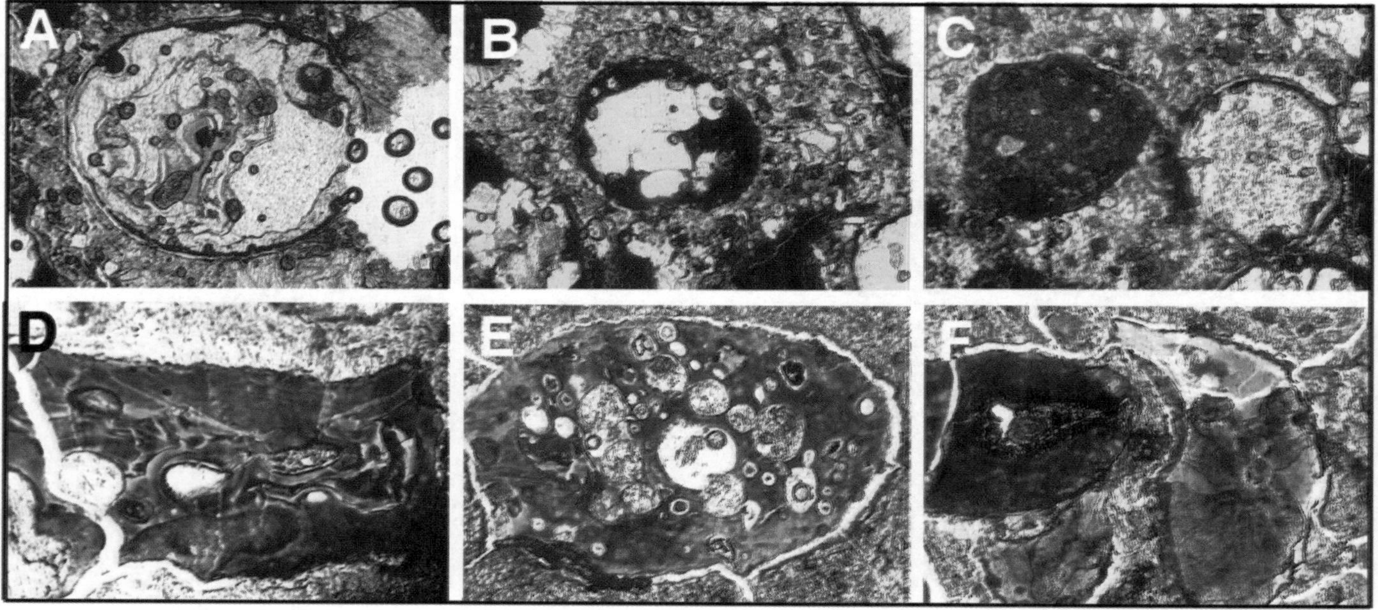

Figure 10. Chicxulub impact spherules from El Peñon, Mexico, as seen in thin sections. A–C: Spherules from the base of the sandstone complex have a matrix of clastic grains, clasts, and shallow-water foraminifera that indicate erosion from shallow areas and transport and redeposition in submarine canyons. D–F: Spherules and shards from the original spherule ejecta layer, ~4 m below the sandstone complex, are embedded in a matrix of calcite cement, which contains no shallow-water debris or clastic grains. Spherules are mostly spherical and range from 2 to 4 mm in size. Most spherules and glass shards contain multiple air cavities, which is characteristic of Chicxulub impact glass. The cavities are generally infilled with diagenetic calcite and the glass altered to Mg-smectite.

Figure 11. Chicxulub impact spherule rock from El Peñon, NE Mexico. Note that the spherulitic texture is largely preserved due to the diagenetic calcite that infills the spherule cavities. Its green color is largely due to glass alteration.

to the overlying K-T boundary (reviews in Smit, 1999; Keller et al., 2003a, 2003b). They infill submarine canyons, as indicated by their lenticular bodies of 100–300 m lateral extent. Laterally, the spherule unit disappears first, followed by the sandstone and finally the topmost calcareous rippled sandstone (see descriptions in Keller et al., 1994b; Stinnesbeck et al., 1996; Smit, 1999). The topmost rippled sandstone unit disappears last and is overlain by the K-T boundary clay, red layer, and Ir anomaly (e.g., El Mimbral, La Parida, La Sierrita) (Keller et al., 1994a; Lopez-Oliva and Keller, 1996; Stinnesbeck et al., 1996). No Chicxulub spherules have been observed in any K-T boundary clay layers in Mexico.

In the late 1990s a group of M.A. and Ph.D. students from Princeton and the Universities of Neuchatel, Switzerland, and Karlsruhe, Germany, undertook detailed field and laboratory investigations of the spherule and sandstone deposits and the underlying Maastrichtian sediments over a 50 km^2 area in NE Mexico (Lopez-Oliva and Keller, 1996; Lindenmaier, 1999; Schulte, 1999, 2003; Schilli, 2000; Affolter, 2000; Harting, 2004). Their studies documented over 45 K-T sequences, many of them with additional spherule layers in pelagic marls of the upper Maastrichtian, and the stratigraphically lowermost spherule layer up to 9 m below the sandstone complex (Fig. 12) (Keller et al., 2002b). This complex stratigraphic distribution of spherule

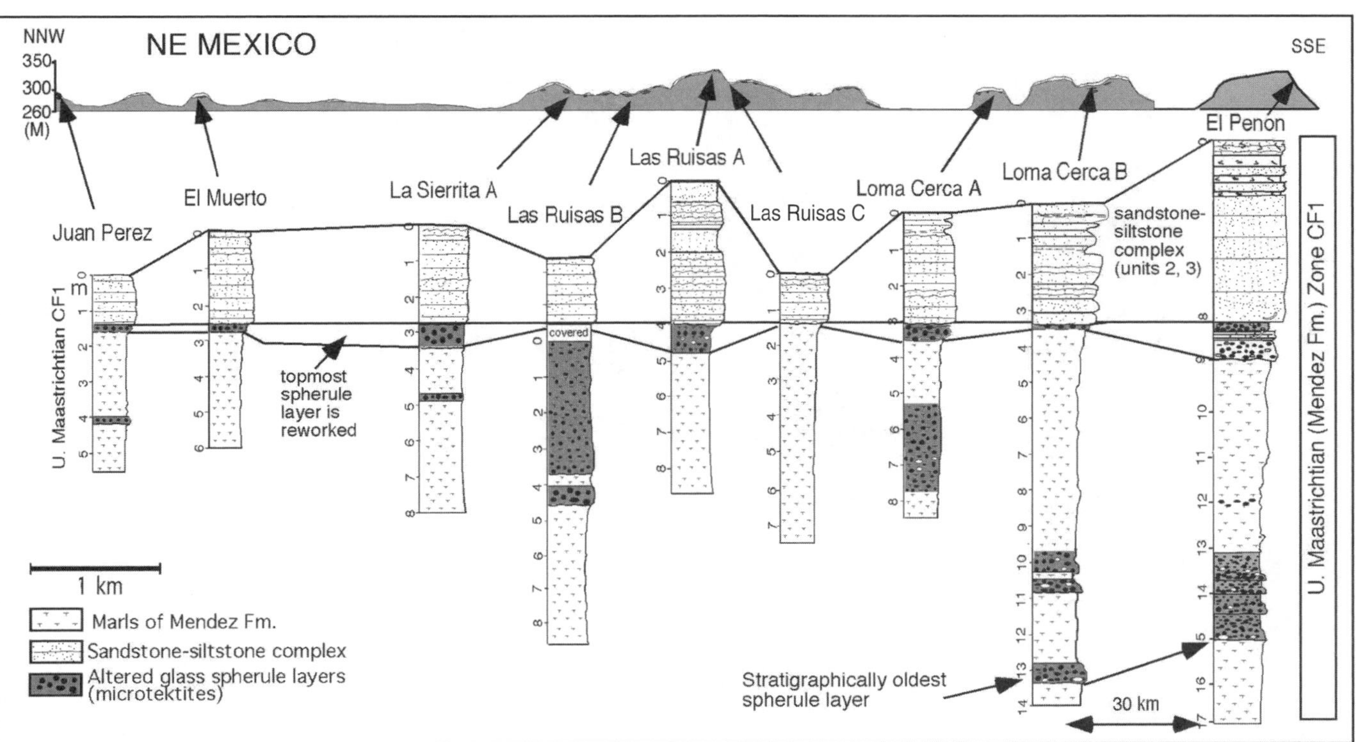

Figure 12. Stratigraphic and lithologic correlation of late Maastrichtian NE Mexico sections with sandstone complexes and altered Chicxulub impact spherule layers from Mesa Juan Perez to Mesa Loma Cerca (~8 km) and El Peñon (~30 km from Mesa Loma Cerca; Fig. 1). The siliciclastic deposits of units 2 and 3 form the top of the mesas as shown by the topographic relief, and the overlying sediments are eroded. In this area, all spherule deposits are within the Maastrichtian marls of the Mendez Formation and within zone CF1, which spans the last 300 k.y. of the Maastrichtian. Modified after Keller et al., 2002b, scale added.

ejecta in upper Maastrichtian sediments suggests a complex post-depositional history of erosion, transport, and redeposition. Thus, an impact ejecta layer cannot be assumed to be the impact time horizon without careful stratigraphic, biostratigraphic, and lithologic examinations of sediments above and below.

Reworked Spherules + Sandstone Complex = Tsunami?

NE Mexico

The most prominent stratigraphic features in K-T sequences of NE Mexico and Texas are thick sandstone complexes with Chicxulub impact glass spherules at the base (Fig. 12). These sandstone deposits underlie the K-T boundary and Ir anomaly. Because of the stratigraphic proximity of the sandstone deposits to the K-T boundary, they have been commonly interpreted as impact-generated tsunami deposits (e.g., Smit et al., 1992, 1996; Smit, 1999; Soria et al., 2001). This hypothesis has not held up under scrutiny and has been the topic of significant controversy. Even before the discovery of the stratigraphically older spherule deposits in upper Maastrichtian sediments, evidence indicated that rapid tsunami deposition was not likely. For example, burrows were first discovered at El Mimbral in the alternating sand, silt, and marl layers by Toni Ekdale during the 1994 NASA-LPI-sponsored field trip (led by Stinnesbeck, Keller, and Adatte) and subsequently documented as discrete burrowing horizons (*Thalassinoides*, *Chondrites*, J-shaped burrows; Fig. 13) in most sections of NE Mexico. These burrowing horizons suggest repeated colonization of the ocean floor during deposition (Keller et al., 1997, 2003a; Ekdale and Stinnesbeck, 1998). In addition, alternating coarse and fine-grained deposition and two volcanic ash (zeolite) layers indicate interrupted deposition over an extended time period (Adatte et al., 1996; Stinnesbeck et al., 1996).

The spherule layer at the base of the sandstone complex yielded further evidence of long-term deposition. In various sequences, most prominently at El Mimbral and El Peñon, a 20-cm-thick sandy limestone with rare spherule-infilled J-shaped burrows separates the spherules below the sandstone into two layers (Fig. 13). This indicates two spherule depositional events separated by sufficient time to deposit the sandy limestone and permit burrowing by marine invertebrates prior to deposition of the second spherule layer. The spherule deposits contain shallow-water debris, including a matrix of sand, clasts, foraminifera, wood, and leaves, transported from nearshore areas and carried by currents downslope to

Figure 13. The sandstone complex at El Peñon, NE Mexico, overlies two reworked spherule layers that are separated by a 20 cm thick sandy limestone. J-shaped burrows infilled with spherules and truncated at the top are present in the limestone and overlying sandstone. Siltstone and marly layers above the sandstone are intensely burrowed by *Chondrites* and *Thalassinoides*. These are not characteristics that would be found in a tsunami deposit, but rather indicate deposition over an extended time period during which the ocean floor was repeatedly colonized by invertebrates. The K-T boundary is eroded at the top of the section.

their resting place at 500–1000 m depths (Keller et al., 1994a; Alegret et al., 2001). These spherules are thus reworked, transported, and redeposited—they do not represent the original Chicxulub ejecta fallout. These data are difficult if not impossible to reconcile with a tsunami interpretation, even without the discovery of the original spherule ejecta deposit in the underlying undisturbed upper Maastrichtian sediments (discussed below).

Texas, USA

In Texas, along the Brazos River of Falls County and its tributaries, Cottonmouth and Darting Minnow Creeks, the sandstone complex that underlies the K-T boundary and Ir anomaly is well developed and up to 2 m thick. These deposits have also been interpreted as of impact tsunami origin (Hansen et al., 1987; Bourgeois et al., 1988; Smit, 1999; Heymann et al., 1998; Schulte et al., 2006, 2008). As in Mexico, this interpretation is controversial and has not held up under scrutiny.

Yancey (1996) described the units of the sandstone complex from the base upward as basal conglomerate, a spherule-rich conglomerate with glauconite, clasts, phosphate and shell hash, a hummocky cross-bedded sandstone, rippled and laminated sandstones, and a calcareous mudstone at the top. He suggested the term "event deposits" for this sandstone complex, though in this study, as in NE Mexico, we will keep the term "sandstone complex" to describe these deposits. Gale (2006) and Keller et al. (2007, 2008) suggested that the stratigraphic discontinuities, multiple burrowed horizons, and alternating low- and high-energy depositional regimes of these deposits suggest seasonal storms, rather than an impact tsunami, and deposition during a low sea level.

An excellent sandstone complex was recovered in the recently drilled new core Mullinax-1, drilled near the classic Brazos-1 section. In this core, the spherules at the base of the sandstone deposit show three upward-fining units rich in impact

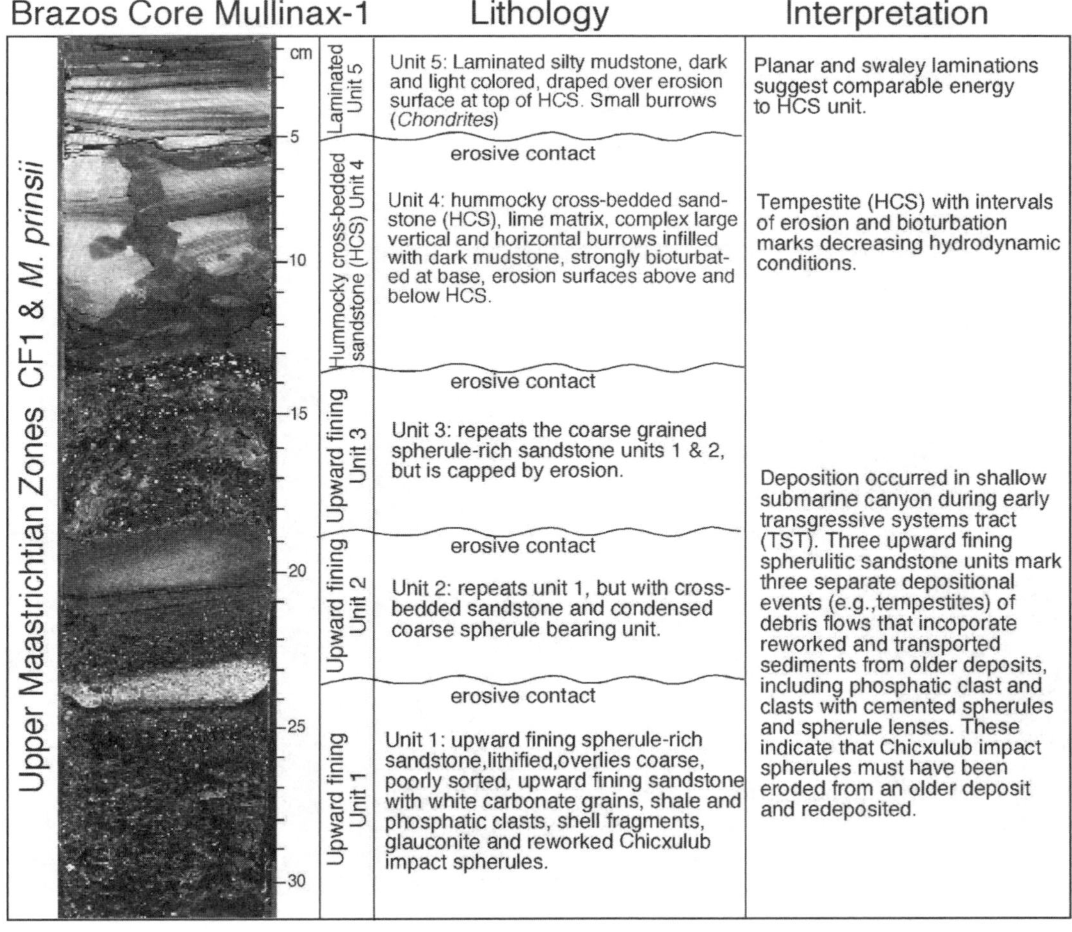

Figure 14. Sandstone complex recovered from core Mullinax-1, drilled near the Brazos River Route 413 bridge. The sandstone deposit shows three upward-fining, spherule-rich units grading into laminated fine sandstone layers. A strongly burrowed hummocky cross-bedded unit overlies the spherule-rich layers. The top of the sandstone deposit consists of a laminated clayey calcareous mudstone. This sequence is difficult to reconcile with an impact tsunami interpretation. Modified after Keller et al., 2007.

spherules, glauconite, phosphate, shell hash, and claystone clasts, which grade into finer laminated sandstones without spherules in units 1 and 2 (Fig. 14). This sequence indicates erosion, transport, and redeposition of spherules rather than the original ejecta fallout. Unit 3 is truncated by erosion and overlain by a hummocky cross-bedded sandstone (HCS) that is strongly bioturbated with large burrows infilled with dark mudstone and truncated at the top. Light and dark gray laminated silty mudstones drape over the erosion surface and contain small burrows of *Chondrites*.

This sequence indicates that deposition occurred in a shallow submarine canyon during the early transgressive systems tract (TST). The three upward-fining spherule-rich sandstone layers suggest a debris flow origin (tempestites) with erosion, transport, and redeposition of older sediments, whereas the HCS and overlying swaley and planar laminations mark decreasing hydrodynamic conditions (Keller et al., 2007).

In outcrops where the sandstone complex is well developed and overlies the scoured base of submarine canyons, a basal conglomerate bed is present of locally derived clasts from the underlying mudstone (Fig. 15A). These lithified mudstone clasts consist of hemipelagic facies and contain latest Maastrichtian planktic foraminiferal assemblages and well-preserved glass spherules (Figs. 15B and 15C). The glass spherules frequently infill fractures, or cracks, which may be rimmed by secondary sparry calcite (Fig. 15B), then truncated and followed by normal sedimentation. This suggests complex diagenetic processes and possible emergence prior to erosion and transport. These clasts with impact spherules provide unequivocal evidence of the existence of an older spherule ejecta layer, which was lithified and subsequently eroded, transported, and redeposited at the base of the sandstone deposit. In contrast, the spherules in the sandstone complex are embedded in a matrix of sand, marl clasts, glauconite, and shell fragments (Figs. 15D–15F).

Original Impact Spherule Layer

NE Mexico

Stratigraphic, lithologic, mineralogic, and paleontologic investigations of the spherule layers at the base of the sandstone complex in NE Mexico revealed these as reworked and redeposited from older sediments. In contrast to Texas, deposition occurred in an upper slope environment at a depth of at least 500 m (Keller et al., 1994a, 1994b; Alegret et al., 2001). Recently, Loma Cerca and El Peñon, which are 30 km apart, revealed the stratigraphically oldest spherule layer 8–10 m and 4–5 m, respectively, below the reworked spherule layers at the base of the sandstone complex (Figs. 1 and 12). The Loma Cerca section was discussed in Keller et al. (2002b), and only a brief summary is given here. The stratigraphically oldest spherule layer, which is between 9.5 and 10 m below the sandstone complex, is dense with spherules, some marl clasts, and foraminifera, but no sedimentary influx from shallow areas. Nearly 2 m

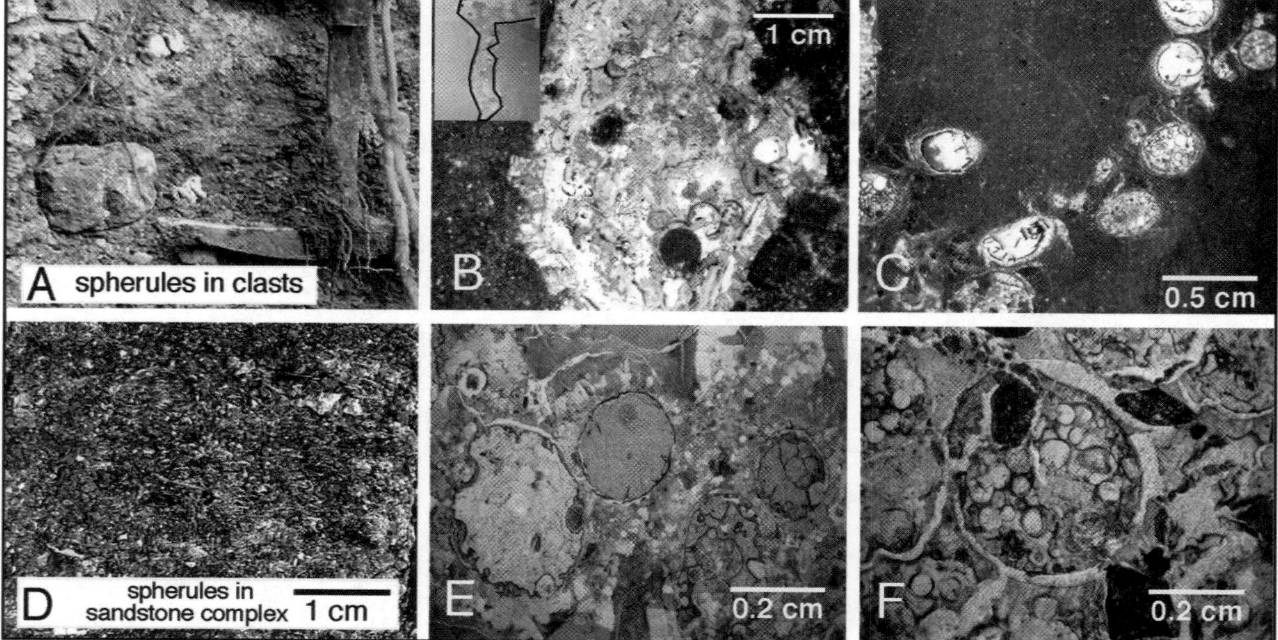

Figure 15. Spherules in mudstone clasts (A) from the basal conglomerate bed of the sandstone complex contain Chicxulub impact spherules. B and C: Some spherules infill cracks rimmed by sparry calcite and are truncated by erosion. Inset in B shows morphology of crack and total length of ~2 cm. In contrast, spherules at the base of the sandstone complex are embedded in a matrix of sand, glauconite, and shell fragments (D–F). These clasts and spherule-rich sandstones reveal the history of Chicxulub ejecta fallout, including lithification well prior to exposure to erosion, transport, and redeposition at the base of the sandstone deposit.

of marls with rare spherules and normal planktic foraminiferal assemblages overlie this spherule layer. Above it are two 50 cm thick spherule layers with upward-decreasing spherule abundance, some angular marl clasts, and anomalous peaks in robust species (e.g., globotruncanids and benthic foraminifera), which suggest reworking and redeposition.

Princeton students discovered the stratigraphically oldest spherule layer at El Peñon in upper Maastrichtian marls more than 4 m below the two spherule layers at the base of the sandstone complex (Fig. 16). Preliminary studies show an erosional base overlain by marl clasts and dense spherules cemented in calcite (10–20 cm thick; Figs. 10C and 10D) and followed by marls with decreasing spherule abundance. Four such units can be recognized in the 1.8 m thick deposit. No shallow-water fossils, wood, or plant debris, similar to the spherule deposits at the base of the sandstone complex, are present, which suggests that reworking and transport was more local. Between the top of this spherule deposit and the two reworked spherule layers at the base of the sandstone complex are ~4 m of normally bedded pelagic marls with no apparent disturbance (Fig. 16). The absence of

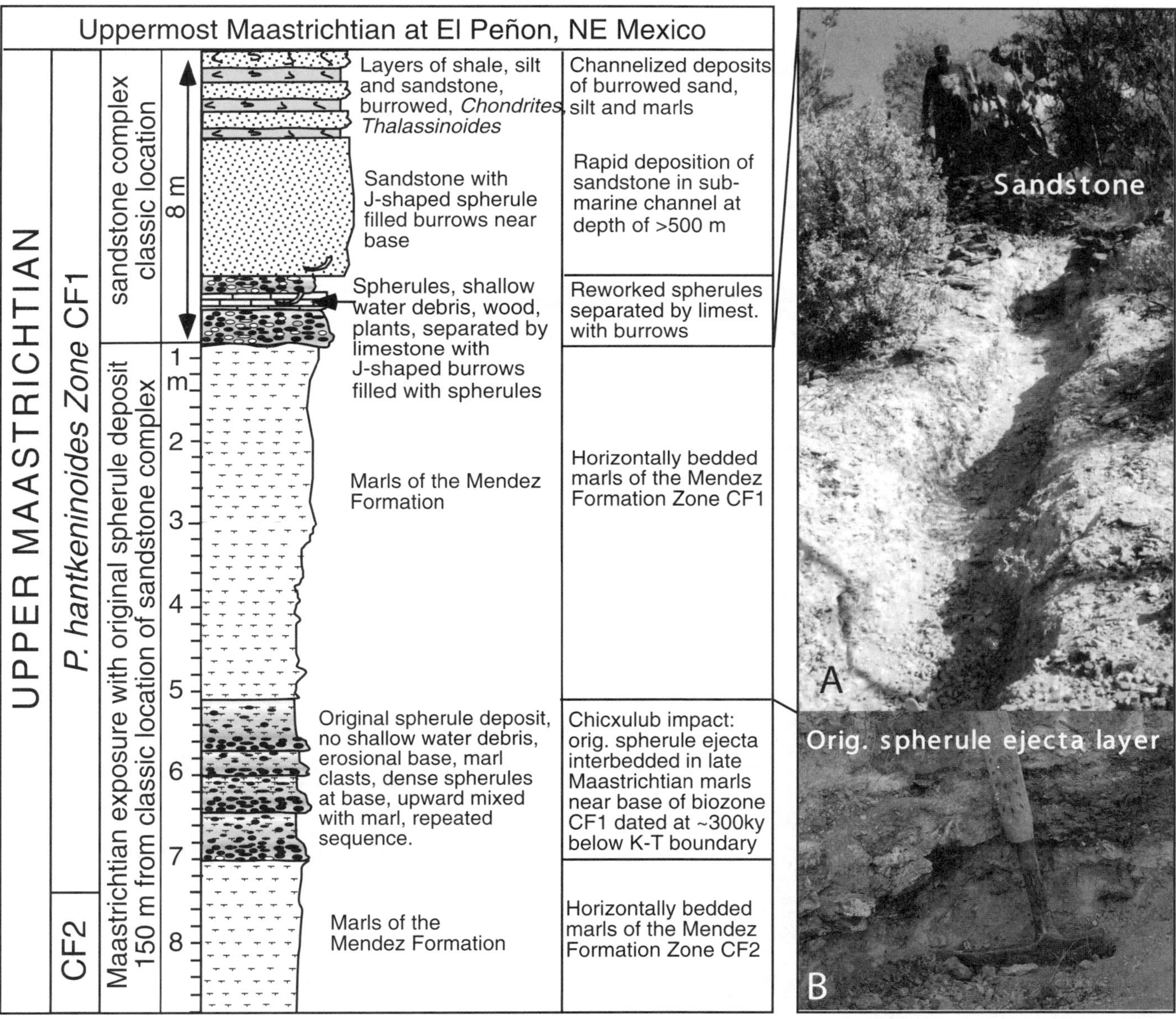

Figure 16. Composite El Peñon litholog showing the classic sandstone deposit with two reworked spherule layers at the base, and 150 m away, the newly discovered original Chicxulub impact spherule deposit near the base of zone CF1 about 4 m below the sandstone unit. A: Trenched sequence from below the sandstone unit to the original spherule deposit. Note that debris cover prevented uncovering the base of the sandstone unit. B: Base of original spherule ejecta layer.

structural disturbance of the sediments and presence of normal latest Maastrichtian microfossil assemblages indicate that these spherule layers represent the stratigraphically oldest and original Chicxulub spherule ejecta. Biostratigraphically, this spherule layer was deposited near the base of zone CF1 and thus predates the K-T boundary by ~300,000 yr.

Some workers suggested that the spherule layers in the upper Maastrichtian marls are due to large-scale earthquake-induced slumps and margin collapse (Soria et al., 2001; Smit et al., 2004) (see also *Chicxulub Debate*, http://www.geolsoc.org.uk/template.cfm?name=NSG1). However, the only evidence for structural disturbance is rare small-scale (<2 m) slumps within the upper reworked spherule layer (Keller and Stinnesbeck, 2002). In the shallow La Popa basin northwest of Monterey, Mexico, Lawton et al. (2005) described valley-like deposits as tsunami backflow, though this speculation finds little support in the sedimentary record (see Keller and Adatte, 2005; Stinnesbeck et al., 2005).

Brazos River, Texas

The original Chicxulub impact spherule layer was recently also discovered along the Cottonmouth Creek (CM) tributary of the Brazos River, Falls County, Texas, where it is best exposed below a small waterfall over the sandstone complex (Fig. 17A). At this locality, the sandstone deposit with its reworked spherules at the base is 40 cm below the K-T boundary. About 40–60 cm below the base of the sandstone deposit with its reworked spherules is a 3 cm thick yellow clay layer that consists of 100% Mg-smectite, or cheto-smectite, which is derived from weathering of glass (Fig. 17B) (Keller et al., 2007). Geochemical analysis of the smectite phases from the spherule-rich sandstone reveals the same cheto-smectite high in SiO_2 (66%–71%), Al_2O_3 (19%–20%), FeO (4.4%–4.8%), and MgO (2.8%–3.3%) with minor K_2O (1%–1.1%) and NaO (<0.5%). This composition is very similar to the altered smectite rims from glass spherules in Haiti, Belize, Guatemala, and Mexico, which indicates a common origin from altered impact glass spherules (Elliott, 1993; Debrabant et al., 1999; Stüben et al., 2002; Keller et al., 2003b). Relic glass fragments within the yellow clay and clay altered glass spherules from the base of the sandstone deposit (Fig. 15) have the same chemical composition, revealing their common origins (Keller et al., 2007, 2008). This provides very strong evidence that the yellow clay

Figure 17 (*continued on following page*). A: In Cottonmouth Creek near the waterfall, outcrops show the stratigraphic separation between the Chicxulub impact spherule ejecta layer now altered to Cheto smectite (yellow clay layer) and the sandstone complex 60 cm above. The K-T boundary is about 40 cm above the top of the sandstone complex.

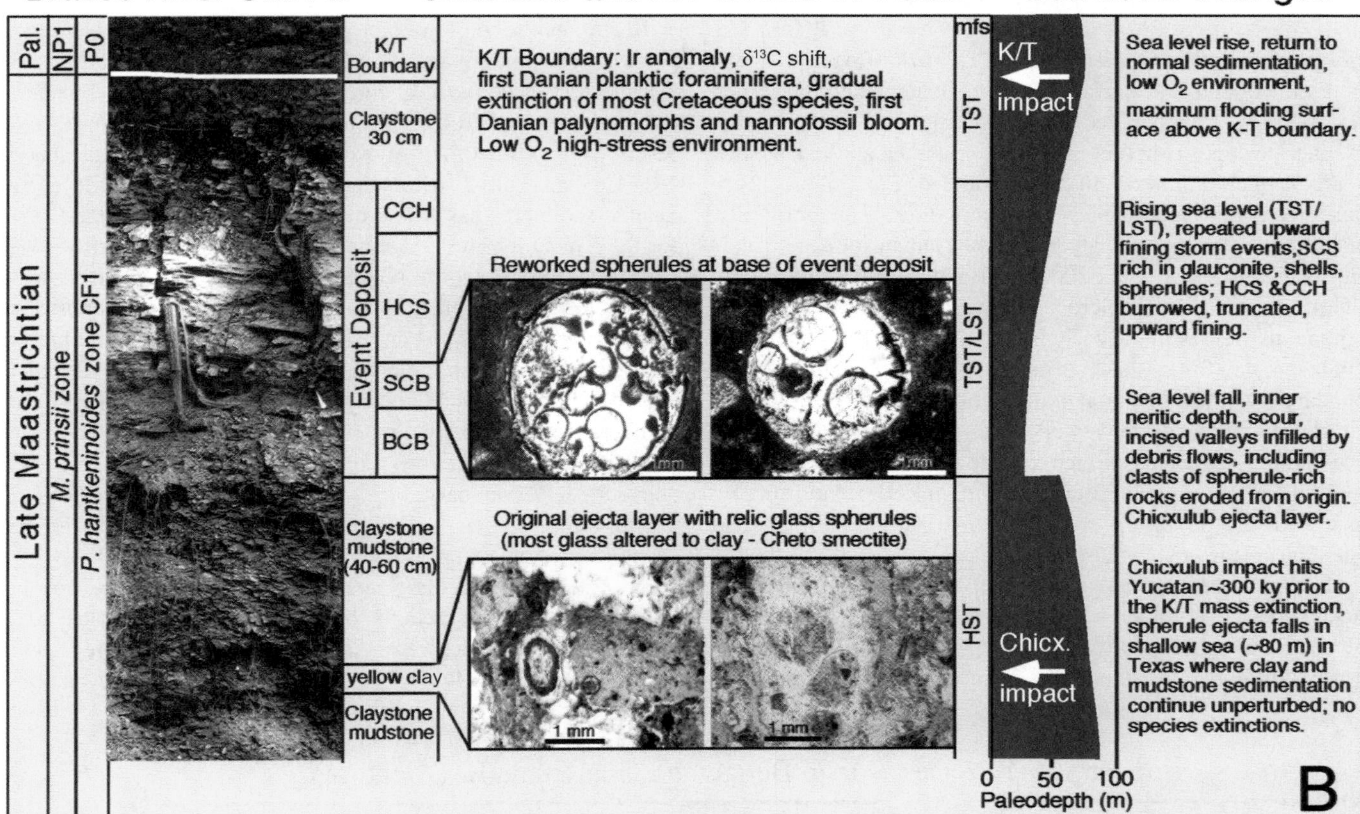

Figure 17 (*continued*). B: Cottonmouth Creek (CM) tributary of the Brazos River, Falls County, Texas. The sandstone deposit and, 40 cm below, the original Chicxulub impact spherule layer, now altered to yellow clay. The K-T boundary is 40 cm above the event deposit. The yellow clay consists of 100% Mg-smectite (cheto), indicative of altered Chicxulub impact glass; relic glass shows Chicxulub amphibolite composition. Abbreviations as in Fig 9A. Modified after Keller et al., 2007.

layer represents the original Chicxulub impact ejecta layer now altered to cheto-smectite. The stratigraphic position of this yellow clay layer near the base of zone CF1 is consistent with observations in NE Mexico and the Chicxulub crater core Yaxcopoil-1 (Keller et al., 2003a, 2004a, 2004b).

Age of Original Spherule Layer

The age of the stratigraphically oldest spherule layer can be determined from biostratigraphy. In the Brazos sections, as well as Loma Cerca and El Peñon in NE Mexico, the oldest spherule layer is near the base of planktic foraminiferal biozone CF1 (*Plummerita hantkeninoides*), which marks the last 300 k.y. of the Maastrichtian (Pardo et al., 1996). Similarly, the impact breccia in the Chicxulub crater core Yaxcopoil-1 underlies biozone CF1 (Keller et al., 2004a, 2004b). The evolutionary first appearance of *P. hantkeninoides* is thus an excellent biomarker for the age of the Chicxulub impact. This biomarker corresponds to within the lower part of nannofossil zone CC26 (*Micula prinsii*), which spans the last 450 k.y. of the Maastrichtian, and the lower part of paleomagnetic Chron 29R. The stratigraphic position and biostratigraphy thus place the Chicxulub impact in the uppermost Maastrichtian, ~300,000 yr below the K-T boundary.

This stratigraphic interval coincides with the rapid global warming that began ~450 k.y. prior to the K-T boundary and is attributed to Deccan volcanism (e.g., Kucera and Malmgren, 1998; Li and Keller, 1998b; Abramovich and Keller, 2003; Nordt et al., 2003). The climate warming follows the maximum cooling of the Maastrichtian, which culminated 500 k.y. prior to the K-T boundary and was associated with a major sea-level fall and widespread erosion. Both of these climatic events are easily identified in the stratigraphic record by paleontologic, stable isotopic, and sedimentologic analyses. Deposition of the sandstone complex coincides with the sea-level fall that followed the end of the global warming, ~100–150 k.y. before the K-T boundary. This sea-level fall resulted in widespread erosion, transport, and redeposition of the spherule layer exposed in nearshore areas.

Chicxulub Spherules in Danian Sediments

Cuba, Haiti, Belize, Guatemala, and Central Mexico

Fieldwork in Cuba, Haiti, Belize, Guatemala, and central Mexico (Fig. 1) has yielded further stratigraphic evidence that the spherule deposits in these areas are reworked and redeposited in early Danian sediments. In central Mexico (Coxquihui) a 2 cm thick altered glass spherule layer truncates the K-T unconformity, but a 60 cm thick spherule layer is in the Danian subzone P1a(1) (Fig. 7) (Stinnesbeck et al., 2002). In Bochil, Chiapas (Mexico), spherules are present in microconglomerates of the upper Maastrichtian, as well as in the early Danian subzone P1a(1) (Fig. 6). In Guatemala (Stinnesbeck et al., 1997) and Belize the K-T boundary is marked by a major unconformity with spherules and late Maastrichtian species reworked into 2–6 m of early Danian zone P1a(1) sediments, which overlie Maastrichtian platform limestone or limestone breccia (summary in Keller et al., 2003b). An Ir anomaly is present in a clayey marl in the early Danian zone P1a(1) (Fig. 18).

In Haiti, tectonic activity resulted in two K-T transitions stacked above each other, with the most commonly studied K-T boundary and spherule layers exposed in road cuts, and the less known, but more complete K-T sequences exposed on the slope ~30 m below the road (Keller et al., 2001). In these outcrops a 5–10 cm thick spherule-rich layer with early Danian zone P1a(1) planktic foraminifera and zone NP1a nannofossils directly overlies the K-T unconformity, and additional 2–4 cm thick spherule layers and spherule clasts are interspersed in the overlying 50 cm with an Ir anomaly above a rippled sandstone (Figs. 5 and 19). This spherule distribution pattern and Ir anomaly in early Danian sediments has led some workers to interpret a K-T age for the Chicxulub impact, arguing that "megaseiches" from the Chicxulub impact and subsequently seiches developed from tectonic adjustments resulted in the observed spherule redeposition (Aguado et al., 2005; Maurrasse et al., 2005). This interpretation is in conflict with the presence of early Danian species that evolved after the K-T boundary event, the apparently long-term and repeated reworking and redeposition pattern through the early Danian, and the distinct Ir anomaly in subzone P1a(1) well above the K-T boundary.

In Cuba, Alegret et al. (2005) documented the Loma Capiro section near Santa Clara. In the same area, Stinnesbeck, Adatte, and Keller analyzed the Santa Clara section where exposure of the breccia is limited to the top 10 m (Fig. 20). The two sequences are complementary, but there are some differences in the breccia and overlying Danian. In the Loma Capiro section,

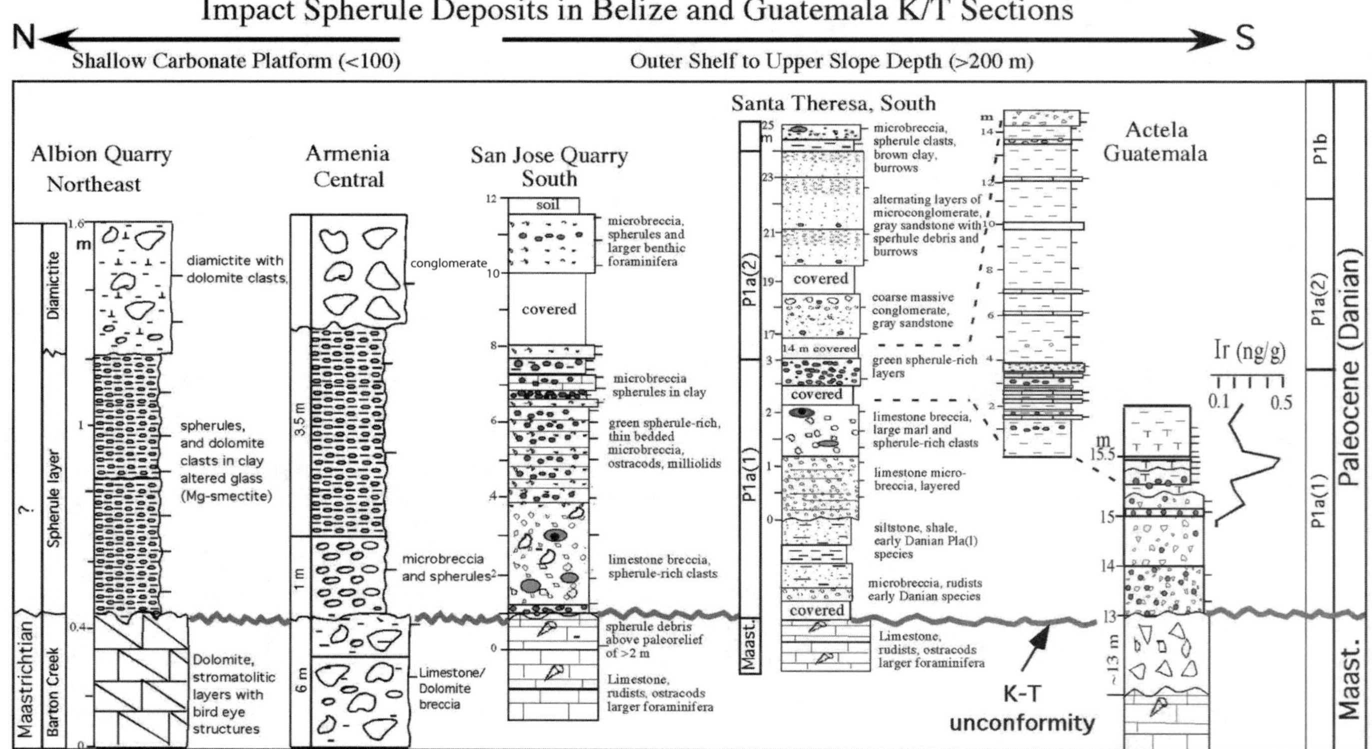

Figure 18. Litho- and biostratigraphy of the Belize and Guatemala K-T sequences from shallow platform environments in the north to deeper outer shelf and upper slope sequences in the south and Guatemala to the west (modified after Keller et al., 2003b). A major unconformity marks the K-T boundary, and all spherule deposits are reworked into early Danian sediments. An Ir anomaly is present in zone P1a(1) in the Actela section of Guatemala.

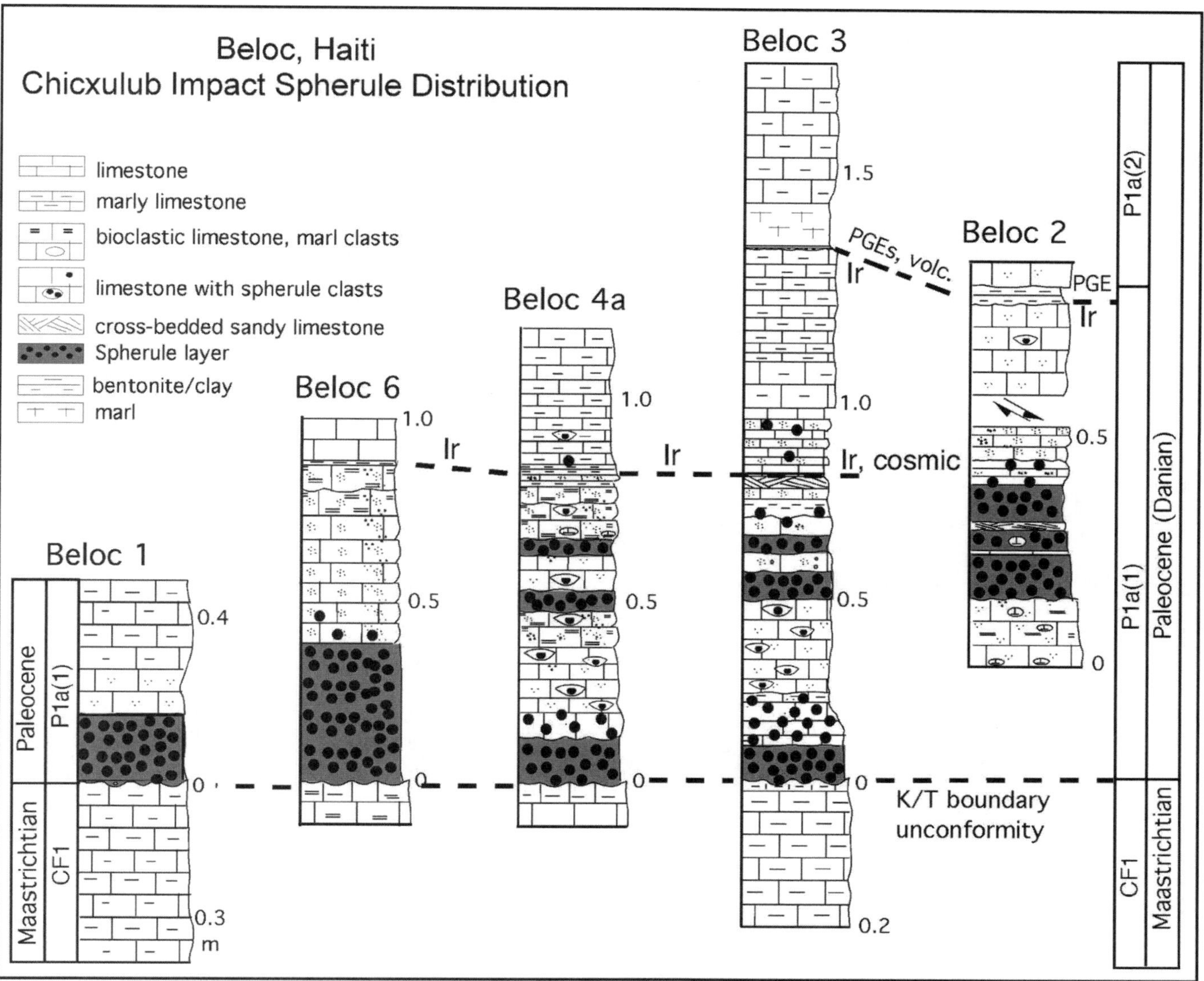

Figure 19. Litho- and biostratigraphy of the Beloc, Haiti, K-T sequences (modified after Keller et al., 2001). All spherule-rich deposits are reworked into early Danian zone P1a(1), as indicated by the presence of early Danian index species and reworked Cretaceous species. An Ir anomaly of cosmic origin is present above the spherule-rich sediments and a Pd anomaly of volcanic origin is present at the P1a(1)-P1a(2) boundary.

the breccia consists of an upward-fining megabreccia, followed by an upward-fining microconglomerate and topped by upward-fining sandstone with spherules at the top. In the Santa Clara section, the upward-fining microconglomerate and sandstone are absent. An undulating erosion surface with the depressions filled by altered glass spherules marks the top of the megabreccia. About 5 cm above this unconformity is a 1 cm thick spherule layer in marlstone. The overlying Danian sequence appears to be the same in both sections, consisting of clayey siltstone or marlstone with sand layers. Alegret et al. (2005) recognized upper zone P1a (or Pα) assemblages above the unconformity and estimated that a hiatus spans zone P0 and most of P1a. At Santa Clara, the presence of rare small *P. eugubina* in an assemblage of *Subbotina triloculinoides* and *Parasubbotina pseudobulloides* marks this interval as the upper part of zone P1a, or P1a(2), in agreement with Alegret et al. (2005).

The breccia contains reworked Maastrichtian species, including the zone CF1 index *Plummerita hantkeninoides*. Alegret et al. (2005) argued that the presence of this species in the breccia excludes a pre-K-T age for the Chicxulub impact. However, this is not the case because in NE Mexico and Texas the original Chicxulub impact ejecta layer is *near the base of zone CF1*—indicating that the impact occurred after the first appearance of *P. hantkeninoides*. This suggests that breccia formation is

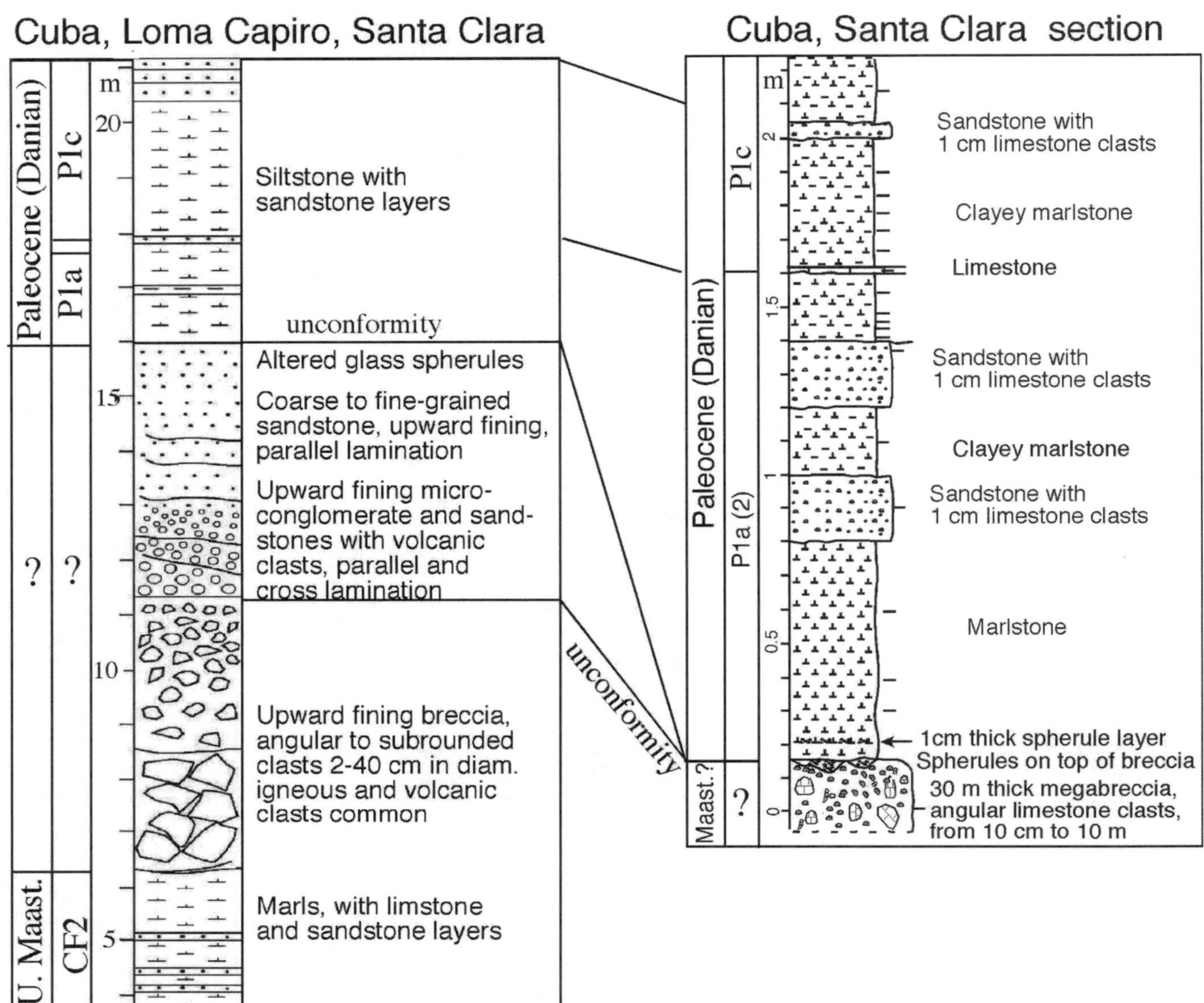

Figure 20. Litho- and biostratigraphy of the Cuba K-T sequences near Santa Clara. The Loma Capiro section (modified after Alegret et al., 2005) has a more expanded breccia complex than the Santa Clara section. In both sections, Chicxulub spherules are present at or above the unconformity between the breccia complex and the overlying Danian sediments. The early Danian zones P0 and P1a(1) are missing at the unconformity.

no older or younger than the base of zone CF1, similar to the age of the impact breccia in Yaxcopoil-1 (Keller et al., 2004a, 2004b). The stratigraphic position of the spherules at the top of the sandstone unit in Loma Capiro and in the Danian zone P1a above the unconformity in the Santa Clara section indicates reworking and redeposition and does not support a K-T age for the Chicxulub impact, as suggested by Alegret et al. (2005).

The Cuba and Haiti spherule distribution patterns are similar to those observed in Belize and Guatemala and represent post-depositional erosion and redeposition during the early Danian. These Central American sequences illustrate the danger of basing interpretations on regional observations that may reflect special marine settings (e.g., shallow platform limestones) or tectonic activities that resulted in incomplete sequences. These sections must be interpreted within the context of the more complete and high-sedimentation deepwater sequences of NE Mexico and the shallow-water sequences in Texas.

Chicxulub Spherules and Ir Anomaly in Stratigraphic Proximity

Advocates for a K-T age of the Chicxulub impact frequently point to the stratigraphic proximity of spherules and an Ir anomaly in terrestrial sequences of the continental United States and some deep-sea marine sequences (e.g., Blake Nose, Bass River, Demerara Rise) as unequivocal evidence that Chicxulub caused the K-T mass extinction (Olsson et al., 1997; Norris et al., 1999, 2000; Klaus et al., 2000; MacLeod et al., 2006). Differential settling of iridium and spherules, estimated at a few hours to days, is often used to explain the proximity of the Ir anomaly at or a few centimeters above the spherule layer in the deep-sea and terrestrial sequences. This argument is plausible only if there is no inverse grading in the spherule layer and no disconformity due to erosion and/or nondeposition—features that are evident at Blake Nose, Bass River, and Demerara Rise.

Stratigraphic proximity of the Ir anomaly and spherule layer alone is no proof for a cause-and-effect relationship, particularly in the deep-sea and terrestrial records, where sedimentary sequences tend to be incomplete due to erosion or condensed sedimentation. Deep-sea marine sequences must be interpreted with caution, as they contain almost invariably less complete sedimentation records than continental shelf sequences. This is because the sedimentation rate is higher on shelves due to terrigenous influx, whereas erosion is higher in the deep sea due to intensified current circulation during climate cooling and lower sea levels, such as during deposition of the latest Maastrichtian zone CF1 sandstone deposits.

Blake Nose, Western North Atlantic

Perhaps the most commonly cited example for a cause-and-effect relationship between impact and Ir anomaly is Blake Nose ODP Site 1049 (Sigurdsson et al., 1997; Norris et al., 1999, 2000; Klaus et al., 2000; Martínez-Ruiz et al., 2002). At this locality, a 15 cm thick spherule layer disconformably overlies slumped carbonate ooze of upper Maastrichtian age (Fig. 21). The spherule layer is graded and contains clasts of limestone, dolomite, chert,

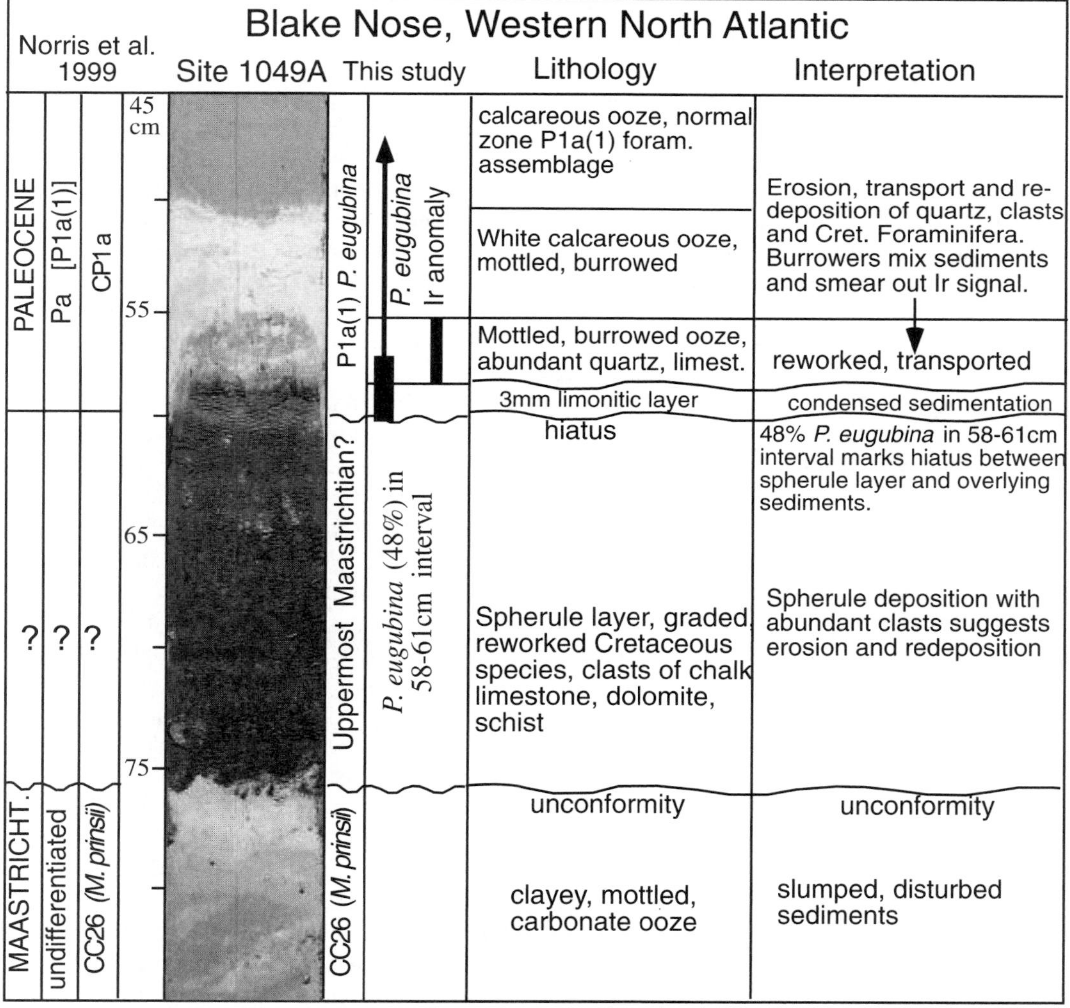

Figure 21. Litho- and biostratigraphy of ODP Site 1049A, Blake Nose, western North Atlantic (modified after Norris et al., 1999). Note that this section contains a highly condensed K-T transition that is disturbed, burrowed, mixed, and incomplete. The juxtaposition of the spherule layer and Ir anomaly in this sequence permits no conclusion as to the precise timing of the Chicxulub impact.

and schist, which indicate reworked and transported sediments. Above it is a 3 mm thick limonitic layer followed by a 3–4 cm thick dark gray mottled and bioturbated ooze that contains abundant quartz, limestone chips, Cretaceous foraminifera, 48% of the Danian species *Parvularugoglobigerina eugubina*, and an Ir anomaly (1.3 ppb). The presence of quartz, limestone chips, and Cretaceous foraminifera indicate that this layer is also reworked. Norris et al. (1999) concluded that the basal Danian zone P0 (e.g., boundary clay) is absent. Nevertheless, they interpreted this mottled and disturbed interval as a complete sedimentary sequence and unequivocal evidence that the Chicxulub impact is the K-T boundary event.

To justify this conclusion, they explained the absence of the boundary clay zone P0 by arguing that it does not exist in the deep sea (Norris et al., 1999). They based this conclusion on their analysis of the K-T clay at the El Kef stratotype section, stating that they found rare tiny *P. eugubina* at the base of the clay layer where no others had detected them. However, the specimen they show is large (80 μm, whereas the first specimens are usually <63 μm; Keller et al., 1995, 2002a), well developed, and known to occur much later in its evolutionary history, which suggests contamination or mixing probably by bioturbation. Without further evidence, Norris et al. (1999) speculate that zone P0 may be an artifact of rare occurrence of *P. eugubina* in shallow-water sequences, whereas in deep waters this species would be more common. They conclude that if P0 is interpreted as a shallow-water biofacies—and hence absent in the deep—then "*the deep-sea sites are biostratigraphically complete and can be used to describe events immediately surrounding the impact event*" (Norris et al., 1999, p. 421). This argument does not take into account that El Kef, deposited in 300–500 m depth, is not a shallow-water sequence, and that the K-T boundary clay, or zone P0, has been documented in deep-sea sequences worldwide beginning with the Gubbio section in 1980 and in NE Mexico localities in the 1990s.

The stratigraphic and sedimentary record of Site 1049A provides evidence for a simpler interpretation (Fig. 21). The close stratigraphic proximity of the spherule layer and Ir anomaly is marked by discontinuity, nondeposition, erosion, and redeposition. For example, after spherule deposition the 3 mm limonitic layer marks a period of condensed sedimentation or nondeposition. Above it, the 3–5 cm burrowed mottled dark and light ooze indicates slow deposition with abundant burrowing organisms during a time of erosion, transport, and redeposition. The high abundance (48%) of *P. eugubina* in the 58–61 cm interval overlying the spherule layer (this study) marks a hiatus with the early part of zone P1a(1) missing. In condensed, burrowed, mixed, and incomplete sequences, it cannot be determined whether the Ir anomaly and well-developed *P. eugubina* represent the K-T or second Danian P1a(1) Ir anomaly of the region (discussed above). The main point here is that the juxtaposition of the spherule layer and Ir anomaly in Site 1049 provides no proof of cause and effect, but appears to be an artifact of a condensed and incomplete stratigraphic record that permits no conclusion as to the precise timing of the Chicxulub and K-T events.

Demerara Rise, Western North Atlantic

ODP Leg 207, Site 1259, on Demerara Rise recovered a K-T transition similar to Site 1049 on Blake Nose. The K-T transition was recovered in four closely spaced holes, each with an ~2 cm thick spherule layer separating Cretaceous chalk from overlying Paleogene claystone. MacLeod et al. (2006) described the lithology of Hole 1259B, but provide Ir and biostratigraphic data from 1259C, where an Ir anomaly of 1.5 ppb coincides with the very top of the spherule layer and high values (0.5–0.7 ppb) persist over 15 cm of the claystone (Fig. 22), which suggests mixing of the original signal. They report very low-diversity (five to ten species) late Maastrichtian zone CF1 assemblages, indicating high-stress conditions. A similar assemblage is present in the spherule layer. A thin section of the basal clay above the spherule layer shows an assemblage dominated by *Guembelitria cretacea*, which is characteristic of high-stress conditions after the mass extinctions, as well as shallow water or restricted basin environments of the latest Maastrichtian (Keller, 2002; Keller and Pardo, 2004). Therefore, this interval can only be questionably assigned to P0 on the basis of abundant *Guembelitria*. The first Danian species are reported 6 cm above the spherule layer and mark the *Parvularugoglobigerina eugubina* P1a(1) zone (MacLeod et al., 2006), though the presence of *Chiloguembelitria crinita* indicates that this is not the basal part of this zone. Thus, the position of the spherule layer stratigraphically separates Cretaceous and Paleogene sediments, but the continuity of sedimentation and the history of spherule and Ir deposition are in doubt.

MacLeod et al. (2006, p. 10) argue that the K-T boundary should be placed at the base of the spherule layer, because this is in agreement with the impact signals (e.g., spherules, shocked quartz, Ir anomaly) at the El Kef stratotype and shows that the Chicxulub impact defines the K-T boundary. They interpret the juxtaposition of the Ir anomaly and spherule layer as unequivocal evidence that the Chicxulub impact defines the K-T boundary (p. 12) and demonstrates "*first-order agreement with the predictions of the impact hypothesis*" recording the history "*within minutes of the impact.*" There are major problems with this interpretation.

For example, there are no Chicxulub-type spherules at the El Kef stratotype, and no Ir anomaly has ever been recorded in association with the Chicxulub spherule layers. The juxtaposition of spherules and Ir anomaly has only been reported from the condensed and apparently incomplete deep-sea sections on Blake Nose, Bass River, and Demerara Rise. In areas of high sedimentation rates and thick (~2 m) spherule layers (e.g., NE Mexico and Brazos River, Texas), the K-T Ir anomaly and Chicxulub spherule layers are well separated by 2–9 m of undisturbed marls or claystones (Figs. 14, 16, 17). This suggests that the juxtaposition of the two impact signals on Demerara Rise is an artifact of a condensed and incomplete sedimentary record. Another potential problem is the inverse grading and soft-sediment deformation (e.g., fine-grained white layers at the base of the spherule layer followed by upward-fining spherules; Fig. 22), which MacLeod et al. (2006) interpreted as dewatering and resuspension of fines

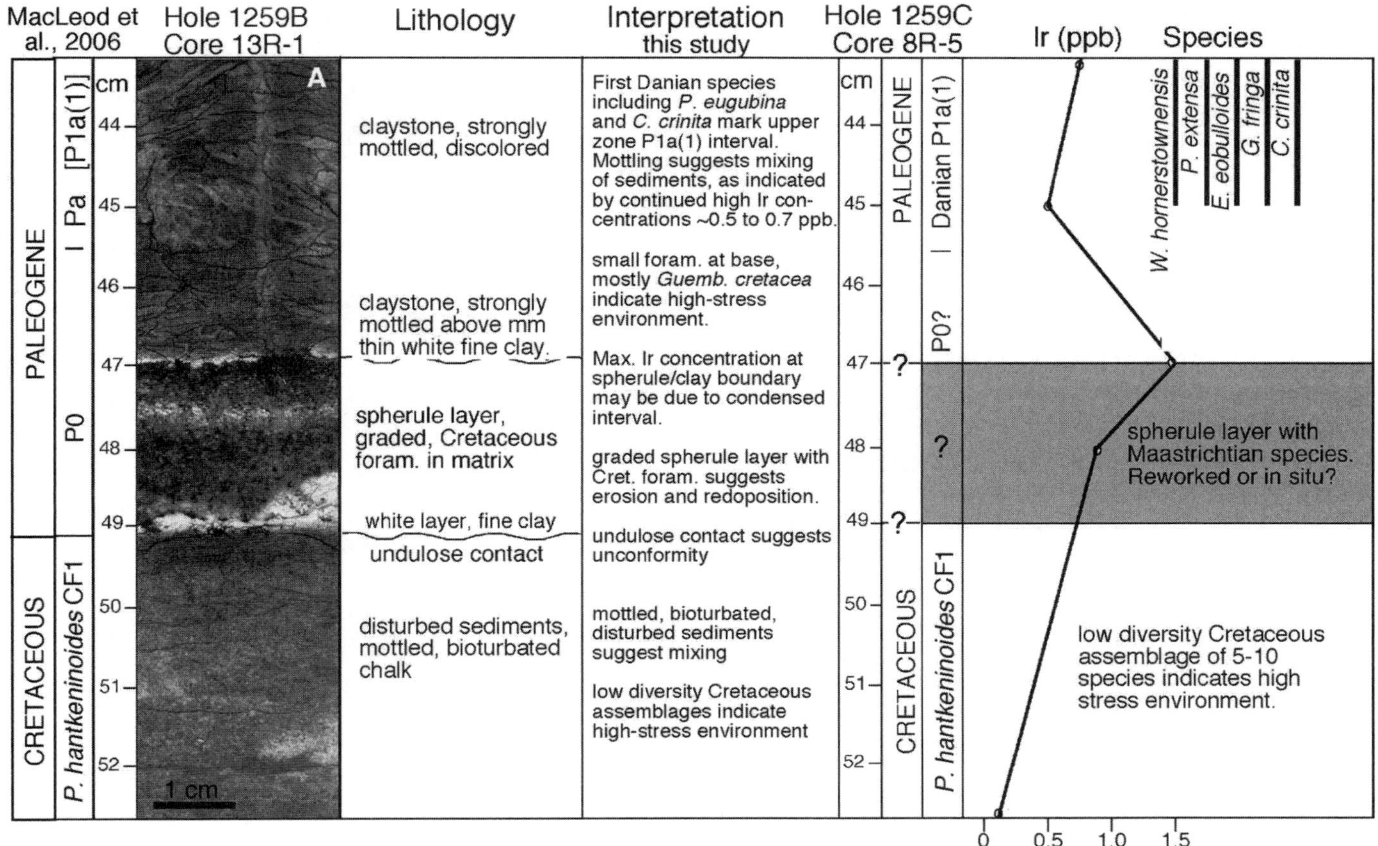

Figure 22. Litho- and biostratigraphy of ODP Site 1259, Demerara Rise, western North Atlantic (modified after MacLeod et al., 2006). Note that this section is very similar to Blake Nose Site 1049, though has the latest Maastrichtian zone CF1 present. Although the spherule layer and maximum Ir concentrations followed by the first Danian species place this impact ejecta at or near the K-T boundary, lithological characteristics indicate condensed sedimentation and redeposition (inverse grading), which allows no conclusion as to the timing of the Chicxulub impact.

as a result of seismic shaking within 15 min of the impact. An alternative explanation is reworking and resettling of sediments during intensified current circulation associated with the end-Maastrichtian global cooling and low sea level that resulted in erosion and redeposition of spherules throughout Central America, the Caribbean, and Texas (Keller et al., 2003a, 2007).

Bass River Core, New Jersey

The Bass River core of New Jersey has also been claimed to contain a complete K-T boundary sequence that provides "unequivocal evidence" that the Chicxulub impact is K-T in age (Olsson et al., 1997). However, this shallow-water (~100 m) sequence is demonstrably incomplete. The Maastrichtian consists of highly bioturbated glauconitic clay with only the top 8 cm indicative of the *M. prinsii* (CC26) zone, which suggests erosion of the uppermost Maastrichtian sediments (Fig. 23). The overlying spherule layer is poorly sorted with altered clay spherules and clasts. Above it is a silty glauconitic clay with numerous clasts with Maastrichtian microfossils that indicate erosion, transport, and redeposition. The presence of an early Danian planktic foraminiferal assemblage with six species, including *P. eugubina*, marks zone P1a as spanning a mere 8 cm interval. This indicates that the boundary clay zone P0 and most of zone P1a are missing in the Bass River core. The juxtaposition of the spherule layer and early Danian is due to a major hiatus, which allows no conclusion as to the timing of the Chicxulub event.

Terrestrial Sequences—U.S. Western Interior

Another frequently used argument in support of Chicxulub as the K-T boundary impact is the close proximity of the spherule and Ir-enriched layers in terrestrial sequences from the U.S. Western Interior (Orth et al., 1982). In these localities a 2 cm thick gray layer of kaolinite and some smectite and hollow goyazite spherules (infilled by kaolinite) mark the Chicxulub spherule layer. Above it is a dark organic-rich thinner layer of laminated smectite and kaolinite clay with shocked minerals and siderophile elements that is believed to mark the K-T boundary "fireball" layer (Bohor, 1990; Izett, 1990; Bohor and Glass, 1995). However, sedimentation in terrestrial sequences is inherently discontinuous, and the close stratigraphic proximity

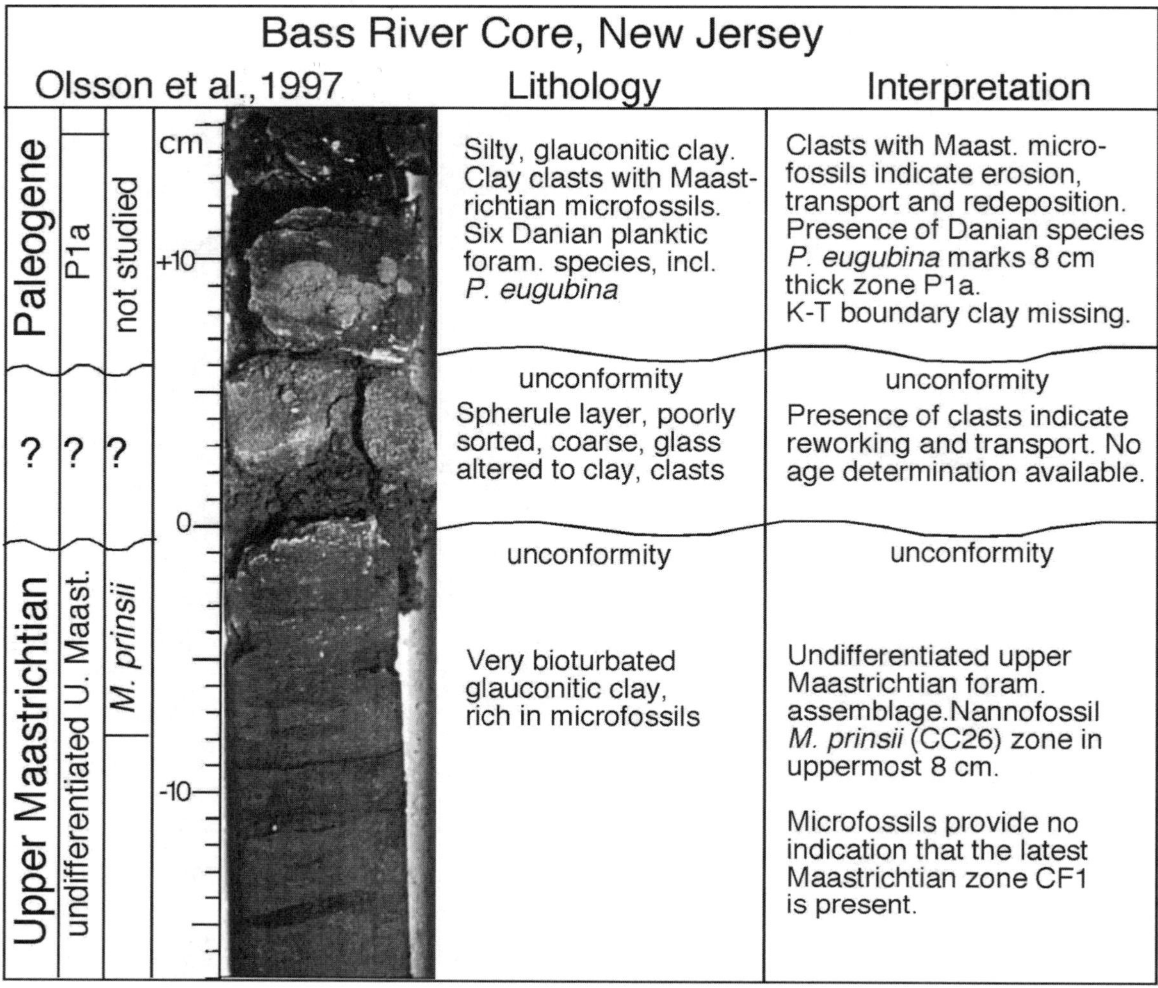

Figure 23. Litho- and biostratigraphy of the Bass River core of New Jersey (modified after Olsson et al., 1997). In this section the juxtaposition of the spherule-rich layer and the overlying 8-cm-thick glauconitic clay of zone P1a marks a major hiatus, which allows no conclusion as to the timing of the Chicxulub event.

of two separate layers alone is not a convincing argument for a cause-and-effect scenario.

DISCUSSION AND SUMMARY

The promise and potential of impact stratigraphy are impact markers as time horizons in correlation and in deciphering the relative order of a closely spaced sequence of events. In principle, impact markers represent an instant in time, and their stratigraphic positions relative to other marker horizons are fixed, providing a time line between younger and older events above and below. But impact markers are subject to the same geological forces as any other marker horizons—erosion, reworking, and redeposition—sometimes leaving multiple impact marker horizons over time from which the postdepositional history must be carefully evaluated.

The global K-T Ir anomaly and more regional Chicxulub impact spherules are the most widespread and consistently present impact markers, and both have been used as unquestionable K-T marker horizons. However, strata with Ir anomalies never contain Chicxulub impact spherules and spherule layers never contain iridium anomalies. Moreover, the stratigraphic position of the oldest spherule layer is in undisturbed upper Maastrichtian sediments. But reworked and redeposited spherule layers are common in the uppermost Maastrichtian and lower Danian, and there is evidence of additional Ir anomalies of both cosmic and volcanic origins in the early Danian (Fig. 24). In the deep sea, the sedimentary records are generally condensed and incomplete due to erosion and nondeposition, which may position the Ir anomaly in close proximity with a thin (1–2 cm) spherule layer. For these reasons, analysis of isolated sequences can lead and has led frequently to erroneous conclusions.

A comprehensive survey of Ir anomaly and spherule impact markers within the Chicxulub proximal and distal ejecta fields from Texas to central Mexico and the Caribbean reveals the main

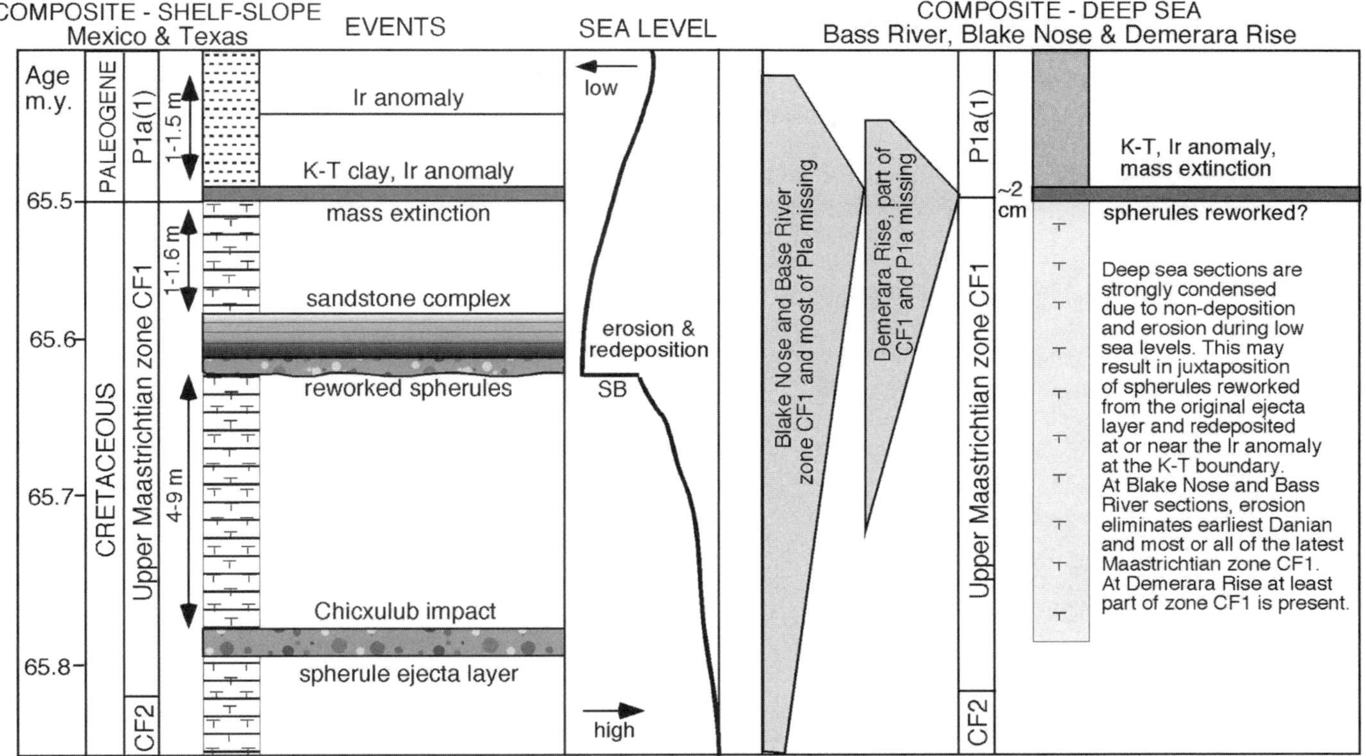

Figure 24. Summary of impact horizons in upper Maastrichtian to lower Danian sediments based on the composite stratigraphy of the most expanded shelf and upper slope sequences in Texas and NE Mexico. The original Chicxulub impact ejecta spherule layer, the reworked spherules at the base of the sandstone complex, and the K-T Ir anomaly are all excellent marker horizons. Comparison with the condensed marine sequences in the northeastern Atlantic indicates that these three marker horizons are condensed into one, juxtaposing the Ir anomaly and spherules at or near the K-T boundary due to erosion and nondeposition.

Ir anomaly at the K-T boundary clay (red layer), consistent with its global occurrence, and a second smaller Ir anomaly of cosmic origin and one of volcanic origin in the early Danian zone P1a(1) (Stüben et al., 2002, 2005). The K-T transition is thus marked by multiple impacts, similar to the late Eocene, the end-Devonian, and the Triassic (Keller, 2005).

The stratigraphy of impact spherules is even more complex than that of Ir anomalies, due to erosion, transport, and redeposition of spherules, and varies on a regional basis, apparently controlled by gravity flows and slumps commonly associated with sea-level falls. For example, impact spherules from predominantly shallow-water sequences of southern Mexico, Guatemala, Belize, Haiti, and Cuba are reworked and redeposited in the early Danian zone P1a(1). An unconformity marks the K-T boundary, and in some localities rare spherules can be found in the underlying breccia (e.g., Chiapas and Guatemala). In NE Mexico and Texas, impact spherules are exclusively found in upper Maastrichtian sediments, with the stratigraphically lowermost and original impact ejecta layer near the base of zone CF1, ~300,000 yr below the K-T boundary.

It is tempting to interpret this pattern of spherule layers as evidence for multiple impacts. But this is not supported by spherule glass geochemistry, which indicates comparable compositions in all spherule layers and is consistent with impact glass in the Chicxulub crater cores (e.g., Schulte et al., 2003; Keller et al., 2007). This indicates that all stratigraphically higher spherule layers are reworked from the stratigraphically oldest and original ejecta deposit. In all areas with abundant spherule deposits, redeposition of impact spherules occurred predominantly in submarine canyons during the latest Maastrichtian sea-level fall (~100–150 k.y. prior to the K-T boundary; Fig. 24), which also deposited a widespread sandstone complex that has been erroneously interpreted as a megatsunami generated by the Chicxulub impact.

Although impacts create tsunamis, which are expected to leave sedimentary records, there is little disturbance associated with the original impact ejecta layer at Brazos River in Texas, and Loma Cerca and El Peñon in NE Mexico, at 500 km and 1000 km, respectively, from the impact crater on Yucatan. The submarine canyon sandstone deposits with reworked spherules at the base were formed much later (probably ~100–150 k.y. prior to the K-T boundary) during a sea-level fall that lowered sea level by ~50–70 m, as estimated from Brazos River sequences. Clasts with spherules from the original deposit and burrowed horizons throughout these sandstone complexes indicate long-term deposition and seasonal storms rather than a tsunami (Yancey 1996; Gale, 2006).

Impact stratigraphy is thus an exceedingly useful tool in stratigraphy that can help evaluate the sequence of events across critical intervals in Earth's history. But it has to be used with caution, applied on a regional scale to avoid isolated artifacts of redeposition, and subject to the basic stratigraphic principle of superposition—that *in any undisturbed sediment sequence the oldest bed is at the base and the youngest at the top.* Observations from this regional evaluation of Ir and spherule impact markers provide some guidance.

Iridium Marker in K-T Sequences

The sharp K-T iridium anomaly is probably the best impact marker, though studies in India suggest that a volcanic origin cannot be excluded (Bajpai and Prasad, 2000). Because of its properties, Ir is unlikely to be reworked and redeposited as a sharp peak anomaly (Sawlowicz, 1993), though it is easily smeared out by bioturbation. In NE Mexico, as in continuous sequences worldwide, the K-T Ir anomaly is only found in the boundary clay or thin red layer, coincident with critical K-T markers such as the negative $\delta^{13}C$ shift, the extinction of all specialized tropical planktic foraminifera, and evolution of the first Danian species in the boundary clay immediately above (Fig. 4). When used together with these bio- and chemomarkers, the Ir anomaly is a reliable K-T indicator. However, it requires some considerations.

1. In NE Mexico the Ir anomaly is in the K-T boundary clay and red layer and stratigraphically always above the sandstone complex with its spherules at the base (e.g., El Mimbral, La Parida, La Sierrita; Figs. 8A and 8B). Along the Brazos River, Texas, the Ir anomaly is also above the sandstone complex with spherules at the base (Figs. 9A and 9B).
2. Ir can be remobilized upward and downward and concentrated at redox boundaries, though in usually much smaller concentrations. Such minor Ir enrichments are present in the sandstone complex below the K-T boundary at El Mimbral and Brazos-1 (Figs. 8A, 8B, 9A, 9B) and have been claimed as support for the impact tsunami hypothesis and to justify the placement of the K-T boundary at the base of these sandstone complexes. Ir anomalies cannot be used in isolation as the sole index for the stratigraphic placement of the K-T boundary.
3. Iridium anomalies are not unique and therefore not infallible K-T markers. At least three impact craters are known during the last several hundred thousand years of the Maastrichtian (Silverpit, Boltysh, Chicxulub). The fourth may be the K-T impact for which no impact crater is yet known. A fifth impact may have occurred during the early Danian zone P1a(1). Ir anomalies are known only from the K-T and Danian impacts.
4. Iridium anomalies may be of cosmic or volcanic origins. In particular, small Ir anomalies should be used with caution and require other PGEs to determine their origin (e.g., cosmic, volcanic, or redox boundaries). For example, early Danian zone P1a(1) sediments in Haiti, Guatemala, and central Mexico contain an Ir anomaly of cosmic origin, but in Haiti a second small Ir anomaly is of volcanic origin (Fig. 5).
5. No iridium enrichment has been observed to date in any impact spherule layers or any other Chicxulub impact ejecta. The Chicxulub bolide may have been a dirty snowball–type comet without Ir enrichment.

Chicxulub Spherule Marker in K-T Sequences

The geographic and stratigraphic distribution of the Chicxulub impact spherules is very complex due to the postdepositional history affecting these deposits (e.g., tectonic, erosion, transport, redeposition, slumps, gravity flows). The multiple stratigraphic occurrences of altered glass spherules show how easily spherules can be transported and redeposited into younger sediments. Their presence in both uppermost Maastrichtian and lower Danian sediments reflects repeated erosion and redeposition of spherules over several hundred thousand years after the impact event. This limits the use of Chicxulub impact spherules as a correlation tool and time horizon without consideration of this temporal and spatial postdepositional history. The following stratigraphic and geographic considerations mitigate this drawback.

1. Only the stratigraphically lowermost (oldest) impact spherule layer in undisturbed sediments provides a marker horizon for the age of the Chicxulub impact. This spherule layer generally contains a calcite matrix and lacks shallow-water debris. In NE Mexico and along the Brazos River in Texas, the stratigraphically lowermost spherule layer is near the base of planktic foraminiferal zone CF1 and predates the K-T boundary by ~300 k.y. This is in agreement with the age of the impact breccia in the Chicxulub crater core Yaxcopoil-1.
2. All stratigraphically higher (younger) spherule layers reflect postdepositional erosion, transport, and redeposition at times of low sea levels, gravity slumps along slopes, or tectonic activity. This is evident in the presence of clastic grains, shallow-water debris, and foraminifera.
3. The first widespread phase of spherule redeposition occurred during the early transgression after the uppermost Maastrichtian sea-level fall (~100–150 k.y. prior to the K-T boundary). At this time, spherules eroded in shallow areas were transported (via gravity flows, slumps) into deep waters via submarine canyons (e.g., NE Mexico and Texas). This phase led to deposition of the spherule layers at the base of the sandstone complex in NE Mexico and Texas, which gave rise to the controversial hypothesis of impact-generated tsunami deposits. The second phase of spherule redeposition occurred during the early Danian zone P1a(1) sea-level low and is mainly found in central Mexico, Guatemala, Belize, and Haiti. A major hiatus marks the K-T boundary in these areas.

4. The stratigraphic proximity of Chicxulub impact spherules and Ir anomaly at or near the K-T boundary in deep-sea sites (e.g., ODP Sites 1049, 1259, and Bass River, New Jersey) does not signify a K-T boundary age for the Chicxulub impact, but may reflect the condensed and often disturbed sedimentary records of the deep sea, which must be carefully evaluated and correlated with the more complete records from NE Mexico and Texas.

CONCLUSIONS

Impact stratigraphy makes use of unique catastrophes in Earth's history and uses their signals in correlation and timing of these events. Impact markers are extremely useful and invaluable tools in stratigraphy with great potential in deciphering a sequence of closely spaced events, such as the timing of the Chicxulub impact and K-T event. But as any sedimentary markers, they are subject to erosion, transport, and redeposition, which can lead to erroneous interpretations. Therefore, the timing of impacts must be deciphered from undisturbed, complete, and continuous sequences with high sedimentation rates. Moreover, the regional stratigraphic distribution is critical to sort out spherule layers that are redeposited from those that represent the original impact ejecta fallout. This study shows that in undisturbed sequences of NE Mexico and Brazos River, Texas, the Chicxulub spherule ejecta and K-T Ir anomaly and red layer are widely separated, with biostratigraphy indicating ~300,000 yr elapsed between these two events. In all other regions examined, the stratigraphic records are demonstrably incomplete, with Chicxulub spherule ejecta redeposited in early Danian sediments (e.g., Cuba, Haiti, Belize, Guatemala, central Mexico), or juxtaposed at unconformities at the K-T boundary (Bass River, New Jersey, Blake Nose, NW Atlantic). For the K-T transition, impact stratigraphy predates the K-T boundary and Ir anomaly. There is also evidence that the environmental upheaval continued during the early Danian with another smaller impact and volcanism.

The story that has emerged to date is that the K-T transition was a time of long-term upheavals including multiple impacts, volcanism, and rapid climate changes that eventually led to the mass extinction, but comprehensive studies are still lacking. Deciphering the full history of the K-T mass extinction will require further detailed regional and global studies. For example, a comprehensive study is needed to evaluate the geochemical similarities and differences between the Chicxulub impact spherules and the rare clay spherules observed in some K-T clay layers, which may reveal their source. Global studies are needed to evaluate the geographic distribution of the Chicxulub impact spherule layer, which currently is restricted largely to Central America and the Caribbean. The absence of this spherule layer in geographically more distant regions suggests that the search has concentrated on the K-T boundary interval, instead of the latest Maastrichtian, or that this impact was too small to have a global distribution and global environmental effects. Perhaps for too long has the search for the cause of the K-T mass extinction concentrated almost exclusively on a large impact while ignoring Deccan volcanism, the other major catastrophe on Earth 65 m.y. ago. Recent studies suggest that the ultimate cause may well have been the combined effects of volcanism and impacts.

ACKNOWLEDGMENTS

This research expository has drawn upon more than 13 years of collaborative studies with Thierry Adatte from the University of Neuchatel, Wolfgang Stinnesbeck, Zsolt Berner, Utz Kramar, and Doris Stueben from the University of Karlsruhe, and numerous former students whose work is acknowledged in the text. I am grateful to all of them for their dedication and hard work in the quest and search for the ultimate answers to the K-T controversy. I am grateful to the reviewers Jere Lipps, Neil Landman, and Kevin Evans for their helpful comments. This material is based upon work supported by the National Science Foundation's Continental Dynamics Program and Sedimentary Geology and Paleobiology Program under NSF Grants EAR-0207407 and EAR-0447171.

REFERENCES CITED

Abramovich, S., and Keller, G., 2003, Planktic foraminiferal response to latest Maastrichtian abrupt warm event: A case study from midlatitude DSDP Site 525: Marine Micropaleontology, v. 48, p. 225–249.

Adatte, T., Stinnesbeck, W., and Keller, G., 1996, Lithostratigraphic and mineralogical correlations of near-K/T boundary clastic sediments in northeastern Mexico: Implications for mega-tsunami or sea level changes?, *in* Ryder, G., et al., eds., The Cretaceous-Tertiary event and other catastrophes in Earth history: Geological Society of America Special Paper 307, p. 197–210.

Affolter, M., 2000, Etude des depots clastiques de la límite Cretace-Tertiaire dans la region de la Sierrita, Nuevo Leon, Mexique [M.S. thesis]: Neuchatel, Switzerland, Geological Institute, University of Neuchatel, 133 p.

Aguado, R., Lamolda, M.L., and Maurrasse, F.J.-M.R., 2005, Nanofósiles del límite Cretácico/Terciario cerca de Beloc (Haití): Bioestratigrafía, composición de las asociaciones e implicaciones paleoclimáticas: Journal of Iberian Geology, v. 31, no. 1, p. 9–24.

Alegret, L., Molina, E., and Thomas, E., 2001, Benthic foraminifera at the Cretaceous-Tertiary boundary around the Gulf of Mexico: Geology, v. 29, p. 891–894, doi: 10.1130/0091-7613(2001)029<0891:BFATCT>2.0.CO;2.

Alegret, L., Arenillas, I., Arz, J.A., Díaz, C., Grajales-Nishimura, J.M., Meléndez, A., Molina, E., Rojas, R., and Soria, A.R., 2005, Cretaceous-Paleogene boundary deposit at Loma Capiro, central Cuba: Evidence for the Chicxulub impact: Geology, v. 33, p. 721–724, doi: 10.1130/G21573.1.

Alvarez, L.W., Alvarez, W., Asaro, F., and Michel, H.V., 1980, Extraterrestrial cause for the Cretaceous-Tertiary extinction: Experimental results and theoretical interpretation: Science, v. 208, p. 1095–1108.

Asaro, H., Michel, H.V., Alvarez, W., Alvarez, L.W., Maddocks, R.F., and Burch, T., 1982, Iridium and other geochemical profiles near the Cretaceous-Tertiary boundary in a Brazos River section in Texas, *in* Maddocks, R.F., ed., Texas Ostracoda: Eighth International Symposium on Ostracoda: Houston, Texas, University of Houston, Department of Geoscience, p. 238–241.

Bajpai, S., and Prasad, G.V.R., 2000, Cretaceous age for Ir-rich Deccan intertrappean deposits: Palaeontological evidence from Anjar, western India: Journal of the Geological Society of London, v. 157, p. 257–260.

Barrera, E., and Keller, G., 1994, Productivity across the Cretaceous-Tertiary boundary in high latitudes: Geological Society of America Bulletin, v. 106, p. 1254–1266.

Berggren, W.A., Kent, D.V., Swisher, C.C., III, and Aubry, M.-P., 1995, A revised Cenozoic geochronology and chronostratigraphy, *in* Berggren, W., et al., eds., Geochronology, time scales and global stratigraphic correlation: Society for Sedimentary Geology Special Publication 54, p. 129–212.

Blum, J.D., Chamberlain, C.P., Hingston, M.P., Koeberl, C., Marin, L.E., Schuraytz, B.C., and Sharpton, V.L., 1993, Isotopic comparison of K-T boundary impact glass with melt rock from the Chicxulub and Manson impact structures: Nature, v. 364, p. 325–327, doi: 10.1038/364325a0.

Bohor, B.F., 1990, Shocked quartz and more: Impact signatures in Cretaceous/Tertiary boundary clays, *in* Sharpton, V.L., and Ward, P., eds., Global catastrophes in Earth history: Geological Society of America Special Paper 247, p. 335–347.

Bohor, B.F., and Glass, B.P., 1995, Origin and diagenesis of the K/T impact spherules from Haiti to Wyoming and beyond: Meteoritics, v. 30, p. 182–198.

Bourgeois, J., Hansen, T.A., Wiberg, P., and Kauffman, E.G., 1988, A tsunami deposit at the Cretaceous-Tertiary boundary in Texas: Science, v. 141, p. 567–570.

Brooks, R.R., Hoek, P.L., Peeves, R.D., Wallace, R.C., Johnston, J.H., Ryan, D.E., Holzbecher, J., and Collen, J.D., 1985, Weathered spheroids in a Cretaceous-Tertiary boundary shale at Woodside Creek, New Zealand: Geology, v. 13, p. 735–740.

Caron, M., 1985, Cretaceous planktonic foraminifera, *in* Bolli, H.M., et al., eds., Plankton stratigraphy: Cambridge, UK, Cambridge University Press, p. 17–86.

Chaussidon, M., Sigurdsson, H., and Metrich, N., 1996, Sulfur and boron isotope study of high-Ca impact glass from the K/T boundary: Constraints on source rocks, *in* Ryder, G., et al., eds., The Cretaceous-Tertiary event and other catastrophes in Earth history: Geological Society of America Special Paper 307, p. 253–262.

Chenet, A.-L., Quidelleur, X., Fluteau, F., and Courtillot, V., 2007, ^{40}K/^{40}Ar dating of the main Deccan large Igneous province: Further evidence of KTB age and short duration: Earth and Planetary Science Letters, v. 263, p. 1–15.

Culver, S.J., 2003, Benthic foraminifera across the Cretaceous-Tertiary (K-T) boundary: A review: Marine Micropaleontology, v. 47, p. 177–226, doi: 10.1016/S0377-8398(02)00117-2.

Dalrymple, B.G., Izett, G.A., Snee, L.W., and Obradovich, J.D., 1993, ^{40}Ar/^{39}Ar age spectra and total fusion ages of tektites from Cretaceous-Tertiary boundary sedimentary rocks in the Beloc formation, Haiti: U.S. Geological Survey Bulletin 2065, 20 p.

Debrabant, P., Fourcade, E.E., Chamley, H., Rocchia, R., Robin, E., Bellier, J.P., Gardin, S., and Thiebolt, F., 1999, Les argiles de la transition Cretace-Tertiaire au Guatemala, temoins d'un impact d'asteroide: Bulletin de la Société Géologique de France, v. 170, p. 643–660.

Ekdale, A.A., and Stinnesbeck, W., 1998, Ichnology of Cretaceous-Tertiary (K/T) boundary beds in northeastern Mexico: Palaios, v. 13, p. 593–602.

Elliott, W.C., 1993, Origin of the Mg smectite at the Cretaceous/Tertiary (K/T) boundary at Stevns Klint, Denmark: Clays and Clay Minerals, v. 41, p. 442–452.

Elliott, W.C., Aronson, J.L., Millard, H.T., and Gierlowski-Kordesch, E., 1989, The origin of clay minerals at the Cretaceous/Tertiary boundary in Denmark: Geological Society of America Bulletin, v. 101, p. 702–710, doi: 10.1130/0016-7606(1989)101<0702:TOOTCM>2.3.CO;2.

Gale, A., 2006, The Cretaceous-Tertiary boundary on the Brazos River, Falls County, Texas: Evidence for impact-induced tsunami sedimentation?: Proceedings of the Geologists' Association, v. 117, p. 1–13.

Ganapathy, R., Gartner, R.S., and Jiang, M.J., 1981, Iridium anomaly at the Cretaceous-Tertiary boundary in Texas: Earth and Planetary Science Letters, v. 54, p. 393–396.

Gilmore, J.S., Knight, J.D., Orth, C.L., Pillmore, C.L., and Tschudy, R.H., 1984, Trace element patterns at a non-marine Cretaceous-Tertiary boundary: Nature, v. 307, p. 224–228.

Glikson, A.Y., and Haines, P.W., 2005, Shoemaker memorial issue on the Australian impact record: 1997–2005 update: Introduction: Australian Journal of Earth Sciences, v. 52, p. 475–476, doi: 10.1080/08120090500170385.

Glikson, A.Y., Allen, C., and Vickers, J., 2004, Multiple 3.47 Ga-old asteroid impact fallout units, Pilbara Craton, Western Australia: Earth and Planetary Science Letters, v. 221, p. 383–396.

Gradstein, F.M., and Ogg, J.G., 2004, Geologic time scale 2004—Why, how, and where next?: Lethaia, v. 37, no. 2, p. 175–181, doi: 10.1080/002411 60410006483.

Grieve, R.A.F., 2005, Planetary and Space Science Centre, 2007, Earth impact database: http://www.unb.ca/passc/ImpactDatabase/ (September 2007).

Hallam, H., and Wignall, P.B., 1997, Mass extinctions and their aftermath: Oxford, UK, Oxford University Press, 320 p.

Hansen, T.A., Farrand, R.B., Montgomery, H.A., Billman, H.G., and Blechschmidt, G.L., 1987, Sedimentology and extinction patterns across the Cretaceous-Tertiary boundary interval in east Texas: Cretaceous Research, v. 8, no. 3, p. 229–252, doi: 10.1016/0195-6671(87)90023-1.

Hansen, T.A., Upshaw, B., III, Kauffman, E.G., and Gose, W., 1993, Patterns of molluscan extinction and recovery across the Cretaceous-Tertiary boundary in east Texas: Report on new outcrops: Cretaceous Research, v. 14, no. 6, p. 685–706, doi: 10.1006/cres.1993.1047.

Harris, R.S., Roden, M.F., Schroeder, P.A., Holland, M.S., Duncan, M.S., and Albin, E.F., 2004, Upper Eocene impact horizon in east-central Georgia: Geology, v. 32, p. 717–720, doi: 10.1130/G20562.1.

Harting, M., 2004, Zum Kreide/Tertiär-Übergang in NE-Mexiko: Geochemische Charakterisierung der Chicxulub-Impaktejekta: Karlsruhe, Germany, Karlsruhe University Press, 568 p.

Heymann, D., Yancey, T.E., Wolbach, W.S., Thiemens, M.H., Johnson, E.A., Roach, D., and Moecker, S., 1998, Geochemical markers of the Cretaceous-Tertiary boundary event at Brazos River, Texas, USA: Geochimica et Cosmochimica Acta, v. 62, p. 173–181, doi: 10.1016/S0016-7037(97)00330-X.

Izett, G.A., 1990, The Cretaceous/Tertiary boundary interval, Raton Basin, Colorado and New Mexico, and its content of shock-metamorphosed minerals: Evidence relevant to the K/T boundary impact-extinction hypothesis: Geological Society of America Special Paper 249, 100 p.

Izett, G.A., Dalrymple, G.B., and Snee, L.W., 1991, ^{40}Ar/^{39}Ar age of K-T boundary tektites from Haiti: Science, v. 252, p. 1539–1543.

Keller, G., 1989, Extended K/T boundary extinctions and delayed populational change in planktic foraminiferal faunas from Brazos River Texas: Paleoceanography, v. 4, no. 3, p. 287–332.

Keller, G., 1993, The Cretaceous-Tertiary boundary transition in the Antarctic Ocean and its global implications: Marine Micropaleontology, v. 21, p. 1–45, doi: 10.1016/0377-8398(93)90010-U.

Keller, G., 2001, The end-Cretaceous mass extinction: Year 2000 assessment: Journal Planetary and Space Science, v. 49, p. 817–830, doi: 10.1016/S0032-0633(01)00032-0.

Keller, G., 2002, *Guembelitria* dominated late Maastrichtian planktic foraminiferal assemblages mimic early Danian in central Egypt: Marine Micropaleontology, v. 47, p. 71–99.

Keller, G., 2005, Impacts, volcanism and mass extinctions: Random coincidence or cause and effect?: Australian Journal of Earth Sciences, v. 52, p. 725–757, doi: 10.1080/08120090500170393.

Keller, G., and Adatte, T., 2005, Basinward transport of Chicxulub ejecta by tsunami-induced backflow, La Popa basin, NE Mexico: Comment: Geology Online, September 2005, p. e88, doi: 10.1130/0091-7613(2005)31<e88:BTOCEB>2.0.CO;2.

Keller, G., and Pardo, A., 2004, Disaster opportunists *Guembelitrinidae*—Index for environmental catastrophes: Marine Micropaleontology, v. 53, p. 83–116, doi: 10.1016/j.marmicro.2004.04.012.

Keller, G., and Stinnesbeck, W., 2002, Slumping and a sandbar deposit at the Cretaceous-Tertiary boundary in the El Tecolote section (northeastern Mexico): An impact-induced sediment gravity flow: Comment: Geology, v. 30, p. 382–383, doi: 10.1130/0091-7613(2002)030<0382:SAASDA>2.0.CO;2.

Keller, G., D'Hondt, S., and Vallier, T.L., 1983, Multiple microtektite horizons in upper Eocene marine sediments: No evidence for mass extinctions: Science, v. 221, p. 150–152.

Keller, G., Stinnesbeck, W., and Lopez-Oliva, J.G., 1994a, Age, deposition and biotic effects of the Cretaceous/Tertiary boundary event at Mimbral, NE Mexico: Palaios, v. 9, p. 144–157.

Keller, G., Stinnesbeck, W., Adatte, T., MacLeod, N., and Lowe, D.R., 1994b, Field guide to Cretaceous-Tertiary boundary sections in northeastern Mexico: Houston, Lunar and Planetary Institute Contribution 827, 110 p.

Keller, G., Li, L., and MacLeod, N., 1995, The Cretaceous/Tertiary boundary stratotype section at El Kef, Tunisia: How catastrophic was the mass extinction?: Palaeogeography, Palaeoclimatology, Palaeoecology, v. 119, p. 221–254, doi: 10.1016/0031-0182(95)00009-7.

Keller, G., Lopez-Oliva, J.G., Stinnesbeck, W., and Adatte, T., 1997, Age, stratigraphy and deposition of near K/T siliciclastic deposits in Mexico: Relation to bolide impact?: Geological Society of America Bulletin, v. 109, p. 410–428, doi: 10.1130/0016-7606(1997)109<0410:ASADON>2.3.CO;2.

Keller, G., Adatte, T., Stinnesbeck, W., Stueben, D., and Berner, Z., 2001, Age, chemo- and biostratigraphy of Haiti spherule-rich deposits: A multi-event K-T scenario: Canadian Journal of Earth Sciences, v. 38, p. 197–227, doi: 10.1139/cjes-38-2-197.

Keller, G., Adatte, T., Stinnesbeck, W., Luciani, V., Karoui-Yaakoub, N., and Zaghbib-Turki, D., 2002a, Paleoecology of the Cretaceous/Tertiary mass extinction in planktonic foraminifera: Palaeogeography, Palaeoclimatology, Palaeoecology, v. 178, p. 257–297.

Keller, G., Adatte, T., Stinnesbeck, W., Affolter, M., Schilli, L., and Lopez-Oliva, J.G., 2002b, Multiple spherule layers in the late Maastrichtian of northeastern Mexico, *in* Koeberl, C., and MacLeod, K.G., eds., Catastrophic events and mass extinctions: Impacts and beyond: Geological Society of America Special Paper 356, p. 145–161.

Keller, G., Stinnesbeck, W., Adatte, T., and Stüben, D., 2003a, Multiple impacts across the Cretaceous-Tertiary boundary: Earth-Science Reviews, v. 62, p. 327–363, doi: 10.1016/S0012-8252(02)00162-9.

Keller, G., Stinnesbeck, W., Adatte, T., Holland, B., Stueben, D., Harting, M., de Leon, C., and de la Cruz, J., 2003b, Spherule deposits in Cretaceous-Tertiary boundary sediments in Belize and Guatemala: Geological Society [London] Journal, v. 160, p. 783–795.

Keller, G., Adatte, T., Stinnesbeck, W., Rebolledo-Vieyra, M., Urrutia Fucugauchi, J., Kramar, U., and Stueben, D., 2004a, Chicxulub crater predates K-T mass extinction: Proceedings of the National Academy of Sciences of the United States of America, v. 101, p. 3753–3758.

Keller, G., Adatte, T., Stinnesbeck, W., Stüben, D., Berner, Z., and Harting, M., 2004b, More evidence that the Chicxulub impact predates the K/T mass extinction: Meteoritics and Planetary Science, v. 39, p. 1127–1144.

Keller, G., Adatte, T., Harting, M., Berner, Z., Baum, G., Prauss, M., Tantawy, A.A., and Stueben, D., 2007, Chicxulub impact predates the K-T boundary: New evidence from Texas: Earth and Planetary Science Letters, v. 255, p. 339–356.

Keller, G., Adatte, T., Baum, G., and Berner, Z., 2008, Reply to "Chicxulub impact predates K-T boundary: New evidence from Brazos, Texas" comment by Schulte et al.: Earth and Planetary Science Letters (in press).

Kelley, P.S., and Gurov, E., 2002, Boltysh, another end Cretaceous impact: Meteoritics and Planetary Science, v. 37, p. 1031–1043.

Klaus, A., Norris, R.D., Kroon, D., and Smit, J., 2000, Impact-induced mass wasting at the K-T boundary: Blake Nose, western North Atlantic: Geology, v. 28, p. 319–322, doi: 10.1130/0091-7613(2000)28<319:IMWATK>2.0.CO;2.

Koeberl, C., Sharpton, V.L., Schuraytz, B.C., Shirley, S.B., Blum, J.D., and Marin, L.E., 1994, Evidence for a meteoric component in impact melt rock from the Chicxulub structure: Geochimica et Cosmochimica Acta, v. 56, p. 2113–2129.

Kucera, M., and Malmgren, B.A., 1998, Terminal Cretaceous warming event in the mid-latitude South Atlantic Ocean: Evidence from poleward migration of *Contusotruncana contusa* (planktonic foraminifera) morphotypes: Palaeogeography, Palaeoclimatology, Palaeoecology, v. 138, p. 1–15.

Lawton, T.F., Shipley, K.W., Aschoff, J.L., Giles, K.A., and Vega, F.J., 2005, Basinward transport of Chicxulub ejecta by tsunami-induced backflow, La Popa basin, northeastern Mexico, and its implication for distribution of impact-related deposits flanking the Gulf of Mexico: Geology, v. 33, p. 81–84, doi: 10.1130/G21057.1.

Li, L., and Keller, G., 1998a, Maastrichtian climate, productivity and faunal turnovers in planktic foraminifera in South Atlantic DSDP Site 525A and 21: Marine Micropaleontology, v. 33, p. 55–86.

Li, L., and Keller, G., 1998b, Abrupt deep-sea warming at the end of the Cretaceous: Geology, v. 26, p. 995–999, doi: 10.1130/0091-7613(1998)026<0995:ADSWAT>2.3.CO;2.

Lindenmaier, F., 1999, Geologie und Geochemie an der Kreide/Tertiär-Grenze im Nordosten von Mexiko: Diplomarbeit, Institut für Regionale Geologie, Karlsruhe, 90 p.

Lopez-Oliva, J.G., and Keller, G., 1996, Age and stratigraphy of near-K/T boundary clastic deposits in NE Mexico, *in* Ryder, G., et al., eds., The Cretaceous-Tertiary event and other catastrophes in Earth history: Geological Society of America Special Paper 307, p. 227–242.

Ma, X.P., and Bai, S.L., 2002, Biological, depositional, microspherule and geochemical records of the Frasnian/Famennian boundary beds, South China: Palaeogeography, Palaeoclimatology, Palaeoecology, v. 181, p. 325–346, doi: 10.1016/S0031-0182(01)00484-9.

MacLeod, K.G., Whitney, D.L., Huber, B.T., and Koeberl, C., 2006, Impact and extinction in remarkably complete Cretaceous-Tertiary boundary sections from Demerara Rise, tropical western North Atlantic: Geological Society of America Bulletin, v. 119, p. 101–115, doi: 10.1130/B25955.1.

MacLeod, N., 1998, Impacts and marine invertebrate extinctions, *in* Grady, M.M., et al., eds., Meteorites: Flux with time and impact effects: Geological Society [London] Special Publication 140, p. 217–246.

Martínez-Ruiz, F., Ortega-Huertas, M., Palomo-Delgado, I., and Smit, J., 2002, Cretaceous-Tertiary boundary at Blake Nose (Ocean Drilling Program Leg 171B): A record of the Chicxulub impact ejecta, *in* Koeberl, C., and MacLeod, K.G., eds., Catastrophic events and mass extinctions: Impacts and beyond: Geological Society of America Special Paper 356, p. 189–200.

Maurrasse, F.J.-M.R., Lamolda, M.A., Aguado, R., Peryt, D., and Sen, G., 2005, Spatial and temporal variations of the Haitian K/T boundary record: Implications concerning the event or events: Journal of Iberian Geology, v. 31, no. 1, p. 113–133.

Méon, H., 1990, Palynologic studies of the Cretaceous/Tertiary boundary interval at El Kef outcrop, northwestern Tunisia: Paleogeographic implications: Review of Palaeobotany and Palynology, v. 65, p. 85–94.

Michel, H.V., Asaro, F., and Alvarez, W., 1991, Geochemical study of the Cretaceous-Tertiary boundary region at Hole 752B, *in* Weissel, J., et al., eds., Proceedings of the Ocean Drilling Program, Scientific Results, Volume 121: College Station, Texas, Ocean Drilling Program, p. 415–422.

Molina, E., Alegret, L., Arenillas, I., Arz, J.A., Gallala, N., Hardenbol, J., von Salis, K., Steurbaut, E., Vandenberghe, N., Zaghbib-Turki, D., 2006, The global boundary stratotype section and point for the base of the Danian stage (Paleocene, Paleogene, "Tertiary," Cenozoic) at El Kef, Tunisia—original definition and revision: Episodes, v. 29, no. 4, p. 263–273.

Montanari, A., and Koeberl, C., 2000, Impact stratigraphy: Lecture notes in earth sciences, 93: Heidelberg, Germany, Springer, 364 p.

Morgan, J., Lana, C., Kearsley, A., Coles, B., Belcher, C., Montanari, S., Díaz-Martínez, E., Barbosa, A., and Neumann, V., 2006, Analyses of shocked quartz at the global K-P boundary indicate an origin from a single, high-angle, oblique impact at Chicxulub: Earth and Planetary Science Letters, v. 251, p. 264–279, doi: 10.1016/j.epsl.2006.09.009.

Napier, W.M., 2001, The influx of comets and their debris, *in* Peucker-Ehrenbrink, B., and Schmitz, B., eds., Accretion of extraterrestrial matter throughout Earth's history: Dordrecht, Kluwer, p. 51–74.

Napier, W.M., 2006, Evidence for cometary bombardment episodes: Monthly Notices of the Royal Astronomical Society, v. 366, p. 977–982.

Nordt, L., Atchley, S., and Dworkin, S., 2003, Terrestrial evidence for two greenhouse events in the latest Cretaceous: GSA Today, v. 13, no. 12, p. 4–9, doi: 10.1130/1052-5173(2003)013<4:TEFTGE>2.0.CO;2.

Norris, R.D., Huber, B.T., and Self-Trail, J., 1999, Synchroneity of the K-T oceanic mass extinction and meteorite impact: Blake Nose, western North Atlantic: Geology, v. 27, p. 419–422, doi: 10.1130/0091-7613(1999)027<0419:SOTKTO>2.3.CO;2.

Norris, R.D., Firth, J., Blusztajn, J.S., and Ravizza, G., 2000, Mass failure of the North Atlantic margin triggered by the Cretaceous-Paleogene bolide impact: Geology, v. 28, p. 1119–1122.

Olsson, R.K., Miller, K.G., Browning, J.V., Habib, D., and Sugarmann, P.J., 1997, Ejecta layer at the Cretaceous-Tertiary boundary, Bass River, New Jersey (Ocean Drilling Program Leg 174AX): Geology, v. 25, p. 759–762, doi: 10.1130/0091-7613(1997)025<0759:ELATCT>2.3.CO;2.

Orth, C.J., Gilmore, J.S., Knight, J.D., Pillmore, C.L., Tschudy, R.H., and Fassett, J.E., 1982, Iridium abundance measurements across the Cretaceous/Tertiary boundary in the San Juan and Raton Basins of northern New Mexico, *in* Silver, L.T., and Schultz, P.H., eds., Geological Implications of Impacts of Large Asteroids and Comets on the Earth: Geological Society of America Special Paper 190, p. 423–433.

Pardo, A., Ortiz, N., and Keller, G., 1996, Latest Maastrichtian and K/T boundary foraminiferal turnover and environmental changes at Agost, Spain, *in* MacLeod, N., and Keller, G., eds., The Cretaceous-Tertiary mass extinction: Biotic and environmental effects: New York, Norton Press, p. 157–191.

Poag, C.W., Plescia, J.B., and Molzer, P.C., 2002, Ancient impact structures on modern continental shelves: The Chesapeake Bay, Montagnais, and Toms Canyon craters, Atlantic margin of North America: Deep Sea Research Part II: Topical Studies in Oceanography, v. 49, no. 6, p. 1081–1102, doi: 10.1016/S0967-0645(01)00144-8.

Rocchia, R., Robin, E., Froget, L., and Gayraud, J., 1996, Stratigraphic distribution of extraterrestrial markers at the Cretaceous-Tertiary boundary in the Gulf of Mexico area: Implications for the temporal complexity of the event, *in* Ryder, G., et al., eds., The Cretaceous-Tertiary event and other catastrophes in Earth history: Geological Society of America Special Paper 307, p. 279–286.

Sawlowicz, Z., 1993, Iridium and other platinum-group elements as geochemical markers in sedimentary environments: Palaeogeography, Palaeoclimatology, Palaeoecology, v. 104, p. 253–270.

Schilli, L., 2000, Etude de la limite K-T dans la région de la Sierrita, Nuevo Leon, Mexique [M.S. thesis]: Neuchatel, Switzerland, Geological Institute, University of Neuchatel, 138 p.

Schulte, P., 1999, Geologisch-sedimentologische Untersuchungen des Kreide/Tertiär (K/T)-Übergangs im Gebiet zwischen La Sierrita und El Toro, Nuevo Leon, Mexiko: Diplomarbeit, Universität Karlsruhe, Institut für Regionale Geologie, Karlsruhe, Germany, 134 p.

Schulte, P., 2003, The Cretaceous-Paleogene transition and Chicxulub impact ejecta in the northwestern Gulf of Mexico: Paleoenvironments, sequence stratigraphic setting, and target lithologies [Ph.D. thesis]: Karlsruhe, Germany, University of Karlsruhe, 204 p.

Schulte, P., Stinnesbeck, W., Steuben, D., Kramar, U., Berner, Z., Keller, G., and Adatte, T., 2003, Multiple slumped? Chicxulub ejecta deposits with iron-rich spherules and quenched carbonates from the K/T transition, La Sierrita, NE Mexico: Journal of International Earth Sciences, v. 92, p. 114–142.

Schulte, P., Speijer, R.P., Mai, H., and Kontny, A., 2006, The Cretaceous-Paleogene (K-P) boundary at Brazos, Texas: Sequence stratigraphy, depositional events and the Chicxulub impact: Sedimentary Geology, v. 184, p. 77–109, doi: 10.1016/j.sedgeo.2005.09.021.

Schulte, P., Spejier, R.P., Brinkhuis, H., Kontny, A., Galeotti, S., and Smit, J., 2008, Comment on the paper "Chicxulub impact predates K-T boundary: New evidence from Brazos, Texas": Earth and Planetary Science Letters (in press).

Sepkoski, J.J., Jr., 1996, Patterns of Phanerozoic extinction: A perspective from global databases, *in* Walliser, O.H., ed., Global events and event stratigraphy: Berlin, Springer-Verlag, p. 35–52.

Sigurdsson, H., Bonté, P., Turpin, L., Chaussidon, M., Metrich, N., Steinberg, M., Pradel, P., and D'Hondt, S., 1991, Geochemical constraints on source region of Cretaceous/Tertiary impact glasses: Nature, v. 353, p. 839–842, doi: 10.1038/353839a0.

Sigurdsson, H., Leckie, R.M., and Acton, G.D., 1997, Caribbean volcanism, Cretaceous/Tertiary impact, and ocean climate history: Synthesis of Leg 165: Proceedings of the Ocean Drilling Program, Initial Reports, Volume 165: College Station, Texas, Ocean Drilling Program, p. 377–400.

Smit, J., 1999, The global stratigraphy of the Cretaceous-Tertiary boundary impact ejecta: Annual Review of Earth and Planetary Sciences, v. 27, p. 75–113.

Smit, J., Montanari, A., Swinburne, N.H.M., Alvarez, W., Hildebrand, A., Margolis, S.V., Claeys, P., Lowerie, W., and Asaro, F., 1992, Tektite-bearing deep-water clastic unit at the Cretaceous-Tertiary boundary in northeastern Mexico: Geology, v. 20, p. 99–104.

Smit, J., Roep, T.B., Alvarez, W., Montanari, A., Claeys, P., Grajales-Nishimura, J.M., and Bermudez, J., 1996, Coarse-grained clastic sandstone complex at the K/T boundary around the Gulf of Mexico: Deposition by tsunami waves induced by the Chicxulub impact, *in* Ryder, G., et al., eds., The Cretaceous-Tertiary event and other catastrophes in Earth history: Geological Society of America Special Paper 307, p. 151–182.

Smit, J., van der Gaast, S., and Lustenhouwer, W., 2004, Is the transition to post-impact rock complete? Some remarks based on XRF scanning electron microprobe and thin section analyses of the Yaxcopoil-1 core: Meteoritics and Planetary Science, v. 39, p. 1113–1126.

Soria, A.R., Llesa, C.L., Mata, M.P., Arz, J.A., Alegret, L., Arenillas, I., and Melendez, A., 2001, Slumping and a sandbar deposit at the Cretaceous-Tertiary boundary in the El Tecolote section (northeastern Mexico): An impact-induced sediment gravity flow: Geology, v. 29, p. 231–234, doi: 10.1130/0091-7613(2001)029<0231:SAASDA>2.0.CO;2.

Stewart, S.A., and Allen, J.P., 2002, A 20-km-diameter multi-ringed impact structure in the North Sea: Nature, v. 418, p. 520–523, doi: 10.1038/nature00914.

Stinnesbeck, W., Barbarin, J.M., Keller, G., Lopez-Oliva, J.G., Pivnik, D., Lyons, J., Officer, C., Adatte, T., Graup, G., Rocchia, R., and Robin, E., 1993, Deposition of channel deposits near the Cretaceous-Tertiary boundary in northeastern Mexico: Catastrophic or "normal" sedimentary deposits?: Geology, v. 21, p. 797–800.

Stinnesbeck, W., Keller, G., Adatte, T., Lopez-Oliva, J.G., and MacLeod, N., 1996, Cretaceous-Tertiary boundary clastic deposits in northeastern Mexico: Impact tsunami or sea level lowstand?, *in* MacLeod, N., and Keller, G., eds., Cretaceous-Tertiary mass extinctions: Biotic and environmental changes: New York, W.W. Norton and Company, p. 471–518.

Stinnesbeck, W., Keller, G., de la Cruz, J., de Leon, C., MacLeod, N., and Whittaker, J.E., 1997, The Cretaceous-Tertiary transition in Guatemala: Limestone breccia deposits from the South Peten basin: Geologische Rundschau, v. 86, p. 686–709.

Stinnesbeck, W., Keller, G., Schulte, P., Stueben, D., Berner, Z., Kramar, U., and Lopez-Oliva, J.G., 2002, The Cretaceous-Tertiary (K/T) boundary transition at Coxquihui, State of Veracruz, Mexico: Evidence for an early Danian impact event?: Journal of South American Research, v. 15, p. 497–509, doi: 10.1016/S0895-9811(02)00079-2.

Stinnesbeck, W., Schafhauser, A., and Götz, S., 2005, Basinward transport of Chicxulub ejecta by tsunami-induced backflow, La Popa basin, NE Mexico: Comment: Geology Online, September 2005, p. e86, doi: 10.1130/0091-7613(2005)31<e86:BTOCEB>2.0.CO;2.

Stüben, D., Kramer, U., Berner, Z., Eckhardt, J.D., Stinnesbeck, W., Keller, G., Adatte, T., and Heide, K., 2002, Two anomalies of platinum group elements above the Cretaceous-Tertiary boundary at Beloc, Haiti: Geochemical context and consequences for the impact scenario, *in* Koeberl, C., and MacLeod, K.G., eds., Catastrophic events and mass extinctions: Impacts and beyond: Geological Society of America Special Paper 356, p. 163–188.

Stüben, D., Kramar, U., Harting, M., Stinnesbeck, W., and Keller, G., 2005, High-resolution geochemical record of Cretaceous-Tertiary boundary sections in Mexico: New constraints on the K/T and Chicxulub events: Geochimica et Cosmochimica Acta, v. 69, p. 2559–2579, doi: 10.1016/j.gca.2004.11.003.

Swisher, C.C., Grajales-Nishimura, J.M., Montanari, A., Margolis, S.V., Claeys, P., Alvarez, W., Renne, P., Cedillo-Pardo, E., Maurrasse, F.J.-M.R., Curtis, G.H., Smit, J., and McWilliams, M.O., 1992, Coeval $^{40}Ar/^{39}Ar$ ages of 65 million years ago from Chicxulub crater melt rock and Cretaceous-Tertiary boundary tektites: Science, v. 257, p. 954–958.

Tantawy, A.A.A., 2003, Calcareous nannofossil biostratigraphy and paleoecology of the Cretaceous-Paleogene transition in the central Eastern Desert of Egypt: Marine Micropaleontology, v. 47, p. 323–356, doi: 10.1016/S0377-8398(02)00135-4.

Tredoux, M., de Wit, M.J., Hart, R.J., Linsay, N.M., Verhagen, B., and Sellschop, J.P.F., 1988, Chemostratigraphy across the Cretaceous-Tertiary boundary and a critical assessment of the iridium anomaly: Journal of Geology, v. 97, p. 585–605.

Tsujita, C., 2001, Coincidence in the geological record: From clam clusters to Cretaceous catastrophe: Canadian Journal of Earth Sciences, v. 38, p. 271–292, doi: 10.1139/cjes-38-2-271.

Yancey, T.E., 1996, Stratigraphy and depositional environments of the Cretaceous/Tertiary boundary complex and basal section, Brazos River, Texas: Gulf Coast Association of Geological Societies Transactions, v. 46, p. 433–442.

Wang, K., Attrep, M., Jr., and Orth, C.J., 1993, Global iridium anomaly, mass extinction and redox change at the Devonian-Carboniferous boundary: Geology, v. 21, p. 1071–1074.

Wignall, P.B., Newton, R., and Brookfield, M.E., 2005, Pyrite framboid evidence for oxygen-poor deposition during the Permian-Triassic crisis in Kashmir: Palaeogeography, Palaeoclimatology, Palaeoecology, v. 216, p. 183–188, doi: 10.1016/j.palaeo.2004.10.009.

Manuscript Accepted by the Society 10 July 2007

Printed in the USA

The Geological Society of America
Special Paper 437
2008

Impact spherule-bearing, Cretaceous-Tertiary boundary sand body, Shell Creek stratigraphic section, Alabama, USA

David T. King Jr.
Department of Geology, Auburn University, Auburn, Alabama 36849-5305, USA

Lucille W. Petruny
Astra-Terra Research, Auburn, Alabama 36831-3323, USA

ABSTRACT

At the Shell Creek stratigraphic section, Wilcox County, Alabama, a 35–75-cm-thick, Cretaceous-Tertiary (K-T) boundary sand body crops out over an area of ~200 m². This sand body consists of (1) a basal impact spherule-bearing, coarse to medium sand and (2) an overlying fine sand with hummocky-type cross-lamination. This K-T boundary sand body probably represents postimpact, shelf sedimentation events involving (1) gravity-driven resedimentation of reworked impact spherule-bearing sands and (2) energetic wave reworking of the impact spherule-bearing, gravity-driven deposits or other subsequently deposited sands. Most impact spherules from Shell Creek are spherically shaped grains (~1 mm in diameter) that are now hollow, or were hollow prior to secondary calcite filling. Most impact spherules from Shell Creek consist of an outer shell, which is composed of smectitic clays, and an inner region of open space or sparry calcite. Most of these impact spherules still retain vesicles that attest to their former molten condition. This stratigraphic section represents the most easterly U.S. Gulf Coastal Plain occurrence of abundant impact spherules in a Cretaceous-Tertiary boundary sand body.

Keywords: impact spherules, Cretaceous-Tertiary boundary, Chicxulub, Alabama, Shell Creek.

INTRODUCTION

Chicxulub impact-related sand bodies (i.e., Cretaceous-Tertiary [K-T] boundary sand bodies) that include impact spherule-bearing beds are known from ~20 localities spanning parts of the U.S. Western Interior, the U.S. Gulf and Atlantic Coastal Plains, the eastern Mexican Coastal Plain, and some DSDP and ODP Atlantic and Gulf of Mexico drill sites (e.g., Sites 536 and 540; Fig. 1). At the Shell Creek stratigraphic section, Wilcox County, Alabama, a K-T boundary sand body that is intermediate in character between sites in the U.S. Atlantic Coastal Plain and the Mexican Coastal Plain crops out. Shell Creek is located on the Catherine, Alabama, 7.5 min U.S. Geological Survey quadrangle map. The exact location of this outcrop, which is located at a ford in the creek, is withheld in this report at the request of the landowners.

King, D.T., Jr., and Petruny, L.W., 2008, Impact spherule-bearing, Cretaceous-Tertiary boundary sand body, Shell Creek stratigraphic section, Alabama, USA, *in* Evans, K.R., Horton, J.W., Jr., King, D.T., Jr., and Morrow, J.R., eds., The Sedimentary Record of Meteorite Impacts: Geological Society of America Special Paper 437, p. 179–187, doi: 10.1130/2008.2437(10). For permission to copy, contact editing@geosociety.org.

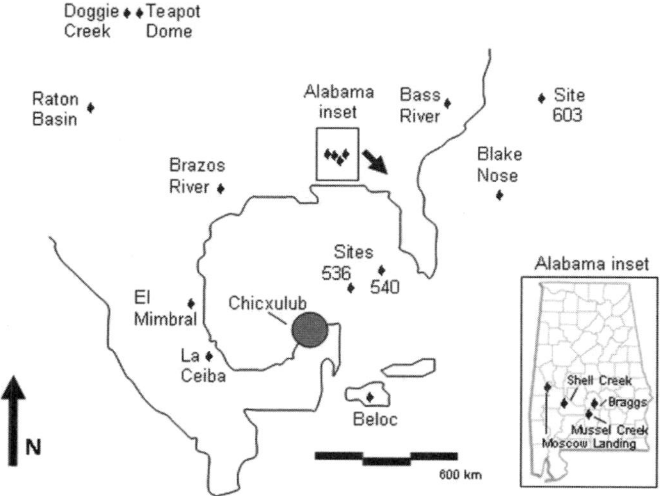

Figure 1. Location of Cretaceous-Tertiary (K-T) boundary sections that contain impact ejecta, including those mentioned in the text, and locations of K-T boundary sections in Alabama (after Smit et al., 1996). The two K-T impact spherule-bearing sites in Alabama (see inset) are Moscow Landing (very rare impact spherules) and Shell Creek (abundant impact spherules). K-T boundary sites with no ejecta are Mussel Creek (noncatastrophic boundary; Savrda, 1993) and Braggs (erosional contact; Mancini et al., 1989).

The Shell Creek K-T sand body is relatively thin (<1 m), which is like the K-T boundary sand bodies at some U.S. Atlantic Coastal Plain sites. However, Shell Creek has a textural dichotomy of two distinctly different sand units, which is like some of the eastern Mexican Coastal Plain sections and the sections at Brazos River, Texas (Smit et al., 1996; Smit, 1999). The high density of impact spherules (i.e., the number of impact spherules per cubic centimeter of rock and/or within the plane of a typical thin section) in the lower bed at Shell Creek is noteworthy because it is much like the density seen in Haitian sections (e.g., the Beloc [Haiti] section; Bohor and Glass, 1995). The preservation state of the Shell Creek impact spherules is also notable. Even though the impact spherules are altered to smectitic clays and thus no longer glassy, the alteration has not obliterated fine details of the former glass such as vesicles and bubble pits.

Shell Creek impact spherules represent the most easterly outcrop occurrence of Chicxulub impact ejecta within the Chicxulub impact spherule-strewn field and, similarly, within the U.S. Gulf Coastal Plain. Coeval U.S. Gulf Coastal Plain impact spherules are reported from several Brazos River sections, Falls County, Texas, which are located ~500 km west of Shell Creek (Smit et al., 1996), and the river sections at Moscow Landing, Sumter County, Alabama, which is located ~28 km west-northwest of Shell Creek (Smit et al., 1996). The size and density of impact spherules as well as the thickness of the host sandy layer are comparable to the K-T boundary section at Beloc, Haiti, which was on about the same paleolongitude as Shell Creek at the time of the Chicxulub impact (Smit et al., 1996; Smit, 1999). This density differs markedly from other basal Clayton sections of the K-T boundary located to the east of Shell Creek, which do not contain any impact spherules or other known K-T ejecta (i.e., Mussel Creek section, Butler County, ~50 km east of Shell Creek, and Braggs section, Lowndes County, ~55 km east of Shell Creek; Savrda, 1993; Smit, 1999). Few known K-T ejecta locales, except perhaps El Mimbral and La Ceiba, both in eastern Mexico, have a higher density of Chicxulub impact spherule material in the rock than what is found at Shell Creek (data in Smit, 1999).

In this paper, the term *impact spherule* is used for the splash-form grains found at Shell Creek, which are interpreted to have been formerly molten droplets of glass, i.e., "pseudomorphs" of impact spherules, but are now entirely or almost entirely altered to more stable mineral phases such as smectitic clays and are, in most instances, internally filled by sparry calcite that occupies a former interior void. We think that most if not all the impact spherules found at Shell Creek were originally microtektites; however, owing to the poor state of preservation of some of these grains, we cannot rule out that some may have been microkrystites (Simonson and Glass, 2004).

In this paper, we use the term *Cretaceous-Tertiary* (or *K-T) boundary sand body* to mean the sand body in a given area that occupies a stratigraphic position between the oldest Danian strata and the youngest Maastrichtian strata in that area, and that contains Chicxulub ejecta and/or sedimentary deposits that are related to K-T ejecta-bearing deposits. Even though a K-T boundary sand body occupies the position between two stratigraphic units and is commonly quite different texturally, it is considered to be part of the overlying Danian unit for nomenclatural and mapping purposes, and is not a separate, formal stratigraphic unit.

SHELL CREEK STRATIGRAPHY

The outcrop at Shell Creek consists of an area of ~200 m^2 of flat-lying rock where one can walk around on the top of the beds and see their physical relationships. They form a series of wide steps leading down from the Danian fine clastic beds, over the K-T boundary sand-body interval, and onto the Maastrichtian chalks. The water in Shell Creek, which can flow very rapidly during sudden, heavy rain events, keeps the outcrop largely swept clean of debris, yet the water has not significantly eroded away the layers during the four-year period that we have visited the site. Despite the presence of other creeks in the area that cross the map level of the K-T boundary, there is only one outcrop of the K-T boundary in the Shell Creek area.

At Shell Creek, the upper surface of the late Maastrichtian Prairie Bluff Chalk, the uppermost Cretaceous unit in the area, is truncated by basal, early Paleocene sands of the Clayton Formation. These basal sands have been interpreted in a sequence-stratigraphic context as basal transgressive lags (e.g., Mancini et al., 1989), but more recent work has shown that the basal Clayton sands in several places in the Alabama Coastal Plain, including

Shell Creek, are part of the global set of K-T boundary sand-body deposits (Smit et al., 1996; Smit, 1999). Pitakpaivan et al. (1994) were the first to recognize that pseudomorphs of some K-T impact spherules were present at Shell Creek, but they did not mention the overlying cross-laminated sand unit. In two brief preliminary reports, we recognized that there are two distinct depositional units within the K-T boundary sand body at Shell Creek, and we noted that they deserve further study (King and Petruny, 2004, 2005).

At Shell Creek, the basal one meter (±) of the youngest Danian unit in the area, the Clayton Formation, consists of a basal sand body, which is made up of (1) an impact spherule-bearing coarse to medium sand bed and (2) an overlying low-angle, cross-laminated fine sand unit (Fig. 2). This K-T boundary sand body ranges in thickness from ~35 to 75 cm and rests directly on top of the Prairie Bluff Chalk. There is minor (i.e., centimeter-scale), irregular relief on the Prairie Bluff's upper surface. In particular, the tops of burrows that were evidently open to the sediment-water interface at the time of emplacement of the basal sand body are filled with K-T boundary impact spherule-bearing sands. The overlying cross-laminated, fine sand unit lies on the basal impact spherule-bearing sand with a nearly flat contact that is not gradational. The upper surface of the fine sand bed has several centimeters of relief due to the presence of irregular sandy hummocks and swales. Above the sandy hummocks and swales of the upper fine sand bed, the typical tan, sandy marls of the Clayton Formation rest conformably. Unfortunately, owing to the topography and vegetation, this sand body cannot be traced laterally beyond the valley of Shell Creek at the stream ford crossing.

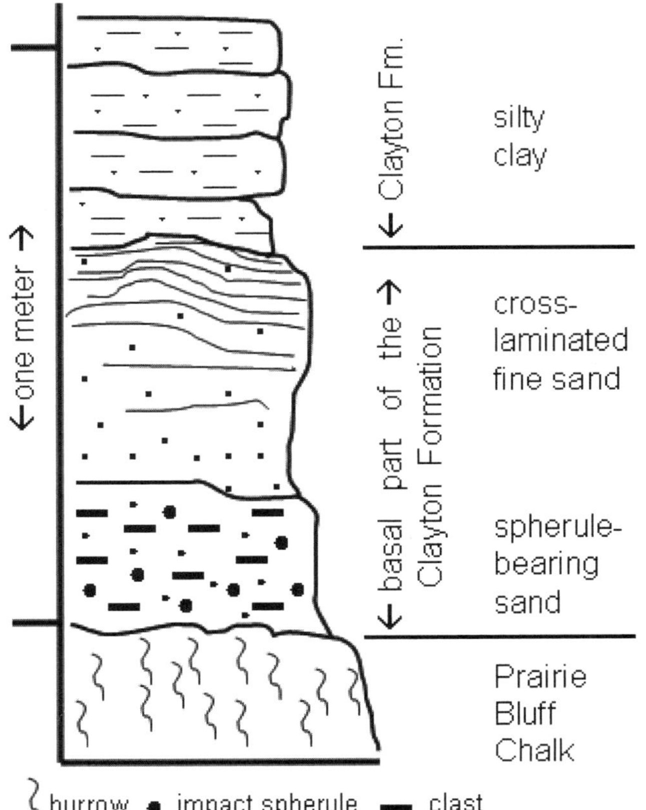

Figure 2. Measured section of the Cretaceous-Tertiary (K-T) boundary at Shell Creek, Wilcox County, Alabama.

The biostratigraphic age of the Prairie Bluff Chalk at Shell Creek is confirmed as uppermost Maastrichtian (i.e., nannofossil zone NC23) by the presence of index nannotaxa *Micula murus* and *Nephrolithus frequens* (Mancini et al., 1989; Pitakpaivan et al., 1994). Further, the local absence of index nannotaxon *Micula prinsii* suggests a minor disconformity at the top of the Prairie Bluff Chalk at Shell Creek (Mancini et al., 1989; Pitakpaivan et al., 1994). A detailed biostratigraphic study of the K-T contact in this area, including the Shell Creek stratigraphic section, reported Danian planktonic foraminifera from the basal Clayton Formation (Mancini et al., 1989). However, these Danian taxa apparently occur above the K-T boundary sand body at Shell Creek. The first sample level above the top of the Prairie Bluff Chalk, as reported by Mancini et al. (1989), contained "very rare" (meaning "one to three individuals") early Danian planktonic foraminifera of the *Subbotina trinidadensis* interval zone (i.e., *Globoconusa daubjergensis*, *Subbotina pseudobulloides*, *Subbotina triloculinoides*, and *Woodringina claytonensis*), which were mixed with ten species of "rare to very rare," reworked Maastrichtian planktonic foraminifera. According to their figure, this sample level is ~1.4 m above the top of the Prairie Bluff, which means it came from the Clayton fine clastics a few centimeters above the top of the cross-laminated sand unit of the K-T boundary sand body described in this paper. These data are not problematic, because finding reworked Maastrichtian taxa in basal strata overlying this K-T sand body would be expected. Basal Clayton sands assigned to the early Danian planktonic foraminiferal *Subbotina pseudobulloides* interval zone overlie the basal Clayton K-T sand body at Shell Creek (Mancini et al., 1989; Pitakpaivan et al., 1994).

Impact Spherule Bed

At Shell Creek, there is a sharp yet slightly irregular contact between the gray, bioturbated shelfal "chalk" (actually a fossiliferous gray marl) of the Prairie Bluff and an overlying gray-green, impact spherule-rich, coarse to medium sand layer (Figs. 3 and 4), which has a typical thickness of ~15–25 cm. The sharp contact is irregular at a fine scale, because of a few centimeters of relief on the surface and because in a few spots, impact spherule-bearing sands occupy burrows within the uppermost Maastrichtian gray marl (Fig. 4). The lowermost few centimeters of the impact spherule-rich layer above the gray marl contain significant amounts of hematite cement (Fig. 5), thus making the rock distinctively red in this zone. However, the upper part of this unit is gray-green and mainly cemented by clay and calcite. The impact spherule-rich layer is not graded in any obvious way and is relatively densely packed with impact spherules, which make up as much as 10 vol% of the grains in some parts of the rock.

Within some low places on top of the Prairie Bluff, a discontinuous, thin conglomeratic zone occurs just at the lower contact. This thin, conglomeratic zone is characterized by abundant, ~3 cm diameter discoid clasts made of Prairie Bluff gray marl (Fig. 6). The orientation of platy mica grains within the gray marl clasts are parallel to the axial plane of the clasts' discoidal shapes, thus indicating that the clasts are ripped up stratal units of the gray marl.

The coarse to medium sands comprising the matrix of the impact spherule bed are highly quartzose (>75%) but also contain small (<5 mm) glauconitic, phosphatic, and carbonate clasts and fragmental fossil debris throughout this unit. Mixed with the quartzose sand is a significant component of feldspar and mica. No shocked silicate grains have been found in thin sections of these sediments. The impact spherule-rich bed contains some irregular rip-up clasts of gray marl (<1 cm) and 1 to 2 cm long, dark phosphatic molds of megafossils (including snails, straight cephalopods, and pelecypods). There are no burrows within the impact spherule-bearing bed, nor do any burrows penetrate into the impact spherule-bearing bed from overlying layers.

These impact spherule-bearing sands and their overlying fine, cross-laminated sands are texturally "out of place" within the interpreted shelfal depositional environment of the gray marl (i.e., the Prairie Bluff Chalk) and overlying fine clastics of

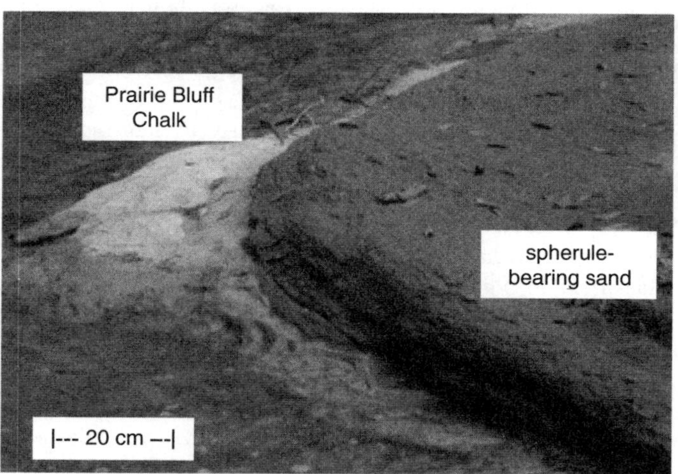

Figure 3. Sharp contact between the gray, bioturbated marl of the Prairie Bluff Chalk and overlying gray-green, impact spherule-bearing, coarse to medium sand.

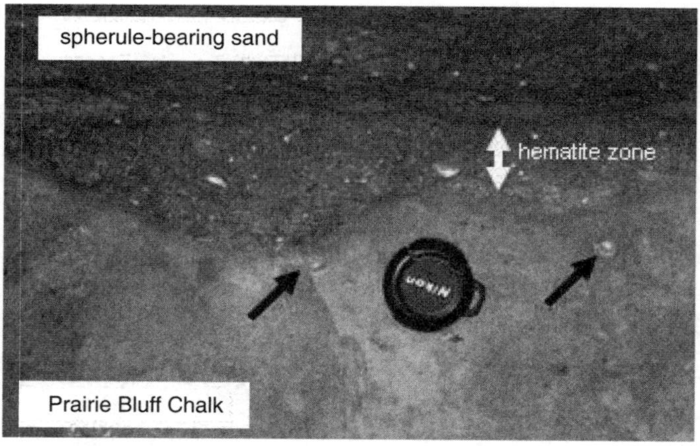

Figure 4. Impact spherule-bearing sands filling burrows (arrows) at the top of the Maastrichtian gray marl. Lens cap diameter is 2.25 cm.

Figure 5. Hematite zone in the basal few centimeters of the impact spherule-bearing sand. Hammer is 20 cm long.

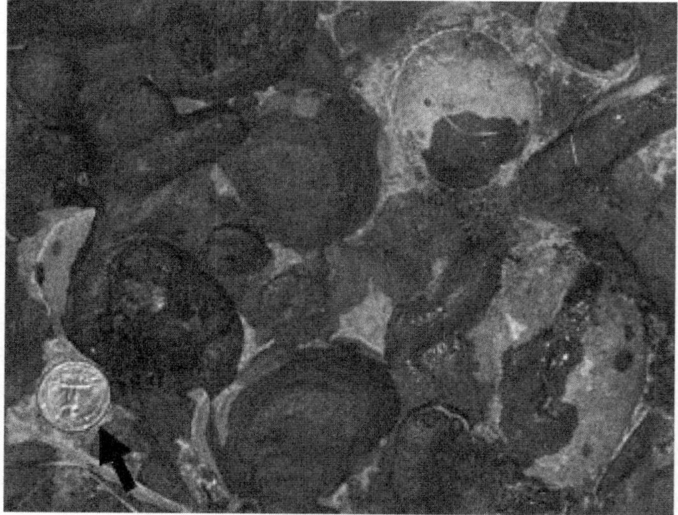

Figure 6. Top view of the thin conglomeratic zone composed of ~3 cm discoid, gray marl clasts (from the Prairie Bluff Chalk), which occurs at the base of the impact spherule-bearing sand. U.S. quarter-dollar coin for scale (arrow).

the Clayton Formation, and thus constitute a separate allochthonous depositional unit (Smit et al., 1996; Smit, 1999; King and Petruny, 2005). This unit is about the same thickness as, and has some of the characteristics of, the Brazos River K-T boundary sand body, namely units C and D, as described by Hansen et al. (1987). Similar characteristics include clasts from underlying layers, mixed impact-spherule and coastal clastic lithologies, and admixed reworked megafossils and phosphatic clasts. The Shell Creek sand body lacks grading, whereas the comparable beds at Brazos River are graded (Hansen et al., 1987).

Impact Spherules

Intact Shell Creek impact spherules are mostly in the size range of ~0.5–3 mm, averaging slightly less than 1 mm, and they have many different shapes, including spheres, spheroids, dumbbells, teardrops, disks, and others (Fig. 7). By far, the most common shape is spherical. Slightly prolate spheroids and teardrops (which may be broken dumbbells) are next most common shapes; other shapes are rare. In thin section, it is evident that some impact spherules are not intact (Fig. 8). They appear

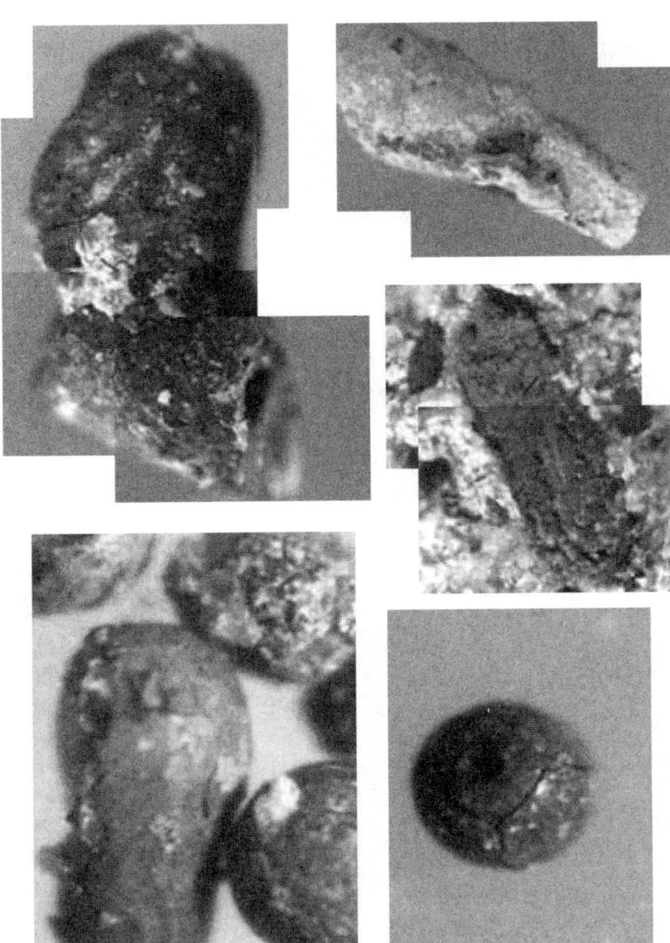

Figure 7. Examples of the variety of shapes of impact spherules seen at Shell Creek, Alabama. Clockwise from upper left: bent and elongate (1.9 mm); teardrop-shaped (1.1 mm); disk-shaped, in matrix (1.5 mm); spherical, with surface depression (0.8 mm); dumbbell-shaped (1.6 mm), shown along with some more nearly spherical forms.

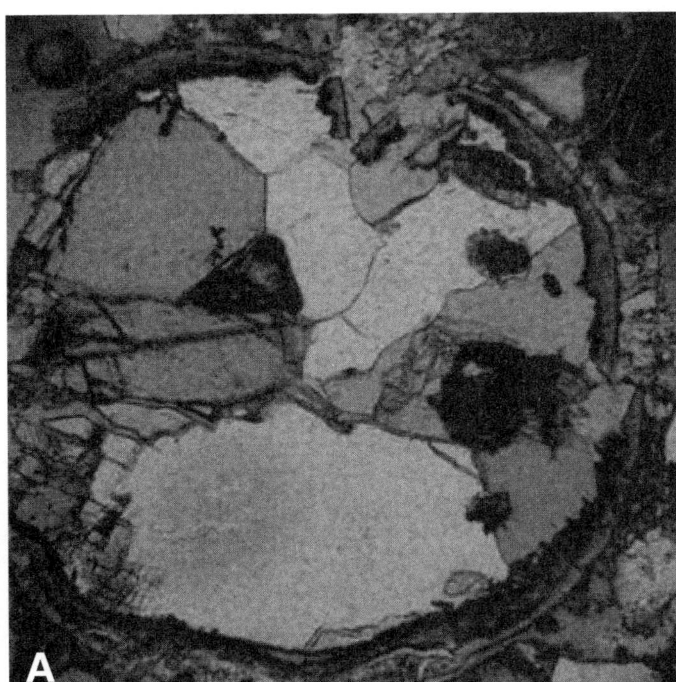

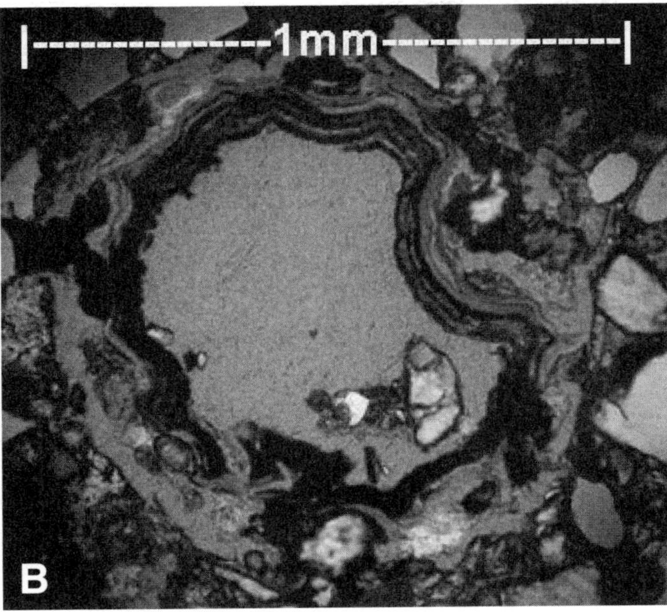

Figure 8. Thin-section views of two spherules from Shell Creek, Alabama. A: Spherically shaped impact spherule showing cracked wall and interior calcite fill. Scale same as B. B: Spherically shaped impact spherule showing laminated wall structure, cracked wall, wall concavities (plastic deformation of the wall), and hollow interior.

either to have been slightly crushed or to have collapsed due to desiccation-induced cracking of the secondary clay minerals (Figs. 7 and 8).

Under a binocular microscope, the exterior of most of the impact spherules appears to be altered entirely to dark, smectitic clays, which have a crackly exterior appearance (Fig. 7). X-ray diffraction work of Pitakpaivan et al. (1994) confirms smectitic clay mineralogy. Colors of impact spherules include black, gray, greenish gray, and light gray. Some impact spherules are encased in a thin shroud of white calcite cement (e.g., see the teardrop impact spherule in Fig. 7), or their interior, once hollow, has been completely or nearly completely filled by colorless calcite. In cross section, many of the freshly broken or cut impact spherules are partially or entirely hollow (i.e., there is no sparry calcite filling). These hollow spherules display features on the inner side of the exterior shell, which range from nearly smooth to quite irregular. The irregular nature of the inner wall is due to intact, small spherically shaped features, which are adhered to or are part of the inner walls of the impact spherule's outer shell. The vesicle-like nature of these features can be seen in thin section, and this is taken as further evidence of a former molten state (Figs. 9 and 10).

In thin section, the outer walls (i.e., shells made of clay minerals) of some spherules display a laminated structure of alternating light and dark layers (Fig. 11). This laminated structure is likely diagenetic (i.e., from devitrification; Bohor and Glass, 1995). The laminated wall structure is also typically deformed. The deformation of impact spherule walls includes concave indentation (plastic deformation) and rim segmentation (brittle breakage; Figs. 8 and 9). The outer walls of some spherules are not laminated, however, and are essentially all one color or have gradational light to dark shading (Fig. 9). Spherules without laminated walls are as likely as those with laminated walls to display hollow and calcite-filled interiors, so it is unclear if there may have been any original compositional difference between spherules with the two wall types.

The outer surfaces of impact spherules commonly show numerous, small indentations that do not appear to be related to plastic deformation (noted above). These are interpreted as primary features resulting from vesicular eruption (i.e., bubble popping) on the surface of the grain or in-flight collision with other fine ejecta (Fig. 10).

The presence of dark, phyllosilicates in the outer rim of these impact spherules has been noted by Smit et al. (1996) in impact spherules from the K-T boundary sands at Moscow Landing, Alabama, and other sites in the Gulf of Mexico region, and by Kohl et al. (2006) in some Late Archean impact spherules from South Africa. The hollowness, i.e., the absence of most of the original interior of these impact spherules, may be due to dissolution of original glassy cores that did not devitrify like the outer walls (Simonson, 2003). We do not favor a possible alternative idea requiring the nearly ubiquitous presence of large vesicles at the center of these impact spherules because we think that would have made the impact spherules

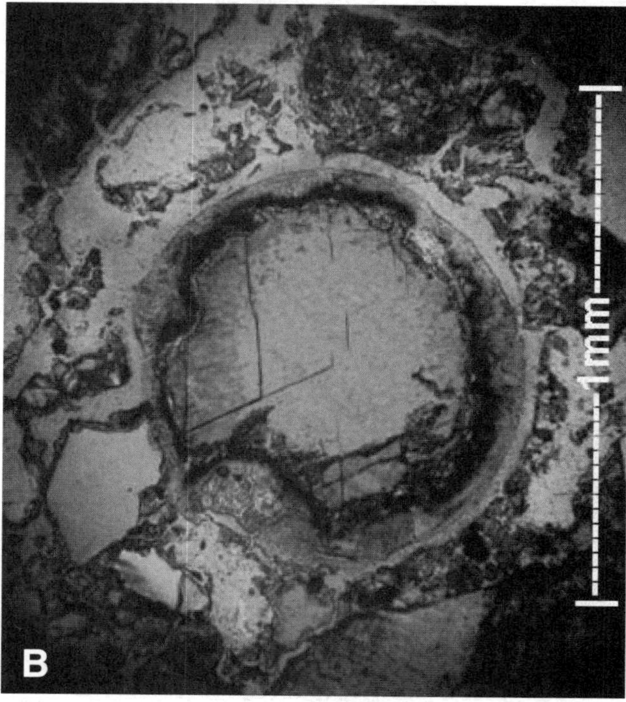

Figure 9. Thin-section views of two spherules from Shell Creek, Alabama. A: Spheroidally shaped impact spherule showing gradational wall structure, vesicles in the wall, wall concavities, and a calcite core with hollow region around it. Scale same as B. B: Spherically shaped impact spherule showing gradational wall structure, calcite filling of central void, and plastic wall deformation.

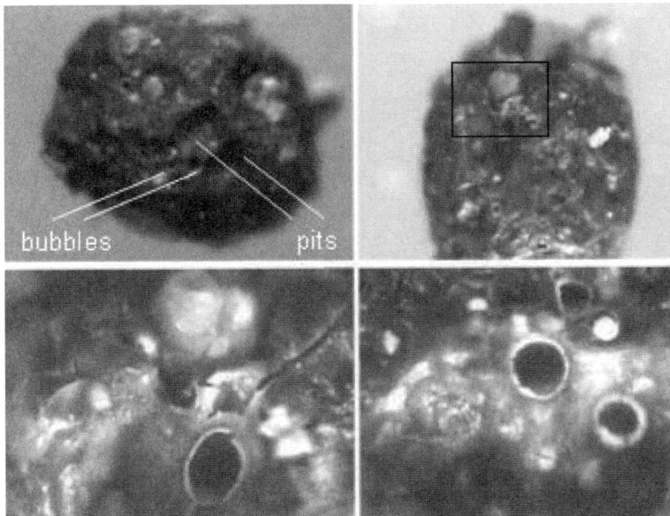

Figure 10. Shell Creek impact spherules displaying surface pits and former bubbles (exterior vesicles). Upper right: box shows enlarged area at lower left. Impact spherules shown range in size from 0.9 to 1.0 mm. Lower right: another close up, same enlargement as lower left, showing additional examples of bubbles on the exterior of the impact spherule.

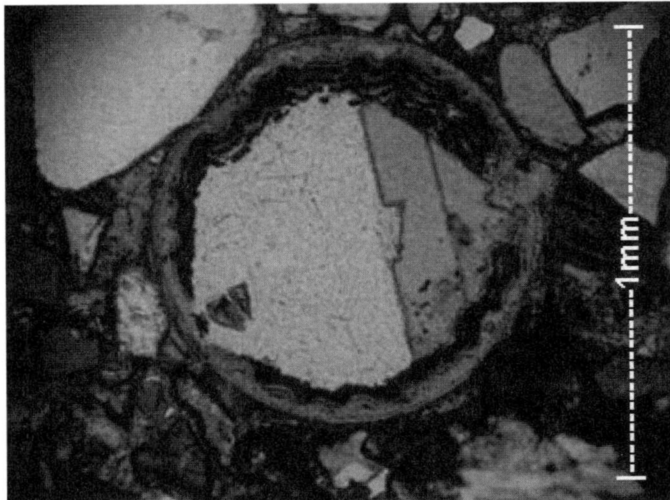

Figure 11. Thin-section view of a spherically shaped impact spherule from Shell Creek, Alabama. Shows multilayered wall structure and internal calcite filling, which partially replaces the wall in one place.

too delicate and too light to be eroded from land and then transported, mainly intact, to the shelfal depositional setting where they were preserved.

Cross-Laminated Sands

The overlying unit of tan, cross-laminated fine quartzose sand (~30–50 cm thick) is much finer than the underlying bed of the sand body. After carefully searching for evidence, we have concluded that the overlying unit contains no impact spherules or other ejecta (or these materials are so rare we cannot find them so far), except at the contact with the underlying unit where some impact spherules were pressed by compaction into the lower sand laminae from the underlying impact spherule-bearing layer. There is a very sharp, flat contact between this overlying fine sand and the previously described, underlying gray, impact spherule-bearing sand layer. At this contact, elongate clasts and fossil molds within the basal few centimeters of the fine sand bed are strongly aligned parallel to one another in the plane of the cross-laminations. In addition, the texture of the base of the cross-laminated unit displays preferential quartz sand-grain alignment, which is parallel to the long axis direction of elongate clasts and fossil molds. Because these observations were made on the bases of detached blocks, not on the outcrop itself, we cannot determine the actual compass direction of flow that produced this alignment.

The main features of these upper fine sands are ubiquitous, thin laminations, which are more nearly horizontal (i.e., parallel to the flat, basal contact) near the layer's base. The thin laminations then grade upward into what we refer to as "hummocky-type cross-laminations" near the top (Fig. 12). This term is used because the sedimentary structure of this bed is superficially similar to hummocky cross-stratification, but is not exactly the same. For example, we do not see any erosional truncation of lamina sets or layers of micas, clays, or macerated plant material between individual lamina (criteria of Dott and Bourgeois, 1982).

The top of the cross-laminated fine sand bed shows original depositional topography of hummocks and swales related to irregularly shaped and distributed ripple-like features. Original relief on these irregular ripple-like features, which are relatively

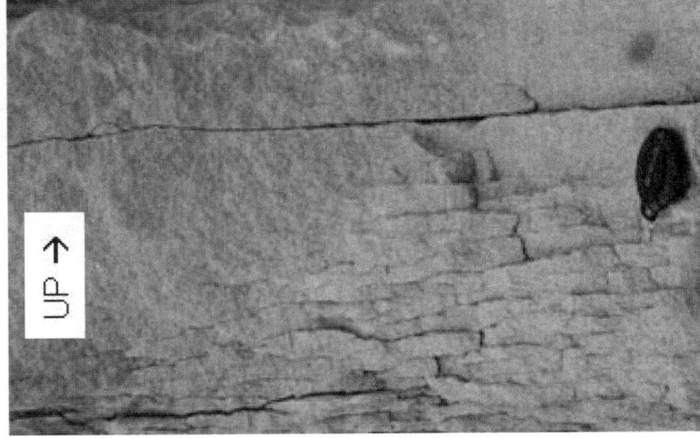

Figure 12. Cross-sectional view of cross-lamination in the upper fine sand at the Shell Creek K-T boundary section. This broken surface, perpendicular to bedding, shows laminations that are nearly horizontal near the base and then grade upward into hummocky-type cross-laminations near the top. Lens cap is 2.25 cm in diameter.

well preserved on outcrop and in some adjacent detached blocks, is ~4–8 cm. As with the underlying impact spherule bed, there are no burrows within the cross-laminated sand bed, nor do burrows penetrate the cross-laminated sand bed from the overlying beds of the Clayton Formation. Multiple flow directions are indicated, but cannot be measured because the block that shows them well has been removed from the creek and the observations were made after movement of the block.

The nature of the upper fine sand bed is similar to, but not exactly like the upper cross-laminated sands seen at other K-T sections. Not present at Shell Creek are the various layers of differently oriented, relatively small-scale ripples and climbing ripples found in the Brazos River sections and in eastern Mexican Coastal Plain sections (Smit et al., 1996; Smit, 1999). The upper cross-laminated sands at Shell Creek have structures akin only to the upper part of unit II in Mexican sections (Smit et al., 1996; Smit, 1999), i.e., hummocky cross-stratification. However, as noted above, we prefer to call the Shell Creek sedimentary structure "hummocky-type cross-lamination" because it is not exactly like the hummocky cross-bedding originally described by Dott and Bourgeois (1982). The structures in the fine layer at Shell Creek may be some kind of large, low-angle climbing ripple structure (Smit et al., 1996; Smit, 1999) or perhaps large interference ripples. True hummocky cross-stratification is the product of deep-acting oscillatory wave motion in a shelfal environment (Dott and Bourgeois, 1982; Bourgeois et al., 1988), which is mainly attributed to water motion effects of cyclonic storm passage. Such conditions may not have prevailed at Shell Creek, as noted below. Unlike Shell Creek, some U.S. Atlantic Coastal Plain and marine U.S. Western Interior sections lack overlying fine sand layers with cross-laminated structure (Smit et al., 1996; Smit, 1999).

Above this upper cross-laminated fine sand bed, bioturbated and rippled, sandy marls of the Clayton Formation crop out. These strata lie directly upon the irregular relief of the ripple-like features at the top of the fine sand bed. We interpret this transition to the basal Clayton Formation as a gradual return to "normal" shallow shelf sedimentation (and thus a lower sedimentation rate), which is characteristic of the lower part of the whole Clayton Formation (King and Petruny, 2005). As noted above, the sample reported by Mancini et al. (1989) from the lower few centimeters of this "normal" Clayton sediment contains a few reworked Maastrichtian benthic foraminifera, which indicates reworking of the upper part of the Shell Creek K-T boundary sand body.

CONCLUSIONS

The Shell Creek impact spherules probably represent pseudomorphs of ballistic melt droplets from the Yucatán deep bedrock (Smit, 1999), which were formed in the Chicxulub impact and subsequently fell upon central Alabama's low-lying coastal plain and adjacent littoral and sublittoral areas soon after impact. This interpretation of the impact spherules seems reasonable considering what is known about Chicxulub ejecta from previous work on comparable K-T boundary sections (Bohor and Glass, 1995). Nevertheless, we know little about how the impact spherules became part of the coarse to medium lower sand unit in the Shell Creek K-T boundary section and the origin of the layer with hummocky-type cross-stratification. In fact, the full explanation may be quite complicated, an observation shared by Schulte et al. (2006) in their analysis of the Texas K-T sand body within Brazos River sections and cores.

Owing to the admixture of coarse to medium quartzose sand plus glauconite grains, intraclasts, and shell fragments, we believe that we can rule out, as others have, the possibility that these are primary ejecta deposits. Therefore, a dynamic process must be involved in generating this ejecta-bearing sedimentary deposit and its overlying cross-laminated unit. Because of the dearth of impact spherules and textural differences between the two beds in the Shell Creek K-T sand body, we attribute these two beds to separate events: (1) resedimentation of reworked impact spherule-bearing sands and (2) subsequent energetic wave reworking of the impact spherule-bearing, gravity-driven deposits or other subsequently deposited sands.

Tsunami-related mixing of coastal and nearshore sand and impact spherules, with subsequent gravity-driven shelf deposition, is a mechanism suggested by Smit (1999) for emplacement of some U.S. Western Interior and U.S. Atlantic Coastal Plain impact spherule beds. While similar processes could be a valid explanation for development of the lower, impact spherule-bearing sand bed in the Shell Creek section, we cannot rule out that other violent sedimentation events, such as the passage of large-scale cyclonic storms, were the mechanism for generating strong currents and debris flows (Schulte et al., 2006). We note that debris flows and associated hummocky-bedded tempestite sands are common features in Upper Cretaceous shelfal sediments of the eastern U.S. Gulf Coastal Plain (King, 1994).

Tsunami-related shelfal currents and sedimentation have been invoked as a possible origin for hummocky-bedded cross-laminated sands in the Brazos River sections (Texas) (Bourgeois et al., 1988) and some K-T boundary sections in the Mexican Coastal Plain (Smit et al., 1996). While similar processes could be a valid explanation for development of the upper cross-laminated sand bed in the Shell Creek section, we cannot rule out the possibility that the cross-laminated unit atop the impact spherule-bearing layer was generated or highly modified by the passage of cyclonic tropical storm waves, which form quite similar deposits (Dott and Bourgeois, 1982). Such cyclonic storms, which may or may not be related to the Chicxulub impact, could have emplaced a separate sand unit above the impact spherule-bearing unit, thus accounting for the textural differences (cf. Simonson et al., 1999).

Regardless of the wave-energy source, the cross-laminated fine sand bed may have been produced either (1) by winnowing of the gravity-emplaced, impact spherule bed (Simonson et al., 1999) or (2) by reworking of a separately emplaced, fine clastic layer with a sharp, planar erosional base. If impact spherules are ever found in the fine, cross-laminated layer, this would favor hypothesis 1; if not, hypothesis 2 remains equally likely.

Because there are no closely adjacent stratigraphic sections that include the subject sand body and no nearby drill cores, we cannot assess whether the Shell Creek K-T sand body is lying in a channel cut into the Prairie Bluff Chalk or in a more sheet-like form of relatively wide extent on top of the Prairie Bluff.

ACKNOWLEDGMENTS

We sincerely thank the landowners at the Shell Creek section for allowing us to work on their property. We thank the following reviewers who helped improve this manuscript: Wright Horton, Glenn Izett, Kevin Pope, and Bruce Simonson. Spherule photomicrographs were made by one of us (L.P.) using Digital Blue™ technology.

REFERENCES CITED

Bohor, B.F., and Glass, B.P., 1995, Origin and diagenesis of K/T impact spherules: From Haiti to Wyoming and beyond: Meteoritics, v. 30, p. 182–198.

Bourgeois, J., Hansen, T.A., Wiberg, P.L., and Kauffman, E.G., 1988, A tsunami deposit at the Cretaceous-Tertiary boundary in Texas: Science, v. 241, p. 567–570, doi: 10.1126/science.241.4865.567.

Dott, R.H., and Bourgeois, J., 1982, Hummocky stratification: Significance of its variable bedding sequences: Geological Society of America Bulletin, v. 93, p. 663–680, doi: 10.1130/0016-7606(1982)93<663:HSSOIV>2.0.CO;2.

Hansen, T.A., Farrand, R.B., Montgomery, H.A., Billman, H.G., and Blechschmidt, G.L., 1987, Sedimentology and extinction patterns across the Cretaceous-Tertiary boundary interval in east Texas: Cretaceous Research, v. 8, p. 229–252, doi: 10.1016/0195-6671(87)90023-1.

King, D.T., Jr., 1994, Upper Cretaceous depositional sequences in the Alabama Gulf Coastal Plain: Their characteristics, origin, and constituent clastic aquifers: Journal of Sedimentary Research, v. B64, p. 258–265.

King, D.T., Jr., and Petruny, L.W., 2004, Cretaceous-Tertiary boundary microtektite-bearing sands and tsunami beds, Alabama Gulf Coastal Plain: Lunar and Planetary Science, v. 35, abstract 1804, CD-ROM.

King, D.T., Jr., and Petruny, L.W., 2005, Cretaceous-Tertiary microtektite-bearing sands and tsunami beds, Alabama Gulf Coastal Plain: Gulf Coast Association of Geological Societies Transactions, v. 55, p. 389–391.

Kohl, I., Simonson, B.M., and Berke, M., 2006, Diagenetic alteration of impact spherules in the Neoarchean Monteville layer, South Africa, *in* Reimold, W.U., and Gibson, R.L., eds., Processes on the early Earth: Geological Society of America Special Paper 405, p. 57–73.

Mancini, E.A., Tew, B.H., and Smith, C.C., 1989, Cretaceous-Tertiary contact, Mississippi and Alabama: Journal of Foraminiferal Research, v. 19, p. 93–104.

Pitakpaivan, K., Byerly, G.R., and Hazel, J.E., 1994, Pseudomorphs of impact spherules from a Cretaceous-Tertiary boundary section at Shell Creek, Alabama: Earth and Planetary Science Letters, v. 124, p. 49–56, doi: 10.1016/0012-821X(94)00077-8.

Savrda, C.E., 1993, Ichnosedimentologic evidence for a noncatastrophic origin of Cretaceous-Tertiary boundary sands in Alabama: Geology, v. 21, p. 1075–1078, doi: 10.1130/0091-7613(1993)021<1075:IEFANO>2.3.CO;2.

Schulte, P., Speijer, R., Mai, H., and Kohnty, A., 2006, The Cretaceous-Paleogene (K-P) boundary at Brazos, Texas: Sequence stratigraphy, depositional events and the Chicxulub impact: Sedimentary Geology, v. 184, p. 77–109, doi: 10.1016/j.sedgeo.2005.09.021.

Simonson, B.M., 2003, Petrographic criteria for recognizing certain types of impact spherules in well-preserved Precambrian successions: Astrobiology, v. 3, p. 49–65, doi: 10.1089/153110703321632417.

Simonson, B.M., and Glass, B.P., 2004, Spherule layers—Records of ancient impacts: Annual Review of Earth and Planetary Sciences, v. 32, p. 329–361, doi: 10.1146/annurev.earth.32.101802.120458.

Simonson, B.M., Hassler, S.W., and Beukes, N.J., 1999, Late Archean impact spherule layer in South Africa that may correlate with a western Australian layer, *in* Dressler, B.O., and Sharpton, V.L., eds., Large meteorite impacts and planetary evolution, II: Geological Society of America Special Paper 339, p. 249–261.

Smit, J., 1999, The global stratigraphy of the Cretaceous-Tertiary boundary impact ejecta: Annual Review of Earth and Planetary Sciences, v. 27, p. 75–113, doi: 10.1146/annurev.earth.27.1.75.

Smit, J., Roep, Th.B., Alvarez, W., Montanari, A., Claeys, P., Grajales-Nishimura, J.M., and Bermudez, J., 1996, Coarse-grained, clastic sandstone complex at the K/T boundary around the Gulf of Mexico: Deposition by tsunami waves induced by the Chicxulub impact?, *in* Ryder, G., et al., eds., The Cretaceous-Tertiary event and other catastrophes in Earth history: Geological Society of America Special Paper 307, p. 151–182.

Manuscript Accepted by the Society 10 July 2007

Printed in the USA

The Geological Society of America
Special Paper 437
2008

Megatsunami deposit in Cretaceous-Paleogene boundary interval of southeastern Missouri

Carl E. Campbell*
Saint Louis Community College—Meramec, Saint Louis, Missouri 63122, USA

Francisca E. Oboh-Ikuenobe*
Tambra L. Eifert
Missouri University of Science and Technology (formerly University of Missouri–Rolla), Rolla, Missouri 65409, USA

ABSTRACT

Crowleys Ridge in southeastern Missouri preserves Cretaceous to Eocene marginal marine sediments deposited in the northwestern portion of the Mississippi Embayment. Sandwiched between the Paleocene Porters Creek Formation and the uppermost Cretaceous Owl Creek Formation is the Paleocene Clayton Formation. Four trenches were excavated and a complete section of Clayton Formation was sampled at a large strip mine in Stoddard County, Missouri. The Clayton at this location consists of 185 cm of graded deposit, the lower part of which includes large Owl Creek rip-up clasts containing layers of microtektites, invertebrate fossils, and abundant terrestrial and marine palynomorphs. Driller's logs and electric logs covering ~9000 km^2 were reviewed. Well data confirm the consistent thickness and lithology of the Clayton Formation in this part of the Mississippi Embayment. Based on sedimentological and palynological data from this study, the Clayton Formation in the northwestern Mississippi Embayment appears to be a megatsunami deposit resulting from post-impact effects (early Paleocene) associated with the Chicxulub impact event.

Keywords: megatsunami, Cretaceous-Paleogene boundary, Mississippi Embayment, Clayton Formation, microtektites.

INTRODUCTION

The Mississippi Embayment has received a fair amount of attention since Alvarez et al. (1980) proposed an asteroid impact at the Cretaceous-Paleogene (K-P) boundary and Hildebrand et al. (1991) located and described the Chicxulub impact crater. Majority of these studies have focused on the Alabama, Mississippi, and Texas portion of the embayment with little attention paid to the Missouri portion. There is good reason for this lack of attention because K-P interval exposures in Missouri are few and weathered. The Clayton Formation in Missouri was first studied in the middle 1930s (Farrar, 1935; Farrar and McManamy, 1937). Exposures were highly weathered and fossils of poor quality. Stephenson (1955) first described Missouri Owl Creek fossils from Crowleys Ridge, and he supplemented his descriptions with photographs from the type area in Tippah County, Mississippi.

*Campbell: cecampbell@stlcc.edu; Oboh-Ikuenobe: Ikuenobe@mst.edu

Campbell, C.E., Oboh-Ikuenobe, F.E., and Eifert, T.L., 2008, Megatsunami deposit in Cretaceous-Paleogene boundary interval of southeastern Missouri, *in* Evans, K.R., Horton, J.W., Jr., King, D.T., Jr., and Morrow, J.R., eds., The Sedimentary Record of Meteorite Impacts: Geological Society of America Special Paper 437, p. 189–198, doi: 10.1130/2008.2437(11). For permission to copy, contact editing@geosociety.org.

The Clayton Formation is a bridge between the Cretaceous and Paleocene sections. Early workers described the Clayton as montmorillonite clay deposited in an outer neritic environment (Pryor and Glass, 1961) with reworked Cretaceous fossils (Stephenson, 1955) but gave little explanation about its deposition. They generally described the Clayton as a regressive sequence (e.g., Farrar, 1935).

REGIONAL GEOLOGICAL SETTING

Missouri's bootheel encompasses the northwestern portion of the Mississippi Embayment (Fig. 1), where the Cretaceous-Paleogene (K-P) boundary interval consists of the Paleocene Clayton Formation sandwiched between the Cretaceous Owl Creek Formation and Paleocene Porters Creek Formation. This set of strata is poorly preserved in a few locations exposed along Crowleys Ridge in southeastern Missouri, where no fresh, unweathered exposures exist.

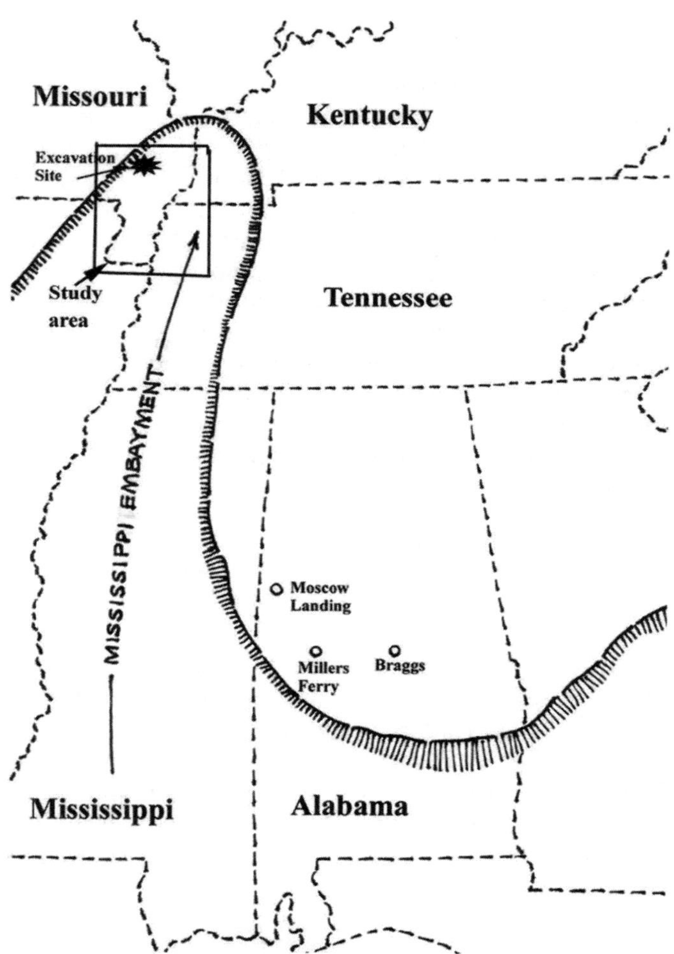

Figure 1. Location map showing excavation site and study area within Gulf Coastal Plain Cretaceous-Paleogene outcrop belt. Microtektites and tsunami deposits have been described at the Alabama locations in recent literature.

Nestlé-Purina Company surface-mines Porters Creek clay at selected locations along the eastern side of Crowleys Ridge in Stoddard County, Missouri. The mine location is in the W/2, S21, T27N, R11E (30°58.100'N, 89°52.300'W), at an elevation of 117 m.

METHODS

Four trenches were excavated to a depth of 5 m, and unweathered Clayton Formation was measured and sampled for analysis. Macrofossils (mostly invertebrates) were collected, and microfossils were sieved from the rock samples, while standard palynological procedures of digesting samples in mineral acids were used for extracting palynomorphs (Traverse, 1988). Transmitted light microscopy was used for routine identification and description of the palynomorphs, and at least 200 specimens per slide were counted in productive samples.

RESULTS

Stratigraphy

The Clayton Formation in the study area is a 185 cm unit resting unconformably on the Upper Cretaceous Owl Creek clay; the contact is undulating (Fig. 2). The overlying Paleocene Porters Creek clay grades vertically from a light gray claystone at the top to a dark gray montmorillonitic clay at its base. The contact between Porters Creek and Clayton often includes large (>100 cm) torpedo-shaped, hollow limonitic concretions (Fig. 3). The Clayton Formation can be divided into four units based on lithology, color, and fossils (Fig. 4).

The lowest unit is a 30 cm thick limestone coquinite containing abundant reworked Cretaceous macro- and microfossils. It is unsorted and is boudined. The coquina appears to be a scour fill (Fig. 5). The scours trend southeast to northwest and are 30 cm to 50 cm in cross section and 30 cm thick. Some are isolated, whereas others are interconnected, and their tops are concordant and nearly flat. Included within this unit are large (3 cm to 20 cm), oblate to spherical rip-up clasts containing layers of microtektites but no fossils. Microtektites are predominantly spherical, but many are teardrop- or dumbbell-shaped, others have been welded together to form doublets and triplets. Their color ranges from dark brown to tan to green. The spherical varieties have a bubbly internal appearance (Figs. 6 and 7). All the microtektites range in size from 0.5 mm to over 3 mm. Most have calcite nuclei, and a few appear to have glass nuclei. Scattered throughout the lower coquinite zone and within the rip-up clasts are tiny ooids (<0.25 mm) with shiny brown coatings. The ooids have either clear, subangular quartz or fecal pellet nuclei. Also scattered randomly throughout the coquinite are floating grains of frosted subrounded quartz sand. The coquinite matrix is biomicrite.

Overlying the coquinite zone is 70 cm of soft, green glauconitic sandy clay containing abundant Upper Cretaceous macrofossil fragments, microfossils, broken limonitic burrows, and

Figure 2. Photograph of trench wall containing complete section of Paleocene Clayton Formation above Cretaceous Owl Creek Formation. Note the deep scour into Owl Creek in lower right corner. Hammer in lower right corner is 29 cm long. White rectangle in lower right is a 15 cm scale bar.

internal casts and molds of mollusks and cephalopods. A few hard concretions of calcified fossils occur within small depressions cut into the lower coquinite and the Owl Creek Formation.

The glauconitic zone grades uniformly into 65 cm of soft, light gray clay containing grains of glauconite and sparse manganese oxide pebbles. The upper portion is burrowed. While this unit contains few macrofossils or foraminifera, it does contain palynomorphs. The uppermost unit consists of 20 cm of claystone. The only difference from the underlying unit is that it is indurated, apparently a hardground. The base of the Porters Creek Formation rests on this upper zone.

Paleontology

Fresh Clayton and Owl Creek exposures are nonexistent in Missouri. However, fossils recovered from this location are abundant, diverse, and exhibit excellent preservation. The highly fossiliferous lower Clayton coquinite zone contains reworked Upper Cretaceous macro- and microfossils along with carbonized wood and seeds. Most of the macrofossils are mollusks with sparse vertebrate specimens such as bony fish, sharks, rays, turtles, and mosasaurs (Campbell and Lee, 2001; Gallagher et al., 2005). We have identified 58 species of invertebrates, of which ~50% are bivalves and 30% are gastropods. Commonly, bivalves are preserved with both valves attached and closed. This zone also contains an abundant and diverse assortment of reworked Upper Cretaceous foraminifera.

The overlying glauconitic zone preserves internal casts and fragments of Upper Cretaceous invertebrates and a rich assortment of foraminifera. Notably, the majority of the ammonites recovered occur in this zone. The glauconitic zone probably contains pelagic remains, whereas the lower coquinite zone contains large benthic fossils such as *Exogyra costata*. The upper gray clay zones lack macrofossils but contain microfossils such as palynomorphs and a few Cretaceous foraminifera.

Palynology

Palynomorphs have been recovered from all the Clayton zones described above, although they vary in type, abundance, and diversity (Fig. 8). Of the 95 taxa of terrestrial palynomorphs (pollen, spores, fungi, algae) identified, ~83% are pollen (mainly angiosperms); the extinct Normapolles accounts for only ~1%. Marine palynomorphs (dinoflagellate cysts and acritarchs) provide a more precise biostratigraphic record than terrestrial palynomorphs, and they increase in abundance and diversity up-section. Late Cretaceous and Paleocene palynomorphs are dominant, and the presence of reworked Cretaceous taxa confirms the macrofossil record discussed above. Important Late Cretaceous–Paleocene dinoflagellate cysts recovered include *Areoligera senonensis*, *Cerodinium diebelii*, *Cerodinium striatum*, *Cordosphaeridium cantharellum*, and *Hafniasphaera graciosa*. *Senoniasphaera inornata* disappeared worldwide during the early Paleocene (Gradstein et al., 2004).

Figure 3. Photograph of limonitic concretion at the base of the Porters Creek Formation. Hammer is 31 cm long.

Well Log Data

The Missouri Department of Natural Resources maintains a library of logs from wells drilled in the state. The Clayton Formation is a good marker on driller's logs and electric logs from water wells and oil test wells in the area (Fig. 9). The transition from soft Porters Creek clay to the Clayton Formation with its hard claystone and carbonate layers is easy to identify. Drillers use terms such as "lime-rock," "marl-stone," and "limestone" on their logs to describe the Clayton. Electric logs are often useful for picking the top of the Clayton in the subsurface. There are ~60 wells with driller's logs and 23 wells with electric logs in the study area. All were reviewed and tops picked for Porters Creek, Clayton, and Owl Creek Formations.

Well log analysis indicates the Clayton Formation dips uniformly 700 m to the southeast over 100 km. This is a typical dip for all the Cretaceous and Paleogene units in this area. The Upper Cretaceous Owl Creek Formation thins from 21 m in the southeast to 5 m in the northwest over the same distance. The overlying Paleocene Porters Creek thins from 190 m to 40 m over the same distance. Based on log data, the overlying Porters Creek and underlying Owl Creek thin from southeast to northwest, while the Clayton Formation appears to have a uniform thickness of ~3–5 m throughout Missouri's bootheel.

In 1978 the U.S. Geological Survey drilled a deep test well (New Madrid Test Well 1-X) in the north-central portion of the Mississippi Embayment in southeastern Missouri and cored several intervals. The complete Clayton interval between a depth of 507 m and 511 m was cored and recovered. Frederiksen et al. (1982) described this unit as "glauconitic shelly marl, grading upward into a glauconitic shelly sand, that grades rather abruptly into glauconitic silty clay of the Porters Creek." The New Madrid test well is located ~100 km southeast of our study area.

DISCUSSION

The Clayton Formation exhibits unusual characteristics at many locations around the embayment margins. Possible impact-related and tsunami-related features such as faulting, chaotic layers, scours, and reworked Maastrichtian sediments have been reported from Mississippi (Lynn Creek) and Alabama (Shell Creek, Millers Ferry, Mussel Creek, and Moscow Landing) (Habib et al., 1996; Olsson et al., 1996, Fig. 15 therein).

Missouri is no exception and may be unique in that it represents the head of the embayment. The Clayton Formation has unusual depositional features, especially in the lower coquinite zone with its lumpy, boudined appearance. It appears as fill within scours carved into the Owl Creek clay. The coquina includes large fossils, some exceeding 6 cm in length, and many rounded to oblate rip-up clasts. Some clasts are more than 10 cm long and 5 cm thick. The clasts are particularly interesting. Many contain layers of microtektites that are morphologically identical to those found at other locations around the Gulf of Mexico. The K-T boundary interval in the New Jersey Coastal Plain contains rip-up clasts with calcite-replaced tektites (Olsson et al., 2000). Furthermore, the Clayton Formation at Shell Creek, Alabama, along the eastern edge of the embayment, has microtektites (Pitakpaivan et al., 1994) identical in size and shape to the ones identified in this study. These microtektites fit Smit's classification as proximal ejecta from the Chicxulub impact (Smit, 1999).

The rip-up clasts seem to be unfossiliferous Owl Creek clay carried into the embayment headland. Many of the microtektites are welded together, thereby suggesting high temperature during their formation. As this cloud of hot ejecta presumably settled into shallow, quiet water it could have formed a "skin" of microtektites mixed with the ooids covering unconsolidated Owl Creek clay. As fast-moving water moved up the embayment, it likely swept up the newly "welded" upper layer of the Owl Creek and formed the rip-up clasts. All of this allochthonous material consisting of intermixed rip-up clasts and fossils,

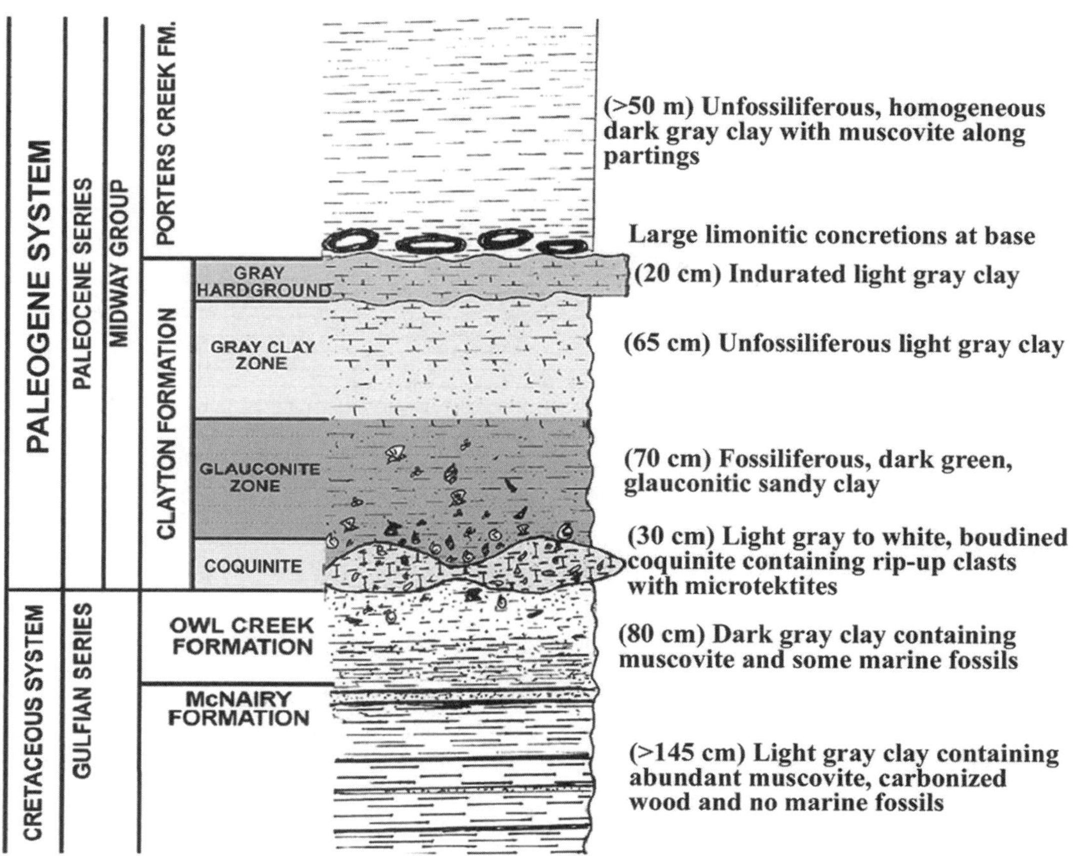

Figure 4. Description of Missouri's K-T boundary interval.

Figure 5. Photographs of coquinite in lower Clayton Formation. A: Coquinite-filled scour in Owl Creek clay. Note boudined shape. B: Coquinite containing rip-up clasts at base of unit.

Figure 6. Photographs of rip-up clast containing microtektites. A: Photograph of portion of rip-up clast containing numerous spherical and splashform microtektites. B: Microphotograph of dumbbell-shaped splashform microtektite next to spherical microtektite exhibiting bubbly internal texture. C: Microphotograph of teardrop-shaped splashform microtektite next to two smaller spherical ones. Arrows locate the two microphotographs on the clast. The two large microtektites are 3 mm long.

some with valves closed, formed the coquinite zone of the Clayton Formation. The coquinite contains reworked Cretaceous invertebrates and foraminifera as well as Late Cretaceous and early Paleocene palynomorphs. Poor preservation may account for the relatively low percentage of marine palynomorphs in the coquinite sample.

The glauconitic clay immediately above the coquinite seems to be rich in animals and material that would float for a short time, such as ammonites, shell fragments, wood, and seeds. The overlying soft, light gray clay is fine-grained and lacks macrofossils but contains abundant and diverse palynomorphs, including marine dinoflagellate cysts. The presence of *Senoniasphaera inornata* with a last appearance datum during the early Paleocene and several other dinoflagellate cysts ranging from Late Cretaceous to Paleocene in age confirms an early Paleocene age for the Clayton Formation.

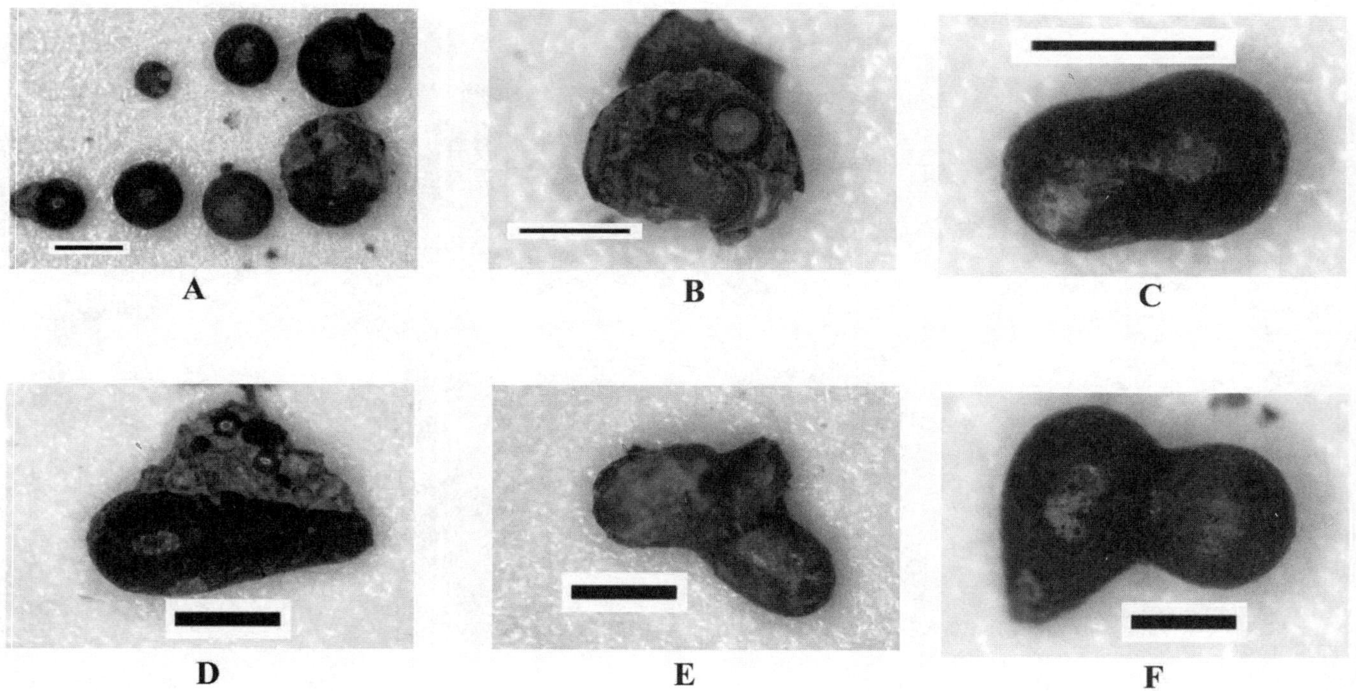

Figure 7. Microphotographs of microtektites. A: Complete spherules. B: Broken spherule exhibiting bubbly internal texture resulting from smaller spherules. C: Small dumbbell-shaped microtektite resulting from two spherules welded together. D: Teardrop-shaped splashform microtektite. E: Three oblong microtektites welded together. F: Spherule welded to a splashform microtektite. Each scale bar is 1 mm.

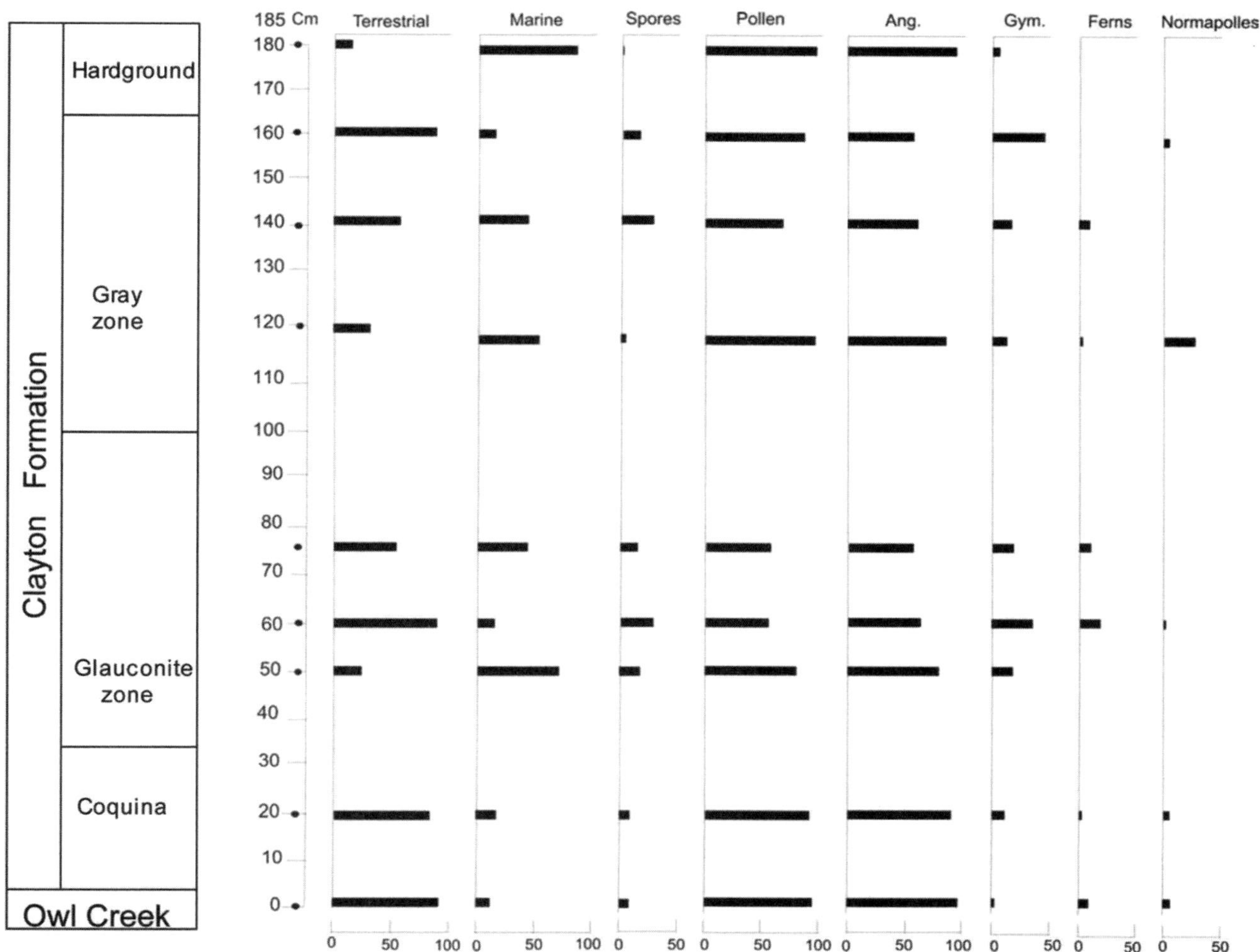

Figure 8. Chart of relative abundances (in percent) of groups of palynomorphs. Ang.—angiosperms; Gym.—gymnosperms.

Available well data in southeastern Missouri demonstrate the Clayton Formation has a rather uniform thickness of 3–5 m throughout the bootheel. It also has a similar description with a hard clay at the top, limestone at the base, and soft clay in between. As expected, overlying Porters Creek clay and underlying Owl Creek clay thin to the northwest from the embayment center. However, the Clayton does not thin much, if at all, toward the paleoshoreline.

The Chicxulub impact event probably generated a large tsunami, not only directly from the Gulf of Mexico rushing into the crater (Matsui et al., 2002), but also indirectly from shelf collapse into the Gulf of Mexico (Bourgeois et al., 1988; Bourgeois, 1994). Postimpact earthquakes likely precipitated shelf collapse. Ejecta could have reached the North American Gulf Coastal Plain within 10 min of the impact event (Olsson et al., 2000). Wave height greater than 200 m with periods of 2 h could have spanned the Gulf of Mexico within 10 h (Matsui et al., 2002). Shelf collapse into deep Gulf of Mexico water could also have generated large waves. Therefore, the resulting megatsunami advancing on the narrow mouth of the Mississippi Embayment probably experienced constructive and destructive interference and generated a tall bore up the embayment (Fig. 10).

Wave heights in the embayment could rise to more than 200 m and wave periods decrease to less than 2 h. Encroaching waves probably ripped up the seafloor and created a suspension cloud covering much of the embayment. Backwash from the embayment head could have added terrestrial material to the mix and the suspension cloud would have taken days to settle out (Smit et al., 1996). Evidently, the heaviest material, consisting of large fossils and rip-up clasts, settled first and filled scours in the Owl Creek seafloor. Lighter material, along with a large amount of marine organic material and some terrestrial material, settled next, forming the glauconitic clay layer. Fine silt and clay devoid of macroorganisms settled out last. A hardground developed, probably during a period of nondeposition, before a return to transgression.

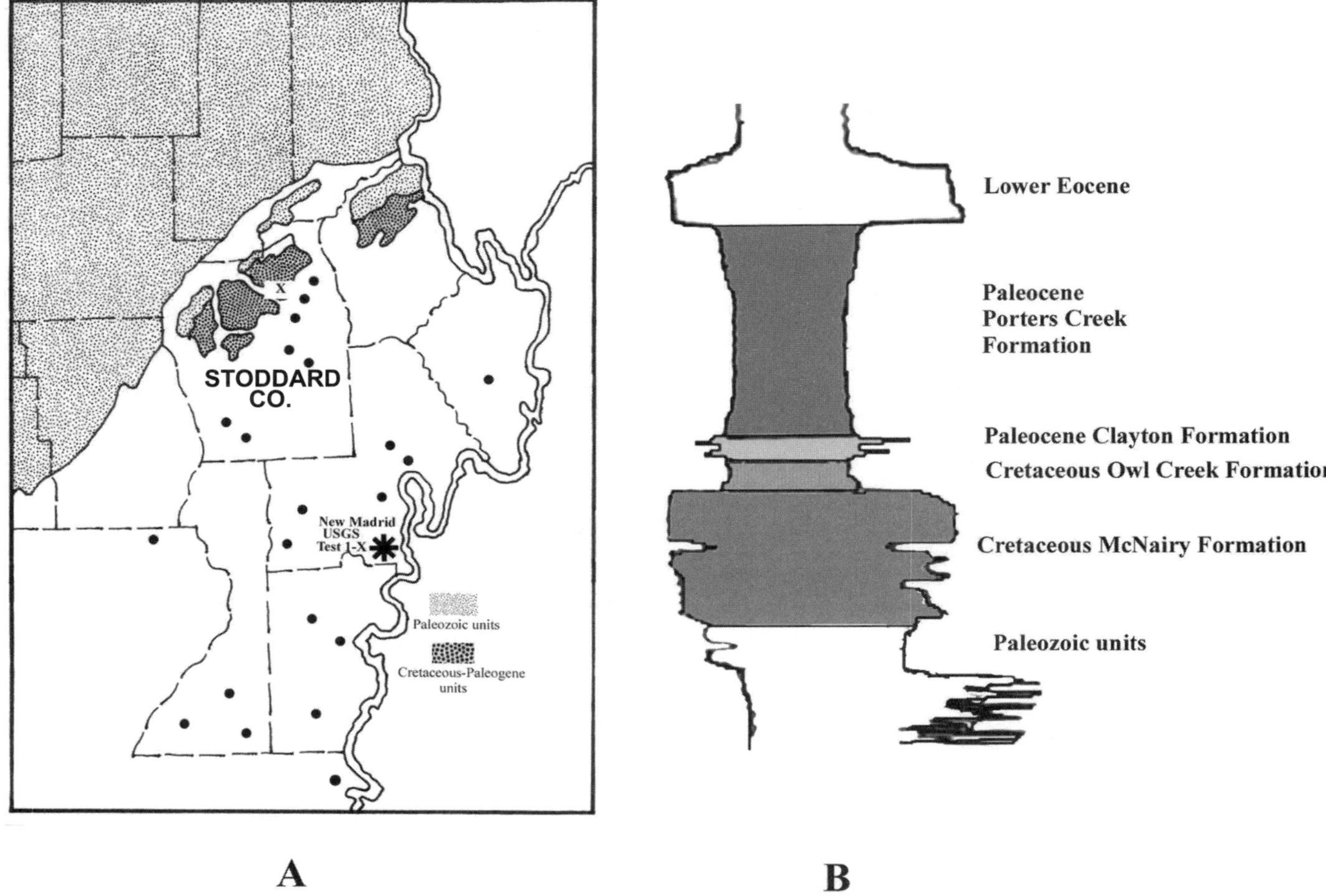

Figure 9. Location map of wells with electric logs (A) and generalized electric log of interval studied (B). Filled dots indicate locations of wells within study area that have electric logs. X marks the excavation site located along southeastern side of Crowleys Ridge. The U.S. Geological Survey (USGS) test well cored and recovered a complete section of Clayton Formation.

Pryor and Glass (1961) noted that the drainage basin of the southern Appalachian Mountains, which extended into the northeastern portion of the embayment, was the source area for micaceous Owl Creek and Porters Creek clays. Muscovite in Clayton clay is rare, and southeast to northwest coquinite-filled scours indicate a possible Clayton source area to the southeast.

CONCLUSIONS

Microtektites in rip-up clasts clearly demonstrate that the Clayton Formation of southeastern Missouri was deposited after the K-P impact. But how soon after the impact: hours, days, weeks, months, or even millions of years? Hot ejecta settling rapidly into shallow, quiet water could have formed slightly lithified layers, almost a welded sedimentary crust several centimeters thick, within a period of hours. Seismic shaking and concomitant shelf collapse into the Gulf of Mexico basin would create a megatsunami merging with large waves from the direct impact. A ramp-like narrow Mississippi Embayment probably focused the tsunami into a bore moving from southeast to northwest, and the tsunami picked up material from the shallow seafloor and generated a suspension cloud. Wave heights greater than 200 m engulfed the headland area in southeastern Missouri and created a backwash. The suspension cloud settled out as one graded unit 185 cm thick. The heaviest material (including reworked fossils) filled the scours on the seabed, depositing the coquinite layer. Organic-rich and lighter elements formed the overlying glauconitic unit, while the finest material settled out as the upper clay unit.

We propose that the Clayton Formation in southeastern Missouri was likely deposited during the early Paleocene as a short-term, isochronous event. It represents a graded section resulting from the megatsunami initiated by the Chicxulub impact event.

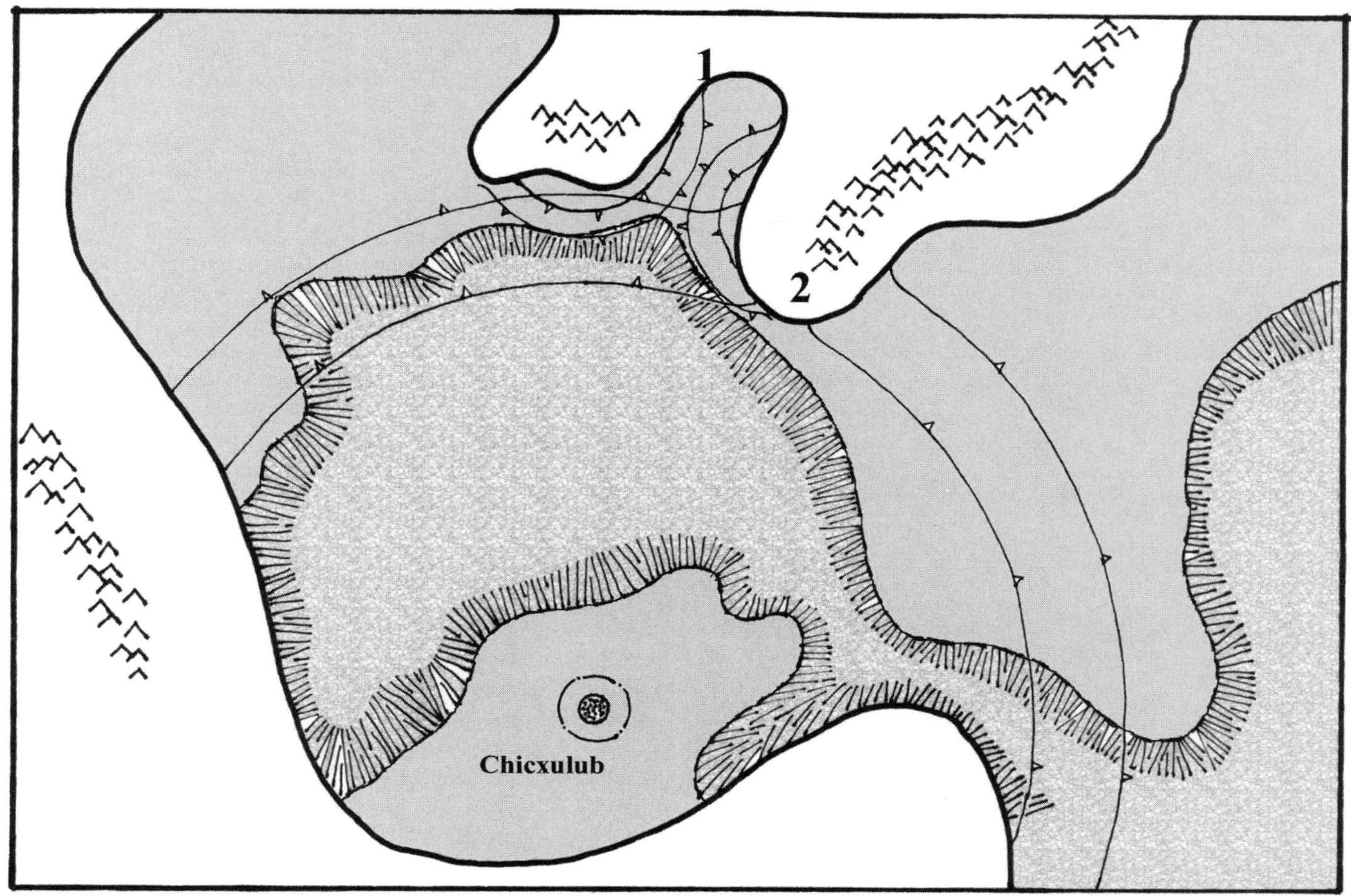

Figure 10. Generalized Cretaceous-Paleogene map of Gulf of Mexico locating Chicxulub impact site. Inner circle represents the transient crater and outer circle the collapse crater. Light gray areas represent shelf and textured area deep ocean. Semicircular lines with arrowheads indicate possible tsunami wave fronts generated from Chicxulub impact and shelf collapse. Upon entering Mississippi Embayment, wave could experience constructive (and destructive) interference. 1—location of study area; 2—location of Alabama sites with microtektites and tsunami deposits.

ACKNOWLEDGMENTS

We thank the Nestlé-Purina Company for allowing us access to its mine over a period of four years, the Missouri Department of Natural Resources for access to well information, Tom Lee and Rick Poropat for financial and physical assistance with this study, and Harold Levin for assistance in foraminifera analysis. Tambra Eifert thanks the American Association of Stratigraphic Palynologists and the Missouri University of Science and Technology's Geology and Geophysics Radcliffe Graduate Scholarship for financial assistance. We also thank reviewers Andrew Rindsberg, Vivi Vajda, and David King for their valuable comments and suggestions. This is Missouri University of Science and Technology Geology and Geophysics Contribution Number 6.

REFERENCES CITED

Alvarez, L.W., Alvarez, W., Asaro, F., and Michel, H.V., 1980, Extraterrestrial cause for the Cretaceous-Tertiary extinction: Science, v. 208, p. 1095–1108, doi: 10.1126/science.208.4448.1095.

Bourgeois, J., 1994, Tsunami deposits and the KT boundary: A sedimentologist's perspective: Lunar and Planetary Institute Contribution 825, p. 16.

Bourgeois, J., Hansen, T.A., Wiberg, P.L., and Kauffman, E.G., 1988, A tsunami deposit at the Cretaceous-Tertiary boundary in Texas: Science, v. 241, p. 567–570, doi: 10.1126/science.241.4865.567.

Campbell, C.E., and Lee, T.L., 2001, "Tails of *Hoffmanni*": Mosasaur fossils in a tsunami deposit at the K/T boundary of southeast Missouri: Journal of Vertebrate Paleontology, v. 21, no. 3, supplement, p. 37A.

Farrar, W., 1935, The Cretaceous and Tertiary geology (of southeast Missouri): Missouri Geological Survey and Water Resources Biennial Report (58th General Assembly), app. 1, pt. 1, p. 1–35.

Farrar, W., and McManamy, L., 1937, The Geology of Stoddard County, Missouri: Missouri Geological Survey and Water Resources Biennial Report (59th General Assembly), app. 6, 92 p., geological map.

Frederiksen, N.O., Bybell, L.M., Christopher, R.A., Crone, A.J., Edwards, L.E., Gibson, T.G., Hazel, J.E., Repetski, J.E., Russ, D.P., Smith, C.C., and Ward, L.W., 1982, Biostratigraphy and paleoecology of lower Paleozoic, Upper Cretaceous, and lower Tertiary rocks in U.S. Geological Survey New Madrid test wells, southeastern Missouri: Tulane Studies in Geology and Paleontology, v. 17, no. 2, p. 23–45.

Gallagher, W.B., Campbell, C.E., Jagt, J.W.M., and Mulder, E.W.A., 2005, Mosasaur (Reptilia, Squamata) material from Cretaceous-Tertiary boundary interval in Missouri: Journal of Vertebrate Paleontology, v. 25, no. 2, p. 473–475, doi: 10.1671/0272-4634(2005)025[0473:MRSMFT]2.0.CO;2.

Gradstein, F.M., Ogg, J.G., and Smith, A.G., editors, 2004, A geologic time scale 2004: Cambridge, UK, Cambridge University Press, 500 p.

Habib, D., Olsson, R.K., Liu, C., and Moshkovitz, S., 1996, High-resolution biostratigraphy of sea-level low, biotic extinction, and chaotic sedimentation at the Cretaceous-Tertiary boundary in Alabama, north of the Chicxulub Crater, *in* Ryder, G., et al., eds., The Cretaceous-Tertiary event and other catastrophes in Earth history: Geological Society of America Special Paper 307, p. 243–252.

Hildebrand, A.E., Penfield, G.T., Kring, D.A., Pilkington, M., Camargo, Z.A., Jacobsen, S.E., and Boynton, W.V., 1991, Chicxulub Crater—A possible Cretaceous Tertiary boundary impact crater on the Yucatan Peninsula, Mexico: Geology, v. 19, p. 867–871, doi: 10.1130/0091-7613(1991)019<0867:CCAPCT>2.3.CO;2.

Matsui, T., Imamura, F., Tajika, E., Nakano, Y., and Fujisawa, Y., 2002, Generation and propagation of a tsunami from the Cretaceous-Tertiary impact event, *in* Koeberl, C., and MacLeod, K.G., eds., Catastrophic events and mass extinctions: Impacts and beyond: Geological Society of America Special Paper 356, p. 69–77.

Olsson, R.K., Liu, C., and van Fossen, M., 1996, The Cretaceous-Tertiary catastrophic event at Millers Ferry, Alabama, *in* Ryder, G., et al., eds., The Cretaceous-Tertiary event and other catastrophes in Earth history: Geological Society of America Special Paper 307, p. 263–277.

Olsson, R.K., Wright, J.D., Miller, K.G., Browning, J.V., and Cramer, B.S., 2000, The Cretaceous-Tertiary boundary events on the New Jersey continental margin: Lunar and Planetary Institute Contribution 1053, Abstracts with Programs, abstract 3130, p. 160.

Pitakpaivan, K., Byerly, G., and Hazel, J.E., 1994, Pseudomorphs of impact spherules from a Cretaceous-Tertiary boundary section at Shell Creek, Alabama: Earth and Planetary Science Letters, v. 124, p. 49–56, doi: 10.1016/0012-821X(94)00077-8.

Pryor, W.A., and Glass, H.D., 1961, Cretaceous-Tertiary clay mineralogy of Upper Mississippi Embayment: Journal of Sedimentary Petrology, v. 31, no. 1, p. 38–51.

Smit, J., 1999, The global stratigraphy of the Cretaceous-Tertiary boundary impact ejecta: Annual Review of Earth and Planetary Sciences, v. 27, p. 75–113, doi: 10.1146/annurev.earth.27.1.75.

Smit, J., Roep, Th.B., Alvarez, W., Montanari, A., Claeys, P., Grajales-Nishimura, J.M., and Bermudez, J., 1996, Coarse-grained, clastic sandstone complex at the K/T boundary around the Gulf of Mexico: Deposition by tsunami waves induced by the Chicxulub impact, *in* Ryder, G., et al., eds., The Cretaceous-Tertiary event and other catastrophes in Earth history: Geological Society of America Special Paper 307, p. 151–182.

Stephenson, L.W., 1955, Owl Creek (Upper Cretaceous) fossils from Crowleys Ridge, southeastern Missouri: U.S. Geological Survey Professional Paper 274-E, 163 p., 11 plates.

Traverse, A., 1988, Paleopalynology: Boston, Unwin Hyman, 600 p.

Manuscript Accepted by the Society 10 July 2007

Printed in the USA

The Geological Society of America
Special Paper 437
2008

Stratigraphic expression of a regionally extensive impactite within the Upper Cretaceous Fox Hills Formation of southwestern South Dakota

Patricia A. Jannett*
Dennis O. Terry Jr.*
Department of Geology, Temple University, Philadelphia, Pennsylvania 19122, USA

ABSTRACT

A zone of intense soft-sediment deformation, with associated spherules and shocked quartz grains, is identified over an area of ~1000 km^2 in southwestern South Dakota. This Disturbed Zone (DZ) is up to 5 m thick and is preserved within distal deltaic deposits of the Upper Cretaceous Fox Hills Formation. Localized structural development caused thinning and eventual subaerial exposure of several sections within Badlands National Park, whereas sections to the north of the park were unaffected. Although previously interpreted as an intense period of soil formation under tropical conditions, the degree of ancient soil overprinting of these sections is minimal, with the exception of bright coloration of the strata, and appears not to have had any effect on ejecta preservation. Biostratigraphic data suggest a middle to late Maastrichtian age for the DZ. When compared to other Cretaceous impactites, our study sections in southwestern South Dakota are most similar to, and may correlate with, the recently documented 68 Ma impactite within the Vermejo Formation of Berwind Canyon in southeastern Colorado. If this correlation is correct, the size of the ejecta within the Fox Hills and Vermejo Formations suggests that the sections in South Dakota represent distal deposits.

Keywords: Fox Hills Formation, impactite, liquefaction, soft-sediment deformation, spherules, shocked quartz, South Dakota, Maastrichtian.

*Present address, Jannett: URS Corporation, 335 Commerce Drive, Suite 300, Fort Washington, Pennsylvania 19034-2623, USA, patricia_jannett@urscorp.com. Corresponding author, Terry: doterry@temple.edu

Jannett, P.A., and Terry, D.O., Jr., 2008, Stratigraphic expression of a regionally extensive impactite within the Upper Cretaceous Fox Hills Formation of southwestern South Dakota, *in* Evans, K.R., Horton, J.W., Jr., King, D.T., Jr., and Morrow, J.R., eds., The Sedimentary Record of Meteorite Impacts: Geological Society of America Special Paper 437, p. 199–213, doi: 10.1130/2008.2437(12). For permission to copy, contact editing@geosociety.org.

INTRODUCTION

A 0.5–5 m thick zone of intense sediment deformation occurs within the Upper Cretaceous Fox Hills Formation of southwestern South Dakota in and around Badlands National Park (Fig. 1). This Disturbed Zone (DZ) was previously interpreted as a distal manifestation of the Chicxulub impact based on the presence of soft-sediment deformation features, ejecta, and fossil data (Chamberlain et al., 2001; Jannett and Terry, 2001; and Terry et al., 2001). However, based on new data, including lithostratigraphic correlation and biostratigraphy (dinoflagellates and calcareous nannoplankton), the DZ is now regarded as a manifestation of a previously unrecognized late Maastrichtian impact event (Palamarczuk et al., 2003; Terry et al., 2004).

Impact layers are usually described from either purely marine or nonmarine environments (Izett, 1990; Smit, 1999). With respect to the Cretaceous-Paleogene (K-Pg) boundary, marine sections are commonly composed of a dispersed unit of clay and silt, enriched in Ir and ejecta, up to 30 cm thick and modified by bioturbation (Kyte and Bostwick, 1995; Kyte, 1998), while in nonmarine sections it is a distinct 2 cm (on average) clay layer with ejecta collected as fallout into marshes and swamps (Nichols and Johnson, 2002). The DZ impactite is preserved within deltaic sediments of the Fox Hills Formation and is variable throughout its extent as a function of geographic position on the delta. In addition, lateral tracing of the DZ to the north reveals that exposures in Badlands National Park were subsequently subjected to ancient pedogenic modification as a function of tectonic uplift (Jannett and Terry, 2001).

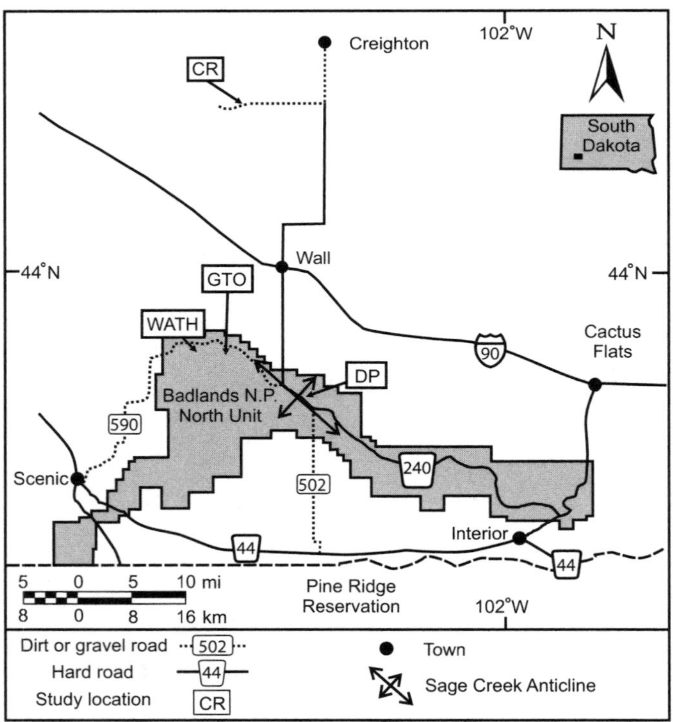

Figure 1. Location map of the study area and measured sections. CR—Creighton; DP—Dillon Pass; GTO—Grassy Tables Overlook; WATH—Wilderness Access Trailhead.

The results of this investigation provide insight into the recognition of impact signatures in deltaic facies and the characterization of an impactite that has been pedogenically modified. In addition, the recognition of a Late Cretaceous impact in the middle of a deltaic setting in this region provides an isochronous marker bed that can be used to clarify chronostratigraphic relationships across the delta. The role of this impact event on Late Cretaceous extinctions is unknown at this time.

Regional Geologic Setting

Western Interior Seaway

The Western Interior Seaway was a warm, shallow (less than 180 m deep), epicontinental sea that was present through the central part of North America during the Cretaceous. At its extreme, the area covered by this seaway extended from the Arctic Ocean to the Gulf of Mexico and from the Wyoming/Idaho border to possibly the Mississippi Valley region (Dean and Arthur, 1998). The position and extent of this sea varied throughout its history, but during the Late Cretaceous North America was effectively split into eastern and western halves (Pagani and Arthur, 1998; Leckie et al., 1998). The Western Interior Seaway gradually vanished through a combination of regression and subsequent filling of the seaway with sediments derived from the Late Cretaceous uplift of the Rocky Mountains (Stoffer, 2003). During the late Maastrichtian, the Western Interior Seaway was in its final retreat, accompanied by the progradation of the deltaic Fox Hills Formation in southwestern South Dakota. The portion of the Fox Hills delta analyzed for this study was along the western margin of the Western Interior Seaway (Lillegraven and Ostresh, 1990).

The Fox Hills Formation Type Areas

The Fox Hills Formation is exposed over a large portion of South Dakota (Fig. 2), but it is manifested differently in various parts of the state due to the diachronous nature of deltaic sedimentation and associated facies. Previous studies of the Fox Hills Formation by Waage (1968) and Pettyjohn (1967) are based on exposures 70 km farther north-northeast in Meade, Zeibach, and Corson Counties, and on exposures along the Montana/Wyoming border (Gill and Cobban, 1973). Our research was conducted in the Badlands National Park region in southwestern South Dakota (Figs. 1 and 2). See Stoffer et al. (2001) and Stoffer (2003) for a detailed background discussion of the stratigraphic nomenclature and type sections of the Fox Hills Formation.

Badlands National Park Area

The stratigraphy and correlation of the Fox Hills Formation is very problematic. Some of the discrepancy is likely the result of several type sections described at various times by different researchers (Pettyjohn, 1967; Waage, 1968; Gill and Cobban,

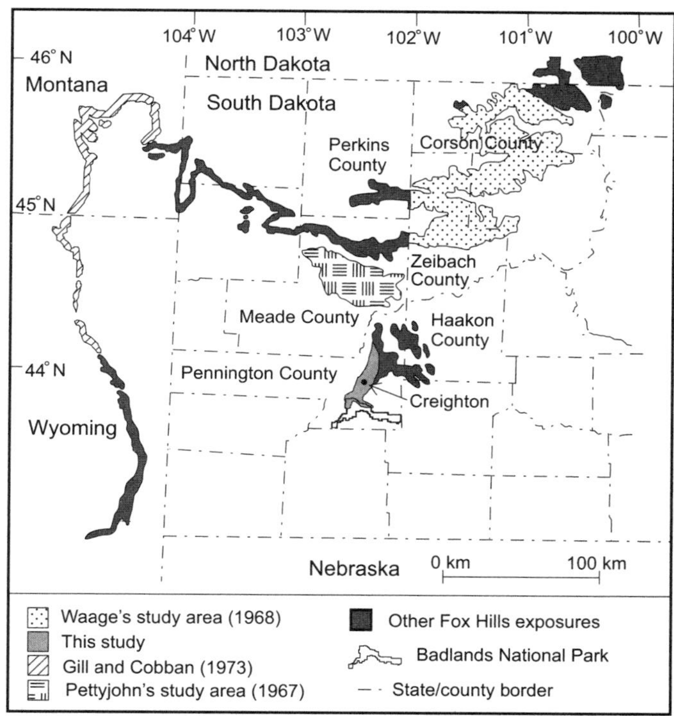

Figure 2. Outcrop map of the Fox Hills Formation. Modified from Chamberlain et al. (2001).

1973). Because the regional assignment of members to facies and lithologies is so variable, the stratigraphic section used for correlation of the badlands exposures is a composite of these different type areas (Fig. 3). Even with the hybridization of these type areas, there is still confusion about the lateral and vertical relationships. According to Stoffer (2003), the simplest lithologic divisions of the formations in the southwestern South Dakota region are as follows: the fossiliferous, marine Pierre Shale (Elk Butte Member) and the deltaic Fox Hills Sandstone, which contains some marine fossils and detrital plant material.

Outcrops of the Fox Hills Formation in our study area do not match any of the previously described type sections because strata are missing in the badlands. The question of why strata are missing was explored by Stoffer (2003), who proposed two explanations: (1) erosion and/or (2) that structural development at the time of deposition, such as the uplift of the Black Hills, and resulting local structures such as the Chadron, Sage Creek, and Creighton arches, possibly prevented an influx of sediments to the badlands region or created condensed sections.

The strata in the Badlands National Park region show transitional characteristics from the marine Pierre Shale to the near-shore deltaic Fox Hills Formation (Fig. 4). There is no evidence of terrestrial facies or coal deposits in the Fox Hills Formation in the badlands area, such as those of the Stoneville Coal facies near Enning (Pettyjohn, 1967), nor does the massive cross-bedded

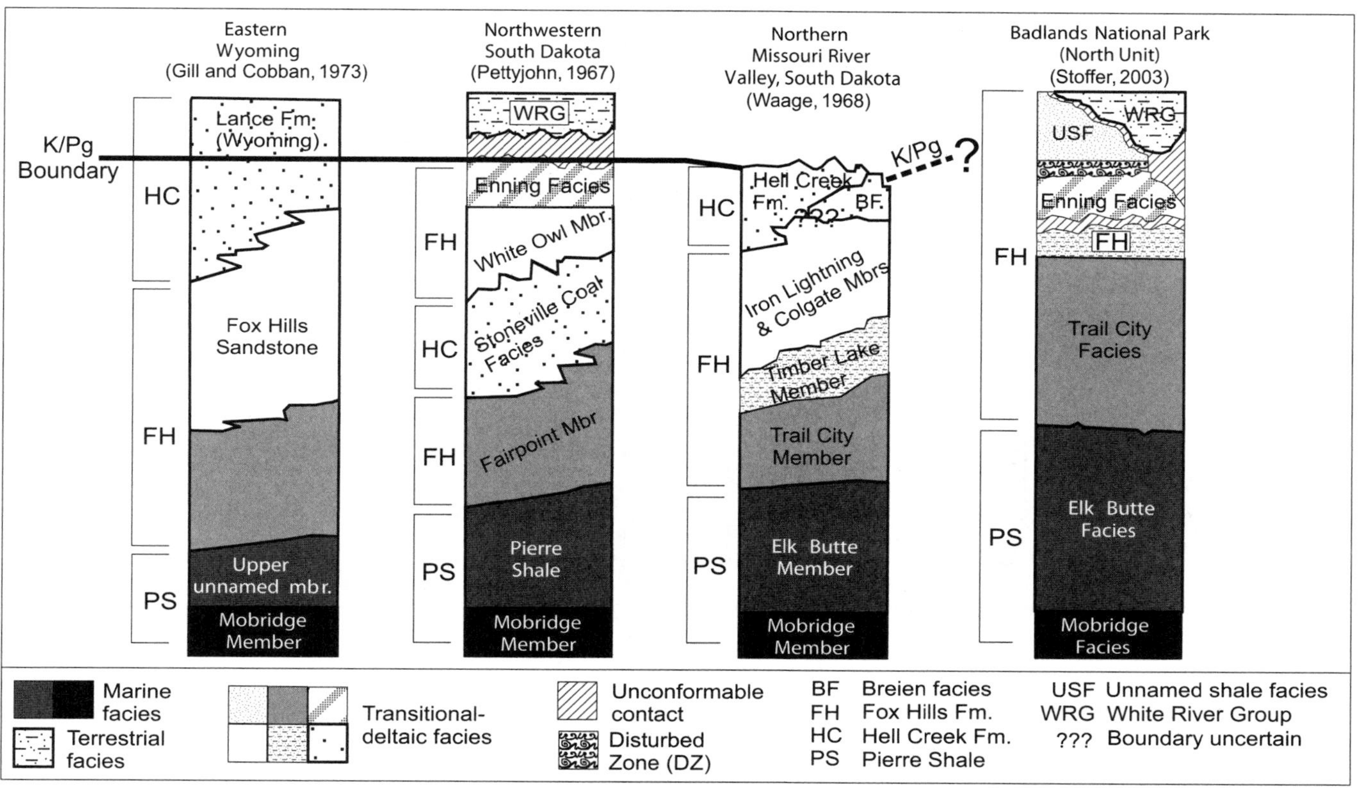

Figure 3. Generalized correlation of the Fox Hills Formation between different type areas based on similarities in lithology and fossil content. Pattern fills do not represent lithologies, but are used instead to show correlations. Time and thickness are not quantified. Modified from Stoffer (2003).

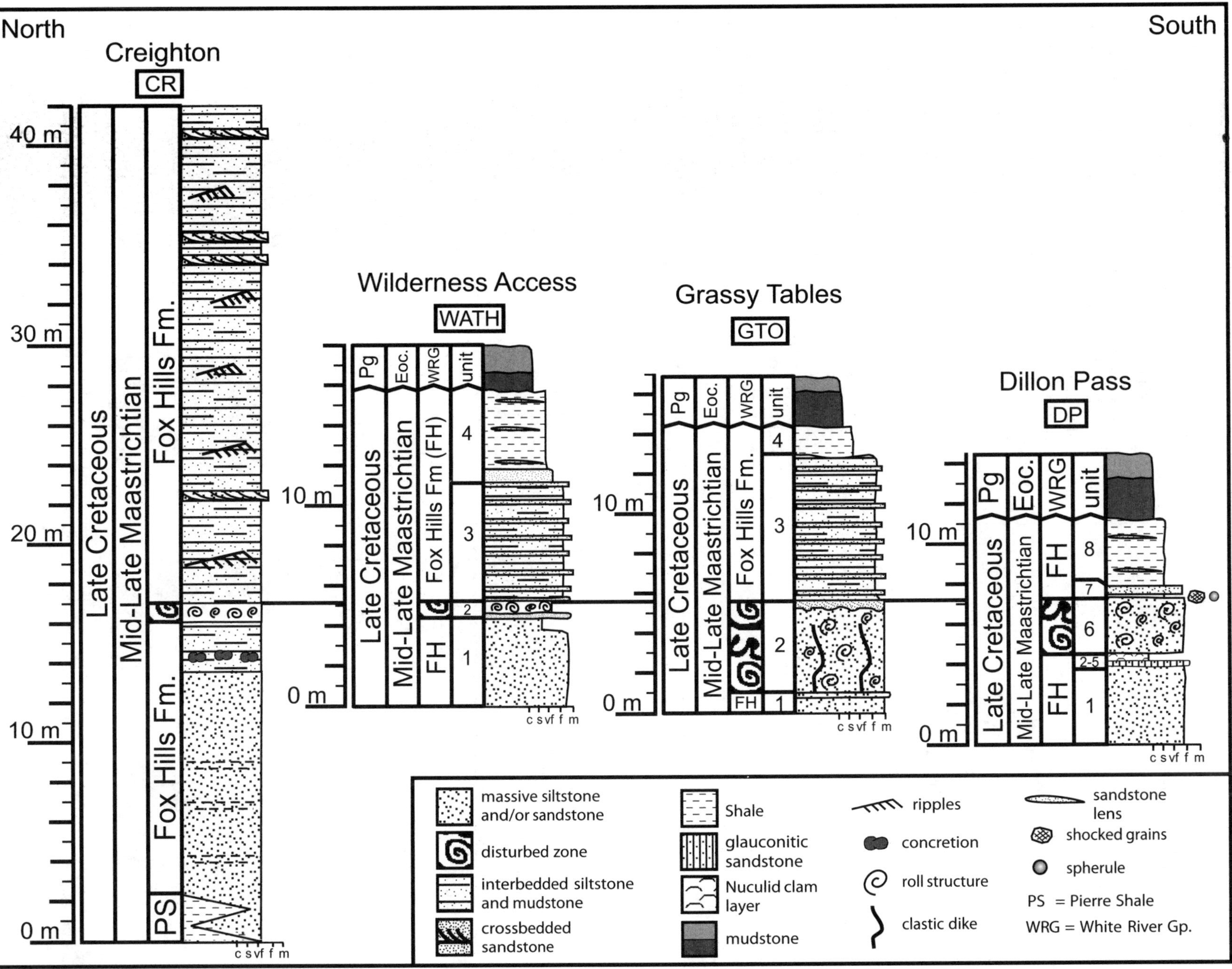

Figure 4. Correlation of measured sections across the study area. Unit numbers correspond to descriptions in text and outcrop photographs in Figure 5. c—clay; f—fine sand; m—medium sand; s—silt; vf—very fine sand.

sandstone of the upper Timber Lake Member occur as is seen in the eastern Wyoming type area (Gill and Cobban, 1973). According to Chamberlain et al. (2001), the lowermost Fox Hills Formation outcrops in the badlands are similar to those in the middle of the type area as described by Pettyjohn (1967), in terms of both lithology and fossil content. However, the age of these deposits in the badlands is currently in question due to the continuous, but variable, progradation over time of individual delta lobes and associated environments. Current palynological data suggest a late Maastrichtian age for the Fox Hills Formation at the Creighton locality (Palamarczuk et al., 2003). How this relates to the type sections in the Enning, Corson County, and Montana/Wyoming border areas is not yet known (Fig. 3).

Within Badlands National Park, the DZ is preserved within bright pink, orange, and yellow strata referred to as the "Interior Zone" underneath the Eocene-Oligocene White River Group (Retallack, 1983). These brightly colored beds, which stretch across the northern Great Plains (Pettyjohn, 1965), have been interpreted in a variety of ways and have held a variety of names throughout the past 30 years (Terry, 1998). The current interpretation in the Badlands National Park area is that the Interior Zone represents at least two separate periods of ancient pedogenesis. The first episode of pedogenesis is the Yellow Mounds Paleosol Series (Retallack, 1983), the yellow sediment that makes up the bottommost part of the Interior Zone (Figs. 5A and 5B). This paleosol formed on Cretaceous Pierre Shale and the Fox Hills Formation (Stoffer et al., 1998). These brightly colored sediments gradually transition downward into unaltered black Pierre Shale and tan Fox Hills Formation. The Yellow Mounds Paleosol Series is similar to modern ultisols: extremely weathered, forest soils with accumulations of subsurface clay, iron, alumina, or humus (Retallack, 1983).

Overlying the Yellow Mounds Paleosol Series is a second episode of pedogenic modification, the Interior Paleosol Series, recognized as the bright red strata at the top of the Interior Zone. This series formed on nonmarine sediments and was likely similar to modern alfisols: fertile forest soils with subsurface clay accumulations (Retallack, 1983). Later work by Terry and Evans (1994) and Evans and Terry (1994) combined the Interior Paleosol Series with the white channel sandstones at the base of the White River Group to form the late Eocene Chamberlain Pass Formation.

Pedogenic modification of our study sections in Badlands National Park is associated with development of the Sage Creek Anticline. The Sage Creek Anticline occurs along the northern boundary of Badlands National Park and is likely related to the Laramide uplift of the Black Hills that began in the Late Cretaceous and lasted into the Eocene (Lisenbee and DeWitt, 1993). This uplift was active at the time of the impact (Stoffer et al., 2001) and was responsible for the eventual exposure, pedogenic modification, and erosion of some of our study sections (Jannett and Terry, 2001), specifically those sections located closer to the arch (Figs. 1 and 4).

The effects of pedogenesis on the preservation potential of impact ejecta (both geochemical and mineralogical phases) are unknown; however, any degree of pedogenesis should have an effect on ejecta preservation. Previous research on paleopedology related to ejecta includes a study of acid trauma associated with the K-Pg impact fallout at Bug Creek, Montana (Retallack, 1996), in which the dual nature of the boundary and overlying impact layers was attributed to the neutralization of acids generated by the impact, and a study by Fastovsky et al. (1989) in which paleopedology was used to interpret the depositional history of the K-Pg boundary layer.

MATERIALS AND METHODS

A vertical trench was excavated through the DZ at each locality to a depth of 30–100 cm in order to expose unweathered sediments. The sediments were described and recorded for color (all fresh unless otherwise noted) using the Munsell soil color charts (Munsell Color, 1975), lithology, thickness, fossil content, bioturbation/pedogenic features, paleocurrents, and soft-sediment deformation characteristics. Reaction of fresh samples to HCl was also recorded for each horizon. Oriented samples were collected for thin-section analysis. Horizontal and vertical thin sections were prepared by dry cutting and polishing to avoid expansion of smectite clays. Ejecta was noted and described based on the criteria of French (1998) and Izett (1990). Paleosol features were described based on the descriptions of Retallack (1990), Fitzpatrick (1993), and Stoops (2003). Samples for microfossils were collected at 1 m intervals spanning the Elk Butte Member of the Pierre Shale, through the DZ, and 27 m above the DZ into the Fox Hills Formation at Creighton. Samples were analyzed for calcareous nannoplankton by Beck Biostrat, Inc., and for palynology by Palamarczuk et al. (2003).

DETAILED LITHOLOGIC DESCRIPTIONS

Localities Within Badlands National Park

Dillon Pass

The Dillon Pass section is located northwest of the Dillon Pass Overlook in the Wall SW 7.5' quadrangle, SW ¼, NE ¼, SE ¼, Section 20, Township 2S, Range 16E, and positioned almost directly on the crest of the Sage Creek Anticline (Figs. 1 and 4). Dillon Pass is thinner than the other sections and is composed of eight distinct units. The section begins with the unaltered grayish black Elk Butte Member of the Maastrichtian Pierre Shale. The shale is overlain by unit 1, a massive, weakly bedded, fine- to medium-grained olive yellow (2.5 YR 6/6) and light greenish gray (GLEY 1 7/1) muddy, clay-rich sandstone that varies from matrix- to clast-supported and is composed of quartz, mica, feldspar, and rare glauconite. Secondary calcite, gypsum, chlorite, and goethite are present as patchy cements. The unit shows slight skelsepic, lattisepic, and clinobimasepic fabrics as well as weak relict bedding. Unit 2 is a pale yellow (2.5 YR 7/3) zone of calcium carbonate nodules of variable thickness ranging from 0 to 5 cm. In thin section the nodules

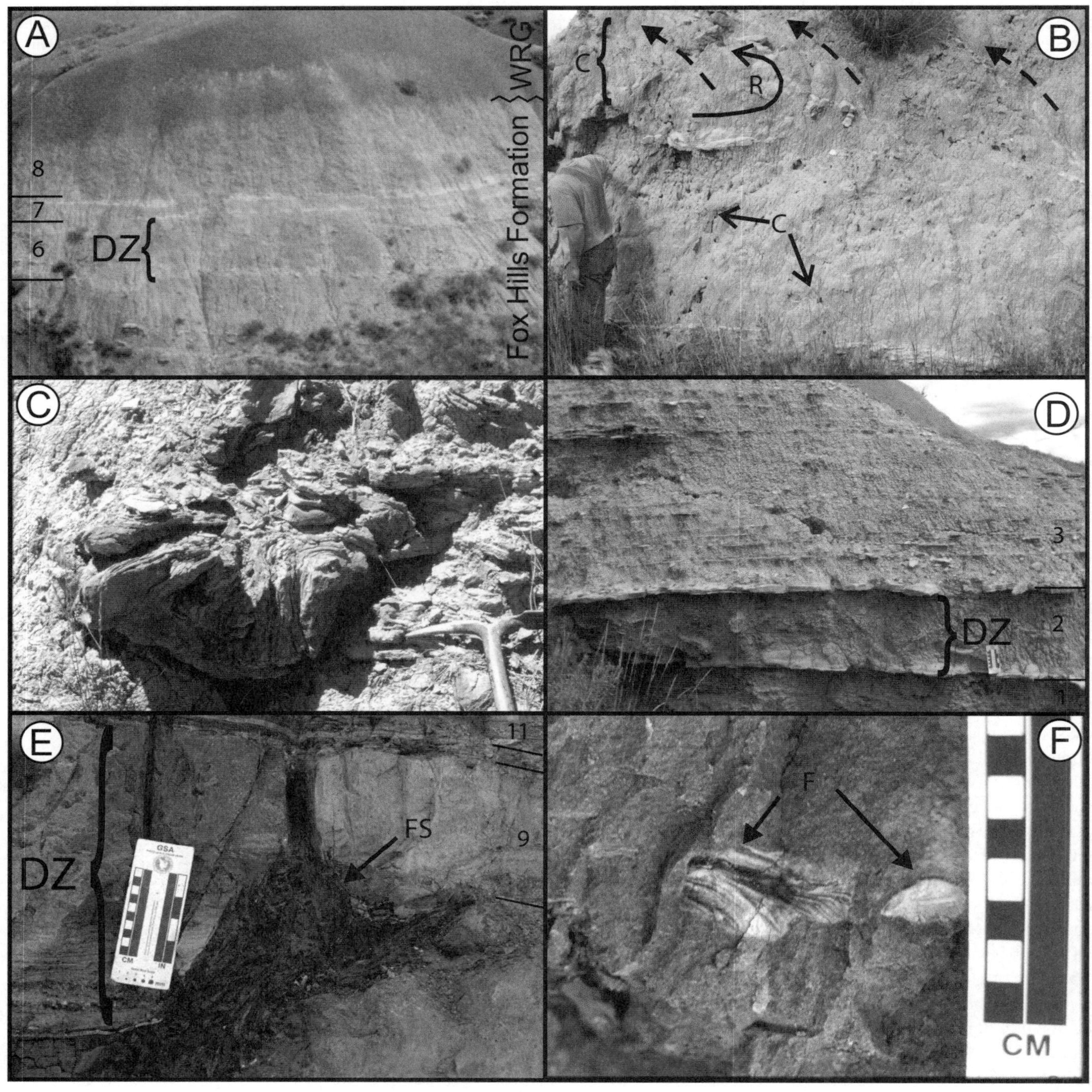

Figure 5. Photographs of the Disturbed Zone (DZ) and associated soft-sediment deformation features. Numbers refer to individual units described in the text and in Figure 4. A: Outcrop photograph at Dillon Pass showing the DZ within the brightly colored Interior Zone and the overlying, unconformable White River Group (WRG). The DZ is 2.5 m thick at this locality. B: Clastic dikes (C) and roll structures (R) at Grassy Tables Overlook. Note that dikes are bent in the same direction as slumping. C: Roll structure at Grassy Tables Overlook. Hoe pick for scale. D: Outcrop photograph of the DZ at the Wilderness Access Trailhead. Note scale at bottom right. E: Outcrop photograph of the DZ at Creighton. Note flame structure (FS). F: Foundered blocks of cross-bedded sandstone (F) within the homogenized DZ at Creighton. See Figure 1 for locations. A, B, and E modified from Terry et al. (2001).

have a crystic fabric and contain common grains of quartz, micas, a few charcoal fragments, and small, rare clusters of ferrihydrite.

Unit 3 is a 30 cm thick pale yellow (5 Y 7/3) interval. The lowest 18 cm of unit 3 resembles the weak laminations of unit 1, whereas the upper 12 cm of unit 3 has strong relict bedding. This unit is a muddy, clay-rich sandstone that varies from matrix- to clast-supported and is composed of quartz, mica, feldspar, and rare glauconite. Secondary calcite, gypsum, chlorite, ferrihydrite, and goethite are present as patchy cements. The unit shows slight skelsepic, lattisepic, and clinobimasepic fabrics as well as weak relict bedding. Unit 4 is 8 cm thick and is composed of a silty, olive yellow (2.5 Y 6/6), matrix-supported sandy mudstone with lenses of greenish gray (5 GY 5/1) glauconite and detrital quartz, feldspars, and micas. A slight clinobimasepic fabric is present.

Unit 5 is 15 cm thick and appears violet in outcrop, but Munsell colors are a mixture of brownish yellow (10 YR 6/8) and light gray (2.5 Y 7/2). This unit is a matrix-supported sandy mudstone composed of quartz, feldspar, mica, and rare glauconite. It contains secondary gypsum and calcite, along with patches of hematite cement. Microfabrics include slight skelsepic plasmic fabric, and relict bedding. This unit also contains weak red (10 R 4/3) clay-infilled burrows, rare nuculid clam fragments, and very rare fish scales.

Unit 6 is the DZ at this locality. It is a 2.5 m thick, severely distorted clayey siltstone to calcite- and iron oxide–cemented fine sandstone composed of quartz, mica, feldspar, and rare charcoal fragments (Fig. 5A). Gypsum is present as secondary void fills and distinct veins up to several millimeters thick. The bottom 40 cm of unit 6 is reddish gray (2.5 YR 7/1) with some disseminated plant fragments. The middle part of the DZ is reddish yellow (7.5 YR 6/6) intermixed with gray (5 YR 6/1) claystone lenses and contains layers of discontinuous carbonate nodules. Thin, discontinuous lenses of glauconite and slickensides are present throughout the unit. The top 10–20 cm of the DZ is pale red (2.5 YR 6.5/2) to light reddish brown (2.5 YR 6/3) and marked by small-scale roll structures at its top.

Unit 7 is interbedded with the bottommost 50 cm of unit 8. Unit 7 consists of two discrete olive gray (5 Y 6.5/2) very fine- to fine-grained sandstone intervals. The sandstones are composed of quartz, feldspar, and mica with rare glauconite grains and charcoal fragments that are calcite-cemented and crosscut by occasional millimeter-wide veins of gypsum. In thin section the sandstones display skelsepic fabric and partially oxidized cubes of secondary pyrite.

Unit 8 is composed of 3.6 m of laminated, thinly bedded claystone and siltstone (shale) that show strong relict bedding and weakly developed clinobimasepic and skelsepic fabrics. The shale is composed of clays, feldspar, mica, quartz, rare glauconite, and charcoal fragments, and abundant secondary iron oxide along relict bedding. The shale is light greenish gray (5 Y 8/1) for the first 60 cm and becomes progressively intermixed with weak red (2.5 YR 5/2) and brownish yellow (10 YR 6/8) shales at higher levels. Unit 8 is unconformable with the overlying late Eocene Chamberlain Pass Formation of the White River Group (Fig. 5A).

Grassy Tables Overlook

The Grassy Tables Overlook section is located ~1 km south of the Sage Creek Rim Road in the SW ¼, Section 12, Township 2S, Range 15E of the Quinn Table NE 7.5' quadrangle (Figs. 1 and 4). This section preserves the greatest thickness of disturbed sediments, up to 5 m, as well as a variety of deformation features not seen elsewhere. The Grassy Tables Overlook section is composed of four discrete units. Only the upper 90 cm of unit 1 is exposed in a stream cut at the base of the section. The bottom 45 cm of unit 1 is massive and composed of weak red to pale red (2.5 YR 5.5/2) fine sandstone with disseminated plant fragments and weak lamination at its base. The contact with the upper half of unit 1 is marked by a zone of discontinuous carbonate nodules up to 30 × 10 cm. The upper part of unit 1 is laminated, mostly shale with infrequent lenticular and glauconitic sandstone bodies up to 1 cm thick.

Unit 2 is the DZ at this locality and is composed of ~5 m of light reddish brown (2.5 YR 6/3) clayey siltstone and fine to medium sandstone that weathers light brown to brownish yellow (7.5 YR 6/4 and 10 YR 6/6). Deformation features are extremely well developed and exposed at this location, including large-scale slump and roll structures, clastic dikes, and a massive homogenized sandstone at the top of the unit (Fig. 5B). The roll structures are commonly preferentially cemented, likely due to enhanced porosity induced by liquefaction, and remain as resistant log-like features within outcrop or on surrounding slopes (Fig. 5C). Clastic dikes range from nearly vertical hairline to centimeter-thick iron-stained traces that pinch out upward and are associated with large-scale roll structures (Fig. 5B). Approximately 2.5 m above the base of the unit is a carbonate nodule layer that is ~10 cm thick. Pedogenic indicators are absent within this unit, although rare detrital plant fragments are present throughout. The top 20 cm of this unit is a light gray bulbous and massive sandstone that protrudes downward. Individual lobes are separated by flame structures.

Unit 3 is 7.25 m thick and is composed of interbedded mudstone, siltstone, fine to medium sandstone, and shale. The sandstone is composed of a light brownish gray (2.5 YR 6/3) and grayish brown (2.5 YR 5/2) resistant, flaggy sandstone and less resistant white (5 Y 8/1) sandstone. The sandstone layers range from 2 to 10 cm thick and contain small-scale trough cross-bedding, lower-phase planar beds, and ripple laminations. Small-scale mud cracks are present in the resistant sandstone layers, as well as burrows perpendicular to bedding and occasional slickensided fracture surfaces. Unit 4 is 150 cm thick and composed primarily of olive brown (2.5 Y 4/4) shale with occasional light gray (5 Y 7/2) and yellowish brown (10 YR 5/6), 1–5 cm thick, indistinct and low-angle cross-bedded sandstone lenses. The unit contains dark brown (10 YR 3/3), dark yellowish brown (10 YR 4/6), and olive gray (5 Y 4/2) clay skins and slickensides, possibly of pedogenic origin, and 0.5 cm diameter dusky red (10 R 3/3) burrows both horizontal and perpendicular to bedding.

The contact with unit 5, the late Eocene Chamberlain Pass Formation, is unconformable, although pedogenic activity appears to increase upward toward this contact. Whether this increased amount of pedogenesis is related to modification of the Chamberlain Pass Formation or an earlier period of pedogenic modification is unknown. The Chamberlain Pass Formation is ~2 m thick at this location.

Wilderness Access Trailhead

The Wilderness Access Trailhead locality is located ~0.32 km south of the Rodeo Point parking lot on the Sage Creek Rim Road within Section 3, Township 2S, Range 15E of the Quinn Table NE 7.5' quadrangle and is accessible by a bison trail (Figs. 1 and 4). This section is on the southwestern flank of the Sage Creek Anticline and is located at the base of a series of normal faults that displace the exposures into several repetitive sections. The section is composed of four units (Fig. 5D). Unit 1 is 4.62 m thick and composed of light brown (7.5 YR 6/3) to yellowish brown (2.5 Y 6/4), medium light greenish yellow (GLEY 1 7/10) to brownish yellow (10 YR 6/8) argillaceous fine to medium sandstone with a yellow (10 YR 7/6) popcorn weathered surface (due to high smectite content). The sandstone also contains detrital quartz, micas, feldspar, and rare organic fragments. Some areas are cemented by iron oxide, whereas others are partially cemented by authigenic chlorite and gypsum. Clays show skelsepic and clinobimasepic fabrics. Thin, discontinuous lenses of glauconite are present throughout the unit as well as a distinct 2 cm thick glauconite layer ~1.8 m above the base of the section. The glauconite contains quartz, feldspar, gypsum, micas, and rare organics, as well as authigenic chlorite and pyrite. Some areas are cemented by goethite. Clays show skelsepic fabric.

Unit 2 is the DZ at this locality (Fig. 5D). It is composed of light gray (5 Y 7/1) and pale red (10 R 6/3) homogenized, matrix-supported clayey siltstone and very fine to fine sandstone with resistant roll structures oriented approximately east-west. The siltstone shows crystic fabric and contains disseminated calcium carbonate fragments, quartz, mica, feldspar, rare glauconite, and rare detrital organics. Some areas are cemented by iron oxide, and there is slight relict bedding present as well. The DZ is 56 cm thick and bounded on top and bottom by light gray (5 Y 7/1) resistant sandstone sheets. The lower sheet is highly deformed, 6–10 cm thick, whereas the upper sandstone sheet is undeformed and 3.5–4 cm thick. The sandstone is calcite-cemented and contains rare organic fragments and glauconite.

Unit 3 is 6.16 m thick and consists of thinly bedded, laminated sandstones and shales identical to unit 3 at the Grassy Tables Overlook locality. The sandstones are light yellow brown (10 YR 6/4) and yellow brown (10 YR 5/8), finely cross-bedded, with some ripple laminations. There are no consistent thicknesses of the laminations, but the sandstones are commonly 10–15 cm apart. The shales are weak red (10 R 5/2) and light olive brown (2.5 Y 5/6). Vertical and horizontal burrows are present, including *Ophiomorpha*, chevron-shaped isopod trails on bedding planes, and straight pencil-like *Planolites* (Chamberlain et al., 2001).

Unit 4 is 4.62 m thick. The bottom 65 cm of this unit is a massive olive yellow (2.5 Y 6/6 and 2.5 Y 6/8) muddy, smectite-rich sandstone composed also of quartz, feldspars, micas, and rare charcoal fragments. It is cemented by calcite and clay with occasional zones of iron oxide along relict bedding. Slickensided fracture surfaces (due to the high smectite content), burrows, and features similar to drab-haloed root traces, although stained light reddish brown (2.5 YR 6/3), are also present. The remainder of unit 4 is multicolored shale, weak red (10 R 4/3), yellowish brown (10 YR 5/6), and dark grayish brown (10 YR 4/2), with quartz, mica, rare to uncommon feldspars and rare glauconite grains. Slickensided fracture surfaces, burrows perpendicular to bedding, and lenticular bodies of very fine sandstone are also present. The contact with the overlying late Eocene Chamberlain Pass Formation is unconformable. The Chamberlain Pass Formation is 65 cm thick and overlain by a thin remnant of the Peanut Peak Member of the Chadron Formation.

Localities Outside Badlands National Park

Creighton

The Creighton section is located 35 km north of the Badlands National Park near the town of Creighton, South Dakota, in the N ½, NW ¼, Section 34, Township 3N, Range 15E of the Wasta NE 7.5' quadrangle. It is within the upper part of the Cheyenne River drainage and is accessed by a gravel road. This section is the farthest away from the crest of the Sage Creek Anticline and has the thickest and most complete section above the DZ (Figs. 1 and 4). Creighton is not pedogenically modified and therefore has pristine bedding features preserved above and below the DZ, as well as the best fossil preservation. The Creighton section begins with the unaltered Elk Butte Member of the Pierre Shale, which is accessible farther down the road at another outcrop. The contact with the overlying Fox Hills Formation is transitional over ~10 m. The DZ is just above this transitional zone. Both the top and bottom of the interval containing the DZ at this locality are arbitrary.

The bottom contact of unit 1 is covered but is at least 16 cm thick. Unit 1 is a very dark gray (10 YR 3/1) and yellowish brown (10 YR 5/8) clayey siltstone with faint yellowish brown laminations and nuculid clams. The very dark gray layers alternate with 1–2 cm thick, olive brown (2.5 Y 4/3) clayey silts. The contact with unit 2 is sharp and planar.

Unit 2 is an 8 cm thick black silty clay-shale, similar to the underlying Elk Butte Member, with the same invertebrate fossils as unit 1. The contact with unit 3 is sharp and planar. Unit 3 is 15 cm thick and similar to unit 1, except unit 3 contains occasional burrow fills ~0.5 cm in diameter. This unit also contains large amounts of carbonized plant fragments, plant fragment impressions, and nuculid clam shell hash. The contact with unit 4 is sharp and planar. Unit 4 is composed of dark olive gray (5 Y 3/2) slightly silty claystone. This 12 cm thick unit also contains silty and sandy burrows, nuculid clams, plant fragment impressions, and carbonized plant fragments. The contact with unit 5 is sharp and planar. Unit 5 is a 3 cm thick, laminated to cross-laminated, pale olive (5 Y 6/3) siltstone with submillimeter dark reddish brown (5 Y 3/2) clay drapes. The unit also contains carbonized and red-stained plant hash and occasional clam fragments. Paleocurrent measurements (N = 2) suggest a SSE flow direction (160°), which is consistent with an overall southeast direction of flow for the Fox Hills Formation in the study area. Laterally, this unit becomes as thin as 2 cm and shows evidence of soft-sediment deformation associated with the DZ. The contact with unit 6 is sharp and slightly wavy. Unit 6 is 7 cm thick and composed of dark olive gray (5 Y 3/2) silty claystone with carbonized plant hash and clams. The contact with unit 7 is very undulatory.

Unit 7 is the lower part of the DZ. It is a 5 cm thick, yellowish brown (10 YR 5/8) and pale olive (5 Y 6/3) massive sandstone. This unit also contains horizontal gypsum veins up to 3 mm wide, and vertical orange-stained (10 R 5/8) cracks. This unit contains abundant lobate soft-sediment deformation with foundering of unit 7 into unit 6. The contact with unit 8 is sharp and undulatory.

Unit 8 is composed of 8 cm of laminated sandstone with undulatory bedding. The basal 2 cm of this unit are a very dark brown (7.5 YR 2.5/2) clay-rich zone with abundant plant hash, films, and carbonized fragments. The middle section of unit 8 is a 3–4 cm thick light yellowish brown (2.5 Y 6/4) zone with abundant very dark brown (7.5 YR 2.5/2) clay laminae and plant fragments. The top portion of unit 8 is a 2 cm thick light yellowish brown (2.5 Y 6/3) and yellowish brown (10 YR 5/8) laminated sandstone. The contact with unit 9 is sharp and undulatory.

Unit 9 is the main DZ sand body. It is a massive olive gray (5 Y 4/2) unit that has been totally liquefied and homogenized, with the exception of occasional coherent blocks of cross-bedded sandstone that foundered into the unit (Figs. 5E and 5F). The DZ sand is 24–25 cm thick at this section, but thickness varies by 10–15 cm along the outcrop. The unit contains nuculid clams, carbonized plant fragments and impressions, scaphite fragments, rare shark teeth, and distinctive iron oxide concretions. The concretions are yellowish red (5 YR 5/8) rounded to dumbbell-shaped bodies composed of iron oxide (goethite and hematite)–cemented grains of variable mineralogy, including quartz and feldspar. They occur up to ~5 mm in size. Some have been replaced entirely by gypsum, while others have a plane of gypsum through their centers. Also, some of the larger bodies have a pyrite center. The contact with unit 10 is undulatory and sharp.

Unit 10 is an undulatory 0.5–2 cm black claystone. It is slightly silty with faint orange (10 R 5/8) laminations. The upper contact with unit 11 is sharp and undulatory. Unit 11 is a 10 cm thick light yellowish brown (2.5 Y 6/3) and yellowish brown (10 YR 5/8) sandstone marked at its base by a 2 cm thick laminated zone of carbonaceous material. The upper 8 cm of this unit is herringbone cross-stratified. Paleocurrents are bidirectional SSE and NNW (170° and 350°, N = 2). The contact with unit 12 is sharp and wavy.

Unit 12 is a 2 cm thick dark grayish brown (2.5 Y 4/2), laminated siltstone and claystone with coalified plant fragments. Silt bands are dark grayish brown (2.5 Y 6/3). The contact with unit 13 is abrupt and slightly wavy. Unit 13 is a 2 cm thick, very dark brown (7.5 YR 2.5/2), light yellowish brown (2.5 Y 6/4), and yellowish brown (10 YR 5/8) cross-laminated sandstone (SSE, 170°, N = 1) with plant fragments at the top of the unit. The contact with unit 14 is sharp and undulatory. Unit 14 is a 24–25 cm thick silty clay with ≤0.5 cm sand lenses. The colors are the same as units 12 and 13. The contact with unit 15 is sharp and slightly wavy.

Unit 15 is a 19–20 cm cross-laminated sandstone (SSE, 170°, N = 1) with 1–2 cm thick zones of laminated, carbonaceous, silty clay at the 5–7 cm and 15–17 cm intervals. The contact with unit 16 is sharp and slightly wavy. Unit 16 is a grayish brown (2.5 Y 5/2) silty clay-shale with sand stringers up to 1–2 cm thick. Colors are the same as unit 15. Abundant plant fragments are present as impressions and carbonized traces, but no invertebrate fossils are observed. This same lithology dominates throughout the rest of the section, with occasional zones of sandstone up to 20 cm thick.

IMPACT EVIDENCE

Ejecta

We identified spherules from the –2 to +6 cm interval at Dillon Pass (DP) and the herringbone cross-stratification zone (unit 11) from +2 to +12 cm above the top of the DZ at Creighton (Figs. 6A–6D). The spherules range in size from 50 to 110 µm. Spherule mineralogy has not been identified, although photomicrographs are very similar to images of feldspar spherules from the Manson impact structure in Iowa (Short and Gold, 1996). The spherules appear almost perfectly round in thin section and exhibit pseudo-uniaxial interference figures, undulatory extinction, and occasional compression fracturing and deformation due to compaction. No attempts were made to determine the concentration of spherules.

Shocked quartz grains ranging in size from 0.1 to 0.25 mm were recovered from the herringbone cross-bedded sandstone +2 to +12 cm above the DZ (unit 11) at Creighton and from 2 cm below to 2 cm above the top of the DZ (unit 6) at Dillon Pass. The grains show evidence of planar deformation features (PDFs) in multiple directions (Figs. 6E and 6F). In addition to PDFs, some grains exhibit undulatory extinction. No attempt was made to determine the concentration of shocked grains or PDF orientations.

Seismicity and Liquefaction

The DZ in the badlands area was originally interpreted by Stoffer et al. (1999) as pro-delta slumping. Later interpretations by Jannett and Terry (2001) and Terry et al. (2001) suggested that the deformation was the result of seismic activity generated by the Chicxulub impact, although this is unlikely given current biostratigraphic data (Palamarczuk et al., 2003). Distinctive soft-sediment deformation features within the DZ include slump and roll structures, flame structures, foundering of the DZ into underlying units,

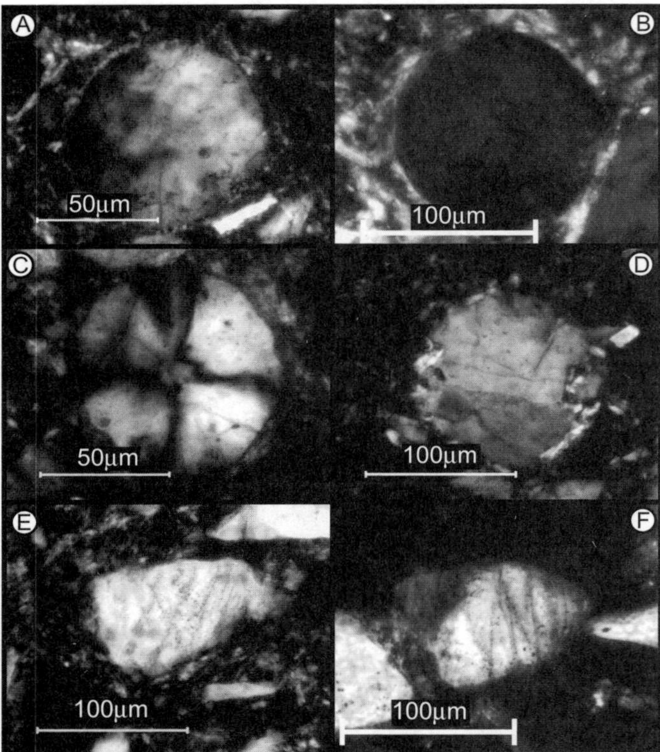

Figure 6. Photomicrographs of spherules (A–D) and shocked quartz (E–F) from the Disturbed Zone (DZ). A–E are from 2 cm below to 6 cm above the top of the DZ at Dillon Pass. F is from unit 11 at Creighton. Refer to Figures 1, 4, and 5E for locations and stratigraphic positions. A is modified from Terry et al. (2001). All photomicrographs in cross-polarized light.

blocks of coherent sediment within homogenized strata, injection dikes, and overall highly contorted bedding over a 1000 km^2 area of southwestern South Dakota (Fig. 5). Roll structures and clastic dikes have a preferred east-west orientation (Terry et al., 2001).

The presence of widespread deformation alone is not enough to make a conclusion of impact origin, due in part to the fact that earthquakes also can produce similar features. However, the DZ is a discrete and distinct unit within the Fox Hills Formation in this region, which suggests that it is not earthquake-related. Stratigraphic sections in seismically active areas tend to contain multiple widespread zones of disruption (Ettensohn et al., 2002). There are no other units in our study sections similar to the DZ.

Localized zones of slumping characteristic of deltaic environments have been noted in the Bullhead facies of the Fox Hills type area (Waage, 1968). The isolated zones of soft-sediment deformation pinch out laterally and are up to 30 m wide and 3 m thick. The DZ is too laterally extensive to be confidently interpreted as typical delta slumping. In addition, the almost total liquefaction and disruption of sedimentary structures and original fabrics in some of our sections suggests intense shaking as opposed to simple slumping (Figs. 5E, F). Similar interpretations of impact seismicity have been proposed for laterally extensive zones of soft-sediment deformation within the Triassic of the United Kingdom (Simms, 2003) and the Fundy Basin (Tanner, 2003). Impact seismicity may also be responsible for the genesis of the Crow Creek Member of the Pierre Shale, a distinct sandy unit containing ejecta within the Pierre Shale that has been linked to the Manson impact (Izett et al., 1993; Katongo et al., 2004).

PALEOSOL OVERPRINTING

Our study sections within Badlands National Park were not pedogenically modified during the Cretaceous, but were marginally affected by late Eocene soil formation. Our initial assumption was that the Dillon Pass, Wilderness Access Trailhead, and Grassy Tables Overlook sections were intensely pedogenically modified. This was due in part to the bright colors and the previous interpretation of ultisols within the Interior Zone (Retallack, 1983). However, upon further examination of the Dillon Pass section, we noticed extremely well-preserved relict bedding and a lack of fossilized root traces. Occasional slickensided surfaces were preserved throughout the section, but only one burrow was found below the DZ.

No roots or burrows were observed in thin section; however, sporadic and inconsistent weakly developed skelsepic, clinobimasepic, and lattisepic soil fabrics were observed throughout the section, even when contorted within the DZ. Strong relict bedding was the most prevalent fabric throughout the section. Preservation of relict bedding suggests that pedoturbation was not excessive, which is counter to the concept of a well-developed soil such as an ultisol.

Carbonate nodules are present in the Dillon Pass section, which would be unlikely if this was an ultisol because $CaCO_3$ would not be favored by intense hydrolysis and leaching. Mineralogically, the Dillon Pass section contains fresh feldspar and glauconite throughout, both of which would have been destroyed under conditions conducive to the formation of ultisols. The preservation of spherules also argues against the formation of an ultisol, which should have been easily leached from the profile. We also noted relict bedding in outcrop and in thin section. In order for relict bedding to be preserved, any soil formation that occurred would have to be weak at best. An ultisol is a well-developed soil, and its formation should destroy any remains of relict bedding. Molecular weathering ratios calculated for the Dillon Pass and Wilderness Access Trailhead sections are similar to those of Creighton, which also suggest that pedogenic modification was minimal (Jannett, 2004).

At the Wilderness Access Trailhead locality, macroscopic evidence of ancient soil formation is rare but includes slickenside ped surfaces and one possible clay-lined root trace near the middle of the measured section. In thin section, sporadic and inconsistent weakly developed skelsepic and clinobimasepic soil fabrics were observed throughout the section. Relict bedding was the most prevalent fabric throughout. At Grassy Tables Overlook, rare slickensides were observed throughout the measured section in addition to occasional horizontal and vertical burrows. No roots were observed in outcrop. The rarity of pedogenic features in combination with extremely well-preserved relict bedding suggests that the Wilderness Access Trailhead and Grassy Tables Overlook localities were not intensely modified by ancient soil formation. At Creighton the bright colors seen within the park are absent, and only relict bedding was observed in outcrop and thin section, both of which support the initial interpretation that the Creighton locality was not pedogenically modified.

By using the DZ as an isochronous, basinwide marker, regional patterns of overall section preservation and subsequent late Eocene pedogenic overprinting can be seen. Sections closer to the axis of the Sage Creek Anticline, such as Dillon Pass on the crest of the anticline, are less complete, whereas the sections farther from the crest of the arch have much more section preserved above the DZ (Figs. 1 and 4). Only 6 m of section are preserved between the DZ and overlying White River Group at Dillon Pass, whereas at Grassy Tables Overlook and Wilderness Access Trailhead there are 9 and 10 m of section, respectively. At Creighton the section continues for at least 27 m before truncation by modern erosion.

Initially, the development of local structure was thought to play a major role in the direct pedologic overprinting of these sections. The sections closer to the Sage Creek Anticline are thinner and brightly colored, possibly due to pedogenic overprinting from the Yellow Mounds and/or the Interior Paleosol Series. Based on the presence of extremely well-preserved relict bedding, it is likely that the upper A and B horizons of the paleosol(s) responsible for pedogenic modification and bright coloration of the study sections in Badlands National Park were removed by erosion as the Sage Creek Anticline continued to develop. What remains of this pre–White River Group period of pedogenesis is the lowermost brightly colored C horizon formed by oxidation. The amount of coloration and overprinting created by the formation of the late Eocene Interior Paleosol Series is unknown, although according to Terry and Evans (1994) this period of pedogenesis was extensive. Lateral tracing of the Interior Zone and DZ may provide sections that were affected by greater amounts of pedogenic modification.

AGE OF THE DZ

Macrofossils

The paleontology of the DZ interval was described by Chamberlain et al. (2001). We make reference to particular fossil groups for the purpose of age control, and report on new finds since Chamberlain et al. (2001). Fossil preservation is extremely variable throughout the study area due, in part, to differing degrees of late Eocene pedogenic overprinting of the Yellow Mounds and Interior Paleosol Series. The best preservation is at Creighton, while the poorest is at Dillon Pass.

Invertebrate fossils are most common, mainly bivalves and cephalopods (Chamberlain et al., 2001). Bivalves are numerous, but diversity is low, with abundant *Nucula cancellata* above and below the DZ, and one possible specimen of the inoceramid *Spyridoceramus tegulatus* from 10 cm below the DZ at Wilderness Access Trailhead (Fig. 4). Cephalopods are rare but are extremely important for biostratigraphic zonations. According to Stoffer et al. (2001), representatives of all Maastrichtian ammonite zones are present, even though the zones are condensed due to syndepositional tectonics of the Sage Creek Anticline. Ammonites and belemnites are present in the Fox Hills Formation, but baculitids are restricted to the underlying Pierre Shale (Chamberlain et al., 2001; Stoffer et al., 2001). This is consistent with the view of Cobban and Kennedy (1992) and Landman and Waage (1993), who assert that baculitids disappear before the end of the Cretaceous.

All ammonite specimens to date were recovered within or below the DZ. The ammonites recovered represent the *Jeletzkytes nebrascensis* ammonite zone, the youngest recognized ammonite zone of the Maastrichtian within the Western Interior Seaway (Kennedy et al., 1998). Reworking of the ammonites from the underlying sediments is unlikely due to the amount of strata present between the Fox Hills Formation and the underlying Pierre Shale. Also, the ammonites are well preserved, which would not be the case if they were reworked. Belemnites are more common and have been found in place from 1.5 to 13 m below the DZ in the vicinity of the Wilderness Access Trailhead (Chamberlain et al., 2001; Terry et al., 2004).

Vertebrates are uncommon in these sections, but include osteichthian scales scattered throughout the section and a few teeth of lamniform sharks that were recovered from within and slightly above the DZ at Creighton (Chamberlain et al., 2001). Trace fossils are common and include *Ophiomorpha*, *Climacodichnus*, and *Diplocraterion*-like burrows, and crawling traces of possible isopods or burrows of amphipods (Chamberlain et al., 2001).

Microfossils

Calcareous nannofossil data suggest that the overall microfossil assemblage of the Creighton section is Maastrichtian in age (Jannett, 2004; Terry et al., 2004). Unfortunately, the low degree of preservation and the likelihood of missing upper Maastrichtian sediments makes the nannofossil data inconclusive. Some microfossils regarded as Danian "survivor species" in other K-Pg sites are present at Creighton, such as *Cyclagelosphaera reinhardtii*, *Thoracosphaera* spp., and *Neocrepidolithus* spp., but here their occurrence is ambiguous as they also occur in the Maastrichtian. *Biscutum magnum* is the only marker species in these samples, which has its last occurrence in the middle Maastrichtian. In addition, the occurrence of this marker above the DZ would not be unusual given the obvious presence of burrowing and extensive reworking. How many, if any, Danian nannoplankton are present is not discernable from these samples.

Palynomorphs from Creighton consist of dinoflagellate cysts, gymnosperm and angiosperm pollen grains, and spores, among others (Palamarczuk et al., 2003). The palynomorphs and sedimentologic data from Creighton indicate a nearshore marine depositional environment for the basal 4 m of the section grading upward into a fluvial/deltaic paleoenvironment (Figs. 1 and 4). Beds above the DZ show a very restricted marine influence. The dinoflagellates indicate that the DZ is middle to early late Maastrichtian in age, although the possibility of reworking cannot be ruled out. The terrestrial palynomorphs have yet to be analyzed.

Based on the fossil data, the DZ and associated sediments of the Fox Hills Formation are likely Maastrichtian in age. The nannoplankton and palynomorphs indicate a middle to late Maastrichtian age; the macrofossils correspond to the *Jeletzkytes nebrascensis* ammonite zone, which spans 67.4–68.1 Ma; and unaltered belemnites (N = 10) recovered from 1.5 m below the DZ have yielded a $^{87}Sr/^{86}Sr$ age of 67.6 ± 0.5 Ma (Stoffer et al., 2001). All of these dates agree and indicate a middle to late Maastrichtian age for the DZ.

DISCUSSION

Lateral Variability of the DZ

Since deltas can encompass very large areas and multiple environments, the preservation potential of the ejecta and degree of expression of the DZ can also vary across the delta. Dispersed impact signatures would be favored by high rates of sedimentation, wave and tidal processes, soils with high porosity and permeability that would favor leaching and translocation of chemical signatures of impact, and oxidizing conditions that promote pedoturbation. Distinct impact signatures would be favored by low rates of sedimentation, calm or waterlogged conditions such as lagoons or swamps, and reducing conditions that retard pedoturbation.

The degree of soft-sediment deformation will be a function of water saturation and position within the delta wedge, with greater degrees of deformation favored by relatively higher water tables and slopes, such as along the delta front or near distributaries. Toward increasingly proximal delta positions, the degree of deformation should become less as water tables become relatively deeper to topset bed deposition.

The sections in our study area represent distal topset to foreset deltaic environments. The degree of slumping is greatest in sections to the south and east, with a decrease in overall thickness of the DZ to the north and west. This would suggest that sections to the south and east were more susceptible to liquefaction. This is in agreement with paleocurrent directions, which suggest progradation of this part of the delta to the southeast (Jannett, 2004).

While overturned beds, slumping, flame structures, and foundering of the DZ into underlying sediments is found in all sections, clastic dikes are found only within the thickest expression of the DZ at Grassy Tables Overlook (Figs. 1, 4, 5B), and complete homogenization is found in the slightly more proximal position at Creighton (Figs. 1, 4, 5E, 5F). Ejecta was recovered from 2 cm below to 6 cm above the DZ at Dillon Pass, while at Creighton it was recovered only from the herringbone cross-stratified unit 2–12 cm above the DZ (Figs. 1 and 6). This lateral variability in thickness of the DZ and ejecta distribution is likely a function of localized environmental processes across the delta. The degree to which the geochemical component of the ejecta was affected is unknown. Preliminary analysis suggests that geochemical signatures are nothing more than a function of lithology, although this requires more investigation (Jannett, 2004).

An Impactite Without a Crater

Our initial working hypothesis was that the DZ was a distal manifestation of the end-Cretaceous impact at Chicxulub, but due to recent biostratigraphic data, this hypothesis has come into question (Palamarczuk et al., 2003; Harries et al., 2002; Terry et al., 2002, 2004). The following discussion of potential impact sites is based on these new data, as well as previous work.

Manson Impact Event, Iowa

Previous studies suggested that meteoritic enrichment and ejecta at the K-Pg boundary could be the result of the Manson impact in north-central Iowa (Izett, 1990). This impact structure is one of the largest known in the United States, measuring ~37 km in diameter. Subsequent radiometric studies suggested a date of ca. 74 Ma (Izett et al., 1993). This date corresponds to the late Campanian, not the late Maastrichtian. In addition, ejecta from the Manson impact has been identified in Nebraska and South Dakota in the Crow Creek Member of the Pierre Shale (Izett et al., 1993; Katongo et al., 2004), which is stratigraphically below our DZ study interval.

End-Cretaceous Impact Events

The Chicxulub impact resulted in the widespread distribution of meteoritic elements, shocked mineral grains, and spherules formed from condensation of the vapor plume. Near the impact site, the evidence of impact is preserved as a chaotic mixture of ejecta and submarine mass wasting deposits that infilled the impact crater. The average size of shocked minerals and spherules decreases with distance from the impact site (Smit, 1999).

Within the Western Interior, the K-Pg boundary is preserved as a 2 cm thick (on average) layer of Ir-enriched clay with shocked quartz and spherules (Izett, 1990). North of the study area, the K-Pg boundary is preserved within the terrestrial Hell Creek Formation in northern South Dakota and North Dakota. From a facies perspective, the Hell Creek Formation represents the topset beds of the prograding Fox Hills delta. This led Terry et al. (2001) to suggest that the Fox Hills Formation in the Badlands National Park area was temporally equivalent to the Hell Creek Formation within northern South Dakota and that the DZ represented the K-Pg boundary within this distal deltaic environment. Based on the middle to late Maastrichtian age of microfossils from the DZ, this scenario is unlikely, although reworking of microfossils cannot be ruled out. Tracing of the DZ to the north into other type areas of the Fox Hills Formation may resolve questions surrounding the stratigraphic relations of the DZ to the Fox Hills and Hell Creek Formations.

While the concept of a large bolide impact at the end of the Cretaceous is widely accepted, some have suggested that the end of the Cretaceous was marked by multiple impacts that span from the Late Cretaceous into the early Paleogene, similar to a comet shower or a period of time when Earth was experiencing an influx of multiple bolides (e.g., Koutsoukos, 1999). More recently, Keller et al. (2004) have proposed that the Chicxulub impact and the K-Pg boundary represent two distinct events separated by ~300 k.y. Keller and her group assert that the Chicxulub impact actually occurred 300 k.y. *earlier* than the K-Pg boundary and that the "real" K-Pg event is an impact of worldwide proportions that caused the recognizable Ir enrichment as well as ejecta. If this is correct, then two zones of meteoritic enrichment should be preserved in distal locations. To date only one zone of enrichment associated with the K-Pg boundary has been documented in distal sections of the Western Interior Seaway, and our DZ is too old to be associated with the 300 k.y. pre-K-Pg impact scenario proposed by Keller et al. (2004).

Berwind Canyon, Colorado

A recently discovered impactite has been described within the late Maastrichtian Vermejo Formation in the Raton Basin of Colorado (Turner et al., 2003). The impactite is ~100 m below the K-Pg boundary. Shocked quartz and quartz/feldspar quench spherules are preserved at this locality. Preliminary $^{40}Ar/^{39}Ar$ dates of feldspars within spherules suggest an age of 68 Ma or younger (middle to late Maastrichtian). This date is comparable with the middle to late Maastrichtian age suggested by the macro- and microfossil data for the DZ. In addition, the stratigraphic succession of Pierre Shale–Trinidad Sandstone–Vermejo Formation is very similar to the Pierre Shale–Fox Hills Formation succession in western South Dakota, both of which record the overall retreat of the Western Interior Seaway at the end of the Cretaceous.

If the Berwind Canyon impactite is related to the DZ, then the crater responsible for both would likely be closer to the Berwind Canyon impactite based on size of ejecta. Assuming that the long axes of roll structures and clastic dike orientations within the DZ are orthogonal to the compressive stresses of the migrating seismic wave (Terry et al., 2001; Simms, 2003), the source crater is likely south of the study region.

CONCLUSIONS

We interpret the DZ as an impactite based on several lines of evidence. The DZ is unique in the study section and is unlike disruption that would be caused by earthquakes or volcanism. The DZ has been stratigraphically linked between all sections in the study area, and several new ones not included in this study (Terry et al., 2004), thus expanding the range of the DZ to 1000 km^2. Simple slumping along delta fronts is usually on the order of tens of meters. Other features suggestive of an impact origin include the presence of spherules and quartz grains with planar deformation features recovered from immediately above the DZ at the Creighton and Dillon Pass sections.

Evidence of ancient soil formation in these sections is very rare but includes some burrows and weakly developed and sporadic microscopic soil fabric. The degree of pedogenesis within these sections was insufficient to affect the preservation of ejecta and seismic disruption.

Although the DZ appears to be an impactite, no crater has been identified. Initially, the DZ was interpreted as a distal manifestation of the K-Pg impact event at Chicxulub (Chamberlain et al., 2001; Jannett and Terry, 2001; Terry et al., 2001), but the recovery of middle to late Maastrichtian microfossils from the study area, including above the DZ (Palamarczuk et al., 2003), strongly suggests that Chicxulub, or the recent interpretation of multiple impacts separated by 300 k.y. at the end of the Cretaceous (Keller et al., 2004), is not responsible for the formation of the DZ and associated ejecta.

According to the biostratigraphic data, the DZ impactite was formed during the Maastrichtian, which spans 70.6–65.5 Ma. Calcareous nannoplankton recovered from above and below the DZ at the Creighton sections suggest an age of middle Maastrichtian; the dinoflagellates recovered above and below the DZ suggest middle to early late Maastrichtian; and the macrofossils (scaphites and belemnites) recovered from only below the DZ also suggest middle to late Maastrichtian. Based on biostratigraphic data, the DZ likely formed within the period spanning 67.4–68.1 Ma. When the known Cretaceous impact locations and impactites are considered, the Berwind Canyon impactite in the Raton Basin appears to be the best match because (1) K-feldspar melt glass spherules from the Berwind Canyon impactite are dated to 68 Ma or younger and (2) both the Berwind Canyon impactite and the DZ impactite are underlain by the Pierre Shale and overlain by progressively shallower deposits. Providing that the Berwind Canyon impactite and the DZ are correlative, the impact was closer to the Berwind Canyon site based on the larger size of ejecta.

Regardless of any eventual correlation of the DZ to a recognized impact crater, the physical expression of an impactite within the deltaic Fox Hills Formation represents an isochronous marker that may eventually help to clarify the bio- and lithostratigraphic complexities associated with this delta system. In addition, the degree and style of preservation of impactites in deltaic settings should be expected to vary across the delta as a function of differences in depositional environments, thus generating a hybrid of the better-known terrestrial and open marine impactites.

ACKNOWLEDGMENTS

We thank Marty Becker, John Chamberlain, Matt Garb, Phil Stoffer, and Paula Messina for many hours of discussion on the finer points of stratigraphy and numerous hours of support during fieldwork. We thank the Jones Family of Quinn, South Dakota, for providing a base of operations during our many field seasons over the past several years. We thank Deborah Beck of Beck Biostrat, Inc., for calcareous nannoplankton data and Susanna Palamarczuk for dinoflagellate data. Thanks also to Rachel Benton and Marianne Mills of Badlands National Park for logistical support. Funding for this research was provided (to Terry) by the Research Incentive Fund program at Temple University. This manuscript was greatly improved by the comments of Emmett Evanoff and Dennis Ruez Jr.

REFERENCES CITED

Chamberlain, J.A., Jr., Terry, D.O., Jr., Stoffer, P.W., and Becker, M., 2001, Paleontology of the K/T boundary, Badlands National Park, South Dakota, *in* Santucci, V.L., and McClelland, L., eds., Proceedings of the 6th Fossil Resource Conference: National Park Service Geological Resource Division Technical Report NPS/NRGRD/GRDTR-01/01, p. 11–22.

Cobban, W.A., and Kennedy, W.J., 1992, The last Western Interior *Baculites* from the Fox Hills Formation of South Dakota: Journal of Paleontology, v. 66, p. 690–692.

Dean, W.E., and Arthur, M.A., 1998, Cretaceous Western Interior Seaway drilling project: An overview, *in* Dean, W.E., and Arthur, M.A., eds., Stratigraphy and paleoenvironments of the Cretaceous Western Interior Seaway, USA: SEPM Concepts in Sedimentology and Paleontology 6, p. 1–10.

Ettensohn, F.R., Rast, N., and Brett, C.E., editors, 2002, Ancient seismites: Geological Society of America Special Paper 359, 200 p.

Evans, J.E., and Terry, D.O., Jr., 1994, The significance of incision and fluvial sedimentation in the basal White River Group (Eocene-Oligocene), badlands of South Dakota, U.S.A.: Sedimentary Geology, v. 90, p. 137–152, doi: 10.1016/0037-0738(94)90021-3.

Fastovsky, D.E., McSweeney, K., and Norton, L.D., 1989, Pedogenic development at the Cretaceous-Tertiary boundary, Garfield County, Montana: Journal of Sedimentary Petrology, v. 59, p. 758–767.

Fitzpatrick, E.A., 1993, Soil microscopy and micromorphology: New York, John Wiley and Sons, 304 p.

French, B.M., 1998, Traces of catastrophe: A handbook of shock-metamorphic effects in terrestrial meteorite impact structures: Houston, Lunar and Planetary Institute Contribution 954, 120 p.

Gill, J.R., and Cobban, W.A., 1973, Stratigraphy and geologic history of the Montana Group and equivalent rocks in Montana, Wyoming, North and South Dakota: U.S. Geologic Survey Professional Paper 776, 36 p.

Harries, P.J., Johnson, K.R., Cobban, W.A., and Nichols, D.J., 2002, Marine Cretaceous-Tertiary boundary section in southwestern South Dakota: Comment: Geology, v. 30, p. 954–955, doi: 10.1130/0091-7613(2002)030<0955:MCTBSI>2.0.CO;2.

Izett, G.A., 1990, The Cretaceous/Tertiary boundary interval, Raton Basin, Colorado and New Mexico, and its content of shock-metamorphosed minerals: Evidence relevant to the K/T boundary impact-extinction theory: Geological Society of America Special Paper 249, 100 p.

Izett, G.A., Cobban, W.A., Obradovich, J.D., and Kunk, M.J., 1993, The Manson impact structure: $^{40}Ar/^{39}Ar$ age and its distal impact ejecta in the Pierre Shale in southeastern South Dakota: Science, v. 262, p. 729–731, doi: 10.1126/science.262.5134.729.

Jannett, P., 2004, An impactite without a crater: Ejecta preservation in the late Maastrichtian Fox Hills Formation of southwestern South Dakota [M.S. thesis]: Philadelphia, Temple University, 245 p.

Jannett, P., and Terry, D.O., Jr., 2001, Ancient pedogenic overprinting of the K-T boundary interval within the Fox Hills Formation of southwestern South Dakota: Geological Society of America Abstracts with Programs, v. 33, no. 7, p. A-202.

Katongo, C., Koeberl, C., Witzke, B.J., Hammond, R.H., and Anderson, R.R., 2004, Geochemistry and shock petrography of the Crow Creek Member, South Dakota, USA: Ejecta from the 74 Ma Manson impact structure: Meteoritics and Planetary Science, v. 39, p. 31–51.

Keller, G., Adatte, T., Stinnesbeck, W., Rebolledo-Vieyra, M., Fucugauchi, J.U., Kramar, U., and Stüben, D., 2004, Chicxulub impact predates the K-T boundary mass extinction: Proceedings of the National Academy of Sciences of the United States of America, v. 101, p. 3753–3758, doi: 10.1073/pnas.0400396101.

Kennedy, W.J., Landman, N.H., Christiansen, W.K., Cobban, W.A., and Hancock, J.M., 1998, Marine connections in North America during the late Maastrichtian: Paleogeographic and paleobiogeographic significance of *Jeletzkytes nebrascensis* zone cephalopod fauna from the Elk Butte Member of the Pierre Shale, SE South Dakota and NE Nebraska: Cretaceous Research, v. 19, p. 745–775, doi: 10.1006/cres.1998.0129.

Koutsoukos, E.A.M., 1999, An extraterrestrial impact event in the early Danian: a secondary K/T boundary event?: Terra Nova, v. 10, no. 2, p. 68–73.

Kyte, F.T., 1998, A meteorite from the Cretaceous/Tertiary boundary: Nature, v. 396, p. 237–239, doi: 10.1038/24322.

Kyte, F.T., and Bostwick, J.A., 1995, Magnesioferrite spinel in Cretaceous/Tertiary boundary sediments of the Pacific basin: Remnants of hot, early ejecta from the Chicxulub impact?: Earth and Planetary Science Letters, v. 132, p. 113–127, doi: 10.1016/0012-821X(95)00051-D.

Landman, N.H., and Waage, K.M., 1993, Morphology and environment of Upper Cretaceous (Maastrichtian) scaphites: Geobios, v. 26, supplement 1, p. 257–265, doi: 10.1016/S0016-6995(06)80380-3.

Leckie, R.M., Yuretich, R.F., West, O.L.O., Finklestein, D., and Schmidt, M., 1998, Paleoceanography of the southwestern Western Interior Sea during the time of the Cenomanian-Turonian boundary (Late Cretaceous), *in* Dean, W.E., and Arthur, M.A., eds., Stratigraphy and paleoenvironments of the Cretaceous Western Interior Seaway, USA: SEPM Concepts in Sedimentology and Paleontology 6, p. 101–126.

Lillegraven, J.A., and Ostresh, L.M., Jr., 1990, Late Cretaceous (earliest Campanian/Maastrichtian) evolution of western shorelines of the North American Western Interior Seaway in relation to known mammalian faunas, *in* Bown, T.M., and Rose, K.D., eds., Dawn of the Age of Mammals in the northern part of the Rocky Mountain interior, North America: Geological Society of America Special Paper 243, p. 1–30.

Lisenbee, A.L., and DeWitt, E., 1993, Laramide evolution of the Black Hills uplift, *in* Snoke, A.W., et al., eds., Geology of Wyoming: Geological Survey of Wyoming Memoir 5, p. 374–412.

Munsell Color, 1975, Munsell color charts: Baltimore, Munsell Color Company, 11 p.

Nichols, D.J., and Johnson, K.R., 2002, Palynology and microstratigraphy of the Cretaceous-Tertiary boundary sections in southwestern North Dakota, *in* Hartman, J.R., et al., eds., The Hell Creek Formation and the Cretaceous-Tertiary boundary in the northern Great Plains: An integrated continental record of the end of the Cretaceous: Geological Society of America Special Paper 361, p. 95–143.

Pagani, M., and Arthur, M.A., 1998, Stable isotopic studies of the Cenomanian-Turonian proximal marine fauna from the U.S. Western Interior Seaway, *in* Dean, W.E., and Arthur, M.A., eds., Stratigraphy and paleoenvironments of the Cretaceous Western Interior Seaway, USA: SEPM Concepts in Sedimentology and Paleontology 6, p. 201–225.

Palamarczuk, S., Chamberlain, J.A., Jr., and Terry, D.O., Jr., 2003, Dinoflagellates of the Fox Hills Formation (Maastrichtian), badlands area of South Dakota: Biostratigraphic and paleoenvironmental implications: Joint Meeting of the American Association of Stratigraphic Palynology, Canadian Association of Palynologists, and the North American Micropaleontology Section of the SEPM, Saint Catharines, Canada, October 5–8, 2003, Programs with Abstracts (not paginated).

Pettyjohn, W.A., 1965, Eocene soil profile in the northern Great Plains: Proceedings of the South Dakota Academy of Science, v. XLIV, p. 80–87.

Pettyjohn, W.A., 1967, New members of the Upper Cretaceous Fox Hills Formation in South Dakota, representing delta deposits: American Association of Petroleum Geologists Bulletin, v. 51, p. 1361–1367.

Retallack, G.J., 1983, Late Eocene and Oligocene paleosols from Badlands National Park, South Dakota: Geological Society of America Special Paper 193, 82 p.

Retallack, G.J., 1990, Soils of the past: An introduction to paleopedology: Boston, Unwin Hyman, 520 p.

Retallack, G.J., 1996, Acid trauma at the Cretaceous-Tertiary boundary in eastern Montana: GSA Today, v. 6, no. 5, p. 1–7.

Short, N.M., and Gold, D.P., 1996, Petrography of shocked rocks from the central peak at the Manson impact structure, *in* Koeberl, C., and Anderson, R.R., eds., The Manson impact structure, Iowa: Anatomy of an impact crater: Geological Society of America Special Paper 302, p. 245–266.

Simms, M.J., 2003, Uniquely extensive seismite from the latest Triassic of the United Kingdom: Evidence for bolide impact?: Geology, v. 31, p. 557–560, doi: 10.1130/0091-7613(2003)031<0557:UESFTL>2.0.CO;2.

Smit, J., 1999, The global stratigraphy of the Cretaceous-Tertiary boundary impact ejecta: Annual Review of Earth and Planetary Sciences, v. 27, p. 75–113, doi: 10.1146/annurev.earth.27.1.75.

Stoffer, P.W., 2003, Geology of Badlands National Park: A preliminary report: U.S. Geological Survey Open-File Report 03-35, 63 p.

Stoffer, P.W., Messina, P., and Chamberlain, J.A., Jr., 1998, Upper Cretaceous stratigraphy of Badlands National Park, South Dakota: Influence of tectonism and sea level change on sedimentation in the Western Interior Seaway: Dakoterra, v. 5, p. 55–62.

Stoffer, P.W., Messina, P., Chamberlain, J.A., Jr., and Terry, D.O., Jr., 1999, Tsunamis in South Dakota: Two asteroid impacts inferred from the Cretaceous/Tertiary (K/T) boundary interval in Badlands National Park: Geological Society of America Abstracts with Programs, v. 31, no. 7, p. A-473.

Stoffer, P.W., Messina, P., Chamberlain, J.A., Jr., and Terry, D.O., 2001, The Cretaceous-Tertiary boundary interval in Badlands National Park, South Dakota: U.S. Geological Survey Open-File Report 01-56, 49 p.

Stoops, G., 2003, Guidelines for analysis and description of soil and regolith thin sections: Madison, Soil Science Society of America, 184 p.

Tanner, L.H., 2003, Far-reaching seismic effects of the Manicouagan impact: Evidence from the Fundy Basin: Geological Society of America Abstracts with Programs, v. 35, no. 6, p. 167.

Terry, D.O., Jr., 1998, Lithostratigraphic revision and correlation of the lower part of the White River Group, South Dakota to Nebraska, *in* Terry, D.O., et al., eds., Geological Society of America Special Paper 325, p. 15–37.

Terry, D.O., Jr., and Evans, J.E., 1994, Pedogenesis and paleoclimatic implications of the Chamberlain Pass Formation, Basal White River Group, badlands of South Dakota: Palaeogeography, Palaeoclimatology, Palaeoecology, v. 110, p. 197–215, doi: 10.1016/0031-0182(94)90084-1.

Terry, D.O., Jr., Chamberlain, J.A., Jr., Stoffer, P.W., Messina, P., and Jannett, P.A., 2001, Marine Cretaceous-Tertiary boundary section in southwestern South Dakota: Geology, v. 29, p. 1055–1058, doi: 10.1130/0091-7613(2001)029<1055:MCTBSI>2.0.CO;2.

Terry, D.O., Jr., Chamberlain, J.A., Jr., Stoffer, P.W., Messina, P., and Jannett, P.A., 2002, Marine Cretaceous-Tertiary boundary section in southwestern South Dakota: Reply: Geology, v. 30, p. 955–956, doi: 10.1130/0091-7613(2002)030<0956:>2.0.CO;2.

Terry, D.O., Jr., Chamberlain, J.A., Jr., Stoffer, P.W., Becker, M., Jannett, P.A., Palamarczuk, S., Garb, M., and Beeney, B., 2004, A widespread zone of soft sediment deformation and ejecta in the Fox Hills Formation of southwestern South Dakota: An impactite without a crater: Geological Society of America Abstracts with Programs, v. 36, no. 2, p. 25.

Turner, P., Sherlock, S.C., Clarke, P., and Cornelius, C., 2003, A new mid-late Maastrichtian impact in the Raton Basin 100 m below the K/T boundary: Houston, Lunar and Planetary Science Institute, Third International Conference on Large Meteorite Impacts, CD-ROM, abstract 4087.

Waage, K.M., 1968, The type Fox Hills Formation, Cretaceous (Maestrichtian), South Dakota, part 1: Stratigraphy and paleoenvironments: Peabody Museum of Natural History Bulletin, v. 27, 171 p.

Manuscript Accepted by the Society 10 July 2007

Printed in the USA